Communications and Control Engineering

Published titles include:

Stability and Stabilization of Infinite Dimensional Systems with Applications
Zheng-Hua Luo, Bao-Zhu Guo and Omer Morgul

Nonsmooth Mechanics (Second edition)
Bernard Brogliato

Nonlinear Control Systems II
Alberto Isidori

L_2-Gain and Passivity Techniques in nonlinear Control
Arjan van der Schaft

Control of Linear Systems with Regulation and Input Constraints
Ali Saberi, Anton A. Stoorvogel and Peddapullaiah Sannuti

Robust and $H\infty$ Control
Ben M. Chen

Computer Controlled Systems
Efim N. Rosenwasser and Bernhard P. Lampe

Dissipative Systems Analysis and Control
Rogelio Lozano, Bernard Brogliato, Olav Egeland and Bernhard Maschke

Control of Complex and Uncertain Systems
Stanislav V. Emelyanov and Sergey K. Korovin

Robust Control Design Using H^∞ Methods
Ian R. Petersen, Valery A. Ugrinovski and Andrey V. Savkin

Model Reduction for Control System Design
Goro Obinata and Brian D.O. Anderson

Control Theory for Linear Systems
Harry L. Trentelman, Anton Stoorvogel and Malo Hautus

Functional Adaptive Control
Simon G. Fabri and Visakan Kadirkamanathan

Positive 1D and 2D Systems
Tadeusz Kaczorek

Identification and Control Using Volterra Models
Francis J. Doyle III, Ronald K. Pearson and Bobatunde A. Ogunnaike

Non-linear Control for Underactuated Mechanical Systems
Isabelle Fantoni and Rogelio Lozano

Robust Control (Second edition)
Jürgen Ackermann

Flow Control by Feedback
Ole Morten Aamo and Miroslav Krstić

Learning and Generalization (Second edition)
Mathukumalli Vidyasagar

Constrained Control and Estimation
Graham C. Goodwin, María M. Seron and José A. De Doná

Randomized Algorithms for Analysis and Control of Uncertain Systems
Roberto Tempo, Giuseppe Calafiore and Fabrizio Dabbene

Switched Linear Systems
Zhendong Sun and Shuzhi S. Ge

Tohru Katayama

Subspace Methods for System Identification

With 66 Figures

Springer

Tohru Katayama, PhD
Department of Applied Mathematics and Physics,
Graduate School of Informatics, Kyoto University, Kyoto 606-8501, Japan

British Library Cataloguing in Publication Data
Katayama, Tohru, 1942-
Subspace methods for system identification : a realization approach. - (Communications and control engineering)
1. System indentification 2. Stochastic analysis
I. Title
003.1

Communications and Control Engineering Series ISSN 0178-5354

ISBN-13 978-1-84996-988-8 e-ISBN-13 978-1-84628-158-7
Springer Science+Business Media
springeronline.com

Printed in Germany
69/3141-543210 Printed on acid-free paper

To my family

Preface

Numerous papers on system identification have been published over the last 40 years. Though there were substantial developments in the theory of stationary stochastic processes and multivariable statistical methods during 1950s, it is widely recognized that the theory of system identification started only in the mid-1960s with the publication of two important papers; one due to Åström and Bohlin [17], in which the maximum likelihood (ML) method was extended to a serially correlated time series to estimate ARMAX models, and the other due to Ho and Kalman [72], in which the deterministic state space realization problem was solved for the first time using a certain Hankel matrix formed in terms of impulse responses. These two papers have laid the foundation for the future developments of system identification theory and techniques [55].

The scope of the ML identification method of Åström and Bohlin [17] was to build single-input, single-output (SISO) ARMAX models from observed input-output data sequences. Since the appearance of their paper, many statistical identification techniques have been developed in the literature, most of which are now comprised under the label of *prediction error methods (PEM)* or *instrumental variable (IV) methods*. This has culminated in the publication of the volumes Ljung [109] and Söderström and Stoica [145]. At this moment we can say that theory of system identification for SISO systems is established, and the various identification algorithms have been well tested, and are now available as MATLAB® programs.

Also, identification of multi-input, multi-output (MIMO) systems is an important problem which is not dealt with satisfactorily by PEM methods. The identification problem based on the minimization of a prediction error criterion (or a least-squares type criterion), which in general is a complicated function of the system parameters, has to be solved by iterative descent methods which may get stuck into local minima. Moreover, optimization methods need canonical parametrizations and it may be difficult to guess a suitable canonical parametrization from the outset. Since no single continuous parametrization covers all possible multivariable linear systems with a fixed McMillan degree, it may be necessary to change parametrization in the course of the optimization routine. Thus the use of optimization criteria and canonical parametrizations can lead to local minima far from the true solution, and to numerically ill-conditioned problems due to poor identifiability, *i.e.*, to near insensitivity of the criterion to the variations of some parameters. Hence it seems that the PEM method has inherent difficulties for MIMO systems.

On the other hand, *stochastic realization theory*, initiated by Faurre [46] and Akaike [1] and others, has brought in a different philosophy of building models from data, which is not based on optimization concepts. A key step in stochastic realization is either to apply the deterministic realization theory to a certain Hankel matrix constructed with sample estimates of the process covariances, or to apply the canonical correlation analysis (CCA) to the future and past of the observed process. These algorithms have been shown to be implemented very efficiently and in a numerically stable way by using the tools of modern numerical linear algebra such as the singular value decomposition (SVD).

Then, a new effort in digital signal processing and system identification based on the QR decomposition and the SVD emerged in the mid-1980s and many papers have been published in the literature [100, 101, 118, 119], *etc.* These realization theory-based techniques have led to a development of various so-called *subspace identification methods*, including [163, 164, 169, 171–173], *etc.* Moreover, Van Overschee and De Moor [165] have published a first comprehensive book on subspace identification of linear systems. An advantage of subspace methods is that we do not need (nonlinear) optimization techniques, nor we need to impose to the system a canonical form, so that subspace methods do not suffer from the inconveniences encountered in applying PEM methods to MIMO system identification.

Though I have been interested in stochastic realization theory for many years, it was around 1990 that I actually resumed studies on realization theory, including subspace identification methods. However, realization results developed for deterministic systems on the one hand, and stochastic systems on the other, could not be applied to the identification of dynamic systems in which both a deterministic test input and a stochastic disturbance are involved. In fact, the deterministic realization result does not consider any noise, and the stochastic realization theory developed up to the early 1990s did address modeling of stochastic processes, or time series, only. Then, I noticed at once that we needed a new realization theory to understand many existing subspace methods and their underlying relations and to develop advanced algorithms. Thus I was fully convinced that a new stochastic realization theory in the presence of exogenous inputs was needed for further developments of subspace system identification theory and algorithms.

While we were attending the MTNS (The International Symposium on Mathematical Theory of Networks and Systems) at Regensburg in 1993, I suggested to Giorgio Picci, University of Padova, that we should do joint work on stochastic realization theory in the presence of exogenous inputs and a collaboration between us started in 1994 when he stayed at Kyoto University as a visiting professor. Also, I successively visited him at the University of Padova in 1997. The collaboration has resulted in several joint papers [87–90, 93, 130, 131]. Professor Picci has in particular introduced the idea of decomposing the output process into deterministic and stochastic components by using a preliminary orthogonal decomposition, and then applying the existing deterministic and stochastic realization techniques to each component to get a realization theory in the presence of exogenous input. On the other hand, inspired by the CCA-based approach, I have developed a method of solving a multi-stage Wiener prediction problem to derive an innovation representation of the stationary process with an observable exogenous input, from which subspace identification methods are successfully obtained.

This book is an outgrowth of the joint work with Professor Picci on stochastic realization theory and subspace identification. It provides an in-depth introduction to subspace methods for system identification of discrete-time linear systems, together with our results on realization theory in the presence of exogenous inputs and subspace system identification methods. I have included proofs of theorems and lemmas as much as possible, as well as solutions to problems, in order to facilitate the basic understanding of the material by the readers and to minimize the effort needed to consult many references.

This textbook is divided into three parts: Part I includes reviews of basic results, from numerical linear algebra to Kalman filtering, to be used throughout this book, Part II provides deterministic and stochastic realization theories developed by Ho and Kalman, Faurre, and Akaike, and Part III discusses stochastic realization results in the presence of exogenous inputs and their adaptation to subspace identification methods; see Section 1.6 for more details. Thus, various people can read this book according to their needs. For example, people with a good knowledge of linear system theory and Kalman filtering can begin with Part II. Also, people mainly interested in applications can just read the algorithms of the various identification methods in Part III, occasionally returning to Part I and/or Part II when needed. I believe that this textbook should be suitable for advanced students, applied scientists and engineers who want to acquire solid knowledge and algorithms of subspace identification methods.

I would like to express my sincere thanks to Giorgio Picci who has greatly contributed to our fruitful collaboration on stochastic realization theory and subspace identification methods over the last ten years. I am deeply grateful to Hideaki Sakai, who has read the whole manuscript carefully and provided invaluable suggestions, which have led to many changes in the manuscript. I am also grateful to Kiyotsugu Takaba and Hideyuki Tanaka for their useful comments on the manuscript. I have benefited from joint works with Takahira Ohki, Toshiaki Itoh, Morimasa Ogawa, and Hajime Ase, who told me about many problems regarding modeling and identification of industrial processes.

The related research from 1996 through 2004 has been sponsored by the Grant-in-Aid for Scientific Research, the Japan Society of Promotion of Sciences, which is gratefully acknowledged.

Tohru Katayama

Kyoto, Japan
January 2005

Contents

1 Introduction 1
1.1 System Identification 1
1.2 Classical Identification Methods 4
1.3 Prediction Error Method for State Space Models 6
1.4 Subspace Methods of System Identification 8
1.5 Historical Remarks 11
1.6 Outline of the Book 13
1.7 Notes and References 14

Part I Preliminaries

2 Linear Algebra and Preliminaries 17
2.1 Vectors and Matrices 17
2.2 Subspaces and Linear Independence 19
2.3 Norms of Vectors and Matrices 21
2.4 QR Decomposition 23
2.5 Projections and Orthogonal Projections 27
2.6 Singular Value Decomposition 30
2.7 Least-Squares Method 33
2.8 Rank of Hankel Matrices 36
2.9 Notes and References 38
2.10 Problems 39

3 Discrete-Time Linear Systems 41
3.1 z-Transform 41
3.2 Discrete-Time LTI Systems 44
3.3 Norms of Signals and Systems 47
3.4 State Space Systems 48
3.5 Lyapunov Stability 50
3.6 Reachability and Observability 51

3.7 Canonical Decomposition of Linear Systems 55
3.8 Balanced Realization and Model Reduction 58
3.9 Realization Theory 65
3.10 Notes and References 70
3.11 Problems 71

4 Stochastic Processes 73
4.1 Stochastic Processes 73
4.1.1 Markov Processes 74
4.1.2 Means and Covariance Matrices 75
4.2 Stationary Stochastic Processes 77
4.3 Ergodic Processes 79
4.4 Spectral Analysis 81
4.5 Hilbert Space and Prediction Theory 87
4.6 Stochastic Linear Systems 95
4.7 Stochastic Linear Time-Invariant Systems 98
4.8 Backward Markov Models 101
4.9 Notes and References 104
4.10 Problems 105

5 Kalman Filter 107
5.1 Multivariate Gaussian Distribution 107
5.2 Optimal Estimation by Orthogonal Projection 113
5.3 Prediction and Filtering Algorithms 116
5.4 Kalman Filter with Inputs 123
5.5 Covariance Equation of Predicted Estimate 127
5.6 Stationary Kalman Filter 128
5.7 Stationary Backward Kalman Filter 131
5.8 Numerical Solution of ARE 134
5.9 Notes and References 136
5.10 Problems 137

Part II Realization Theory

6 Realization of Deterministic Systems 141
6.1 Realization Problems 141
6.2 Ho-Kalman's Method 142
6.3 Data Matrices 149
6.4 LQ Decomposition 155
6.5 MOESP Method 157
6.6 N4SID Method 161
6.7 SVD and Additive Noises 166
6.8 Notes and References 169
6.9 Problems 170

7 Stochastic Realization Theory (1) ... 171
7.1 Preliminaries ... 171
7.2 Stochastic Realization Problem ... 174
7.3 Solution of Stochastic Realization Problem ... 176
7.3.1 Linear Matrix Inequality ... 177
7.3.2 Simple Examples ... 180
7.4 Positivity and Existence of Markov Models ... 183
7.4.1 Positive Real Lemma ... 183
7.4.2 Computation of Extremal Points ... 189
7.5 Algebraic Riccati-like Equations ... 192
7.6 Strictly Positive Real Conditions ... 194
7.7 Stochastic Realization Algorithm ... 198
7.8 Notes and References ... 199
7.9 Problems ... 200
7.10 Appendix: Proof of Lemma 7.4 ... 201

8 Stochastic Realization Theory (2) ... 203
8.1 Canonical Correlation Analysis ... 203
8.2 Stochastic Realization Problem ... 207
8.3 Akaike's Method ... 209
8.3.1 Predictor Spaces ... 209
8.3.2 Markovian Representations ... 212
8.4 Canonical Correlations Between Future and Past ... 216
8.5 Balanced Stochastic Realization ... 217
8.5.1 Forward and Backward State Vectors ... 217
8.5.2 Innovation Representations ... 219
8.6 Reduced Stochastic Realization ... 223
8.7 Stochastic Realization Algorithms ... 227
8.8 Numerical Results ... 230
8.9 Notes and References ... 232
8.10 Problems ... 233
8.11 Appendix: Proof of Lemma 8.5 ... 234

Part III Subspace Identification

9 Subspace Identification (1) – ORT ... 239
9.1 Projections ... 239
9.2 Stochastic Realization with Exogenous Inputs ... 241
9.3 Feedback-Free Processes ... 243
9.4 Orthogonal Decomposition of Output Process ... 245
9.4.1 Orthogonal Decomposition ... 245
9.4.2 PE Condition ... 246
9.5 State Space Realizations ... 248
9.5.1 Realization of Stochastic Component ... 248

9.5.2 Realization of Deterministic Component . . . 249
9.5.3 The Joint Model . . . 253
9.6 Realization Based on Finite Data . . . 254
9.7 Subspace Identification Method – ORT Method . . . 256
9.7.1 Subspace Identification of Deterministic Subsystem . . . 256
9.7.2 Subspace Identification of Stochastic Subsystem . . . 259
9.8 Numerical Example . . . 261
9.9 Notes and References . . . 265
9.10 Appendix: Proofs of Theorem and Lemma . . . 265
9.10.1 Proof of Theorem 9.1 . . . 265
9.10.2 Proof of Lemma 9.7 . . . 268

10 Subspace Identification (2) – CCA . . . 271
10.1 Stochastic Realization with Exogenous Inputs . . . 271
10.2 Optimal Predictor . . . 275
10.3 Conditional Canonical Correlation Analysis . . . 278
10.4 Innovation Representation . . . 282
10.5 Stochastic Realization Based on Finite Data . . . 286
10.6 CCA Method . . . 288
10.7 Numerical Examples . . . 292
10.8 Notes and References . . . 296

11 Identification of Closed-loop System . . . 299
11.1 Overview of Closed-loop Identification . . . 299
11.2 Problem Formulation . . . 301
11.2.1 Feedback System . . . 301
11.2.2 Identification by Joint Input-Output Approach . . . 303
11.3 CCA Method . . . 304
11.3.1 Realization of Joint Input-Output Process . . . 304
11.3.2 Subspace Identification Method . . . 307
11.4 ORT Method . . . 309
11.4.1 Orthogonal Decomposition of Joint Input-Output Process . . . 309
11.4.2 Realization of Closed-loop System . . . 311
11.4.3 Subspace Identification Method . . . 312
11.5 Model Reduction . . . 315
11.6 Numerical Results . . . 317
11.6.1 Example 1 . . . 318
11.6.2 Example 2 . . . 321
11.7 Notes and References . . . 323
11.8 Appendix: Identification of Stable Transfer Matrices . . . 324
11.8.1 Identification of Deterministic Parts . . . 324
11.8.2 Identification of Noise Models . . . 325

Appendix

A Least-Squares Method 329
A.1 Linear Regressions 329
A.2 LQ Decomposition 334

B Input Signals for System Identification 337

C Overlapping Parametrization 343

D List of Programs 349
D.1 Deterministic Realization Algorithm 349
D.2 MOESP Algorithm 350
D.3 Stochastic Realization Algorithms 351
D.4 Subspace Identification Algorithms 353

E Solutions to Problems 357

Glossary 377

References 379

Index 389

1

Introduction

In this introductory chapter, we briefly review the classical prediction error method (PEM) for identifying linear time-invariant (LTI) systems. We then discuss the basic idea of subspace methods of system identification, together with the advantages of subspace methods over the PEM as applied to multivariable dynamic systems.

1.1 System Identification

Figure 1.1 shows a schematic diagram of a dynamic system with input u, output y and disturbance v. We can observe u and y but not v; we can directly manipulate the input u but not y. Even if we do not know the inside structure of the system, the measured input and output data provide useful information about the system behavior. Thus, we can construct mathematical models to describe dynamics of the system of interest from observed input-output data.

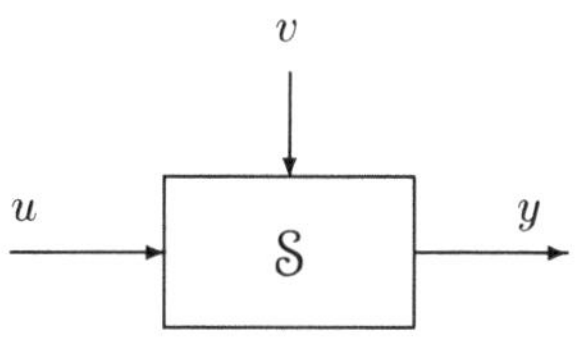

Figure 1.1. A system with input and disturbance

Dynamic models for prediction and control include transfer functions, state space models, time-series models, which are parametrized in terms of finite number of parameters. Thus these dynamic models are referred to as parametric models. Also used are non-parametric models such as impulse responses, and frequency responses, spectral density functions, *etc.*

System identification is a methodology developed mainly in the area of automatic control, by which we can choose the best model(s) from a given model set based on the observed input-output data from the system. Hence, the problem of system identification is specified by three elements [109]:

- A data set $\mathcal{D}$ obtained by input-output measurements.
- A model set $\mathcal{M}$, or a model structure, containing candidate models.
- A criterion, or loss, function $\mathcal{L}$ to select the best model(s), or a rule to evaluate candidate models, based on the data.

The input-output data $\mathcal{D}$ are collected through experiment. In this case, we must design the experiment by deciding input (or test) signals, output signals to be measured, the sampling interval, *etc.*, thereby systems characteristics are well reflected in the observed data. Thus, to obtain useful data for system identification, we should have some *a priori* information, or some physical knowledge, about the system. Also, there are cases where we cannot perform open-loop experiments due to safety, some technical and/or economic reasons, so that we can use data only measured under normal operating conditions.

A choice of model set $\mathcal{M}$ is a difficult issue in system identification, but usually several class of discrete-time linear time-invariant (LTI) systems are used. Since these models do not necessarily reflect the knowledge about the structure of the system, they are referred to as black-box models. One of the most difficult problems is to find a good model structure, or to fix orders of the models, based on the given input-output data. A solution to this problem is given by the Akaike information criterion (AIC) [3].

Also, by using some physical principles, we can construct models that contain several unknown parameters. These models are called gray-box models because some basic laws from physics are employed to describe the dynamics of a system or a phenomenon.

The next step is to find a model in the model set $\mathcal{M}$, by which the experimental data is best explained. To this end, we need a criterion to measure the distance between a model and a real system, so that the criterion should be of physical meaning and simple enough to be handled mathematically. In terms of the input u, the output y of a real system, and the model output y_M, the criterion is usually defined as

$$V_N = \sum_{t=0}^{N-1} l(y(t),\, y_M(t),\, u(t))$$

where $l(\cdot)$ is a nonnegative loss function, and N the number of data. If the model set is parametrized as $\mathcal{M} = \{M_\alpha,\, \alpha \in A\}$, then the identification in narrow sense reduces to an optimization problem minimizing the criterion V_N with respect to α.

Given three basic elements in system identification, we can in principle find the best model $M^* \in \mathcal{M}$. In this case, we need

- A condition for the existence of a model that minimizes the criterion.
- An algorithm of computing models.
- A method of model validation.

In particular, model validation is to determine whether or not an identified model should be accepted as a suitable description that explains the dynamics of a system. Thus, model validation is based on the way in which the model is used, *a priori* information on the system, the fitness of the model to real data, *etc.* For example, if we identify the transfer function of a system, the quality of an identified model is evaluated based on the step response and/or the pole-zero configuration. Furthermore, if the ultimate goal is to design a control system, then we must evaluate control performance of a system designed by the identified model. If the performance is not satisfactory, we must go back to some earlier stages of system identification, including the selection of model structure, or experiment design, *etc.* A flow diagram of system identification is displayed in Figure 1.2, where we see that the system identification procedure has an inherent iterative or feedback structure.

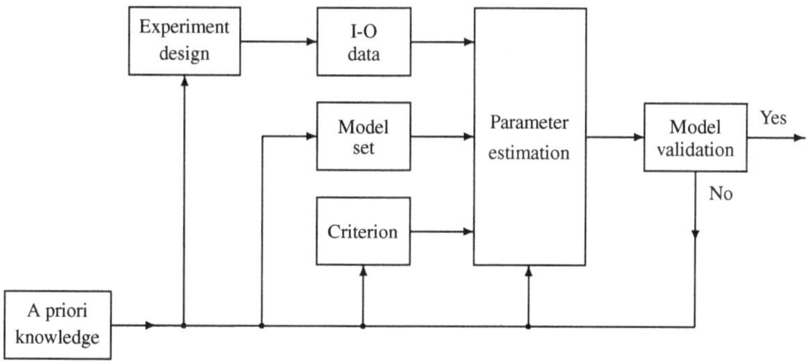

Figure 1.2. A flow diagram of system identification [109, 145]

Models obtained by system identification are valid under some prescribed conditions, *e.g.*, they are valid for a certain neighborhood of working point, and also do not provide a physical insight into the system because parameters in the model have no physical meaning. It should be noted that it is engineering skills and deep insights into the systems, shown as *a priori* knowledge, that help us to construct mathematical models based on ill-conditioned data. As shown in Figure 1.2, we cannot get a desired model unless we iteratively evaluate identified models by trying several model structures, model orders, *etc.* At this stage, the AIC plays a very important role in that it can automatically select the best model based on the given input-output data in the sense of maximum likelihood (ML) estimation.

It is well known that real systems of interest are nonlinear, time-varying, and may contain delays, and some variables or signals of central importance may not be measured. It is also true that LTI systems are the simplest and most important class of dynamic systems used in practice and in the literature [109]. Though they are nothing but idealized models, our experiences show that they can well approximate many industrial processes. Besides, control design methods based on LTI models often lead to good results in many cases. Also, it should be emphasized that system identification is a technique of approximating real systems by means of our models since there is no "true" system in practical applications [4].

1.2 Classical Identification Methods

Let the "true" system be represented by

$$(\mathcal{S}) \qquad y(t) = P_0(z)u(t) + v_0(t)$$

where $P_0(z)$ is the "true" plant, and $v_0(t)$ is the output disturbance. Suppose that we want to fit a stochastic single-input, single-output (SISO) LTI model (Figure 1.3)

$$(\mathcal{M}) \qquad y(t) = P(z, \theta)u(t) + H(z, \theta)e(t)$$

to a given set of input-output data, where e is a white noise with mean 0 and variance σ^2, and $\theta \in \mathbb{R}^d$ contains all unknown parameters other than the noise variance. It may also be noted that the noise v includes the effect of unmeasurable disturbances, modeling errors, *etc.*

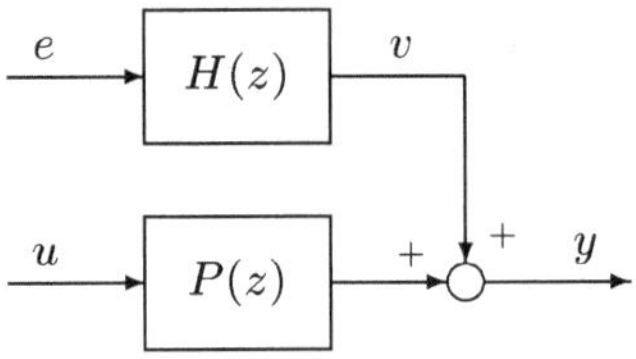

Figure 1.3. An SISO transfer function model

The transfer function of the plant model is usually given by

$$P(z, \theta) = \frac{B(z, \theta)}{A(z, \theta)} = \frac{b_1 z^{-1} + \cdots + b_m z^{-m}}{1 + a_1 z^{-1} + \cdots + a_n z^{-n}}, \qquad n \geq m$$

where, if the plant has a delay, then the parameters $b_1, \cdots, b_l$ with $l \geq 1$ reduce to zero. Also, the transfer function of the noise model is

$$H(z, \theta) = \frac{C(z, \theta)}{D(z, \theta)} = \frac{1 + c_1 z^{-1} + \cdots + c_p z^{-p}}{1 + d_1 z^{-1} + \cdots + d_q z^{-q}} \tag{1.1}$$

where $H(z, \theta)$ is of minimal phase with $H(\infty, \theta) = 1$.

Suppose that we have observed a sequence of input-output data. Let the input-output data up to time $t-1$ be defined by

$$Z^{t-1} := \{u(k),\ y(k),\ k = 0,\ 1,\ \cdots,\ t-1\}$$

Then, it can be shown [109] that the one-step predicted estimate of the output $y(t)$ based on Z^{t-1} is given by

$$\hat{y}(t,\theta) = H^{-1}(z,\theta)P(z,\theta)u(t) + [1 - H^{-1}(z,\theta)]y(t)$$

Moreover, we define the one-step prediction error $\varepsilon(t,\theta) := y(t) - \hat{y}(t,\theta)$. Then, it can be expressed as

$$\varepsilon(t,\theta) = H^{-1}(z,\theta)[P_0(z) - P(z,\theta)]u(t) + H^{-1}(z,\theta)v_0(t) \tag{1.2}$$

Suppose that we have a set of data Z^{N-1}. If we specify a particular value to the parameter θ, then from (1.2), we can obtain a sequence of prediction errors

$$\{\varepsilon(t,\theta),\ t = 0,\ 1,\ \cdots,\ N-1\}$$

where the initial conditions $\{\varepsilon(t,\theta),\ t = -1,\ \cdots,\ -p\}$ should be given. When we fit a model to the data Z^{N-1}, a principle of estimation is to select $\hat{\theta}$ that produces the minimum variance of prediction error. Thus the criterion function is given by

$$V_N(\theta) = \frac{1}{N}\sum_{t=0}^{N-1} \varepsilon^2(t,\theta) \tag{1.3}$$

A schematic diagram of the prediction error method (PEM) is displayed in Figure 1.4. Thus, a class of algorithms designed so that a function of prediction errors is minimized is commonly called the PEM. Since the performance criterion of (1.3) is in general a complicated function of the system parameters, the problem has to be solved by iterative descent methods, which may get stuck in local minima.

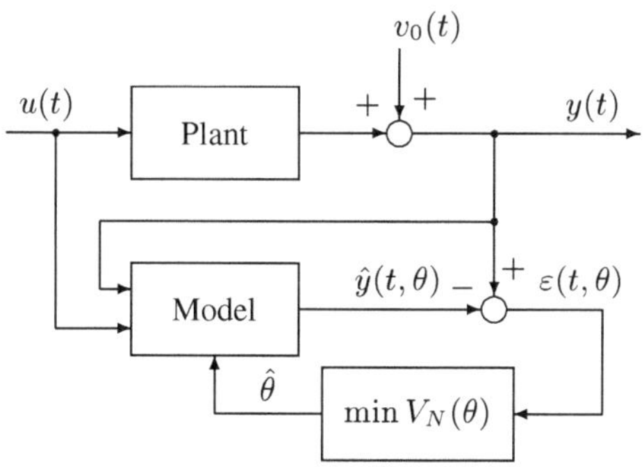

Figure 1.4. Prediction error method

Example 1.1. Let $H(z) = C(z)/A(z)$ in (1.1). Then we get the ARMAX[1] model of the form

[1] ARMAX= AutoRegressive Moving Average with eXogenous input.

$$A(z)y(t) = B(z)u(t) + C(z)e(t) \tag{1.4}$$

where the unknown parameters are $\theta = (a_1 \ \cdots \ a_n \ b_1 \ \cdots \ b_m \ c_1 \ \cdots \ c_p)^{\mathrm{T}}$ and the noise variance σ^2. Then, the one-step prediction error for the ARMAX model of (1.4) is expressed as

$$\varepsilon(t,\theta) = [C(z,\theta)]^{-1}[A(z,\theta)y(t) - B(z,\theta)u(t)] \tag{1.5}$$

Obviously, the polynomial $C(z,\theta)$ should be stable in order to get a sequence of prediction errors. Substituting (1.5) into (1.3) yields

$$V_N(\theta) = \frac{1}{N}\sum_{t=0}^{N-1}\left[\frac{A(z,\theta)}{C(z,\theta)}y(t) - \frac{B(z,\theta)}{C(z,\theta)}u(t)\right]^2$$

Thus, in this case, the PEM reduces to a nonlinear optimization problem of minimizing the performance index $V_N(\theta)$ with respect to the parameter vector θ under the constraint that $C(z,\theta)$ is stable. □

For the detailed exposition of the PEM, including a frequency domain interpretation of the PEM and the analysis of convergence of the estimate, see [109, 145].

1.3 Prediction Error Method for State Space Models

Consider an innovation representation of a discrete-time LTI system of the form

$$x(t+1) = Ax(t) + Bu(t) + Ke(t) \tag{1.6a}$$

$$y(t) = Cx(t) + Du(t) + e(t) \tag{1.6b}$$

where $y \in \mathbb{R}^p$ is the output vector, $u \in \mathbb{R}^m$ the input vector, $x \in \mathbb{R}^n$ the state vector, $e \in \mathbb{R}^p$ the innovation vector with mean zero and covariance matrix $R > 0$, and $(A,\ B,\ C,\ D,\ K)$ are matrices of appropriate dimensions. The unknown parameters in the state space model are contained in these system matrices and covariance matrix R of the innovation process.

Consider the application of the PEM to the multi-input multi-output (MIMO) model (1.6). In view of Theorem 5.2, the prediction error $\varepsilon(t,\theta)$ is computed by a linear state space model with inputs $u(t)$, $y(t)$ of the form

$$\hat{x}(t+1,\theta) = [A(\theta) - K(\theta)C]\hat{x}(t,\theta) + B(\theta)u(t) + K(\theta)y(t)$$
$$\varepsilon(t,\theta) = -C\hat{x}(t,\theta) - D(\theta)u(t) + y(t)$$

with the initial condition $\hat{x}(0,\theta) = 0$. Then, in terms of $\varepsilon(t,\theta)$, the performance index is given by

$$V_N(\theta) = \frac{1}{N}\sum_{t=0}^{N-1}\|\varepsilon(t,\theta)\|^2$$

Thus the PEM estimates are obtained by minimizing $V_N(\theta)$ with respect to θ, and the covariance matrix R of e is estimated by computing the sample covariance matrix of $\varepsilon(t), t = 0, 1, \cdots, N-1$.

If we can evaluate the gradient $\partial V_N / \partial \theta$, we can in principle compute a (local) minimum of the criterion $V_N(\theta)$ by utilizing a (conjugate) gradient method. Also optimization methods need canonical parametrizations and it may be difficult to guess a suitable canonical parametrization from the outset. Since no single continuous parametrization covers all possible multivariable linear systems with a fixed McMillan degree, it may be necessary to change parametrization in the course of the optimization routine.

Even if this difficulty can be tackled by using overlapping parametrizations or pseudo-canonical forms, sensible complications in the algorithm in general result. Thus the use of optimization criteria and canonical parametrizations can lead to local minima far from the true solution, to complicated algorithms for switching between canonical forms, and to numerically ill-conditioned problems due to poor identifiability, *i.e.*, to near insensitivity of the criterion to the variations of some parameters. Hence it seems that the PEM method has inherent difficulties for MIMO systems.

It is well known that for a given triplet (n, m, p), there does not exist a global canonical MIMO linear state space form [57, 67]. But there are some interests in deriving a convenient parametrization for MIMO systems called an overlapping parametrization, or pseudo-canonical form [54, 68].

Example 1.2. Consider the state space model of (1.6). An MIMO observable pseudo-canonical form with $(p = 3,\ m = 4,\ n = 9)$ can be given by

$$A = \left[\begin{array}{ccc|cccc|cc} 0 & 1 & 0 & 0 & 0 & 0 & 0 & 0 & 0 \\ 0 & 0 & 1 & 0 & 0 & 0 & 0 & 0 & 0 \\ \times & \times & \times & \times & \times & \times & \times & \times & \times \\ \hline 0 & 0 & 0 & 0 & 1 & 0 & 0 & 0 & 0 \\ 0 & 0 & 0 & 0 & 0 & 1 & 0 & 0 & 0 \\ 0 & 0 & 0 & 0 & 0 & 0 & 1 & 0 & 0 \\ \times & \times & \times & \times & \times & \times & \times & \times & \times \\ \hline 0 & 0 & 0 & 0 & 0 & 0 & 0 & 0 & 1 \\ \times & \times & \times & \times & \times & \times & \times & \times & \times \end{array}\right], \quad B = \left[\begin{array}{cccc} \times & \times & \times & \times \\ \times & \times & \times & \times \\ \times & \times & \times & \times \\ \hline \times & \times & \times & \times \\ \times & \times & \times & \times \\ \times & \times & \times & \times \\ \times & \times & \times & \times \\ \hline \times & \times & \times & \times \\ \times & \times & \times & \times \end{array}\right], \quad K = \left[\begin{array}{ccc} \times & \times & \times \\ \times & \times & \times \\ \times & \times & \times \\ \hline \times & \times & \times \\ \times & \times & \times \\ \times & \times & \times \\ \times & \times & \times \\ \hline \times & \times & \times \\ \times & \times & \times \end{array}\right]$$

$$C = \left[\begin{array}{ccc|cccc|cc} 1 & 0 & 0 & 0 & 0 & 0 & 0 & 0 & 0 \\ \hline 0 & 0 & 0 & 1 & 0 & 0 & 0 & 0 & 0 \\ \hline 0 & 0 & 0 & 0 & 0 & 0 & 0 & 1 & 0 \end{array}\right], \quad D = \left[\begin{array}{cccc} \times & \times & \times & \times \\ \times & \times & \times & \times \\ \times & \times & \times & \times \end{array}\right]$$

where $\times$ indicates independent parameters. See Appendix C, where an overlapping parametrization is derived for a stochastic system.

The pair $(C,\ A)$ is observable by definition, but the reachability of pairs $(A,\ B)$ and $(A,\ K)$ depends on the actual values of parameters. We see that A has pn independent parameters, and all the elements in $(B,\ D,\ K)$ are independent parameters, but the parameters in C are fixed. Thus the number of unknown parameters in the overlapping parametrization above is $N_{\text{ovlap}} = n(2p + m) + pm$. On the other hand, the total number of parameters in (A, B, C, D, K) is $N_{\mathrm{T}} = n^2 + n(2p+m) + pm$, so that we can save n^2 parameters by using the above overlapping parametrization. □

Recently, data driven local coordinates, which is closely related to the overlapping parametrizations, have been introduced in McKelvey *et al.* [114].

1.4 Subspace Methods of System Identification

In this section, we glance at some basic ideas in subspace identification methods. For more detail, see Chapter 6.

Basic Idea of Subspace Methods

Subspace identification methods are based on the following idea. Suppose that an estimate of a sequence of state vectors of the state space model of (1.6) are somehow constructed from the observed input-output data (see below). Then, for $t = 0, 1, \cdots, N-1$, we have

$$\begin{bmatrix} \bar{x}(t+1) \\ y(t) \end{bmatrix} = \begin{bmatrix} A & B \\ C & D \end{bmatrix} \begin{bmatrix} \bar{x}(t) \\ u(t) \end{bmatrix} + \begin{bmatrix} \eta(t) \\ \nu(t) \end{bmatrix} \tag{1.7}$$

where $\bar{x} \in \mathbb{R}^n$ is the estimate of state vector, $u \in \mathbb{R}^m$ the input, $y \in \mathbb{R}^p$ the output, and η, ν are residuals. It may be noted that since all the variables are given, (1.7) is a regression model for system parameters $\Theta := \begin{bmatrix} A & B \\ C & D \end{bmatrix} \in \mathbb{R}^{(n+p)\times(n+m)}$. Thus the least-squares estimate of Θ is given by

$$\Theta = \left(\sum_{t=0}^{N-1} \begin{bmatrix} \bar{x}(t+1) \\ y(t) \end{bmatrix} \begin{bmatrix} \bar{x}^{\mathrm{T}}(t) & u^{\mathrm{T}}(t) \end{bmatrix} \right) \left(\sum_{t=0}^{N-1} \begin{bmatrix} \bar{x}(t) \\ u(t) \end{bmatrix} \begin{bmatrix} \bar{x}^{\mathrm{T}}(t) & u^{\mathrm{T}}(t) \end{bmatrix} \right)^{-1}$$

This class of approaches are called the *direct N4SID methods* [175]. We see that this estimate uniquely exists if the rank condition

$$\operatorname{rank} \begin{bmatrix} \bar{x}(0) & \bar{x}(1) & \cdots & \bar{x}(N-1) \\ u(0) & u(1) & \cdots & u(N-1) \end{bmatrix} = n + m$$

is satisfied. This condition, discussed some 30 years ago by Gopinath [62], plays an important role in subspace identification as well; see Section 6.3.

Moreover, the covariance matrices of the residuals are given by

$$\begin{bmatrix} Q & S \\ S^{\mathrm{T}} & R \end{bmatrix} = \frac{1}{N} \sum_{t=0}^{N-1} \begin{bmatrix} \eta(t) \\ \nu(t) \end{bmatrix} \begin{bmatrix} \eta^{\mathrm{T}}(t) & \nu^{\mathrm{T}}(t) \end{bmatrix}$$

Thus, by solving a certain algebraic Riccati equation, we can derive a steady state Kalman filter (or an innovation model) of the form

$$\begin{bmatrix} \hat{x}(t+1) \\ y(t) \end{bmatrix} = \begin{bmatrix} A & B \\ C & D \end{bmatrix} \begin{bmatrix} \hat{x}(t) \\ u(t) \end{bmatrix} + \begin{bmatrix} K \\ I_p \end{bmatrix} \hat{e}(t)$$

where K is the steady state Kalman gain, $\hat{x}$ is the estimate of state vector, and $\hat{e}$ is the estimate of innovation process.

Computation of State Vectors

We explain how we compute the estimate of state vectors by the LQ decomposition; this is a basic technique in subspace identification methods (Section 6.6). Suppose that we have an input-output data from an LTI system. Let the block Hankel matrices be defined by

$$U_{0|k-1} = \begin{bmatrix} u(0) & u(1) & \cdots & u(N-1) \\ u(1) & u(2) & \cdots & u(N) \\ \vdots & \vdots & \ddots & \vdots \\ u(k-1) & u(k) & \cdots & u(N+k-2) \end{bmatrix} \in \mathbb{R}^{km \times N}$$

and

$$Y_{0|k-1} = \begin{bmatrix} y(0) & y(1) & \cdots & y(N-1) \\ y(1) & y(2) & \cdots & y(N) \\ \vdots & \vdots & \ddots & \vdots \\ y(k-1) & y(k) & \cdots & y(N+k-2) \end{bmatrix} \in \mathbb{R}^{kp \times N}$$

where $k > n$ and N is sufficiently large.

For notational convenience, let p and f denote the past and future, respectively. Then, we define the past data as $U_p := U_{0|k-1}$ and $Y_p := Y_{0|k-1}$. Similarly, we define the future data as $U_f := U_{k|2k-1}$ and $Y_f := Y_{k|2k-1}$. Let the LQ decomposition be given by

$$\begin{bmatrix} U_f \\ W_p \\ Y_f \end{bmatrix} = \begin{bmatrix} R_{11} & 0 & 0 \\ R_{21} & R_{22} & 0 \\ R_{31} & R_{32} & R_{33} \end{bmatrix} \begin{bmatrix} Q_1^{\mathrm{T}} \\ Q_2^{\mathrm{T}} \\ Q_3^{\mathrm{T}} \end{bmatrix}$$

where $R_{11} \in \mathbb{R}^{km \times km}$, $R_{22} \in \mathbb{R}^{k(m+p) \times k(m+p)}$ and $R_{33} \in \mathbb{R}^{kp \times kp}$ are upper triangular, and $Q_i,\ i = 1, 2, 3$ are orthogonal matrices. Then, from Theorem 6.3, we see that the oblique projection of the future Y_f onto the joint past $W_p := \begin{bmatrix} U_p \\ Y_p \end{bmatrix}$ along the future U_f is given by

$$\xi := \hat{E}_{\|U_f}\{Y_f | W_p\} = R_{32} R_{22}^{\dagger} W_p$$

where $(\cdot)^{\dagger}$ denotes the pseudo-inverse. We can show that ξ can be factored as a product of the extended observability matrix $\mathcal{O}_k$ and the future state vector $X_f := [x(k) \ \cdots \ x(k+N-1)] \in \mathbb{R}^{n \times N}$. It thus follows that

$$\xi = \mathcal{O}_k X_f = R_{32} R_{22}^{\dagger} W_p$$

Suppose that the SVD of ξ be given by $\xi = U \Sigma V^{\mathrm{T}}$ with $\mathrm{rank}(\Sigma) = n$. Thus, we can take the extended observability matrix as

$$\mathcal{O}_k = U \Sigma^{1/2} \tag{1.8}$$

Hence, it follows that the state vector is given by $X_f = \mathcal{O}_k^{\dagger} \xi = \Sigma^{1/2} V^{\mathrm{T}}$.

Alternatively, by using a so-called shift invariant property of the extended observability matrix of (1.8), we can respectively compute matrices A and C as

$$A = \mathcal{O}_{k-1}^{\dagger} \mathcal{O}_k(p+1 : pk, 1 : n), \qquad C = \mathcal{O}_k(1 : p, 1 : n)$$

This class of approaches are called the *realization-based N4SID methods* [175]. For detail, see the MOESP method in Section 6.5.

Summarizing, under certain assumptions, we can reconstruct the estimate of a sequence of state vectors and the extended observability matrix from given input-output data. Numerical methods of obtaining the state estimates and extended observability matrix of LTI systems will be explained in detail in Chapter 6. Once this "trick" is understood, subspace identification methods in the literature can be understood without any difficulty.

Why Subspace Methods?

Although modern control design techniques have evolved based on the state space approach, the classical system identification methods have been developed in the input-output framework until the mid-1980s. It is quite recent that the state concept was introduced in system identification, thereby developing many subspace methods based on classical (stochastic) realization theory.

From Figure 1.5, we see some differences in the classical and subspace methods of system identification, where the left-hand side is the subspace method, and the right-hand side is the classical optimization-based method. It is interesting to observe the difference in the flow of two approaches; in the classical method, a transfer function model is first identified, and then a state space model is obtained by using some realization technique; from the state space model, we can compute state vectors, or the Kalman filter state vectors. In subspace methods, however, we first construct the state estimates from given input-output data by using a simple procedure based on tools of numerical linear algebra, and a state space model is obtained by solving a least-squares problem as explained above, from which we can easily compute a transfer matrix if necessary. Thus an important point of the study of subspace methods is to understand the key point of how the Kalman filter state vectors and the extended observability matrix are obtained by using tools of numerical linear algebra.

To recapitulate, the advantage of subspace methods, being based on reliable numerical algorithms of the QR decomposition and the SVD, is that we do not need (nonlinear) optimization techniques, nor do we need to impose onto the system a

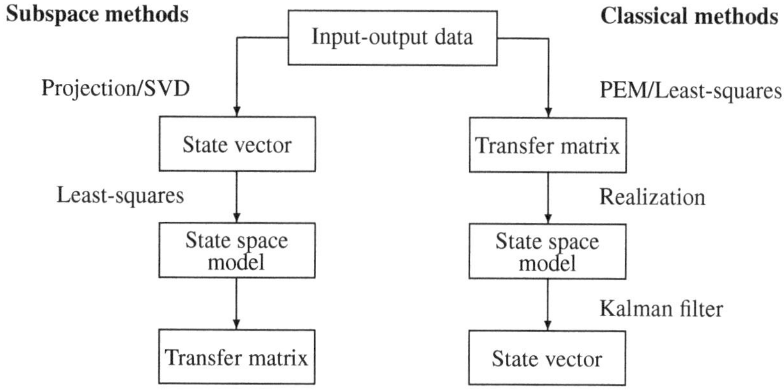

Figure 1.5. Subspace and classical methods of system identification ([165])

canonical form. This implies that subspace algorithms can equally be applicable to MIMO as well as SISO system identification. In other words, subspace methods do not suffer from the inconveniences encountered in applying PEM methods to MIMO system identification.

1.5 Historical Remarks

The origin of subspace methods may date back to multivariable statistical analysis [96], in particular to the principal component analysis (PCA) and canonical correlation analysis (CCA) due to Hotelling [74, 75] developed nearly 70 years ago. It is, however, generally understood that the concepts of subspace methods have spread to the areas of signal processing and system identification with the invention of the MUSIC (MUltiple SIgnal Classification) algorithm due to Schmidt [140]. We can also observe that the MUSIC is an extension of harmonic decomposition method of Pisarenko [133], which is in fact closely related to the classical idea of Levin [104] in the mid-1960s. For more detail, see the editorial of two special issues on Subspace Methods (Parts I and II) of Signal Processing [176], and also [150, 162].

Canonical Correlation Analysis

Hotelling [75] has developed the CCA technique to analyze linear relations between two sets of random variables. The CCA has been further developed by Anderson [14]. The predecessor of the concept of canonical correlations is that of canonical angles between two subspaces; see [21]. In fact, the ith canonical correlation ρ_i between two sets of random variables is related to the ith canonical angle θ_i between two Hilbert spaces generated by them via $\rho_i = \cos\theta_i$.

Gel'fand and Yaglom [52] have introduced mutual information between two stationary random processes in terms of canonical correlations of the two processes. Björck and Golub [21] have solved the canonical correlation problem by using the SVD. Akaike [2, 3] has analyzed the structure of the information interface between the future and past of a stochastic process by means of the CCA, and thereby developed a novel stochastic realization theory. Pavon [126] has studied the mutual information for a vector stationary process, and Desai *et al.* [42, 43] have developed a theory of stochastic balanced realization by using the CCA. Also, Jonckheere and Helton [77] have solved the spectral reduction problem by using the CCA and explored its relation to the Hankel norm approximation problem.

Hannan and Poskit [68] have derived conditions under which a vector ARMA process has unit canonical correlations. More recently, several analytical formulas for computing canonical correlations between the past and future of stationary stochastic processes have been developed by De Cock [39].

Stochastic Realization

Earlier results on the stochastic realization are due to Anderson [9] and Faurre [45]. Also, related spectral factorization results based on the state space methods are given by Anderson [7, 8]. By using the deterministic realization theory together with the LMI and algebraic Riccati equations, Faurre [45–47] has made a fundamental contribution to the stochastic realization theory. In Akaike [1], a stochastic interpretation of various realization algorithms, including the algorithm of Ho and Kalman [72], is provided. Moreover, Aoki [15] has derived stochastic realization algorithm based on the CCA and deterministic realization theory. Subspace methods of identifying state space models have been developed by De Moor *et al.* [41], Larimore [100, 101] and Van Overschee and De Moor [163]. Lindquist and Picci [106] have analyzed state space identification algorithms in the light of geometric theory of stochastic realization. Also, the conditional canonical correlations have been defined and employed to develop a stochastic realization theory in the presence of exogenous inputs by Katayama and Picci [90].

Subspace Methods

A new approach to system identification based on the QR decomposition and the SVD has emerged and many papers have been published in the literature in the late-1980s, *e.g.* De Moor [41], Moonen *et al.* [118, 119]. Then, these new techniques have led to a development of various subspace identification methods, including Verhaegen and Dewilde [172, 173], Van Overschee and De Moor [164], Picci and Katayama [130], *etc.* In 1996, a first comprehensive book on subspace identification of linear systems is published by Van Overschee and De Moor [165]. Moreover, some recent developments in the asymptotic analysis of N4SID methods are found in Jansson and Wahlberg [76], Bauer and Jansson [19], and Chiuso and Picci [31, 32]. Frequency domain subspace identification methods are also developed in McKelvey *et al.* [113] and Van Overschee *et al.* [166]. Among many papers on subspace identification of continuous-time systems, we just mention Ohsumi *et al.* [120], which is based on a mathematically sound distribution approach.

1.6 Outline of the Book

The primary goal of this book is to provide an in-depth knowledge and algorithms for the subspace methods for system identification to advanced students, engineers and applied scientists. The plan of this book is as follows.

Part I is devoted to reviews of some results frequently used throughout this book. More precisely, Chapter 2 introduces basic facts in numerical linear algebra, including the QR decomposition, the SVD, the projection and orthogonal projection, the least-squares method, the rank of Hankel matrices, *etc*. Some useful matrix formulas are given at the end of chapter as problems.

Chapter 3 deals with the state space theory for linear discrete-time systems, including the reachability, observability, realization theory, and model reduction method, *etc*.

In Chapter 4, we introduce stochastic processes, spectral analysis, and discuss the Wold decomposition theorem in a Hilbert space of a second-order stationary stochastic process. We also present a stochastic state space model, together with forward and backward Markov models for a stationary process.

Chapter 5 considers the minimum variance state estimation problem based on the orthogonal projection, and then derives the Kalman filter algorithm and discrete-time Riccati equations. Also derived are forward and backward stationary Kalman filters, which are respectively called forward and backward innovation models for a stationary stochastic process.

Part II provides a comprehensive treatment of the theories of deterministic and stochastic realization. In Chapter 6, we deal with the classical deterministic realization result due to Ho and Kalman [72] based on the SVD of Hankel matrix formed by impulse responses. By defining the future and past of the data, we explain how the LQ decomposition of the data matrix is utilized to retrieve the information about the extended observability matrix of a linear system. We then derive the MOESP method [172] and N4SID method [164, 165] in deterministic setting. The influence of white noise on the SVD of a wide rectangular matrix is also discussed, and some numerical results are included.

Chapter 7 is addressed to the stochastic realization theory due to Faurre [46] by using the LMI and spectral factorization technique, and to the associated algebraic Riccati equation (ARE) and algebraic Riccati inequality (ARI). The positive realness of covariance matrices is also proved with the help of AREs.

In Chapter 8, we present the stochastic realization theory developed by Akaike [2]. We discuss the predictor spaces for stationary stochastic processes. Then, based on the canonical correlations of the future and past of a stationary process, balanced and reduced stochastic realizations of Desai *et al.* [42, 43] are derived by using the forward and backward Markov models.

Part III presents our stochastic realization results and their adaptation to subspace identification methods. Chapter 9 considers a stochastic realization theory in the presence of an exogenous input based on Picci and Katayama [130]. We first review projections in a Hilbert space and consider feedback-free conditions between the joint input-output process. We then develop a state space model with a natural block structure of such processes based on a preliminary orthogonal decomposition of the output process into the deterministic and stochastic components. By adapting it to the finite input-output data, subspace identification algorithms, called the ORT, are derived based on the LQ decomposition and the SVD.

In Chapter 10, based on Katayama and Picci [90], we consider the same stochastic realization problem treated in Chapter 9. By formulating it as a multi-stage Wiener prediction problem and introducing the conditional canonical correlations, we extend the Akaike's stochastic realization theory to a stochastic system with an exogenous input, deriving a subspace stochastic identification method called the CCA method. Some comparative numerical studies are included.

Chapter 11 is addressed to closed-loop subspace identification problems in the framework of the joint input-output approach. Based on our results [87, 88], two methods are derived by applying the ORT and CCA methods, and some simulation results are included. Also, under the assumption that the system is open-loop stable, a simple method of identifying the plant, controller and the noise model based on the ORT method is presented [92].

Finally, Appendix A reviews the classical least-squares method for linear regression models and its relation to the LQ decomposition. Appendix B is concerned with input signals for system identification and the PE condition for deterministic as well as stationary stochastic signals. In Appendix C, we derive an overlapping parametrization of MIMO linear stochastic systems. Appendix D presents some of MATLAB® programs used for simulation studies in this book. Solutions to problems are also provided in Appendix E.

1.7 Notes and References

Among many books on system identification, we just mention Box and Jenkins [22], Goodwin and Payne [61], Ljung [109], Söderström and Stoica [145], and a recent book by Pintelon and Schoukens [132], which is devoted to a frequency domain approach. The book by Van Overschee and De Moor [165] is a first comprehensive book on subspace identification of linear systems, and there are some sections dealing with subspace methods in [109, 132]. Also, Mehra and Lainiotis [116], as a research oriented monograph, includes collections of important articles for system identification in the mid-1970s.

Part I

Preliminaries

2

Linear Algebra and Preliminaries

In this chapter, we review some basic results in numerical linear algebra, which are repeatedly used in later chapters. Among others, the QR decomposition and the singular value decomposition (SVD) are the most valuable tools in the areas of signal processing and system identification.

2.1 Vectors and Matrices

Let $\mathbb{R}$ be the set of real numbers, $\mathbb{R}^n$ the set of n-dimensional real vectors, and $\mathbb{R}^{m\times n}$ the set of $m \times n$ real matrices. The lower case letters $x,\ y,\ \cdots$ denote vectors, and capital letters $A,\ B,\ C,\ \cdots\ ;\ X,\ Y,\ Z,\ \cdots$ denote matrices. Transpositions of a vector x and a matrix A are denoted by x^{T} and A^{T}, respectively. The determinant of a square matrix A is denoted by $|A|$, or $\det(A)$, and the trace by $\mathrm{trace}(A)$.

The $n \times n$ identity matrix is denoted by I_n. If there is no confusion, we simply write I, deleting the subscript denoting the dimension. The inverse of a square matrix A is denoted by A^{-1}. We also use $A^{-\mathrm{T}}$ to denote $(A^{-1})^{\mathrm{T}} = (A^{\mathrm{T}})^{-1}$. A matrix satisfying $A^{\mathrm{T}} = A$ is called a symmetric matrix. If a matrix $A \in \mathbb{R}^{m\times n}$ with $m \geq n$ satisfies $A^{\mathrm{T}}A = I_n$, it is called an orthogonal matrix. Thus, for an orthogonal matrix $A = [a_1\ a_2\ \cdots\ a_n],\ a_i \in \mathbb{R}^m,\ i = 1,\ \cdots\ n$, we have $a_i^{\mathrm{T}} a_j = \delta_{ij},\ i,\ j = 1,\ \cdots,\ n$, where δ_{ij} is the Kronecker delta defined by

$$\delta_{ij} = \begin{cases} 1, & i = j \\ 0, & i \neq j \end{cases}$$

For vectors $x,\ y \in \mathbb{R}^n$, the inner product is defined by

$$(x,\ y) = x^{\mathrm{T}} y = \sum_{i=1}^{n} x_i y_i = y^{\mathrm{T}} x$$

Also, for $A \in \mathbb{R}^{n\times n}$ and $x \in \mathbb{R}^n$, we define the quadratic form

$$x^{\mathrm{T}}Ax = (x,\ Ax) = \sum_{i,j=1}^{n} a_{ij}x_ix_j \tag{2.1}$$

Define $\bar{A} = (A + A^{\mathrm{T}})/2$. Then, we have $x^{\mathrm{T}}\bar{A}x = x^{\mathrm{T}}Ax$. Thus it is assumed without loss of generality that A is symmetric in defining a quadratic form.

If $x^{\mathrm{T}}Ax > 0,\ x \neq 0$, then A is positive definite, and is written as $A > 0$. If $x^{\mathrm{T}}Ax \geq 0$ holds, A is called nonnegative definite, and is written as $A \geq 0$. Moreover, if $A - B > 0$ (or ≥ 0) holds, then we simply write $A > B$ (or $A \geq B$).

The basic facts for real vectors and matrices mentioned above can carry over to complex vectors and matrices. Let $\mathbb{C}$ be the set of complex numbers, $\mathbb{C}^n$ the set of n-dimensional complex vectors, and $\mathbb{C}^{m\times n}$ the set of $m \times n$ complex matrices. The complex conjugate of $\lambda \in \mathbb{C}$ is denoted by $\bar{\lambda}$, and similarly the complex conjugate transpose of $A = (a_{ij}) \in \mathbb{C}^{m\times n}$ is denoted by $A^{\mathrm{H}} = (\bar{a}_{ji})$. We say that $A \in \mathbb{C}^{n\times n}$ is Hermitian if $A^{\mathrm{H}} = A$, and unitary if $A^{\mathrm{H}}A = I_n$.

The inner product of $x,\ y \in \mathbb{C}^n$ is defined by

$$x^{\mathrm{H}}y = \sum_{i=1}^{n} \bar{x}_iy_i = \overline{y^{\mathrm{H}}x}$$

As in the real case, the quadratic form $x^{\mathrm{H}}Ax,\ x \in \mathbb{C}^n$ is defined for a Hermitian matrix A. We say that A is positive definite if $x^{\mathrm{H}}Ax > 0, x \neq 0$, and nonnegative definite if $x^{\mathrm{H}}Ax \geq 0$; being positive (nonnegative) definite is written as $A > 0$ $(A \geq 0)$.

The characteristic polynomial for $A \in \mathbb{R}^{n\times n}$ is defined by

$$\varphi_A(z) := \det(zI - A) = z^n + \alpha_1z^{n-1} + \cdots + \alpha_{n-1}z + \alpha_n \tag{2.2}$$

The n roots of $\varphi_A(z) = 0$ are called the eigenvalues of A. The set of eigenvalues of A, denoted by $\lambda(A)$, is called the spectrum of A. The ith eigenvalue is described by $\lambda_i(A)$. Since $\varphi_A(z)$ has real coefficients, if $\lambda \in \mathbb{C}$ is an eigenvalue, so is $\bar{\lambda} \in \mathbb{C}$. If $\lambda \in \lambda(A)$, there exists a vector $v \in \mathbb{C}^n$ satisfying

$$Av = \lambda v, \qquad v \neq 0$$

In this case, $v \in \mathbb{C}^n$ is called an eigenvector corresponding to the eigenvalue λ. It may be noted that, since the eigenvalues are complex, the corresponding eigenvectors are also complex.

Let the characteristic polynomial of $A \in \mathbb{R}^{n\times n}$ be given by (2.2). Then the following matrix polynomial equation holds:

$$\varphi_A(A) := A^n + \alpha_1A^{n-1} + \cdots + \alpha_{n-1}A + \alpha_n\,I = 0 \tag{2.3}$$

where the right-hand side is the zero matrix of size $n \times n$. This result is known as the Cayley-Hamilton theorem.

We see that the eigenvalues $\lambda_i,\ i = 1,\ \cdots,\ n$ of a symmetric nonnegative definite matrix $A \in \mathbb{R}^{n\times n}$ are nonnegative. Thus by means of an orthogonal matrix T, we can transform A into a diagonal form, *i.e.*,

$$T^{-1}AT = \begin{bmatrix} \lambda_1 & & & \\ & \lambda_2 & & \\ & & \ddots & \\ & & & \lambda_n \end{bmatrix} = \mathrm{diag}(\lambda_1,\ \lambda_2,\ \cdots,\ \lambda_n)$$

Define $\mu_i = \sqrt{\lambda_i}$, $i = 1,\ \cdots,\ n$. Then we have

$$A = T \begin{bmatrix} \mu_1 & & & \\ & \mu_2 & & \\ & & \ddots & \\ & & & \mu_n \end{bmatrix} \begin{bmatrix} \mu_1 & & & \\ & \mu_2 & & \\ & & \ddots & \\ & & & \mu_n \end{bmatrix} T^{-1}$$

Also, let B be given by

$$B = \begin{bmatrix} \mu_1 & & & \\ & \mu_2 & & \\ & & \ddots & \\ & & & \mu_n \end{bmatrix} T^{-1}$$

Then it follows that $A = B^{\mathrm{T}}B$, so that B is called a square root matrix of A, and is written as $\sqrt{A}$ or $A^{1/2}$. For any orthogonal matrix Q, we see that $B_1 = QB$ satisfies $A = B_1^{\mathrm{T}}B_1$ so that B_1 is also a square root matrix of A, showing that a square root matrix is not unique.

Suppose that $A = (a_{ij}) \in \mathbb{R}^{m\times n}$. Then, $A(p : q, r : s)$ denotes the submatrix of A formed by p, $p+1, \cdots$, q rows and r, $r+1, \cdots$, s columns, *e.g.*,

$$A(2:4;3:6) = \begin{bmatrix} a_{23} & a_{24} & a_{25} & a_{26} \\ a_{33} & a_{34} & a_{35} & a_{36} \\ a_{43} & a_{44} & a_{45} & a_{46} \end{bmatrix}$$

In particular, $A(p : q, :)$ means the submatrix formed by p, $p+1, \cdots$, q rows, and similarly $A(:, r : s)$ the submatrix formed by r, $r+1, \cdots$, s columns. Also, $A(i, :)$ and $A(:, j)$ respectively represent the ith row and jth column of A.

2.2 Subspaces and Linear Independence

In the following, we assume that scalars are real; but all the results are extended to complex scalars.

Consider a set $\mathcal{V}$ which is not void. For $x, y \in \mathcal{V}$, and for a scalar $\alpha \in \mathbb{R}$, the sum $x + y$ and product αx are defined. Suppose that the set $\mathcal{V}$ satisfies the axiom of "linear space" with respect to the addition and product defined above. Then $\mathcal{V}$ is called a linear space over $\mathbb{R}$. The set of n-dimensional vectors $\mathbb{R}^n$ and $\mathbb{C}^n$ are linear spaces over $\mathbb{R}$ and $\mathbb{C}$, respectively.

Suppose that $\mathcal{W}$ be a subset of a linear space $\mathcal{V}$. For any $w_1,\ w_2 \in \mathcal{W}$ and α_1, $\alpha_2 \in \mathbb{R}$, if $\alpha_1 w_1 + \alpha_2 w_2 \in \mathcal{W}$ holds, then $\mathcal{W}$ is called a subspace of $\mathcal{V}$, and this fact is simply expressed as $\mathcal{W} \subset \mathcal{V}$.

For a set of vectors $\{x_1,\ \cdots,\ x_n\}$ in $\mathbb{R}^m$, if there exist scalars $\alpha_1,\ \cdots,\ \alpha_n$ with $\alpha_i \neq 0$ for at least an i such that

$$\sum_{j=1}^{n} \alpha_j x_j = 0$$

holds, then $\{x_1,\ \cdots,\ x_n\}$ are called linearly dependent. Conversely, if we have

$$\sum_{j=1}^{n} \alpha_j x_j = 0 \quad \Rightarrow \quad \alpha_1 = \cdots = \alpha_n = 0$$

then $\{x_1,\ \cdots,\ x_n\}$ are called linearly independent.

All the linear combinations of vectors $\{w_1,\ \cdots,\ w_p\}$ in $\mathbb{R}^m$ form a subspace of $\mathbb{R}^n$, which is written as

$$\mathcal{W} = \mathrm{span}\{w_1,\ \cdots,\ w_p\} = \left\{ \sum_{j=1}^{p} \alpha_j w_j \;\middle|\; \alpha_1,\ \cdots,\ \alpha_p \in \mathbb{R} \right\}$$

If $\{w_1,\ \cdots,\ w_p\}$ are linearly independent, they are called a basis of the space $\mathcal{W}$.

Suppose that $\mathcal{V}$ is a subspace of $\mathbb{R}^m$. Then there exists a basis $\{v_1,\ \cdots,\ v_d\}$ in $\mathcal{V}$ such that

$$\mathcal{V} = \mathrm{span}\{v_1,\ \cdots,\ v_d\}$$

Hence, any $x \in \mathcal{V}$ can be expressed as a linear combination of the form

$$x = \sum_{j=1}^{d} \beta_j v_j, \qquad \beta_1,\ \cdots,\ \beta_d \in \mathbb{R}$$

where $\beta_1,\ \cdots,\ \beta_d$ are components of x with respect to the basis $\{v_1,\ \cdots,\ v_d\}$. Choice of basis is not unique, but the number of the elements of any basis is unique. The number is called the dimension of $\mathcal{V}$, which is denoted by $\dim(\mathcal{V})$.

For a matrix $A \in \mathbb{R}^{m \times n}$, the image of A is defined by

$$\mathrm{Im}(A) = \{y \in \mathbb{R}^m \mid y = Ax,\ x \in \mathbb{R}^n\} = A\mathbb{R}^n$$

This is a subspace of $\mathbb{R}^m$, and is also called the range of A. If $A = [a_1 \ \cdots \ a_n]$, then we have $\mathrm{Im}(A) = \mathrm{span}\{a_1,\ \cdots,\ a_n\}$. Moreover, the set of vectors mapped to zero are called the kernel of A, which is written as

$$\mathrm{Ker}(A) = \{x \in \mathbb{R}^n \mid Ax = 0\}$$

This is also called the null space of A, a subspace of $\mathbb{R}^n$.

The rank of $A \in \mathbb{R}^{m \times n}$ is defined by dim(Im A) and is expressed as rank(A). We see that rank(A) $= r$ if and only if the maximum number of independent vectors among the column vectors $a_1, \cdots, a_n$ of A is r. This is also equal to the number of independent vectors in row vectors $\tilde{a}_1^{\mathrm{T}}, \cdots, \tilde{a}_m^{\mathrm{T}}$ of A. Thus it follows that rank(A) = rank(A^{T}).

It can be shown that for $A \in \mathbb{R}^{m \times n}$,

$$\dim(\mathrm{Im}\, A) + \dim(\mathrm{Ker}\, A) = n \tag{2.4}$$

Hence, if $m = n$ holds, the following are equivalent:

(i) A : nonsingular (ii) $\mathrm{Ker}(A) = \{0\}$ (iii) $\mathrm{rank}(A) = n$

Suppose that $x,\ y \in \mathbb{R}^n$. If $x^{\mathrm{T}} y = 0$, or if the vectors are mutually orthogonal, we write $x \perp y$. If $y^{\mathrm{T}} x = 0$ holds for all $x \in \mathcal{V} \subset \mathbb{R}^n$, we say that y is orthogonal to $\mathcal{V}$, which is written as $y \perp \mathcal{V}$. The set of $y \in \mathbb{R}^n$ satisfying $y \perp \mathcal{V}$ is called the orthogonal complement, which is expressed as

$$\mathcal{V}^{\perp} = \{y \in \mathbb{R}^n \mid y^{\mathrm{T}} x = 0,\ \forall\, x \in \mathcal{V}\}$$

The orthogonal complement $\mathcal{V}^{\perp}$ is a subspace whether or not $\mathcal{V}$ is a subspace.

Let $\mathcal{V},\ \mathcal{W} \subset \mathbb{R}^n$ be two subspaces. If $v^{\mathrm{T}} w = 0$ holds for any $v \in \mathcal{V}$ and $w \in \mathcal{W}$, then we say that $\mathcal{V}$ and $\mathcal{W}$ are orthogonal, so that we write $\mathcal{V} \perp \mathcal{W}$. Also, the vector sum of $\mathcal{V}$ and $\mathcal{W}$ is defined by

$$\mathcal{V} \vee \mathcal{W} = \{v + w \mid\ v \in \mathcal{V},\ w \in \mathcal{W}\}$$

It may be noted that this is not the union $\mathcal{V} \cup \mathcal{W}$ of the two subspaces. Moreover, if $\mathcal{V} \cap \mathcal{W} = \{0\}$ holds, the vector sum is called the direct sum, and is written as $\mathcal{V} + \mathcal{W}$. Also, if $\mathcal{V} \perp \mathcal{W}$ holds, then it is called the direct orthogonal sum, and is expressed as $\mathcal{V} \oplus \mathcal{W}$.

For a subspace $\mathcal{V} \subset \mathbb{R}^n$, we have a unique decomposition

$$\mathbb{R}^n = \mathcal{V} \oplus \mathcal{V}^{\perp} \tag{2.5}$$

This implies that $x \in \mathbb{R}^n$ has a unique decomposition $x = v + w$, $v \in \mathcal{V},\ w \in \mathcal{V}^{\perp}$.

Let $\mathcal{V} \subset \mathbb{R}^n$ be a subspace, and $A \in \mathbb{R}^{n \times n}$ a linear transform. Then, if

$$x \in \mathcal{V} \Rightarrow Ax \in \mathcal{V} \quad (A\mathcal{V} \subset \mathcal{V})$$

holds, $\mathcal{V}$ is called an A-invariant subspace. The spaces spanned by eigenvectors, and Im(A), Ker(A) are all important A-invariant subspaces of $\mathbb{R}^n$.

2.3 Norms of Vectors and Matrices

Definition 2.1. *A vector norm* ($\|\cdot\|$) *has the following properties.*

(i) $||x|| \geq 0; \quad ||x|| = 0 \ \Leftrightarrow \ x = 0$

(ii) $||\lambda x|| = |\lambda| \, ||x||, \quad \lambda$ *: scalar*

(iii) $||x + y|| \leq ||x|| + ||y||$ *(triangular inequality)* □

For a vector $x = (x_1, \cdots, x_n)^{\mathrm{T}} \in \mathbb{R}^n$, the 2-norm (or Euclidean norm) is defined by

$$||x||_2 = \left(|x_1|^2 + \cdots + |x_n|^2\right)^{1/2}$$

and the infinity-norm is defined by

$$||x||_\infty = \max(|x_1|, \cdots, |x_n|)$$

Since $A \in \mathbb{R}^{m \times n}$ can be viewed as a vector in $\mathbb{R}^{mn}$, the definition of a matrix norm should be compatible with that of the vector norm. The most popular matrix norms are the Frobenius norm and the 2-norm. The former norm is given by

$$||A||_F = \sqrt{\sum_{i=1}^{m} \sum_{j=1}^{n} a_{ij}^2} = \sqrt{\mathrm{trace}(A^{\mathrm{T}} A)} \tag{2.6}$$

The latter is called an operator norm, which is defined by

$$||A||_2 = \sup_{x \neq 0} \frac{||Ax||_2}{||x||_2} \tag{2.7}$$

We have the following inequalities for the above two norms:

$$||Ax||_2 \leq ||A||_2 \, ||x||_2, \quad ||AB||_\alpha \leq ||A||_\alpha \, ||B||_\alpha, \quad \alpha = 2, \ F$$

If Q is orthogonal, *i.e.*, $Q^{\mathrm{T}} Q = I$, we have $||Qx||_2^2 = x^{\mathrm{T}} Q^{\mathrm{T}} Q x = ||x||_2^2$. Moreover, it follows that $||QA||_\alpha = ||A||_\alpha$ for $\alpha = 2, F$. Thus we see that the 2-norm and Frobenius norm are invariant under orthogonal transforms. We often write the 2-norm of x as $||x||$, suppressing the subscript.

For a complex vector $x \in \mathbb{C}^n$, and a complex matrix $A \in \mathbb{C}^{m \times n}$, their norms are defined similarly to the real cases.

Lemma 2.1. *For $A \in \mathbb{R}^{n \times n}$, the spectral radius is defined by*

$$\rho(A) = \max\{|\lambda_i(A)| \mid i = 1, \cdots, n\} \tag{2.8}$$

Then, $\rho(A) \leq ||A||_\alpha$ holds.

Proof. Clearly, there exists an eigenvalue λ for which $|\lambda| = \rho(A)$. Let $Ax = \lambda x$, $x \neq 0$. Let $X := [x \ \ x \ \ \cdots \ \ x] \in \mathbb{C}^{n \times n}$, and consider $AX = \lambda X$. Then, for any matrix norm $|| \cdot ||_\alpha$, we have

$$|\lambda| \cdot ||X||_\alpha = ||\lambda X||_\alpha = ||AX||_\alpha \leq ||A||_\alpha \cdot ||X||_\alpha, \qquad ||X||_\alpha \neq 0$$

and hence $|\lambda| = \rho(A) \leq ||A||_\alpha$. □

More precisely, the above result holds for many matrix norms [73].

2.4 QR Decomposition

In order to consider the QR decomposition, we shall introduce an orthogonal transform, called the Householder transform (see Figure 2.1).

Lemma 2.2. *Consider two vectors $x \neq y \in \mathbb{R}^n$ with $\|x\| = \|y\|$. Then there exists a vector $u \in \mathbb{R}^n$ such that*

$$(I - 2uu^{\mathrm{T}})x = y, \qquad \|u\| = 1 \tag{2.9}$$

The vector u is defined uniquely up to signature by

$$u = \pm \frac{x - y}{\|x - y\|} \tag{2.10}$$

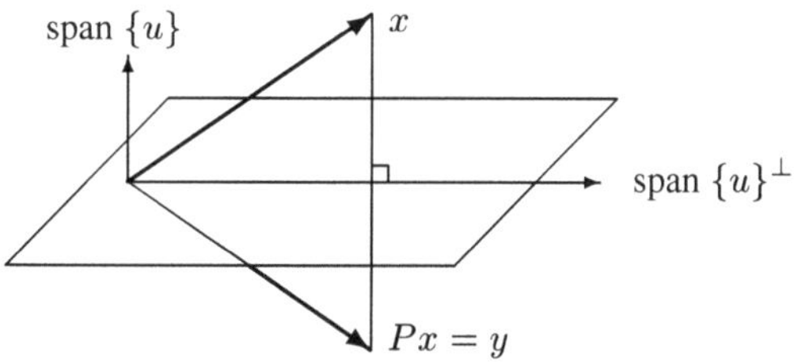

Figure 2.1. Householder transform

Proof. By using (2.10) and the fact that $\|x\| = \|y\|$ and $x^{\mathrm{T}}y = y^{\mathrm{T}}x$, we compute the left-hand side of (2.9) to get

$$\begin{aligned}(I - 2uu^{\mathrm{T}})x &= x - \frac{2(x-y)(x-y)^{\mathrm{T}}}{(x-y)^{\mathrm{T}}(x-y)}x = x - \frac{2(x-y)(x^{\mathrm{T}}x - y^{\mathrm{T}}x)}{x^{\mathrm{T}}x - x^{\mathrm{T}}y - y^{\mathrm{T}}x + y^{\mathrm{T}}y} \\ &= x - \frac{2(x-y)(x^{\mathrm{T}}x - y^{\mathrm{T}}x)}{2(x^{\mathrm{T}}x - y^{\mathrm{T}}x)} = y\end{aligned}$$

Suppose now that a vector $v \in \mathbb{R}^n$ also satisfies the condition of this lemma. Then we have $\|v\| = 1$, and hence

$$y = (I - 2uu^{\mathrm{T}})x = (I - 2vv^{\mathrm{T}})x \;\Rightarrow\; u(u^{\mathrm{T}}x) = v(v^{\mathrm{T}}x), \quad \forall\, x$$

Putting $x = u$ (or $x = v$) yields $v^{\mathrm{T}}u = \pm 1$. Thus it follows that $v = \pm u$, showing the uniqueness of the vector u up to signature. □

The matrix $P := I - 2uu^{\mathrm{T}}$ of Lemma 2.2, called the Householder transform, is symmetric and satisfies

$$P^2 = (I - 2uu^{\mathrm{T}})^2 = I - 4uu^{\mathrm{T}} + 4u(u^{\mathrm{T}}u)u^{\mathrm{T}} = I$$

Thus it follows that $P^{-1} = P = P^{\mathrm{T}}$, implying that P is an orthogonal transform, and that $Px = y$, $Py = x$ hold.

Let $a, b \in \mathbb{R}^n$ with $||a|| = ||b||$. We consider a problem of transforming the vector a into the vector b, of which the first element is nonzero, but the other elements are all zeros. More precisely, we wish to find a transform that performs the following reduction

$$a = \begin{bmatrix} a_1 \\ a_2 \\ \vdots \\ a_n \end{bmatrix} \longrightarrow b = \begin{bmatrix} b_1 \\ 0 \\ \vdots \\ 0 \end{bmatrix}, \quad ||a|| = ||b|| = |b_1|$$

Since $||a|| = ||b||$, we see that $b_1 = \pm||a||$. It follows that

$$\tilde{a} := a - b = \begin{bmatrix} a_1 - b_1 \\ a_2 \\ \vdots \\ a_n \end{bmatrix}, \quad ||\tilde{a}||^2 = 2b_1(b_1 - a_1) = -2b^{\mathrm{T}}\tilde{a}$$

It should be noted that if the sign of b_1 is chosen as the same as that of a_1, there is a possibility that $|a_1 - b_1|$ has a small value, so that a large relative error may arise in the term $u^{\mathrm{T}}u = \tilde{a}^{\mathrm{T}}\tilde{a}$. This difficulty can be simply avoided by choosing the sign of b_1 opposite to that of a_1.

We now define

$$P := I - \frac{2\tilde{a}\tilde{a}^{\mathrm{T}}}{||\tilde{a}||^2} = I + \frac{\tilde{a}\tilde{a}^{\mathrm{T}}}{b^{\mathrm{T}}\tilde{a}} = I + \frac{\tilde{a}\tilde{a}^{\mathrm{T}}}{b_1\tilde{a}_1} \tag{2.11}$$

Noting that $a^{\mathrm{T}}a = b_1^2$ and $b^{\mathrm{T}}a = b_1a_1$, we have

$$Pa = \left[I + \frac{(a-b)(a-b)^{\mathrm{T}}}{b_1\tilde{a}_1}\right]a = a - \frac{(a-b)(a^{\mathrm{T}}a - b^{\mathrm{T}}a)}{b_1(b_1 - a_1)} = b$$

Hence, by knowing $\tilde{a} = a - b$ and b_1, the vector a can be transformed into the vector b with the specified form. It is well known that this method is very efficient and reliable, since only the first component of a is modified in the computation. In the following, $\tilde{a}$ plays the role of the vector u in the Householder transform, though the norm of $\tilde{a}$ is not unity.

Now we introduce the QR decomposition, which is quite useful in numerical linear algebra. We assume that matrices are real, though the QR decomposition is applicable to complex matrices.

Lemma 2.3. *A tall rectangular matrix $A \in \mathbb{R}^{m\times n}$, $m \geq n$ is decomposed into a product of two matrices:*

$$A = QR \tag{2.12}$$

where $Q \in \mathbb{R}^{m\times n}$ is an orthogonal matrix with $Q^{\mathrm{T}}Q = I_n$, and $R \in \mathbb{R}^{n\times n}$ is an upper triangular matrix. The right-hand side of (2.12) is called the QR decomposition of A.

Proof. The decomposition is equivalent to $Q^{\mathrm{T}}A = R$, so that Q^{T} is an orthogonal matrix that transforms a given matrix A into an upper triangular matrix. In the following, we give a method of performing this transform by means of the Householder transforms.

Let $a^{(1)} = A(:,1)$, the first column vector of A. By computing $u^{(1)} := \tilde{a}$ and $b_1^{(1)}$, we perform the following transform:

$$a^{(1)} := \begin{bmatrix} a_{11} \\ a_{21} \\ \vdots \\ a_{m1} \end{bmatrix} \longrightarrow u^{(1)} = \begin{bmatrix} a_{11} - b_1^{(1)} \\ a_{21} \\ \vdots \\ a_{m1} \end{bmatrix}, \quad b^{(1)} = \begin{bmatrix} b_1^{(1)} \\ 0 \\ \vdots \\ 0 \end{bmatrix}$$

where $b_1^{(1)} = \pm\|a^{(1)}\|$. According to (2.11), let

$$P^{(1)} := I + u^{(1)}(u^{(1)})^{\mathrm{T}}/(b^{(1)})^{\mathrm{T}}u^{(1)}$$

and $P^{(1)}A := A^{(1)}$. Then we get

$$P^{(1)}A = A^{(1)} = \begin{bmatrix} b_1^{(1)} & a_{12}^{(1)} & a_{13}^{(1)} & \cdots & a_{1n}^{(1)} \\ 0 & a_{22}^{(1)} & a_{23}^{(1)} & \cdots & a_{2n}^{(1)} \\ \vdots & \vdots & \vdots & & \vdots \\ 0 & a_{m2}^{(1)} & a_{m3}^{(1)} & \cdots & a_{mn}^{(1)} \end{bmatrix}$$

Thus the first column vector of $A^{(1)}$ is reduced to the vector $b^{(1)}$, where the column vectors $a_2, \cdots, a_n$ are subject to effects of $P^{(1)}$. But, in the transforms that follow, the vector $b^{(1)} := A^{(1)}(:,1)$ is intact, and this becomes the first column of R.

Next we consider the transforms of the second column vector of $A^{(1)}$. We define $a^{(2)}$, $u^{(2)}$ and $b^{(2)}$ as

$$a^{(2)} := \begin{bmatrix} 0 \\ a_{22}^{(1)} \\ a_{32}^{(1)} \\ \vdots \\ a_{m2}^{(1)} \end{bmatrix} \longrightarrow u^{(2)} = \begin{bmatrix} 0 \\ a_{22}^{(1)} - b_2^{(2)} \\ a_{32}^{(1)} \\ \vdots \\ a_{m2}^{(1)} \end{bmatrix}, \quad b^{(2)} = \begin{bmatrix} 0 \\ b_2^{(2)} \\ 0 \\ \vdots \\ 0 \end{bmatrix}$$

where $b_2^{(2)} = \pm\|a^{(2)}\|$. Let $P^{(2)}$ be defined by

$$P^{(2)} := I + u^{(2)}(u^{(2)})^{\mathrm{T}}/(b^{(2)})^{\mathrm{T}}u^{(2)} = \left[\begin{array}{c|ccc} 1 & 0 & \cdots & 0 \\ \hline 0 & P_{22}^{(2)} & \cdots & P_{2m}^{(2)} \\ \vdots & \vdots & \ddots & \vdots \\ 0 & P_{m2}^{(2)} & \cdots & P_{mm}^{(2)} \end{array}\right]$$

We see that $P^{(2)}$ is an orthogonal matrix, for which all the elements of the first row and column are zero except for $(1,1)$-element. Thus pre-multiplying $A^{(1)}$ by $P^{(2)}$ yields

$$P^{(2)}A^{(1)} = P^{(2)}P^{(1)}A = A^{(2)} = \begin{bmatrix} b_1^{(1)} & a_{12}^{(1)} & a_{13}^{(2)} & \cdots & a_{1n}^{(2)} \\ & b_2^{(2)} & a_{23}^{(2)} & \cdots & a_{2n}^{(2)} \\ & & a_{33}^{(2)} & \cdots & a_{3n}^{(2)} \\ & & \vdots & & \vdots \\ 0 & & a_{m3}^{(2)} & \cdots & a_{mn}^{(2)} \end{bmatrix}$$

where we note that the first row and column of $A^{(2)}$ are the same as those of $A^{(1)}$ due to the form of $P^{(2)}$.

Repeating this procedure until the nth column, we get an upper triangular matrix $A^{(n)}$ of the form

$$P^{(n)}P^{(n-1)}\cdots P^{(1)}A = A^{(n)} = \begin{bmatrix} R \\ 0 \end{bmatrix} \tag{2.13}$$

Since each component $P^{(j)}$, $j = 1, \cdots, n$ is orthogonal and symmetric, we get

$$A = P^{(1)}P^{(2)}\cdots P^{(n)}A^{(n)} = Q\begin{bmatrix} R \\ 0 \end{bmatrix}$$

where $R \in \mathbb{R}^{n\times n}$ is upper triangular and $Q = [q_1, \cdots, q_n] \in \mathbb{R}^{m\times n}$ is orthogonal. This completes a proof of lemma. □

The QR decomposition is quite useful for computing an orthonormal basis for a set of vectors. In fact, it is a matrix realization of the Gram-Schmidt orthogonalization process. Suppose that $A \in \mathbb{R}^{m\times n}$ and $\text{rank}(A) = n$. Let the QR decomposition of A be given by

$$A = [Q_A \;\; Q_A^{\perp}]\begin{bmatrix} R \\ 0 \end{bmatrix} = Q_A R, \qquad Q_A \in \mathbb{R}^{m\times n} \tag{2.14}$$

Since R is nonsingular, we have $\text{Im}(A) = \text{Im}(Q_A)$, *i.e.*, the column vectors of Q_A form an orthonormal basis of $\text{Im}(A)$, and those of $Q_A^{\perp}$ forms an orthonormal basis of the orthogonal complement $(\text{Im}\, A)^{\perp}$.

It should, however, be noted that if $\text{rank}(A) = r < n$, the QR decomposition does not necessarily gives an orthonormal basis for $\text{Im}(A)$, since some of the diagonal elements of R become zero. For example, consider the following QR decomposition

$$A = [a_1 \;\; a_2 \;\; a_3] = [q_1 \;\; q_2 \;\; q_3]\begin{bmatrix} 1 & 2 & 1 \\ 0 & 0 & 1 \\ 0 & 0 & 1 \end{bmatrix}$$

Though we see that $\text{rank}(A) = 2$, it is impossible to span $\text{Im}(A)$ by any two vectors from q_1, q_2, q_3. But, for $\text{rank}(A) = r < n$, it is easy to modify the QR decomposition algorithm so that the r column vectors $(q_1, \cdots, q_r)$ form an orthonormal basis of $\text{Im}(A)$ with column pivoting; see [59].

2.5 Projections and Orthogonal Projections

Definition 2.2. *Suppose that $\mathbb{R}^n$ be given by a direct sum of subspaces $\mathcal{V}$ and $\mathcal{W}$, i.e.,*

$$\mathbb{R}^n = \mathcal{V} + \mathcal{W}, \qquad \mathcal{V} \cap \mathcal{W} = \{0\}$$

Then, $x \in \mathbb{R}^n$ can be uniquely expressed as

$$x = v + w, \qquad v \in \mathcal{V}, \qquad w \in \mathcal{W} \tag{2.15}$$

where v is the projection of x onto $\mathcal{V}$ along $\mathcal{W}$, and w is the projection of x onto $\mathcal{W}$ along $\mathcal{V}$. The uniqueness follows from $\mathcal{V} \cap \mathcal{W} = \{0\}$. □

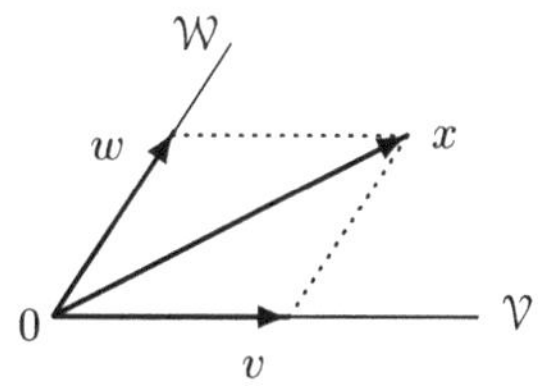

Figure 2.2. Oblique (or parallel) projection

The projection is often called the oblique (or parallel) projection, see Figure 2.2. We write the projection operator that transforms x onto $\mathcal{V}$ along $\mathcal{W}$ as $P_{\|\mathcal{W}}^{\mathcal{V}}$. Then, we have $v = P_{\|\mathcal{W}}^{\mathcal{V}}(x)$ and $w = P_{\|\mathcal{V}}^{\mathcal{W}}(x)$, and hence the unique decomposition of (2.15) is written as

$$x = P_{\|\mathcal{W}}^{\mathcal{V}}(x) + P_{\|\mathcal{V}}^{\mathcal{W}}(x)$$

We show that the projection is a linear operator. For $x,\ y \in \mathbb{R}^n$, we have the following decompositions

$$x = v + w, \qquad y = u + z, \qquad v,\ u \in \mathcal{V}, \qquad w,\ z \in \mathcal{W}$$

Since $x + y = (v + u) + (w + z)$, $u + v \in \mathcal{V}$, $w + z \in \mathcal{W}$, we see that $v + u$ is the oblique projection of $x + y$ onto $\mathcal{V}$ along $\mathcal{W}$. Hence, we have

$$P_{\|\mathcal{W}}^{\mathcal{V}}(x + y) = v + u = P_{\|\mathcal{W}}^{\mathcal{V}}(x) + P_{\|\mathcal{W}}^{\mathcal{V}}(y)$$

Moreover, for any α, we get $\alpha x = \alpha v + \alpha w$, $\alpha v \in \mathcal{V}$, $\alpha w \in \mathcal{W}$, so that αv is the oblique projection of αx onto $\mathcal{V}$ along $\mathcal{W}$, implying that

$$P_{\|\mathcal{W}}^{\mathcal{V}}(\alpha x) = \alpha v = \alpha P_{\|\mathcal{W}}^{\mathcal{V}}(x)$$

From the above, we see that the projection $P_{\|\mathcal{W}}^{\mathcal{V}}$ is a linear operator on $\mathbb{R}^n$, so that it can be expressed as a matrix.

Lemma 2.4. *Suppose that $P \in \mathbb{R}^{n\times n}$ is idempotent, i.e.,*

$$P^2 = P \tag{2.16}$$

Then, we have

$$\mathrm{Ker}(P) = \mathrm{Im}(I_n - P) \tag{2.17}$$

and vice versa.

Proof. Let $x \in \mathrm{Ker}(P)$. Then, since $Px = 0$, we get $x = (I - P)x \in \mathrm{Im}(I - P)$, implying that $\mathrm{Ker}(P) \subset \mathrm{Im}(I - P)$. Also, for any $x \in \mathbb{R}^n$, we see that $P(I - P)x = 0$, showing that $\mathrm{Im}(I - P) \subset \mathrm{Ker}(P)$. This proves (2.17). Conversely, for any $z \in \mathbb{R}^n$, let $x = (I - P)z$. Then, we have $x \in \mathrm{Ker}(P)$, so that $0 = Px = P(I - P)z$ holds for any $z \in \mathbb{R}^n$, implying that $P^2 = P$. □

Corollary 2.1. *Suppose that (2.16) holds. Then, we have*

$$\mathbb{R}^n = \mathrm{Im}(P) + \mathrm{Ker}(P) \tag{2.18}$$

Proof. Since any $x \in \mathbb{R}^n$ can be written as $x = Px + (I - P)x$, we see from (2.17) that

$$\mathbb{R}^n = \mathrm{Im}(P) \vee \mathrm{Im}(I - P) = \mathrm{Im}(P) \vee \mathrm{Ker}(P) \tag{2.19}$$

Now let $x \in \mathrm{Im}(P) \cap \mathrm{Ker}(P)$. Then we have $x = Py$, $y \in \mathbb{R}^n$ and $Px = 0$. From (2.16), we get $0 = Px = P^2 y = Py = x$ and hence $\mathrm{Im}(P) \cap \mathrm{Ker}(P) = \{0\}$. Thus the right-hand side of (2.19) is expressed as the direct sum. □

We now provide a necessary and sufficient condition such that P is a matrix that represents an oblique projection.

Lemma 2.5. *A matrix $P \in \mathbb{R}^{n\times n}$ is the projection matrix onto* $\mathrm{Im}(P)$ *along* $\mathrm{Ker}(P)$ *if and only if (2.16) holds.*

Proof. We prove the necessity. Since, for any $x \in \mathbb{R}^n$, $v = Px \in \mathrm{Im}(P)$, we have $P(Px) = Pv = v = Px$ for all x, implying that $P^2 = P$ holds. Conversely, to prove the sufficiency, we define

$$\mathcal{V} := \{v \mid v = Px,\ x \in \mathbb{R}^n\}, \qquad \mathcal{W} := \{w \mid w = (I - P)x,\ x \in \mathbb{R}^n\}$$

Since $\mathcal{V} \cap \mathcal{W} = \{0\}$, Lemma 2.4 implies that $x \in \mathbb{R}^n$ is decomposed uniquely as

$$x = Px + (I - P)x = v + w, \qquad v \in \mathcal{V}, \qquad w \in \mathcal{W}$$

From Definition 2.2, we see that P is the projection matrix onto $\mathcal{V} = \mathrm{Im}(P)$ along $\mathcal{W} = \mathrm{Ker}(P)$. □

Example 2.1. It can be shown that $P \in \mathbb{R}^{n\times n}$ is a projection if and only if P is expressed as

$$P = T\Delta_r T^{-1} \tag{2.20}$$

where T is a nonsingular matrix, and Δ_r is given by

$$\Delta_r = \text{diag}(\underbrace{1, \cdots, 1}_{r}, 0, \ldots, 0) \tag{2.21}$$

In fact, it is obvious that P of (2.20) satisfies $P^2 = P$. Conversely, suppose that $P^2 = P$ holds. Let

$$\text{Im}(P) = \text{span}\{t_1, \cdots, t_r\}, \qquad \text{Ker}(P) = \text{span}\{t_{r+1}, \cdots, t_n\}$$

Noting that $x \in \text{Im}(P) \Leftrightarrow Px = x$ and that $x \in \text{Ker}(P) \Leftrightarrow Px = 0$, we get

$$P[t_1 \ \cdots \ t_r \ t_{r+1} \ \cdots \ t_n] = [t_1 \ \cdots \ t_r \ t_{r+1} \ \cdots \ t_n] \begin{bmatrix} I_r & 0 \\ 0 & 0 \end{bmatrix}$$

From Corollary 2.1, $T = [t_1 \ \cdots \ t_n]$ is nonsingular, showing that (2.20) holds.

Thus it follows from (2.20) that if $P^2 = P$, then $\text{rank}(P) = \text{trace}(P)$. □

Definition 2.3. *Suppose that $\mathcal{V} \subset \mathbb{R}^n$. Then, any $x \in \mathbb{R}^n$ can uniquely be decomposed as*

$$x = v + w, \qquad v \in \mathcal{V}, \qquad w \in \mathcal{V}^\perp \tag{2.22}$$

This is a particular case with $\mathcal{W} = \mathcal{V}^\perp$ in Definition 2.2, and v is called the orthogonal projection of x onto $\mathcal{V}$. See Figure 2.3 below. □

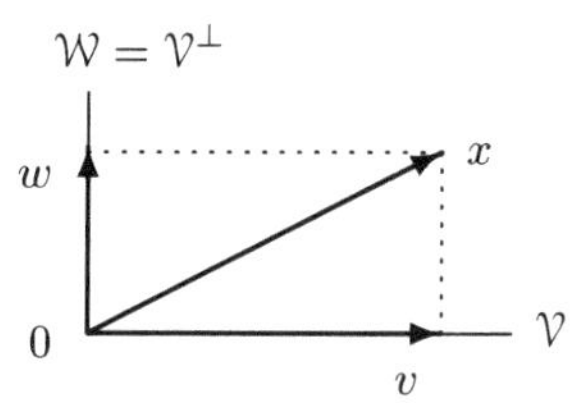

Figure 2.3. Orthogonal projection

For $x, y \in \mathbb{R}^n$, we consider the orthogonal decompositions $x = v_1 + w_1$ and $y = v_2 + w_2$, where $v_1, v_2 \in \mathcal{V}$ and $w_1, w_2 \in \mathcal{V}^\perp$. Let P be the orthogonal projection onto $\mathcal{V}$ along $\mathcal{V}^\perp$. Then, $v_1 = Px$, $v_2 = Py$. Since $v_2 \perp w_1$, $v_1 \perp w_2$,

$$\begin{aligned}(x, Py) &= (v_1 + w_1, v_2) = (v_1, v_2) = (v_1, v_2 + w_2) \\ &= (Px, y) = (x, P^{\mathrm{T}} y)\end{aligned}$$

holds for any x, y, so that we have $P = P^{\mathrm{T}}$. The next lemma provides a necessary and sufficient condition such that P is an orthogonal projection.

Lemma 2.6. *The matrix $P \in \mathbb{R}^{n\times n}$ is the orthogonal projection onto $\text{Im}(P)$ if and only if the following two conditions hold.*

$$(i) \quad P^2 = P \qquad\qquad (ii) \quad P^{\mathrm{T}} = P \tag{2.23}$$

Proof. (Necessity) It is clear from Lemma 2.5 that $P^2 = P$ holds. The fact that $P^{\mathrm{T}} = P$ is already proved above.

(Sufficiency) It follows from Lemma 2.5 that the condition (i) implies that P is the projection matrix onto $\mathrm{Im}(P)$ along $\mathrm{Ker}(P)$. Condition (ii) implies that $\mathrm{Ker}(P) = \mathrm{Ker}(P^{\mathrm{T}}) = (\mathrm{Im}\ P)^{\perp}$. This means that the sufficiency part holds. □

Let $A \in \mathbb{R}^{n \times r}$ with $\mathrm{rank}(A) = r$ and $\mathrm{Im}(A) = \mathcal{A} \subset \mathbb{R}^n$. Let the QR decomposition of A be given by (2.14). Then, it follows that $\mathrm{Im}(Q_A) = \mathcal{A}$. Also, define

$$P_A = Q_A Q_A^{\mathrm{T}} \in \mathbb{R}^{n \times n} \tag{2.24}$$

It is clear that $P_A^{\mathrm{T}} = P_A$ and $P_A^2 = P_A$, so that the conditions (i) and (ii) of Lemma 2.6 are satisfied. Therefore, if we decompose $z \in \mathbb{R}^n$ as

$$z = x + y, \qquad x \in \mathcal{A}, \qquad y \in \mathcal{A}^{\perp} \tag{2.25}$$

then we get $x = P_A z$ and $y = (I - P_A)z$. Hence, P_A and $I - P_A$ are orthogonal projections onto $\mathcal{A}(= \mathrm{Im}\ A)$ and $\mathcal{A}^{\perp}$, respectively.

Lemma 2.7. *Suppose that $\mathcal{A}$ is a subspace of $\mathbb{R}^n$. Then, for any $z \in \mathbb{R}^n$, $P_A z$ is the unique vector satisfying the following*

$$\min_{x \in \mathcal{A}} \|z - x\| = \|z - P_A z\|$$

Proof. If $z \in \mathcal{A}$, then $P_A z = z$. Now suppose that $z \notin \mathcal{A}$. For any $x \in \mathcal{A}$, we have $x - P_A z \in \mathcal{A}$, but $(I - P_A)z$ is orthogonal to $\mathcal{A}$. Thus it follows that $x - P_A z \perp (I - P_A)z$. Hence,

$$\|z - x\|^2 = \|(I - P_A)z - (x - P_A z)\|^2 = \|(I - P_A)z\|^2 + \|x - P_A z\|^2$$

The right-hand side is minimized by $x = P_A z$, which is unique. □

2.6 Singular Value Decomposition

Though the singular value decomposition (SVD) can be applied to complex matrices, it is assumed here that matrices are real.

Lemma 2.8. *Suppose that the rank of $A \in \mathbb{R}^{m \times n}$ is $r \le \min(n, m)$. Then, there exist orthogonal matrices $U \in \mathbb{R}^{m \times m}$ and $V \in \mathbb{R}^{n \times n}$ such that*

$$A = U \begin{bmatrix} \Sigma_+ & 0 \\ 0 & 0 \end{bmatrix} V^{\mathrm{T}}, \qquad \Sigma_+ = \begin{bmatrix} \sigma_1 & & & \\ & \sigma_2 & & \\ & & \ddots & \\ & & & \sigma_r \end{bmatrix} \tag{2.26}$$

where $U^{\mathrm{T}}U = I_m$, $V^{\mathrm{T}}V = I_n$, and

$$\sigma_1 \geq \sigma_2 \geq \cdots \geq \sigma_r > \sigma_{r+1} = \cdots = \sigma_p = 0, \quad p = \min(m, n)$$

We say that $\sigma_1, \cdots, \sigma_p$ are the singular values of A, and that (2.26) is the singular value decomposition (SVD).

Proof. Suppose that we know the eigenvalue decomposition of a nonnegative definite matrix. Since $A^{\mathrm{T}}A \in \mathbb{R}^{n\times n}$ is nonnegative definite, it can be diagonalized by an orthogonal transform $V \in \mathbb{R}^{n\times n}$. Let the eigenvalues of $A^{\mathrm{T}}A$ be given by $\lambda_1, \lambda_2, \cdots, \lambda_n$, and let the corresponding eigenvectors be given by $v_1, v_2, \cdots, v_n \in \mathbb{R}^n$. Thus we have $A^{\mathrm{T}}Av_i = \lambda_i v_i$, $i = 1, \cdots, n$. However, since $\mathrm{rank}(A) = r$, we have $\lambda_1 \geq \lambda_2 \geq \cdots \geq \lambda_r > \lambda_{r+1} = \cdots = \lambda_n = 0$. Define $\sigma_i = \sqrt{\lambda_i}$, $i = 1, \cdots, n$, and $V = [V_r \;\; \tilde{V}_r]$, where

$$V_r = [v_1 \; v_2 \; \cdots \; v_r], \qquad \tilde{V}_r = [v_{r+1} \; v_{r+2} \; \cdots \; v_n]$$

It then follows that $V^{\mathrm{T}}V = I_n$ and that

$$A^{\mathrm{T}}Av_i = \sigma_i^2 v_i, \qquad i = 1, \cdots, r \tag{2.27a}$$

$$A^{\mathrm{T}}Av_i = 0, \qquad i = r+1, \cdots, n \tag{2.27b}$$

Also we define $U_r := AV_r\Sigma_+^{-1} \in \mathbb{R}^{m\times r}$. We see from (2.27a) that $A^{\mathrm{T}}AV_r = V_r\Sigma_+^2$ holds and

$$U_r^{\mathrm{T}}U_r = \Sigma_+^{-1}V_r^{\mathrm{T}}A^{\mathrm{T}}AV_r\Sigma_+^{-1} = \Sigma_+^{-1}V_r^{\mathrm{T}}(V_r\Sigma_+^2)\Sigma_+^{-1} = I_r \tag{2.28}$$

In other words, the column vectors in U_r form a set of orthonormal basis.

Now we choose $\tilde{U}_r \in \mathbb{R}^{m\times(m-r)}$ so that

$$U = [U_r \;\; \tilde{U}_r] \in \mathbb{R}^{m\times m}$$

is an orthogonal matrix, *i.e.*, $U^{\mathrm{T}}U = I_m$. Then it follows that

$$U^{\mathrm{T}}AV = \begin{bmatrix} U_r^{\mathrm{T}} \\ \tilde{U}_r^{\mathrm{T}} \end{bmatrix} A\,[V_r \;\; \tilde{V}_r] = \begin{bmatrix} U_r^{\mathrm{T}}AV_r & U_r^{\mathrm{T}}A\tilde{V}_r \\ \tilde{U}_r^{\mathrm{T}}AV_r & \tilde{U}_r^{\mathrm{T}}A\tilde{V}_r \end{bmatrix}$$

We see from (2.28) that the $(1,1)$-block element of the right-hand side of the above equation is Σ_+. From (2.27b), we get $A\tilde{V}_r = 0$. Thus $(1,2)$- and $(2,2)$-block elements are zero matrices. Also, since $\tilde{U}_r$ and U_r are orthogonal, $\tilde{U}_r^{\mathrm{T}}AV_r\Sigma_+^{-1} = 0$, implying that $(2,1)$-block element is also zero matrix. Thus we have shown that

$$U^{\mathrm{T}}AV = \begin{bmatrix} \Sigma_+ & 0 \\ 0 & 0 \end{bmatrix} = \Sigma$$

This completes the proof. □

It is clear that (2.26) can be expressed as

$$A = U\Sigma V^{\mathrm{T}} = U_r\Sigma_+V_r^{\mathrm{T}} \tag{2.29}$$

where $U_r \in \mathbb{R}^{m\times r}$ and $V_r \in \mathbb{R}^{n\times r}$. Note that in the following, we often write (2.29) as $A = U\Sigma_+ V^{\mathrm{T}}$, which is called the reduced SVD.

Let $\sigma(A)$ be the set of singular values of A, and $\sigma_i(A)$ ith singular value. As we can see from the above proof, the singular values of A are equal to the positive square roots of the eigenvalues of $A^{\mathrm{T}}A$, *i.e.*, for $A \in \mathbb{R}^{m\times n}$,

$$\sigma_i(A) = \sqrt{\lambda_i(A^{\mathrm{T}}A)}, \qquad i = 1, \cdots, n$$

Also, the column vectors of U, the left singular vectors of A, are the eigenvectors of AA^{T}, and the column vectors of V, the right singular vectors of A, are the eigenvectors of $A^{\mathrm{T}}A$. From (2.29), we have $AV_r = U_r\Sigma_+$ and $A^{\mathrm{T}}U_r = V_r\Sigma_+$, so that the ith right singular vector and the ith left singular vector are related by

$$Av_i = \sigma_i u_i, \qquad A^{\mathrm{T}}u_i = \sigma_i v_i, \qquad i = 1, \cdots, r$$

In the following, $\overline{\sigma}(A)$ and $\underline{\sigma}(A)$ denote the maximum and the minimum singular values, respectively.

Lemma 2.9. *Suppose that* $\operatorname{rank}(A) = r \le \min(m, n)$. *Then, the following properties (i)∼(v) hold.*

(i) Images and kernels of A and A^{T}:

$$\operatorname{Im}(A) = \operatorname{Im}(U_r), \qquad \operatorname{Ker}(A) = \operatorname{Im}(\tilde{V}_r)$$

$$\operatorname{Im}(A^{\mathrm{T}}) = \operatorname{Im}(V_r), \qquad \operatorname{Ker}(A^{\mathrm{T}}) = \operatorname{Im}(\tilde{U}_r)$$

(ii) The dyadic decomposition of A:

$$A = \sum_{i=1}^{r} \sigma_i u_i v_i^{\mathrm{T}} \ (= U\Sigma V^{\mathrm{T}})$$

(iii) The Frobenius norm and 2-norm:

$$||A||_F = \sqrt{\sigma_1^2 + \cdots + \sigma_r^2}, \qquad ||A||_2 = \sigma_1$$

(iv) Equivalence of norms:

$$||A||_2 \le ||A||_F \le \sqrt{p}\,||A||_2, \qquad p = \min(m, n)$$

(v) The approximation by a lower rank matrix: Define the matrix A_k by

$$A_k = \sum_{i=1}^{k} \sigma_i u_i v_i^{\mathrm{T}}, \qquad k < r$$

Then, we have $\operatorname{rank}(A_k) = k$, *and*

$$\min_{\operatorname{rank}(B)=k} ||A - B||_2 = ||A - A_k||_2 = \sigma_{k+1}$$

where $B \in \mathbb{R}^{m\times n}$.

Proof. For a proof, see [59]. We prove only (v). Since

$$A - A_k = U\text{diag}(0, \cdots, 0, \sigma_{k+1}, \cdots, \sigma_p)V^{\mathrm{T}}$$

we have $||A - A_k||_2 = \sigma_{k+1}$. Let $B \in \mathbb{R}^{m\times n}$ be a matrix with rank k. Then, it suffices to show that $||A - B||_2 \geq \sigma_{k+1}$. Let $\{x_i \in \mathbb{R}^n, i = 1, \cdots, n-k\}$ be orthonormal vectors such that $\mathrm{Ker}(B) = \mathrm{span}\{x_1, \cdots, x_{n-k}\}$. Define also $\mathcal{V}_{k+1} := \mathrm{span}\{v_1, \cdots, v_{k+1}\}$. We see that $\dim \mathrm{Ker}(B) = n - k$ and $\dim(\mathcal{V}_{k+1}) = k+1$. But $\mathrm{Ker}(B)$ and $\mathcal{V}_{k+1}$ are subspaces of $\mathbb{R}^n$, so that $\mathrm{Ker}(B) \cap \mathcal{V}_{k+1} \neq \{0\}$.

Let $z \in \mathrm{Ker}(B) \cap \mathcal{V}_{k+1} \subset \mathbb{R}^n$ be a vector with $||z|| = 1$. Then it follows that $Bz = 0$ and

$$Az = \sum_{i=1}^{p} \sigma_i u_i(v_i^{\mathrm{T}} z) = \sum_{i=1}^{k+1} \sigma_i (v_i^{\mathrm{T}} z) u_i$$

Since $(v_i^{\mathrm{T}} z)^2 \leq ||v_i||^2 ||z||^2 = 1$, we have

$$||A - B||_2^2 \geq ||(A - B)z||^2 = ||Az||^2 = \sum_{i=1}^{k+1} \sigma_i^2 (v_i^{\mathrm{T}} z)^2 \geq \sigma_{k+1}^2$$

as was to be proved. □

Finding the rank of a matrix is most reliably done by the SVD. Let the SVD of $A \in \mathbb{R}^{m\times n}$ be given by (2.26). Let E be a perturbation to the matrix A, and $\{\tilde{\sigma}_i, i = 1, \cdots, p\}$ be the singular values of the perturbed matrix $A + E$. Then, it follows from [59] that

$$|\sigma_i - \tilde{\sigma}_i| < ||E||_2 = \overline{\sigma}(E), \qquad i = 1, \cdots, p \tag{2.30}$$

This implies that the singular values are not very sensitive to perturbations.

We now define a matrix B with rank $r - 1$ as

$$B = U\text{diag}(\sigma_1, \cdots, \sigma_{r-1}, 0, \cdots, 0)V^{\mathrm{T}}$$

Then we have $||A - B||_2 = \sigma_r$. Thus, for any matrix B satisfying $||A - B||_2 < \sigma_r$, the rank of B is greater than or equal to r. Hence, as a "zero threshold," if we can choose a number $\delta < \sigma_r$, we can say that A has numerical rank r. Thus the smallest nonzero singular value plays a significant role in determining numerical rank of matrices.

2.7 Least-Squares Method

In this section, we consider the least-squares problem:

$$\min_{x\in\mathbb{R}^n} ||Ax - b||, \qquad A \in \mathbb{R}^{m\times n}, \qquad b \in \mathbb{R}^m \tag{2.31}$$

where $m \geq n$. Suppose that $\mathrm{rank}(A) = n$, and let the QR decomposition of A be given by

$$A = Q\begin{bmatrix} R \\ 0 \end{bmatrix}, \qquad Q \in \mathbb{R}^{m\times m}, \qquad R \in \mathbb{R}^{n\times n}$$

Since the 2-norm is invariant under orthogonal transforms, we have

$$||Ax - b||^2 = ||Q^{\mathrm{T}}(Ax - b)||^2 = \left\| \begin{bmatrix} R \\ 0 \end{bmatrix} x - \begin{bmatrix} b_1 \\ b_2 \end{bmatrix} \right\|^2, \quad Q^{\mathrm{T}} b = \begin{bmatrix} b_1 \\ b_2 \end{bmatrix}$$

where $b_1 \in \mathbb{R}^n$, $b_2 \in \mathbb{R}^{m-n}$. Hence, it follows that

$$||Ax - b||^2 = ||Rx - b_1||^2 + ||b_2||^2$$

Since the second term $||b_2||^2$ in the right-hand side is independent of x, the least-squares problem is reduced to

$$Rx = b_1 \quad \Rightarrow \quad \begin{bmatrix} r_{11} & r_{12} & \cdots & r_{1n} \\ & r_{22} & \cdots & r_{2n} \\ & & \ddots & \vdots \\ \text{0} & & & r_{nn} \end{bmatrix} \begin{bmatrix} x_1 \\ x_2 \\ \vdots \\ x_n \end{bmatrix} = \begin{bmatrix} \beta_1 \\ \beta_2 \\ \vdots \\ \beta_n \end{bmatrix}$$

Since R is upper triangular, the solutions $x_n, x_{n-1}, \cdots, x_1$ are recursively computed by back substitution, starting from $x_n = \beta_n / r_{nn}$.

If the rank of $A \in \mathbb{R}^{m\times n}$ is less than n, some of the diagonal elements of R are zeros, so that the solution of the least-squares problem is not unique. But, putting the additional condition that the norm $||x||$ is minimum, we can obtain a unique solution. In the following, we explain a method of finding the minimum norm solution of the least-squares problem by means of the SVD.

Lemma 2.10. *Suppose that the rank of $A \in \mathbb{R}^{m\times n}$ is $r < \min(m, n)$, and the SVD is given by $A = U\Sigma_+ V^{\mathrm{T}}$, where $U \in \mathbb{R}^{m\times r}$ and $V \in \mathbb{R}^{n\times r}$. Then, there exists a unique solution X satisfying the Moore-Penrose conditions:*

$$\text{(i)} \quad AXA = A \qquad\qquad \text{(iii)} \quad (AX)^{\mathrm{T}} = AX$$
$$\text{(ii)} \quad XAX = X \qquad\qquad \text{(iv)} \quad (XA)^{\mathrm{T}} = XA$$

The unique solution is given by

$$X = V\Sigma_+^{-1}U^{\mathrm{T}} =: A^\dagger \tag{2.32}$$

In this case, $X = A^\dagger$ is called the Moore-Penrose generalized inverse, or the pseudo-inverse, of A.

Proof. [83] It is easy to see that $A^\dagger$ of (2.32) satisfies the above four conditions. To prove the uniqueness, suppose that both X and Y satisfy the conditions. Then, it follows that

$$\begin{aligned} X &= XAX = (XA)^{\mathrm{T}}X = A^{\mathrm{T}}X^{\mathrm{T}}X = A^{\mathrm{T}}Y^{\mathrm{T}}A^{\mathrm{T}}X^{\mathrm{T}}X \\ &= A^{\mathrm{T}}Y^{\mathrm{T}}(XA)X = A^{\mathrm{T}}Y^{\mathrm{T}}X = YAX = (YAY)(AX) \\ &= YY^{\mathrm{T}}A^{\mathrm{T}}X^{\mathrm{T}}A^{\mathrm{T}} = YY^{\mathrm{T}}A^{\mathrm{T}} = YAY = Y \end{aligned}$$

as was to be proved. Note that all four conditions are used in the proof. □

In the above lemma, if $\mathrm{rank}(A) = n$, then $A^\dagger = (A^\mathrm{T}A)^{-1}A^\mathrm{T}$ and $A^\dagger A = I_n$. If $\mathrm{rank}(A) = m$, then we have $A^\dagger = A^\mathrm{T}(AA^\mathrm{T})^{-1}$ and $AA^\dagger = I_m$.

Lemma 2.11. *Suppose that the rank of $A \in \mathbb{R}^{m\times n}$ is $r < n$. Then, a general solution of the least-squares problem*

$$\min_{x\in\mathbb{R}^n} \|Ax - b\|, \qquad A \in \mathbb{R}^{m\times n}, \qquad b \in \mathbb{R}^m$$

is given by

$$x = A^\dagger b + (I_n - A^\dagger A)z, \qquad \forall\, z \in \mathbb{R}^n \tag{2.33}$$

Moreover, $x = A^\dagger b$ is the unique minimum norm solution.

Proof. It follows from Lemma 2.7 that the minimizing vector x should satisfy $Ax = P_A b$, where P_A is the orthogonal projection onto $\mathrm{Im}(A)$, which is given by $UU^\mathrm{T} = AA^\dagger$. Since $A(A^\dagger b) = P_A b$, we see that $x = A^\dagger b$ is a solution of the least-squares problem. We now seek a general solution of the form $x = A^\dagger b + y$, where y is to be determined. Since

$$Ay = A(x - A^\dagger b) = Ax - P_A b = 0$$

we get $y \in \mathrm{Ker}(A)$. By using $A = U\Sigma_+ V^\mathrm{T}$,

$$A^\dagger A = V\Sigma_+^{-1}U^\mathrm{T}U\Sigma_+V^\mathrm{T} = VV^\mathrm{T}$$

Since $VV^\mathrm{T} = A^\dagger A$ is the orthogonal projection onto $\mathrm{Im}(A^\mathrm{T}) = (\mathrm{Ker}\ A)^\perp$, the orthogonal projection onto $\mathrm{Ker}(A)$ is given by $I_n - VV^\mathrm{T} = I_n - A^\dagger A$. Thus $y \in \mathrm{Ker}(A)$ is expressed as

$$y = (I_n - A^\dagger A)z, \qquad z \in \mathbb{R}^n$$

This proves (2.33). Finally, since $A^\dagger b$ and $(I_n - A^\dagger A)z$ are orthogonal, we get

$$\|x\|^2 = \|A^\dagger b\|^2 + \|(I_n - A^\dagger A)z\|^2 \geq \|A^\dagger b\|^2$$

where the equality holds if and only if $z = 0$. This completes the proof. □

Lemma 2.12. *A general solution of the least-squares problem*

$$\min_{X\in\mathbb{R}^{n\times p}} \|AX - B\|_F, \qquad A \in \mathbb{R}^{m\times n}, \qquad B \in \mathbb{R}^{m\times p}$$

is given by

$$X = A^\dagger B + (I_n - A^\dagger A)Z, \qquad \forall\, Z \in \mathbb{R}^{n\times p} \tag{2.34}$$

Proof. A proof is similar to that of Lemma 2.11. □

The minimum norm solution defined by (2.33) is expressed as

$$x = A^\dagger b = V\Sigma_+^{-1}U^\mathrm{T}b = \sum_{i=1}^{r} \frac{u_i^\mathrm{T} b}{\sigma_i} v_i$$

This indicates that if the singular values are small, then small changes in A and b may result in large changes in the solution x. From Lemma 2.9 (iv), $\|A - A_{r-1}\|_2 = \sigma_r$. Since the smallest singular value σ_r equals the distance from A to a set of matrices with ranks less than $r - 1$, it has the largest effect on the solution x. But, since the singular values are scale dependent, the normalized quantity, called the condition number,

$$\kappa(A) = \|A\|_2 \cdot \|A^\dagger\|_2 = \frac{\sigma_1}{\sigma_r}$$

is used as the sensitivity measure of the minimum norm solution to the data.

By definition, the condition number satisfies $\kappa(A) \geq 1$. If $\kappa(A)$ is very large, then A is called ill-conditioned. If $\kappa(A)$ is not very large, we say that A is well-conditioned. Obviously, the condition number of any orthogonal matrix is one, so that orthogonal matrices are perfectly conditioned.

2.8 Rank of Hankel Matrices

In this section, we consider the rank of Hankel matrices [51]. We assume that the sequence $h_1, h_2, \cdots$ below are real, but results are valid for complex sequences.

Definition 2.4. *Consider the infinite matrix*

$$H = \begin{bmatrix} h_1 & h_2 & h_3 & \cdots \\ h_2 & h_3 & h_4 & \cdots \\ h_3 & h_4 & h_5 & \cdots \\ \vdots & \vdots & \vdots & \ddots \end{bmatrix} \tag{2.35}$$

where (i, j)-element is given by h_{i+j}. This is called an infinite Hankel matrix, or Hankel operator. It should be noted that H has the same element along anti-diagonals. Also, define the matrix formed by the first k rows and l columns of H by

$$H_{k,l} = \begin{bmatrix} h_1 & h_2 & h_3 & \cdots & h_l \\ h_2 & h_3 & h_4 & \cdots & h_{l+1} \\ h_3 & h_4 & h_5 & \cdots & h_{l+2} \\ \vdots & \vdots & \vdots & & \vdots \\ h_k & h_{k+1} & h_{k+2} & \cdots & h_{k+l-1} \end{bmatrix} \tag{2.36}$$

This is called a finite Hankel matrix. □

Lemma 2.13. *Consider the finite Hankel matrix $H_{n,n}$ of order n. Suppose that the first l row vectors are linearly independent, but the first $l+1$ row vectors are linearly dependent. Then, it follows that* $\det H_{l,l} \neq 0$.

Proof. Let the first $l + 1$ row vectors of $H_{n,n}$ be given by $R_1, R_2, \cdots, R_l, R_{l+1}$. Since, from the assumption, $R_1, \cdots, R_l$ are linearly independent, we see that R_{l+1} is expressed as

$$R_{l+1} = \sum_{k=1}^{l} \alpha_k R_{l-k+1}$$

In particular, we have

$$h_i = \sum_{k=1}^{l} \alpha_k h_{i-k}, \qquad i = l+1, \cdots, l+n \tag{2.37}$$

Then the matrix formed by the first l row vectors $R_1, R_2, \cdots, R_l$ is given by

$$H_{l,n} = \begin{bmatrix} h_1 & h_2 & \cdots & h_n \\ h_2 & h_3 & \cdots & h_{n+1} \\ \vdots & \vdots & \ddots & \vdots \\ h_l & h_{l+1} & \cdots & h_{l+n-1} \end{bmatrix} \in \mathbb{R}^{l \times n} \tag{2.38}$$

where the rank of this matrix is l.

Now consider the column vectors of $H_{l,n}$. It follows from (2.37) that all the column vectors are expressed as a linear combination of the l preceding column vectors. Hence, in particular, the $(l+1)$th column vector is linearly dependent on the first l column vectors. But since the matrix of (2.38) has rank l, the first l column vectors are linearly independent, showing that $\det H_{l,l} \neq 0$. □

Example 2.2. Consider a finite symmetric Hankel matrix

$$H_{n,n} = \begin{bmatrix} h_1 & h_2 & \cdots & h_n \\ h_2 & h_3 & \cdots & h_{n+1} \\ \vdots & \vdots & \ddots & \vdots \\ h_n & h_{n+1} & \cdots & h_{2n-1} \end{bmatrix} \in \mathbb{R}^{n \times n}$$

Define the anti-diagonal matrix (or the backward identity)

$$J_n = \begin{bmatrix} 0 & & & 1 \\ & & 1 & \\ & \iddots & & \\ & 1 & & \\ 1 & & & 0 \end{bmatrix} \in \mathbb{R}^{n \times n} \tag{2.39}$$

Then it is easy to see that

$$J_n H_{n,n} = \begin{bmatrix} h_n & h_{n+1} & \cdots & h_{2n-1} \\ h_{n-1} & h_n & \cdots & h_{2n-2} \\ \vdots & \vdots & \ddots & \vdots \\ h_1 & h_2 & \cdots & h_n \end{bmatrix} =: T_n$$

where the matrix T_n is called a Toeplitz matrix with $t_{ij} = h_{n-i+j}$, *i.e.*, elements are constant along each diagonal. Also, from $J_n = J_n^{\mathrm{T}} = J_n^{-1}$, we see that $J_n T$ is a Hankel matrix for any Toeplitz matrix $T \in \mathbb{R}^n$. □

Lemma 2.14. *The infinite Hankel matrix of (2.35) has finite rank r if and only if there exist r real numbers $a_1, a_2, \cdots, a_r$ such that*

$$h_i = \sum_{k=1}^{r} a_k h_{i-k}, \qquad i = r+1,\, r+2,\, \cdots \tag{2.40}$$

Moreover, r is the least number with this property.

Proof. Suppose that $\text{rank}(H) = r$ holds. Then the first $r+1$ rows $R_1, R_2, \cdots, R_{r+1}$ are linearly dependent. Hence, there exists an $l\ (\leq r)$ such that $R_1, \cdots, R_l$ are linearly independent, and R_{l+1} is expressed as their linear combination

$$R_{l+1} = \sum_{k=1}^{l} a_k R_{l-k+1}$$

Now consider the row vectors $R_{i+1}, R_{i+2}, \cdots, R_{i+l+1}$, where i is an arbitrary nonnegative integer. From the structure of H, these vectors are obtained by removing the first i elements from $R_1, R_2, \cdots, R_{l+1}$, respectively. Thus we have

$$R_{i+l+1} = \sum_{k=1}^{l} a_k R_{i+l-k+1}, \qquad i = 0,\, 1,\, \cdots \tag{2.41}$$

It therefore follows that any row vector of H below the $(l+1)$th row can be expressed in terms of a linear combination of the l preceding row vectors, and hence in terms of linearly independent first l row vectors. Replacing l by r in (2.41), we have (2.40).

Conversely, suppose that (2.40) holds. Then, all the rows (columns) of H are expressed in terms of linear combinations of the first r rows (columns). Thus all the minors of H whose orders are greater than r are zero, and H has rank r at most. But the rank cannot be smaller than r; otherwise (2.40) is satisfied with a smaller value of r. This contradicts the second condition of the lemma. □

The above result is a basis for the realization theory due to Ho and Kalman [72], to be discussed in Chapter 3, where a matrix version of Lemma 2.14 will be proved.

2.9 Notes and References

- In this chapter, we have presented basic facts related to numerical linear algebra which will be needed in later chapters, including the QR decomposition, the orthogonal and oblique projections, the SVD, the least-squares method, the rank of Hankel matrices. Problems at the end of chapter include some useful formulas and results to be used in this book.
- Main references used are Golub and Van Loan [59], Gantmacher [51], Horn and Johnson [73], and Trefethen and Bau [157]. Earlier papers that have dealt with the issues of numerical linear algebra in system theory are [94] and [122]; see also the preprint book [125].

- For the history of SVD and related numerical methods, see [60, 148, 165]. The early developments in statistics, including the least-squares and the measurement of uncertainties, are covered in [149].

2.10 Problems

2.1 Prove the following by using the SVD, where $A \in \mathbb{R}^{m\times n}$, $B \in \mathbb{R}^{n\times p}$.

(a) $\mathrm{Im}(A) \oplus \mathrm{Ker}(A^{\mathrm{T}}) = \mathbb{R}^m, \quad \mathrm{Im}(A^{\mathrm{T}}) \oplus \mathrm{Ker}(A) = \mathbb{R}^n$

(b) $\mathrm{Ker}(A^{\mathrm{T}}) = (\mathrm{Im}\ A)^{\perp}, \quad \mathrm{Im}(A^{\mathrm{T}}) = (\mathrm{Ker}\ A)^{\perp}$

(c) $\mathrm{Im}(A) = A\mathbb{R}^n = \mathrm{Im}(AA^{\mathrm{T}}), \quad A\,\mathrm{Im}(B) = \mathrm{Im}(AB)$

2.2 Prove the following matrix identities

$$\begin{bmatrix} A & B \\ C & D \end{bmatrix} = \begin{bmatrix} I & BD^{-1} \\ 0 & I \end{bmatrix} \begin{bmatrix} A - BD^{-1}C & 0 \\ 0 & D \end{bmatrix} \begin{bmatrix} I & 0 \\ D^{-1}C & I \end{bmatrix}$$

$$= \begin{bmatrix} I & 0 \\ CA^{-1} & I \end{bmatrix} \begin{bmatrix} A & 0 \\ 0 & D - CA^{-1}B \end{bmatrix} \begin{bmatrix} I & A^{-1}B \\ 0 & I \end{bmatrix}$$

where it is assumed that A^{-1} and D^{-1} exist.

2.3 (a) Using the above results, prove the determinant of the block matrix.

$$\det \begin{bmatrix} A & B \\ C & D \end{bmatrix} = \det(A)\det(D - CA^{-1}B)$$

$$= \det(D)\det(A - BD^{-1}C)$$

(b) Defining $A = I_n$ and $D = I_m$, show that

$$\det(I_m - CB) = \det(I_n - BC)$$

(c) Prove the formulas for the inverses of block matrices

$$\begin{bmatrix} A & B \\ C & D \end{bmatrix}^{-1} = \begin{bmatrix} A^{-1} + A^{-1}B\Delta^{-1}CA^{-1} & -A^{-1}B\Delta^{-1} \\ -\Delta^{-1}CA^{-1} & \Delta^{-1} \end{bmatrix}$$

$$= \begin{bmatrix} \Pi^{-1} & -\Pi^{-1}BD^{-1} \\ -D^{-1}C\Pi^{-1} & D^{-1} + D^{-1}C\Pi^{-1}BD^{-1} \end{bmatrix}$$

where $\Delta := D - CA^{-1}B$, $\Pi := A - BD^{-1}C$. For $C = 0$, we get

$$\begin{bmatrix} A & B \\ 0 & D \end{bmatrix}^{-1} = \begin{bmatrix} A^{-1} & -A^{-1}BD^{-1} \\ 0 & D^{-1} \end{bmatrix}$$

(d) Prove the matrix inversion lemma.

$$[A + BD^{-1}C]^{-1} = A^{-1} - A^{-1}B[D + CA^{-1}B]^{-1}CA^{-1}$$

2.4 Show without using the result of Example 2.1 that if P is idempotent ($P^2 = P$), then all the eigenvalues are either zero or one.

2.5 For $P \in \mathbb{R}^{n\times n}$, show that the following statements are equivalent.

(a) $P^2 = P$

(b) $\mathrm{Im}(P) + \mathrm{Im}(I_n - P) = \mathbb{R}^n$

(c) $\mathrm{rank}(P) + \mathrm{rank}(I_n - P) = n$

2.6 Suppose that $Z = [T \quad U] \in \mathbb{R}^{n\times n}$ is nonsingular, where $T \in \mathbb{R}^{n\times r}$, $U \in \mathbb{R}^{n\times(n-r)}$. Let the inverse matrix of Z be given by

$$Z^{-1} = \begin{bmatrix} L \\ V \end{bmatrix}, \quad L \in \mathbb{R}^{r\times n}, \quad V \in \mathbb{R}^{(n-r)\times n}$$

Then it follows that $TL + UV = I_n$ and

$$\begin{bmatrix} L \\ V \end{bmatrix} [T \quad U] = \begin{bmatrix} LT & LU \\ VT & VU \end{bmatrix} = \begin{bmatrix} I_r & 0 \\ 0 & I_{n-r} \end{bmatrix}$$

Show that $P := TL$ is the oblique projection onto $\mathrm{Im}(T)$ along $\mathrm{Ker}(L)$, and that $Q := UV$ is the oblique projection onto $\mathrm{Ker}(L)$ [$= \mathrm{Im}(U)$] along $\mathrm{Im}(T)$ [$= \mathrm{Ker}(V)$].

2.7 In the above problem, define

$$T = \begin{bmatrix} I_r \\ 0 \end{bmatrix}, \qquad U = \begin{bmatrix} -X \\ I_{n-r} \end{bmatrix}$$

Compute the projection $P = TL$ by means of L and V. Show that (2.16) is satisfied if P has the following representation

$$P = \begin{bmatrix} I_r & X \\ 0 & 0 \end{bmatrix}, \qquad X \in \mathbb{R}^{r\times n}$$

2.8 By using (2.29), prove the following.

(a) $V_r V_r^{\mathrm{T}}$: the orthogonal projection from $\mathbb{R}^n$ onto $\mathrm{Im}(A^{\mathrm{T}})$

(b) $\tilde{V}_r \tilde{V}_r^{\mathrm{T}}$: the orthogonal projection from $\mathbb{R}^n$ onto $\mathrm{Ker}(A)$

(c) $U_r U_r^{\mathrm{T}}$: the orthogonal projection from $\mathbb{R}^m$ onto $\mathrm{Im}(A)$

(d) $\tilde{U}_r \tilde{U}_r^{\mathrm{T}}$: the orthogonal projection from $\mathbb{R}^m$ onto $\mathrm{Ker}(A^{\mathrm{T}})$

2.9 For $A \in \mathbb{R}^{m\times n}$, show that $A^{\mathrm{T}}(AA^{\mathrm{T}})^{\dagger} = A^{\dagger}$ and $(A^{\mathrm{T}}A)^{\dagger}A^{\mathrm{T}} = A^{\dagger}$.

2.10 Let $A \in \mathbb{R}^{m\times n}$ with $m \geq n$. By using the SVD, show that there exist an orthogonal matrix $Q \in \mathbb{R}^{m\times n}$ and a nonnegative matrix $\Pi \in \mathbb{R}^{n\times n}$ such that $A = Q\Pi$.

3

Discrete-Time Linear Systems

This chapter reviews discrete-time LTI systems and related basic results, including the stability, norms of signals and systems, state space equations, the Lyapunov stability theory, reachability, and observability, *etc.* Moreover, we consider canonical structure of linear systems, balanced realization, model reduction, and realization theory.

3.1 z-Transform

Let $f = (f(0),\ f(1),\ \cdots)$ be a one-sided infinite sequence, or a one-sided signal. Let z be the complex variable, and define

$$F(z) = \sum_{k=0}^{\infty} f(k) z^{-k} \tag{3.1}$$

It follows from the theory of power series that there exists $\rho > 0$ such that $F(z)$ absolutely converges for $|z| > \rho$, but diverges for $|z| < \rho$. Then, ρ is called the radius of convergence, and $\rho = |z|$ is the circle of convergence. If the power series in the right-hand side of (3.1) converges, $F(z)$ is called the one-sided z-transform of f, and is written as

$$F(z) = \mathfrak{Z}[f](z) \tag{3.2}$$

Also, let $f = (\cdots,\ f(-1),\ f(0),\ f(1),\ \cdots)$ be a two-sided infinite sequence, or a two-sided signal. Then, if

$$F(z) = \sum_{k=-\infty}^{\infty} f(k) z^{-k} \tag{3.3}$$

does converge, then $F(z)$ is called a two-sided z-transform of f. If the two-sided transform exists, it converges in an annular domain $\rho_1 < |z| < \rho_2$.

It is obvious that the one-sided z-transform is nothing but a two-sided transform of a sequence f with $f(k) = 0,\ k = -1,\ -2,\ \cdots$. Thus both transforms are expressed as in (3.2).

Lemma 3.1. *Let $A,\ r > 0$. If the one-sided signal f satisfies*

$$|f(k)| \le Ar^k, \qquad k = 0,\ 1,\ \cdots$$

Then the z-transform $\mathfrak{Z}[f](z)$ is absolutely convergent for $|z| > r$, and is analytic therein.

Proof. The absolute convergence is clear from

$$\sum_{k=0}^{\infty} \left| f(k) z^{-k} \right| \le \sum_{k=0}^{\infty} Ar^k |z|^{-k} = \frac{A}{1 - r|z|^{-1}}, \qquad r/|z| < 1$$

A proof of analyticity is omitted [36]. □

Similarly, if the two-sided signal $f = (\cdots,\ f(-1),\ f(0),\ f(1),\ \cdots)$ satisfies

$$|f(k)| \le \begin{cases} Ar_1^k, & k = 0,\ 1,\ \cdots \\ Ar_2^k, & k = -1,\ -2,\ \cdots;\ 0 < r_1 < r_2 \end{cases}$$

then the two-sided transform $F(z)$ is absolutely convergent for $r_1 < |z| < r_2$, and is analytic therein.

Example 3.1. (a) Consider the step function defined by

$$1(k) = \begin{cases} 1, & k = 0,\ 1,\ \cdots \\ 0, & k = -1,\ -2,\ \cdots \end{cases}$$

Then the z-transform of $1(k)$ is given by

$$\mathfrak{Z}[1(k)](z) = \sum_{k=0}^{\infty} z^{-k} = \frac{1}{1 - z^{-1}} = \frac{z}{z - 1}, \qquad |z| > 1$$

(b) For the (one-sided) exponential function $f(k) = \alpha^k,\ \ k = 0,\ 1,\ \cdots$,

$$\mathfrak{Z}[f](z) = \sum_{k=0}^{\infty} \alpha^k z^{-k} = \frac{1}{1 - \alpha z^{-1}} = \frac{z}{z - \alpha}, \qquad |z| > |\alpha|$$

(c) Let the two-sided exponential function f be defined by

$$f(k) = \begin{cases} a^k, & k = 0,\ 1,\ \cdots \\ b^k, & k = -1,\ -2,\ \cdots \end{cases}$$

where $0 < a < 1 < b$. Then, the two-sided transform is given by

$$\mathfrak{Z}[f](z) = \sum_{k=0}^{\infty} a^k z^{-k} + \sum_{k=-\infty}^{-1} b^k z^{-k} = \frac{(a-b)z}{(z-a)(z-b)}, \quad a < |z| < b \qquad \square$$

Lemma 3.2. *The inverse transform of $F(z)$ is given by the formula*

$$f(k) = \frac{1}{2\pi j} \oint_C F(z) z^{k-1} dz, \qquad k = 0, \pm 1, \cdots \tag{3.4}$$

where C denotes a closed curve containing all the poles a_i, $i = 1, \cdots, p$ of $F(z)$. Thus $f(k)$ is also obtained by

$$f(k) = \sum_{i=1}^{p} \mathrm{Res}[F(z) z^{k-1},\ z = a_i], \qquad k = 0, \pm 1, \cdots$$

which is the sum of residues of $F(z)z^{k-1}$ at poles contained in $C \subset \mathbb{C}$.

Proof. See [98, 121]. □

Lemma 3.3. *(Properties of z-transform)*

(i) *(Linearity)*

$$\mathfrak{Z}[\alpha f + \beta g](z) = \alpha \mathfrak{Z}[f](z) + \beta \mathfrak{Z}[g](z), \qquad \alpha,\ \beta : \text{scalars}$$

(ii) *(Time shift) Let f be a one-sided signal with $f(k) = 0$, $k = -1, -2, \cdots$. Let σ be a shift operator defined by $(\sigma f)(k) = f(k+1)$. Then, the z-transform of $\sigma^l f$ is given by*

$$\mathfrak{Z}[\sigma^l f](z) = \begin{cases} z^l F(z), & l = 0, -1, \cdots \\ z^l \left[F(z) - \sum_{k=0}^{l-1} f(k) z^{-k} \right], & l = 1, 2, \cdots \end{cases}$$

It should be noted that for the two-sided case, the term consisting of finite sum $\sum_{k=0}^{l-1} f(k) z^{-k}$ does not appear in the above formula.

(iii) *(Convolution) Consider the convolution of two-sided signals f and g, i.e.,*

$$h(k) = \sum_{l=-\infty}^{\infty} f(l) g(k-l) = \sum_{l=-\infty}^{\infty} f(k-l) g(l)$$

Let z-transforms of f and g be absolutely convergent and respectively be given by

$$F(z) = \sum_{k=-\infty}^{\infty} f(k) z^{-k}, \qquad \rho_1 < |z| < \rho_2$$

and

$$G(z) = \sum_{k=-\infty}^{\infty} g(k) z^{-k}, \qquad \rho_3 < |z| < \rho_4$$

Then the z-transform of h is absolutely convergent and is given by

$$H(z) = F(z) G(z), \qquad \rho^- < |z| < \rho^+ \tag{3.5}$$

where $\rho^- := \max(\rho_1, \rho_3)$ and $\rho^+ := \min(\rho_2, \rho_4)$.

(iv) Let the partial sum of a one-sided signal f be given by $g(k) := f(0) + f(1) + \cdots + f(k)$. Then the z-transform of g has the form

$$\mathfrak{Z}[g] = \frac{1}{1 - z^{-1}} F(z) \tag{3.6}$$

(v) For the difference of f, i.e., $g(k) := f(k) - f(k-1)$, we have

$$\mathfrak{Z}[(1 - \sigma^{-1}) f](z) = (1 - z^{-1}) F(z) \tag{3.7}$$

Proof. See [98, 121]. □

In the following, it is necessary to consider the case where

$$f = (\cdots,\ f(-1),\ f(0),\ f(1),\ \cdots)$$

is a sequence of vectors or matrices. For a vector or matrix case, the z-transform is defined componentwise. For example, let $A(t) = (a_{ij}(t))$, $t = 0, \pm 1, \cdots$, where $i = 1, \cdots, m$ and $j = 1, \cdots, n$. Then, the z-transform of the matrix function $A(t)$ is defined by

$$\mathfrak{Z}[A(t)](z) = \begin{bmatrix} \mathfrak{Z}[a_{11}(t)](z) & \cdots & \mathfrak{Z}[a_{1n}(t)](z) \\ \vdots & \ddots & \vdots \\ \mathfrak{Z}[a_{m1}(t)](z) & \cdots & \mathfrak{Z}[a_{mn}(t)](z) \end{bmatrix}$$

3.2 Discrete-Time LTI Systems

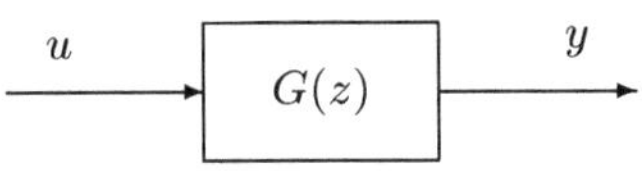

Figure 3.1. Discrete-time LTI system

Consider a single-input, single-output (SISO) discrete-time LTI system shown in Figure 3.1, where u is the input and y the output. We assume that the system is at rest for $t = -1, -2, \cdots$, *i.e.*, $u(t) = 0,\ y(t) = 0,\ t < 0$. Then the output is expressed as a convolution of the form

$$y(t) = \sum_{k=0}^{t} g(k) u(t-k), \qquad t = 0, 1, \cdots \tag{3.8}$$

where $g = (g(0),\ g(1),\ \cdots)$ is the impulse response of the system. An impulse response sequence satisfying $g(t) = 0,\ t = -1, -2, \cdots$ is called physically realizable, or causal, because physical systems are causal.

Let the z-transform of the impulse response g be given by

$$G(z) = \sum_{k=0}^{\infty} g(k) z^{-k}, \qquad |z| > \rho \tag{3.9}$$

We say that $G(z)$ is the transfer function from u to y. Let the z-transforms of u and y be defined by $u(z)$ and $y(z)$, respectively[1]. It then follows from (3.8) and Lemma 3.3 (iii) that

$$y(z) = G(z)u(z) \tag{3.10}$$

Example 3.2. Consider a difference equation of the form

$$\begin{aligned} y(t) + a_1 y(t-1) + a_2 y(t-2) + a_3 y(t-3) \\ = b_1 u(t-1) + b_2 u(t-2) + b_3 u(t-3) \end{aligned}$$

Taking the z-transform of the above equation under the assumption that all the initial values are zero, we get the transfer function of the form

$$G(z) = \frac{b_1 z^2 + b_2 z + b_3}{z^3 + a_1 z^2 + a_2 z + a_3}$$

This is a rational function in z, so that $G(z)$ is called a rational transfer function. □

Most transfer functions treated in this book are rational, so that $G(z)$ is expressed as a ratio of two polynomials

$$G(z) = \frac{b(z)}{a(z)}, \qquad \deg b(z) \le \deg a(z) \tag{3.11}$$

where $a(z)$ and $b(z)$ are polynomials in z. We say that the transfer function $G(z)$ with $\deg b(z) \le \deg a(z)$ is proper. It should be noted that since $g(t) = 0,\ t < 0$, the transfer function $G(z)$ of (3.9) is always proper.

Definition 3.1. *Consider the discrete-time LTI system with the transfer function $G(z)$ shown in Figure 3.1. We say that the system is bounded-input bounded-output (BIBO) stable if for any bounded signal u, the output y is bounded. In this case, we simply say that the system is stable, or $G(z)$ is stable.* □

Theorem 3.1. *The discrete-time LTI system shown in Figure 3.1 is stable if and only if the impulse response is absolutely summable, i.e.,*

$$\sum_{l=0}^{\infty} |g(l)| < \infty \tag{3.12}$$

Proof. (Sufficiency) Let u be a bounded input with $|u(t)| \le M$. Then it follows from (3.8) that

[1] For simplicity, we do not use the hat notation like $\hat{u}(z)$ and $\hat{y}(z)$ in this book.

$$|y(t)| \le \sum_{l=0}^{t} |g(l)| \cdot |u(t-l)| \le M \sum_{l=0}^{\infty} |g(l)| < \infty$$

(Necessity) Suppose that the absolute sum in (3.12) diverges. Let $M_k,\ k = 1,\ 2,\ \cdots$ be a divergent sequence. Then, there exists a divergent sequence $t_k,\ k = 1,\ 2,\ \cdots$ such that $\sum_{l=0}^{t_k} |g(l)| \ge M_k,\ k = 1,\ 2,\ \cdots$. Define $\tilde{u}$ as

$$\tilde{u}(t_k - l) = \begin{cases} 1, & g(l) \ge 0 \\ -1, & g(l) < 0 \end{cases}$$

Then we have

$$y(t_k) = \sum_{l=0}^{t_k} g(l)\tilde{u}(t_k - l) = \sum_{l=0}^{t_k} |g(l)| \ge M_k, \qquad k = 1,\ 2,\ \cdots$$

This implies that if the absolute sum of impulse response diverges, we can make the output diverge by using the bounded input $\tilde{u}$, so that the system is unstable. This completes a proof of the theorem. □

In the following, a number $\lambda \in \mathbb{C}$ is called a pole of $G(z)$ if $G(\lambda) = \infty$. It is also called a zero of $G(z)$ if $G(\lambda) = 0$.

Theorem 3.2. *A discrete-time LTI system with a proper transfer function $G(z)$ is stable if and only if all the poles of the transfer function lie inside the unit disk.*

Proof. Let $a_1,\ \cdots,\ a_p$ be poles of $G(z)$. We assume for simplicity that they are distinct. Partial fraction expansion of the right-hand side of (3.11) yields

$$\frac{G(z)}{z} = \frac{A_0}{z} + \frac{A_1}{z - a_1} + \cdots + \frac{A_p}{z - a_p}$$

Since the right-hand side is absolutely convergent for $|z| > \max_i |a_i|$, the inverse z-transform is given by

$$g(t) = A_0\delta_{t0} + A_1(a_1)^t + \cdots + A_p(a_p)^t, \qquad t = 0,\ 1,\ \cdots$$

Now suppose that $|a_i| < 1,\ i = 1,\ \cdots,\ p$. Then we have

$$\sum_{t=0}^{\infty} |g(t)| \le |A_0| + \sum_{i=1}^{p} \sum_{t=0}^{\infty} |A_i||a_i|^t < \infty$$

Thus it follows from Theorem 3.1 that the system is stable. Conversely, suppose that at least one a_i is outside of the unit disk. Without loss of generality, we assume that $|a_1| \ge 1$ and $|a_i| < 1,\ i = 2,\ \cdots,\ p$. Then, it follows that

$$|g(t)| \ge |A_1||(a_1)^t| - \sum_{i=2}^{p} |A_i||(a_i)^t| - |A_0|\delta_{t0}$$

Note that the sum of the first term in the right-hand side of the above inequality diverges, and that those of the second and the third terms converge. Thus, we have $\sum_{t=0}^{\infty} |g(t)| = \infty$, showing that the system is unstable. □

3.3 Norms of Signals and Systems

We begin with norms of signals. Let $u(t),\ t = 0,\ \pm 1,\ \cdots$ be m-dimensional vectors. Define $u = (\cdots,\ u(-1),\ u(0),\ u(1),\ \cdots)$ be a two-sided signal. The 2-norm of u is then defined by

$$||u||_2 = \sqrt{\sum_{t=-\infty}^{\infty} ||u(t)||^2}$$

where $||\cdot||$ denotes the Euclidean norm of a vector. The set of signals u with finite 2-norm is a Hilbert space denoted by

$$l_2(-\infty,\ \infty) = \{u \mid ||u||_2 < \infty\}$$

If the signal is one-sided, *i.e.*, $u(t) = 0,\ t < 0$, then the space is denoted by $l_2[0,\ \infty)$.

For $u \in l_2(-\infty,\ \infty)$, the Fourier transform, or the two-sided z-transform, is defined by

$$u(z) = \sum_{t=-\infty}^{\infty} u(t) z^{-t}, \qquad z = e^{j\omega}$$

Then, the 2-norm of u in the frequency domain is expressed as

$$||u||_2 = \left(\frac{1}{2\pi}\int_{-\pi}^{\pi} u^{\mathrm{H}}(j\omega)u(j\omega)d\omega\right)^{1/2} = \left(\frac{1}{2\pi}\int_{-\pi}^{\pi} ||u(j\omega)||^2 d\omega\right)^{1/2}$$

where $u^{\mathrm{H}}(j\omega) = u^{\mathrm{T}}(-j\omega)$ denotes the complex conjugate transpose.

We now consider a stable discrete-time LTI system with the input $u \in \mathbb{R}^m$ and the output $y \in \mathbb{R}^p$. Let $G(z)$ be a $p \times m$ transfer matrix, and $G_{ij}(z)$ the (i, j)-element of $G(z)$. Then, if all the elements $G_{ij}(z)$ are BIBO stable, we simply say that $G(z)$ is stable.

Definition 3.2. *For a stable $p \times m$ transfer matrix $G(z)$, two different norms can be defined:*

(i) H_2-norm:

$$||G||_2 = \left(\frac{1}{2\pi}\int_{-\pi}^{\pi} \mathrm{trace}[G^{\mathrm{H}}(e^{j\omega})G(e^{j\omega})]d\omega\right)^{1/2}$$

where trace $[\cdot]$ *denotes the trace of a matrix.*

(ii) H_∞-norm:

$$||G||_\infty = \sup_{-\pi < \omega \le \pi} \overline{\sigma}[G(e^{j\omega})]$$

where $\overline{\sigma}[\cdot]$ denotes the maximum singular value of a matrix. Also, the H_∞-norm can be expressed as

$$||G||_\infty = \sup_{u \neq 0} \frac{||Gu||_2}{||u||_2} = \sup_{u \neq 0} \frac{||y||_2}{||u||_2}, \qquad u \in l_2[0,\ \infty)$$

This is called the l_2-induced norm. □

Lemma 3.4. *Suppose that $G(z)$ is stable, and satisfies $\|G\|_\infty < \infty$.*

(i) If $u \in l_2(-\infty, \infty)$, then the output satisfies $y \in l_2(-\infty, \infty)$.

(ii) Let the z-transforms of u and y be given by $u(z)$ and $y(z)$, respectively. The inner product (y, u) is expressed as

$$\sum_{k=-\infty}^{\infty} y^{\mathrm{T}}(k)u(k) = \frac{1}{2\pi j}\int_{|z|=1} u^{\mathrm{T}}(z)G^{\mathrm{T}}(z)u(z^{-1})\frac{dz}{z} \tag{3.13a}$$

$$= \frac{1}{2\pi}\int_{-\pi}^{\pi} u^{\mathrm{T}}(e^{j\omega})G^{\mathrm{T}}(e^{j\omega})u(e^{-j\omega})d\omega \tag{3.13b}$$

Proof. (i) Since $y = G(z)u$, it follows that

$$\begin{aligned}
\|y\|_2^2 &= \frac{1}{2\pi}\int_{-\pi}^{\pi} y^{\mathrm{H}}(e^{j\omega})y(e^{j\omega})d\omega \\
&= \frac{1}{2\pi}\int_{-\pi}^{\pi} u^{\mathrm{H}}(e^{j\omega})G^{\mathrm{H}}(e^{j\omega})G(e^{j\omega})u(e^{j\omega})d\omega \\
&\le \|G\|_\infty^2 \frac{1}{2\pi}\int_{-\pi}^{\pi} u^{\mathrm{H}}(e^{j\omega})u(e^{j\omega})d\omega = \|G\|_\infty^2\|u\|_2^2
\end{aligned}$$

Thus we get $\|y\|_2 \le \|G\|_\infty\|u\|_2 < \infty$.

(ii) From item (i), if $u \in l_2(-\infty, \infty)$, the inner product $y^{\mathrm{T}}u$ is bounded, so that the sum in the left-hand side of (3.13a) converges. It follows from the inversion formula of Lemma 3.2 that

$$\begin{aligned}
\sum_{k=-\infty}^{\infty} y^{\mathrm{T}}(k)u(k) &= \sum_{k=-\infty}^{\infty}\left(\frac{1}{2\pi j}\int_{|z|=1} y(z)\, z^{k}\frac{dz}{z}\right)^{\mathrm{T}} u(k) \\
&= \frac{1}{2\pi j}\int_{|z|=1} y^{\mathrm{T}}(z)\left(\sum_{k=-\infty}^{\infty} u(k)z^{k}\right)\frac{dz}{z} \\
&= \frac{1}{2\pi j}\int_{|z|=1} y^{\mathrm{T}}(z)\, u(z^{-1})\,\frac{dz}{z}
\end{aligned}$$

Since $y(z) = G(z)u(z)$, we get (3.13a). Letting $z = e^{j\omega}$ $(-\pi < \omega < \pi)$ gives (3.13b). □

3.4 State Space Systems

Consider an m-input, p-output discrete-time LTI system described by

$$x(t+1) = Ax(t) + Bu(t) \tag{3.14a}$$

$$y(t) = Cx(t) + Du(t), \qquad t = 0, 1, \cdots \tag{3.14b}$$

where $x \in \mathbb{R}^n$ is the state vector, $u \in \mathbb{R}^m$ the input vector, and $y \in \mathbb{R}^p$ the output vector. The matrices $A \in \mathbb{R}^{n\times n}$, $B \in \mathbb{R}^{n\times m}$, $C \in \mathbb{R}^{p\times n}$, $D \in \mathbb{R}^{p\times m}$ are constant. Given the initial condition $x(0)$ and the inputs $u(t),\ t = 0,\ 1,\ \cdots$, we see that the state vectors $x(t),\ t = 1,\ 2,\ \cdots$ are recursively obtained, and hence the outputs $y(t),\ t = 0,\ 1,\ \cdots$ are determined. In the following, we simply write $\Sigma = (A,\ B,\ C,\ D)$ for the LTI system described by (3.14).

By solving (3.14),

$$y(t) = CA^t x(0) + Du(t) + \sum_{i=0}^{t-1} CA^{t-1-i}Bu(i), \qquad t = 0, 1, \cdots$$

If $u(t) = 0,\ t = 0, 1, \cdots$, the above equation reduces to

$$y(t) = CA^t x(0), \qquad t = 0, 1, \cdots \tag{3.15}$$

This equation is called the zero-input response. Also, if $x(0) = 0$, we have

$$y(t) = Du(t) + \sum_{i=0}^{t-1} CA^{t-1-i}Bu(i), \qquad t = 0, 1, \cdots \tag{3.16}$$

which is the response due to the external input $u(t)$, and is called the zero-state response. Thus the response of a linear state space system can always be expressed as the sum of the zero-input response and the zero-state response.

In connection with the zero state response, we define the $p \times m$ matrices as

$$G_t = \begin{cases} D, & t = 0 \\ CA^{t-1}B, & t = 1,\ 2,\ \cdots \end{cases} \tag{3.17}$$

The $(G_0,\ G_1,\ \cdots)$ is called the impulse response, or the Markov parameters, of the LTI system $\Sigma = (A,\ B,\ C,\ D)$.

Taking the z-transform of the impulse response, we have

$$G(z) := \left[\begin{array}{c|c} A & B \\ \hline C & D \end{array}\right] = D + C(zI - A)^{-1}B \tag{3.18}$$

which is called the transfer matrix of the LTI system $\Sigma = (A,\ B,\ C,\ D)$.

As shown in Figure 3.1, we can directly access the input and output vectors u and y from the outside of the system, so these vectors are called external vectors. Hence, the transfer matrix $G(z)$ relating the input vector to the output vector is an external description of the system Σ. On the other hand, we cannot directly access the state vector appearing in (3.14), since it is inside the system. Thus (3.14) is called an internal description of the LTI system Σ with the state vector x.

We easily observe that if an internal description of the system $\Sigma = (A,\ B,\ C,\ D)$ is given, the transfer matrix and impulse response matrices are calculated by means of (3.18) and (3.17), respectively. But, for a given external description $G(z)$, there exist infinitely many internal descriptions that realize the external description. In fact, let $T \in \mathbb{R}^{n\times n}$ be an arbitrary nonsingular matrix, and define

$$\bar{A} = T^{-1}AT, \qquad \bar{B} = T^{-1}B, \qquad \bar{C} = CT, \qquad \bar{D} = D \tag{3.19}$$

Then, a simple computation shows that

$$\begin{aligned}\bar{G}(z) &= \bar{D} + \bar{C}(zI - \bar{A})^{-1}\bar{B} \\ &= D + CT(zI - T^{-1}AT)^{-1}T^{-1}B \\ &= D + C(zI - A)^{-1}B = G(z)\end{aligned}$$

Thus the two internal descriptions $\bar{\Sigma} = (\bar{A},\ \bar{B},\ \bar{C},\ D)$ and $\Sigma = (A,\ B,\ C,\ D)$ have the same external description. The LTI systems Σ and $\bar{\Sigma}$ that represent the same input-output relation are called input-output equivalent.

This implies that models we obtain from the input-output data by using system identification techniques are necessarily external representations of systems. To get a state space model from a given external representation, we need to specify a coordinate of the state space.

3.5 Lyapunov Stability

Let $u = 0$ in (3.14). Then we have a homogeneous system

$$x(t+1) = Ax(t), \qquad x(0) = x_0 \tag{3.20}$$

A set $\{x \mid x = Ax\}$ of state vectors are called the equilibrium points. It is clear that the origin $x = 0$ is an equilibrium point of (3.20). If $\det(I - A) \neq 0$, then $x = 0$ is the unique equilibrium point.

Definition 3.3. *If for any $x(0) \in \mathbb{R}^n$ the solution $x(t)$ converges to 0, then the origin of the system (3.20) is asymptotically stable. In this case, we say that the system (3.20) is asymptotically stable. Moreover, A is simply called stable.* □

We now prove the Lyapunov stability theorem.

Theorem 3.3. *The following are equivalent conditions such that the homogeneous system (3.20) is asymptotically stable.*

(i) The absolute values of all the eigenvalues of A are less than 1, i.e.

$$|\lambda_i(A)| < 1, \qquad i = 1, \cdots, n \tag{3.21}$$

It may be noted that this is simply written as $\rho(A) < 1$.

(ii) For any $Q > 0$, there exists a unique solution $P > 0$ that satisfies

$$P = A^{\mathrm{T}}PA + Q \tag{3.22}$$

The above matrix equation is called a Lyapunov equation for a discrete-time LTI system.

Proof. (i) From $x(t) = A^t x(0)$, we see that (3.21) is a necessary and sufficient condition of the asymptotic stability of (3.20).

(ii) (Necessity) Suppose that (3.21) holds. Then, the sum

$$P = \sum_{i=0}^{\infty} (A^{\mathrm{T}})^i Q A^i = Q + A^{\mathrm{T}} \left(\sum_{i=1}^{\infty} (A^{\mathrm{T}})^{i-1} Q A^{i-1} \right) A \tag{3.23}$$

converges. It is easy to see that P defined above is a solution of (3.22), and that $P > 0$ since $Q > 0$. To prove the uniqueness of P, suppose that P_1 and P_2 are two solutions of (3.22). Then we have

$$P_1 - P_2 = A^{\mathrm{T}}(P_1 - P_2)A = (A^{\mathrm{T}})^k (P_1 - P_2) A^k$$

Since A is stable, $P_1 = P_2$ follows taking the limit $k \to \infty$.

(Sufficiency) Suppose that the solution of (3.22) is positive definite, *i.e.*, $P > 0$, but A is not stable. Then, there exist an eigenvalue λ_0 and an eigenvector $\xi \in \mathbb{C}^n$ such that

$$A\xi = \lambda_0 \xi, \qquad |\lambda_0| \geq 1, \qquad \xi \neq 0 \tag{3.24}$$

Pre-multiplying (3.22) by ξ^{H} and post-multiplying by ξ yield

$$\xi^{\mathrm{H}} P \xi = \xi^{\mathrm{H}} A^{\mathrm{T}} P A \xi + \xi^{\mathrm{H}} C^{\mathrm{T}} C \xi = |\lambda_0|^2 \xi^{\mathrm{H}} P \xi + \xi^{\mathrm{H}} Q \xi$$

Thus it follows that $(|\lambda_0|^2 - 1)\xi^{\mathrm{H}} P \xi + \xi^{\mathrm{H}} Q \xi = 0$. Since $|\lambda_0| \geq 1$, the two terms in the left-hand side of this equation should be zero. In particular, we have $Q\xi = 0$, so that $\xi = 0$, a contradiction. Thus A is stable. □

3.6 Reachability and Observability

In this section, we present basic definitions and theorems for reachability and observability of the discrete-time LTI system $\Sigma = (A,\ B,\ C,\ D)$.

Definition 3.4. *Consider a discrete-time LTI system Σ. If the initial state vector $x(0) = 0$ can be transferred to any state $\xi \in \mathbb{R}^n$ at time n, i.e., $x(n) = \xi$, by means of a sequence of control vectors $u(0),\ u(1),\ \cdots,\ u(n-1)$, then the system is called reachable. Also, if any state $x(0) \in \mathbb{R}^n$ can be transferred to the zero state by means of a sequence of control vectors, the system is called controllable.* □

We simply say that $(A,\ B)$ is reachable (or controllable), since the reachability (or controllability) is related to the pair $(A,\ B)$ only.

Theorem 3.4. *The following are necessary and sufficient conditions such that the pair $(A,\ B)$ is reachable.*

(i) Define the reachability matrix as

$$\mathcal{C} = [B\ \ AB\ \ \cdots\ \ A^{n-1}B] \in \mathbb{R}^{n \times nm} \tag{3.25}$$

Then $\operatorname{rank}(\mathcal{C}) = n$ *holds, or* $\operatorname{Im}(\mathcal{C}) = \mathbb{R}^n$.

(ii) For any $\lambda \in \mathbb{C}$, $\operatorname{rank}[A - \lambda I \;\; B] = n$ *holds.*

(iii) The eigenvalues of $A + BK$ *are arbitrarily assigned by a suitable choice of* $K \in \mathbb{R}^{m \times n}$.

Proof. We prove item (i) only. By using (3.14) and (3.25), the state vector at time n is described by

$$\begin{aligned} x(n) &= A^n x(0) + A^{n-1}Bu(0) + \cdots + ABu(n-2) + Bu(n-1) \\ &= A^n x(0) + \mathcal{C} \begin{bmatrix} u(n-1) \\ u(n-2) \\ \vdots \\ u(0) \end{bmatrix} \end{aligned}$$

Let $x(0) = 0$. Then, we see that the vector $x(n)$ takes arbitrary values in $\mathbb{R}^n$ if and only if item (i) holds. For items (ii) and (iii), see Kailath [80]. □

From Definition 3.4, (A, B) is controllable if and only if there exists a sequence of control vectors that transfers the state to zero at n. This is equivalent to

$$A^n x(0) \in \operatorname{Im}(\mathcal{C}), \qquad \forall x(0) \in \mathbb{R}^n$$

Thus if A is nonsingular, the above condition is reduced to $\operatorname{Im}(\mathcal{C}) = \mathbb{R}^n$, which is equivalent to item (i) of Theorem 3.4. Hence if A is nonsingular, we see that the reachability and controllability of (A, B) are equivalent.

Theorem 3.5. *Suppose that the pair* (A, B) *is not reachable, and let* $\operatorname{rank}(\mathcal{C}) = n_c < n$. *Then, there exists a nonsingular matrix* T *such that* A *and* B *are decomposed as*

$$T^{-1}AT = \begin{bmatrix} A_{11} & A_{12} \\ 0 & A_{22} \end{bmatrix}, \qquad T^{-1}B = \begin{bmatrix} B_1 \\ 0 \end{bmatrix} \tag{3.26}$$

where $A_{11} \in \mathbb{R}^{n_c \times n_c}$, $B_1 \in \mathbb{R}^{n_c \times m}$, *and where* (A_{11}, B_1) *are reachable.*

Proof. For a proof, see Kailath [80]. □

Definition 3.5. *We say that* (A, B) *is stabilizable, if there exists a matrix* $K \in \mathbb{R}^{m \times n}$ *such that* $A + BK$ *is stable, i.e.,* $\rho(A + BK) < 1$. *This is equivalent to the fact that the system* Σ *is stabilized by a state feedback control* $u = Kx$. □

Theorem 3.6. *The following are necessary and sufficient conditions such that the pair* (A, B) *is stabilizable.*

(i) For any $\lambda \in \mathbb{C}$ *with* $|\lambda| \geq 1$, *we have* $\operatorname{rank}[A - \lambda I \;\; B] = n$.

(ii) Suppose that A *and* B *are decomposed as in (3.26). Then,* A_{22} *is stable, i.e.* $\rho(A_{22}) < 1$ *holds.*

Proof. For a proof, see Kailath [80]. □

We introduce the observability for the discrete-time LTI system, which is the dual of the reachability.

Definition 3.6. *Let $u \equiv 0$ in (3.14). We say that the system is observable, if the initial state $x(0)$ is completely recoverable from n output observations $y(0)$, $y(1)$, $\cdots$, $y(n-1)$. In this case, we say that (C, A) is observable. This is equivalent to the fact that if both the input and output are zero, i.e., $u(t) = 0$, $y(t) = 0$ for $t = 0, 1, \cdots, n-1$, then we can say that the initial state is $x(0) = 0$.* □

Theorem 3.7. *The following are necessary and sufficient conditions such that the pair (C, A) is observable.*

(i) Define the observability matrix as

$$\mathcal{O} = \begin{bmatrix} C \\ CA \\ \vdots \\ CA^{n-1} \end{bmatrix} \in \mathbb{R}^{np \times n} \tag{3.27}$$

Then, we have $\operatorname{rank}(\mathcal{O}) = n$, *or* $\operatorname{Ker}(\mathcal{O}) = \{0\}$.

(ii) For any $\lambda \in \mathbb{C}$, it follows that $\operatorname{rank} \begin{bmatrix} A - \lambda I \\ C \end{bmatrix} = n$ *holds.*

(iii) All the eigenvalues of $A + LC$ are specified arbitrarily by a suitable choice of $L \in \mathbb{R}^{n \times p}$.

Proof. For a proof, see [80]. □

Theorem 3.8. *Suppose that (C, A) is not observable, and define* $\operatorname{rank}(\mathcal{O}) = n_o < n$. *Then, there exists a nonsingular matrix T such that A and C are decomposed as*

$$T^{-1}AT = \begin{bmatrix} A_{11} & 0 \\ A_{21} & A_{22} \end{bmatrix}, \qquad CT = [C_1 \;\; 0] \tag{3.28}$$

where $A_{11} \in \mathbb{R}^{n_o \times n_o}$, $C_1 \in \mathbb{R}^{p \times n_o}$ with the pair (C_1, A_{11}) observable. □

Now we provide the definition of detectability, which is weaker than the observability condition stated above.

Definition 3.7. *Let $u = 0$ in (3.14). If $\lim_{t \to \infty} y(t) = 0$ implies that $\lim_{t \to \infty} x(t) = 0$, then (C, A) is called detectable.* □

Theorem 3.9. *The following are necessary and sufficient conditions such that the pair (C, A) is detectable.*

(i) There exists a matrix $L \in \mathbb{R}^{n \times p}$ such that $A + LC$ is stabilized.

(ii) For any $\lambda \in \mathbb{C}$ with $|\lambda| \geq 1$, $\operatorname{rank} \begin{bmatrix} A - \lambda I \\ C \end{bmatrix} = n$ *holds.*

(iii) Suppose that A and C are decomposed as in (3.28). Then $\rho(A_{22}) < 1$ holds.

Proof. According to Definition 3.7, we show item (iii). Define $x = T\bar{x}$. It then follows from (3.28) that

$$\bar{x}_1(t+1) = A_{11}\bar{x}_1(t) \tag{3.29a}$$
$$\bar{x}_2(t+1) = A_{21}\bar{x}_1(t) + A_{22}\bar{x}_2(t) \tag{3.29b}$$
$$y(t) = C_1\bar{x}_1(t) \tag{3.29c}$$

From (3.29a) and (3.29c),

$$\begin{bmatrix} y(t) \\ y(t+1) \\ \vdots \\ y(t+n_o-1) \end{bmatrix} = \begin{bmatrix} C_1 \\ C_1A_{11} \\ \vdots \\ C_1(A_{11})^{n_o-1} \end{bmatrix} \bar{x}_1(t)$$

Since $(C_1,\ A_{11})$ is observable, the observability matrix formed by $(C_1,\ A_{11})$ has full rank. Thus we see that $\lim_{t\to\infty} y(t) = 0$ implies that $\lim_{t\to\infty} \bar{x}_1(t) = 0$. Hence it suffices to consider the condition so that $\bar{x}_2(t)$ converges to zero as $\bar{x}_1(t)$ tends to zero. From (3.29b) it follows that

$$\bar{x}_2(t) = (A_{22})^t\bar{x}_2(0) + \sum_{i=0}^{t-1}(A_{22})^{t-i-1}A_{21}\bar{x}_1(k)$$

It can be shown that $\lim_{t\to\infty} \bar{x}_2(t) = 0$ holds, if A_{22} is stable (see Problems 3.6 and 3.7). This shows that the detectability of $(C,\ A)$ requires that unobservable modes are stable. □

Theorem 3.10. *Suppose that $(C,\ A)$ is detectable (observable). Then A is stable if and only if the Lyapunov equation*

$$P = A^{\mathrm{T}}PA + C^{\mathrm{T}}C \tag{3.30}$$

has a unique nonnegative (positive) definite solution P.

Proof. (Sufficiency) If A is stable, then the solution of (3.30) is given by

$$P = \sum_{i=0}^{\infty}(A^{\mathrm{T}})^iC^{\mathrm{T}}CA^i$$

Since $Q = C^{\mathrm{T}}C \geq 0$, we have $P \geq 0$. ($P > 0$ if and only if (C, A) is observable.) For uniqueness of P, see the proof in Theorem 3.3.

(Necessity) Suppose that A is not stable. Then, there exists an unstable eigenvalue λ_0 and a nonzero vector $\xi \in \mathbb{C}^n$ such that

$$A\xi = \lambda_0\xi, \qquad |\lambda_0| \geq 1, \qquad \xi \neq 0 \tag{3.31}$$

Pre-multiplying (3.30) by ξ^{H} and post-multiplying ξ yield

$$\xi^{\mathrm{H}} P \xi = \xi^{\mathrm{H}} A^{\mathrm{T}} P A \xi + \xi^{\mathrm{H}} C^{\mathrm{T}} C \xi = |\lambda_0|^2 \xi^{\mathrm{H}} P \xi + \xi^{\mathrm{H}} C^{\mathrm{T}} C \xi$$

so that we get $(|\lambda_0|^2 - 1)\xi^{\mathrm{H}} P \xi + \xi^{\mathrm{H}} C^{\mathrm{T}} C \xi = 0$. Since $|\lambda_0| \geq 1$, both terms in the left-hand side are nonnegative. Thus we have $C\xi = 0$, which together with (3.31) shows that

$$A\xi = \lambda_0 \xi, \qquad C\xi = 0, \qquad |\lambda_0| \geq 1, \qquad \xi \neq 0$$

It follows from the item (ii) of Theorem 3.9 that this implies that $(C,\ A)$ is not detectable, a contradiction. This completes the proof. □

3.7 Canonical Decomposition of Linear Systems

We consider a finite-dimensional block Hankel matrix defined by

$$H_{n,n} = \begin{bmatrix} CB & CAB & CA^2B & \cdots & CA^{n-1}B \\ CAB & CA^2B & \cdots & \cdots & CA^nB \\ CA^2B & & & & \vdots \\ \vdots & & & & \vdots \\ CA^{n-1}B & \cdots & \cdots & \cdots & CA^{2n-2}B \end{bmatrix} \in \mathbb{R}^{pn \times mn} \tag{3.32}$$

This is called the Hankel matrix associated with the system $\Sigma = (A,\ B,\ C,\ D)$, so that its elements are formed by the impulse response matrices of the discrete-time LTI system Σ.

In terms of the observability matrix $\mathcal{O}$ of (3.25) and the reachability matrix $\mathcal{C}$ of (3.27), the block Hankel matrix is decomposed as $H_{n,n} = \mathcal{O}\mathcal{C}$. Thus we see that $\operatorname{rank}(H_{n,n}) = n_h \leq \min(n_o,\ n_r) \leq n$ (see Lemma 3.11 below).

The following is the canonical decomposition theorem due to Kalman [82].

Theorem 3.11. *(Canonical decomposition) By means of a nonsingular transform, the system* $\Sigma = (A,\ B,\ C,\ D)$ *can be reduced to* $\bar{\Sigma} = (\bar{A},\ \bar{B},\ \bar{C},\ \bar{D})$ *of the form*

$$\begin{bmatrix} \bar{x}_{c\bar{o}}(t+1) \\ \bar{x}_{co}(t+1) \\ \bar{x}_{\bar{c}\bar{o}}(t+1) \\ \bar{x}_{\bar{c}o}(t+1) \end{bmatrix} = \begin{bmatrix} \bar{A}_{11} & \bar{A}_{12} & \bar{A}_{13} & \bar{A}_{14} \\ 0 & \bar{A}_{22} & 0 & \bar{A}_{24} \\ 0 & 0 & \bar{A}_{33} & \bar{A}_{34} \\ 0 & 0 & 0 & \bar{A}_{44} \end{bmatrix} \begin{bmatrix} \bar{x}_{c\bar{o}}(t) \\ \bar{x}_{co}(t) \\ \bar{x}_{\bar{c}\bar{o}}(t) \\ \bar{x}_{\bar{c}o}(t) \end{bmatrix} + \begin{bmatrix} \bar{B}_1 \\ \bar{B}_2 \\ 0 \\ 0 \end{bmatrix} u(t) \tag{3.33a}$$

$$y(t) = \begin{bmatrix} 0 & \bar{C}_2 & 0 & \bar{C}_4 \end{bmatrix} \begin{bmatrix} \bar{x}_{c\bar{o}}(t) \\ \bar{x}_{co}(t) \\ \bar{x}_{\bar{c}\bar{o}}(t) \\ \bar{x}_{\bar{c}o}(t) \end{bmatrix} + Du(t) \tag{3.33b}$$

where the vector $\bar{x}_{c\bar{o}}(t)$ *is reachable but not observable;* $\bar{x}_{co}(t)$ *reachable and observable;* $\bar{x}_{\bar{c}\bar{o}}(t)$ *not reachable and not observable;* $\bar{x}_{\bar{c}o}(t)$ *observable but not reachable. Also, it follows that* $\dim \bar{x}_{c\bar{o}}(t) = n_c - n_h$, $\dim \bar{x}_{co}(t) = n_h$; $\dim \bar{x}_{\bar{c}\bar{o}}(t) = n - n_o - n_c + n_h$; $\dim \bar{x}_{\bar{c}o}(t) = n_o - n_h$. □

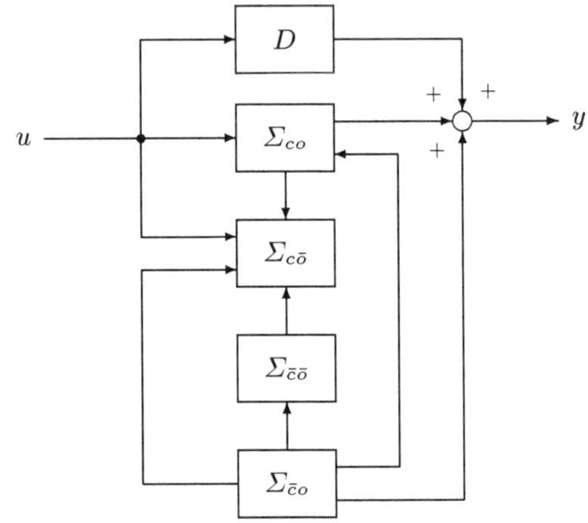

Figure 3.2. Canonical decomposition of LTI system

Figure 3.2 shows the canonical structure of the linear system Σ, where $\Sigma_{c\bar{o}}$, Σ_{co}, $\Sigma_{\bar{c}\bar{o}}$, $\Sigma_{\bar{c}o}$ respectively denote subsystems whose state vectors are $\bar{x}_{c\bar{o}}(t)$, $\bar{x}_{co}(t)$, $\bar{x}_{\bar{c}\bar{o}}(t)$, $\bar{x}_{\bar{c}o}(t)$, and the arrows reflect the block structure of system matrices in $\bar{\Sigma}$.

As mentioned in Section 3.4, the external description of the system Σ is invariant under nonsingular transforms, so that the transfer matrices and impulse response matrices of Σ and $\bar{\Sigma}$ are the same; they are given by

$$G(z) = \bar{C}_2(zI - \bar{A}_{22})^{-1}\bar{B}_2 + \bar{D}$$

and

$$G_t = \bar{C}_2(\bar{A}_{22})^{t-1}\bar{B}_2, \qquad t = 1, 2, \cdots$$

Hence we see that $\Sigma_2 := (\bar{A}_{22},\, \bar{B}_2,\, \bar{C}_2,\, \bar{D})$, $\bar{\Sigma}$ and Σ are all equivalent. This implies that the transfer matrix of a system is related to the subsystem Σ_2 only. In other words, models we obtain from the input-output data by using system identification techniques are necessarily those of the subsystem Σ_2.

Given a transfer matrix $G(z)$, the system $\Sigma = (A,\, B,\, C,\, D)$ is called a realization of $G(z)$. As shown above, the realizations are not unique. A realization with the least dimension is referred to as a minimal realization, which is unique up to nonsingular transforms. In fact, we have the following theorem.

Theorem 3.12. *A triplet $(A,\, B,\, C)$ is minimal if and only if $(A,\, B)$ is reachable and $(C,\, A)$ is observable. Moreover, if both $\Sigma_1 = (A_1,\, B_1,\, C_1,\, D_1)$ and $\Sigma_2 = (A_2,\, B_2,\, C_2,\, D_2)$ are minimal realizations of $G(z)$, then the relation of (3.19) holds for some nonsingular transform T.*

Proof. The first part is obvious from Theorem 3.11. We show the second part. Define the reachability and observability matrices as

$$\mathcal{C}_1 = [B_1\ A_1B_1\ \cdots\ A_1^{n-1}B_1], \qquad \mathcal{C}_2 = [B_2\ A_2B_2\ \cdots\ A_2^{n-1}B_2]$$

$$\mathcal{O}_1 = \begin{bmatrix} C_1 \\ C_1 A_1 \\ \vdots \\ C_1 A_1^{n-1} \end{bmatrix}, \qquad \mathcal{O}_2 = \begin{bmatrix} C_2 \\ C_2 A_2 \\ \vdots \\ C_2 A_2^{n-1} \end{bmatrix}$$

Then, from the hypothesis, we have $D_1 = D_2$ and

$$C_1(zI - A_1)^{-1} B_1 = C_2(zI - A_2)^{-1} B_2$$

By using a series expansion of the above relation, it can easily be shown that

$$C_1 A_1^l B_1 = C_2 A_2^l B_2, \qquad l = 0, 1, \cdots$$

This implies that $\mathcal{O}_1 \mathcal{C}_1 = \mathcal{O}_2 \mathcal{C}_2$. Since $\mathcal{O}_1$ and $\mathcal{C}_1$ have full rank, we define two matrices

$$T_1 = \mathcal{C}_2 \mathcal{C}_1^{\mathrm{T}} (\mathcal{C}_1 \mathcal{C}_1^{\mathrm{T}})^{-1}, \qquad T_2 = (\mathcal{O}_1^{\mathrm{T}} \mathcal{O}_1)^{-1} \mathcal{O}_1^{\mathrm{T}} \mathcal{O}_2$$

It can be shown that $T_2 T_1 = I_n$, implying that both T_1 and T_2 are nonsingular with $T_2 = T_1^{-1}$. Also, we have $T_2 \mathcal{C}_2 = \mathcal{C}_1$ and $\mathcal{O}_2 T_1 = \mathcal{O}_1$. Therefore it follows that

$$\mathcal{O}_1 A_1 \mathcal{C}_1 = \mathcal{O}_2 A_2 \mathcal{C}_2 = \mathcal{O}_1 T_1^{-1} A_2 T_1 \mathcal{C}_1$$

Since $\operatorname{rank}(\mathcal{O}_1) = n$ and $\operatorname{rank}(\mathcal{C}_1) = n$, it follows that $A_1 = T_1^{-1} A_2 T_1$. Hence, comparing the first block columns of $T_1^{-1} \mathcal{C}_2 = \mathcal{C}_1$ yields $T_1^{-1} B_2 = B_1$. Similarly, from $\mathcal{O}_2 T_1 = \mathcal{O}_1$, we have $C_2 T_1 = C_1$. This completes the input-output equivalence of (A_1, B_1, C_1) and (A_2, B_2, C_2). □

Example 3.3. Consider the transfer function of an SISO system

$$G(z) = \frac{b_1 z^{n-1} + \cdots + b_n}{z^n + a_1 z^{n-1} + \cdots + a_n}$$

It is easy to see that both

$$\Sigma = \left(\begin{bmatrix} 0 & 1 & & 0 \\ & & \ddots & \\ & & & 1 \\ -a_n & -a_{n-1} & \cdots & -a_1 \end{bmatrix}, \begin{bmatrix} 0 \\ \vdots \\ 0 \\ 1 \end{bmatrix}, \begin{bmatrix} b_n & b_{n-1} & \cdots & b_1 \end{bmatrix} \right)$$

and

$$\tilde{\Sigma} = \left(\begin{bmatrix} 0 & & -a_n \\ 1 & & -a_{n-1} \\ & \ddots & \vdots \\ & 1 & -a_1 \end{bmatrix}, \begin{bmatrix} b_n \\ b_{n-1} \\ \vdots \\ b_1 \end{bmatrix}, \begin{bmatrix} 0 & \cdots & 0 & 1 \end{bmatrix} \right)$$

are realizations of $G(z)$. It is clear that (A, B) is reachable and $(\tilde{C}, \tilde{A})$ is observable, and that $\tilde{A} = A^{\mathrm{T}}$, $\tilde{B} = C^{\mathrm{T}}$, $\tilde{C} = B^{\mathrm{T}}$ hold. □

3.8 Balanced Realization and Model Reduction

For a given linear system, there exist infinitely many realizations; among others, the balanced realization described below is quite useful in modeling and system identification. First we give the definition of two Gramians associated with a discrete-time LTI system.

Definition 3.8. *Let a realization be given by (A, B, C) with A stable. Consider two Lyapunov equations defined by*

$$P = APA^{\mathrm{T}} + BB^{\mathrm{T}} \tag{3.34}$$

and

$$Q = A^{\mathrm{T}}QA + C^{\mathrm{T}}C \tag{3.35}$$

Then, the solutions P and Q are respectively called reachability Gramian and observability Gramian, where they are nonnegative definite. Also, the square roots of the eigenvalues of PQ are called the Hankel singular values of (A, B, C). □

Lemma 3.5. *Suppose that A is stable. Then, we have*

$$(A, B) : \textit{reachable} \Leftrightarrow P > 0; \quad (C, A) : \textit{observable} \Leftrightarrow Q > 0$$

Proof. (Necessity) Since A is stable, the solution of (3.34) is given by

$$P = \sum_{i=0}^{\infty} A^i BB^{\mathrm{T}} (A^{\mathrm{T}})^i \geq \sum_{i=0}^{n-1} A^i BB^{\mathrm{T}} (A^{\mathrm{T}})^i$$

Thus, if (A, B) is reachable, $P > 0$ follows.

(Sufficiency) Suppose that P is positive definite, but (A, B) is not reachable. Then there exist $\eta \in \mathbb{C}^n$ and $\lambda \in \mathbb{C}$ such that

$$\eta^{\mathrm{H}} A = \lambda \eta^{\mathrm{H}}, \qquad \eta^{\mathrm{H}} B = 0, \qquad \eta \neq 0$$

where $|\lambda| < 1$. Pre-multiplying (3.34) by η^{H} and post-multiplying η yield

$$\eta^{\mathrm{H}} P\eta = \eta^{\mathrm{H}} APA^{\mathrm{T}}\eta + \eta^{\mathrm{H}} BB^{\mathrm{T}}\eta = |\lambda|^2 \eta^{\mathrm{H}} P\eta \quad \Rightarrow \quad (1 - |\lambda|^2)\eta^{\mathrm{H}} P\eta = 0$$

Since $1 - |\lambda|^2 \neq 0$, we get $\eta^{\mathrm{H}} P\eta = 0$, implying that P is not positive definite, a contradiction. This proves the first half of this lemma. The assertion for the Gramian Q is proved similarly. □

Definition 3.9. *Let $G(z) = (A, B, C, D)$ be a minimal realization. Then it is called a balanced realization if the following conditions (i) and (ii) hold.*

(i) The matrix A is stable, i.e., $\rho(A) < 1$.

(ii) *The Gramians P and Q are equal, and diagonal, i.e., there exists a diagonal matrix*

$$\Sigma = \begin{bmatrix} \sigma_1 & & & \\ & \sigma_2 & & \\ & & \ddots & \\ & & & \sigma_n \end{bmatrix}, \qquad \sigma_1 \geq \sigma_2 \geq \cdots \geq \sigma_n > 0$$

satisfying

$$\Sigma = A\Sigma A^{\mathrm{T}} + BB^{\mathrm{T}}, \qquad \Sigma = A^{\mathrm{T}}\Sigma A + C^{\mathrm{T}}C \tag{3.36}$$

Note that $\sigma_1, \sigma_2, \cdots \sigma_n$ are Hankel singular values of (A, B, C). □

Lemma 3.6. *If (A, B, C, D) is a balanced realization, then the 2-norm of A, the maximum singular value, satisfies $\|A\|_2 = \overline{\sigma}(A) \leq 1$. Moreover, if all the elements of Σ are different, we have $\|A\|_2 < 1$.*

Proof. We prove the first part of the lemma. Pre-multiplying the first equation of (3.36) by A^{T} and post-multiplying A, and then adding the resultant equation to the second equation yield

$$A^{\mathrm{T}}A\Sigma A^{\mathrm{T}}A - \Sigma = -(A^{\mathrm{T}}BB^{\mathrm{T}}A + C^{\mathrm{T}}C) \tag{3.37}$$

Let $\lambda \geq 0$ be an eigenvalue of $A^{\mathrm{T}}A$, and $v \in \mathbb{R}^n$ a corresponding eigenvector. Then, we have $A^{\mathrm{T}}Av = \lambda v$, $v \neq 0$. Pre-multiplying (3.37) by v^{T} and post-multiplying v yield

$$(\lambda^2 - 1)v^{\mathrm{T}}\Sigma v = -(v^{\mathrm{T}}A^{\mathrm{T}}BB^{\mathrm{T}}Av + v^{\mathrm{T}}C^{\mathrm{T}}Cv) \leq 0$$

Since $v^{\mathrm{T}}\Sigma v > 0$, we have $\lambda^2 \leq 1$. But $|\lambda|$ is a singular value of A, so that we get $\|A\|_2 \leq 1$. For a proof of the latter half, see [127]. □

Partition Σ into $\Sigma_1 = \mathrm{diag}(\sigma_1, \cdots, \sigma_r)$ and $\Sigma_2 = \mathrm{diag}(\sigma_{r+1}, \cdots, \sigma_n)$, and accordingly write A, B, C as

$$A = \begin{bmatrix} A_{11} & A_{12} \\ A_{21} & A_{22} \end{bmatrix}, \qquad B = \begin{bmatrix} B_1 \\ B_2 \end{bmatrix}, \qquad C = [C_1 \;\; C_2] \tag{3.38}$$

where $A_1 \in \mathbb{R}^{r\times r}$, $B_1 \in \mathbb{R}^{r\times m}$ and $C_1 \in \mathbb{R}^{p\times r}$.

Lemma 3.7. *Suppose that (A, B, C, D) is a balanced realization with A stable. From (3.38), we define a reduced order model*

$$G_r(z) = (A_{11}, B_1, C_1, D) \tag{3.39}$$

Then, the following (i) ∼ (iii) hold[2].

(i) *The model $G_r(z)$ is stable.*

---

[2]Unlike continuous-time systems, the discrete-time model $G_r(z)$ is not balanced. If, however, we relax the definition of balanced realization by using the Riccati inequalities rather than Riccati equations, then $G_r(z)$ may be called a balanced realization [185].

(ii) If $\sigma_r > \sigma_{r+1}$, then $G_r(z)$ is a minimal realization.

(iii) For any $r = 1, \cdots, n-1$, the following bound holds.

$$\|G(e^{j\omega}) - G_r(e^{j\omega})\|_\infty \le 2(\sigma_{r+1} + \cdots + \sigma_n) \tag{3.40}$$

Proof. First we show (i). From (3.38), the first Lyapunov equation of (3.36) is rewritten as

$$\begin{bmatrix} \Sigma_1 & 0 \\ 0 & \Sigma_2 \end{bmatrix} = \begin{bmatrix} A_{11} & A_{12} \\ A_{21} & A_{22} \end{bmatrix} \begin{bmatrix} \Sigma_1 & 0 \\ 0 & \Sigma_2 \end{bmatrix} \begin{bmatrix} A_{11} & A_{12} \\ A_{21} & A_{22} \end{bmatrix}^{\mathrm{T}} + \begin{bmatrix} B_1 \\ B_2 \end{bmatrix} [B_1^{\mathrm{T}} \ B_2^{\mathrm{T}}]$$

Thus we have

$$\Sigma_1 = A_{11}\Sigma_1 A_{11}^{\mathrm{T}} + A_{12}\Sigma_2 A_{12}^{\mathrm{T}} + B_1 B_1^{\mathrm{T}} \tag{3.41a}$$
$$\Sigma_2 = A_{22}\Sigma_2 A_{22}^{\mathrm{T}} + A_{21}\Sigma_1 A_{21}^{\mathrm{T}} + B_2 B_2^{\mathrm{T}} \tag{3.41b}$$
$$0 = A_{11}\Sigma_1 A_{21}^{\mathrm{T}} + A_{12}\Sigma_2 A_{22}^{\mathrm{T}} + B_1 B_2^{\mathrm{T}} \tag{3.41c}$$

Let $\lambda \in \mathbb{C}$ be a non-zero eigenvalue of A_{11}^{T}, and $v \in \mathbb{C}^n$ be a corresponding eigenvector, *i.e.*, $A_{11}^{\mathrm{T}} v = \lambda v$. Pre-multiplying (3.41a) by v^{H} and post-multiplying v yield

$$(1 - |\lambda|^2) v^{\mathrm{H}} \Sigma_1 v = v^{\mathrm{H}} A_{12}\Sigma_2 A_{12}^{\mathrm{T}} v + v^{\mathrm{H}} B_1 B_1^{\mathrm{T}} v \tag{3.42}$$

Since the right-hand side of the above equation is nonnegative, and since $v^{\mathrm{H}}\Sigma_1 v > 0$, we get $|\lambda| \le 1$.

Now suppose that $|\lambda| = 1$. Then the left-hand side of (3.42) becomes 0. But, noting that $\Sigma_2 > 0$, we have

$$v^{\mathrm{H}} A_{12} = 0, \qquad v^{\mathrm{H}} B_1 = 0$$

Since $v^{\mathrm{H}} A_{11} = \bar{\lambda} v^{\mathrm{H}}$, it follows that

$$[v^{\mathrm{H}} \ 0] \begin{bmatrix} A_{11} & A_{12} \\ A_{21} & A_{22} \end{bmatrix} = \bar{\lambda}[v^{\mathrm{H}} \ 0], \qquad [v^{\mathrm{H}} \ 0] \begin{bmatrix} B_1 \\ B_2 \end{bmatrix} = 0$$

But from Theorem 3.4 (ii), this contradicts the reachability of (A, B). Hence, we have $|\lambda| \ne 1$, so that $|\lambda| < 1$. A similar proof is also applicable to the second equation of (3.36).

Now we show item (ii). From the second equation of (3.36),

$$\begin{bmatrix} \Sigma_1 & 0 \\ 0 & \Sigma_2 \end{bmatrix} = \begin{bmatrix} A_{11} & A_{12} \\ A_{21} & A_{22} \end{bmatrix}^{\mathrm{T}} \begin{bmatrix} \Sigma_1 & 0 \\ 0 & \Sigma_2 \end{bmatrix} \begin{bmatrix} A_{11} & A_{12} \\ A_{21} & A_{22} \end{bmatrix} + \begin{bmatrix} C_1^{\mathrm{T}} \\ C_2^{\mathrm{T}} \end{bmatrix} [C_1 \ C_2]$$

Hence, the $(1,1)$-block of the above equation gives

$$A_{11}^{\mathrm{T}}\Sigma_1 A_{11} + A_{21}^{\mathrm{T}}\Sigma_2 A_{21} - \Sigma_1 = -C_1^{\mathrm{T}} C_1 \tag{3.43}$$

Suppose that (C_1, A_{11}) is not observable. Then, there exist an eigenvalue $\lambda \in \mathbb{C}$ and an eigenvector $v \in \mathbb{C}^n$ of A_{11} such that

$$A_{11}v = \lambda v, \qquad C_1 v = 0 \tag{3.44}$$

We assume without loss of generality that $||v|| = 1$. Pre-multiplying (3.43) by v^{H} and post-multiplying v yield

$$(|\lambda|^2 - 1)v^{\mathrm{T}}\Sigma_1 v + v^{\mathrm{H}} A_{21}^{\mathrm{T}} \Sigma_2 A_{21} v = 0$$

Since

$$\underline{\sigma}(\Sigma_1) \leq v^{\mathrm{H}}\Sigma_1 v, \qquad v^{\mathrm{H}} A_{21}^{\mathrm{T}} \Sigma_2 A_{21} v \leq ||A_{21}v||^2 \overline{\sigma}(\Sigma_2)$$

hold, we see that

$$(1 - |\lambda|^2)\underline{\sigma}(\Sigma_1) \leq ||A_{21}v||^2 \overline{\sigma}(\Sigma_2)$$

From Lemma 3.6, it follows that $||A||_2 \leq 1$, so that the norm of any submatrix of A is less than 1. Hence, we get

$$\left\| \begin{bmatrix} A_{11} \\ A_{21} \end{bmatrix} v \right\| \leq 1 \quad \Leftrightarrow \quad ||A_{11}v||^2 + ||A_{21}v||^2 \leq 1$$

Since, from (3.44), $||A_{11}v||^2 = |\lambda|^2$, it follows that $||A_{21}v||^2 \leq 1 - |\lambda|^2$, implying that

$$(1 - |\lambda|^2)\underline{\sigma}(\Sigma_1) \leq (1 - |\lambda|^2)\overline{\sigma}(\Sigma_2)$$

Since $|\lambda|^2 < 1$, we have $\underline{\sigma}(\Sigma_1) \leq \overline{\sigma}(\Sigma_2)$. But this contradicts the assumption that $\sigma_r > \sigma_{r+1}$. Hence we conclude that $(C_1,\ A_{11})$ is observable. Similarly, we can show that $(A_{11},\ B_1)$ is reachable.

For a proof of (iii), see [5, 71, 185]. □

Similarly to the proof of Lemma 3.7 (i), we can show that the subsystem $(A_{22},\ B_2,\ C_2)$ is also stable. Thus, since $|\lambda(A_{22})| < 1$, we see that $\alpha I - A_{22}$ is nonsingular, where $|\alpha| = 1$, $\alpha \in \mathbb{C}$. Hence we can define (A_r, B_r, C_r, D_r) as

$$A_r = A_{11} + A_{12}(\alpha I - A_{22})^{-1} A_{21} \tag{3.45a}$$
$$B_r = B_1 + A_{12}(\alpha I - A_{22})^{-1} B_2 \tag{3.45b}$$
$$C_r = C_1 + C_2(\alpha I - A_{22})^{-1} A_{21} \tag{3.45c}$$
$$D_r = D + C_2(\alpha I - A_{22})^{-1} B_2 \tag{3.45d}$$

Then, we have the following lemma.

Lemma 3.8. *Suppose that $(A,\ B,\ C,\ D)$ is a balanced realization. Then, we have the following (i) ∼ (iii).*

(i) The triplet $(A_r,\ B_r,\ C_r)$ defined by (3.45) satisfies the following Lyapunov equations

$$\Sigma_1 = A_r \Sigma_1 A_r^{\mathrm{H}} + B_r B_r^{\mathrm{H}}, \qquad \Sigma_1 = A_r^{\mathrm{H}} \Sigma_1 A_r + C_r^{\mathrm{H}} C_r$$

where $\Sigma_1 = \mathrm{diag}(\sigma_1, \cdots, \sigma_r)$. Hence $(A_r,\ B_r,\ C_r)$ is a balanced realization with $\overline{\sigma}(A_r) \leq 1$.

(ii) If the Gramians Σ_1 and Σ_2 have no common elements, or $\sigma_r > \sigma_{r+1}$, then A_r is stable, and $G_r(z) = (A_r,\ B_r,\ C_r, D_r)$ is a minimal realization.

(iii) The approximation error is the same as that of (3.40), i.e.,

$$\|G(e^{j\omega}) - G_r(e^{j\omega})\|_\infty \le 2(\sigma_{r+1} + \cdots + \sigma_n)$$

Proof. We show (i). For simplicity, define $\Phi = \alpha I - A_{22}$. Then by the definitions of A_r and B_r,

$$\begin{aligned} J := &\ A_r \Sigma_1 A_r^{\mathrm{H}} + B_r B_r^{\mathrm{H}} \\ = &\ (A_{11} + A_{12}\Phi^{-1}A_{21})\Sigma_1(A_{11} + A_{12}\Phi^{-1}A_{21})^{\mathrm{H}} \\ &+ (B_1 + A_{12}\Phi^{-1}B_2)(B_1 + A_{12}\Phi^{-1}B_2)^{\mathrm{H}} \end{aligned}$$

We show that J equals Σ_1 by a direct calculation. Expanding the above equation gives

$$\begin{aligned} J = &\ A_{11}\Sigma_1 A_{11}^{\mathrm{T}} + A_{11}\Sigma_1 A_{21}^{\mathrm{T}}\Phi^{-\mathrm{H}}A_{12}^{\mathrm{T}} + A_{12}\Phi^{-1}A_{21}\Sigma_1 A_{11}^{\mathrm{T}} \\ &+ A_{12}\Phi^{-1}A_{21}\Sigma_1 A_{21}^{\mathrm{T}}\Phi^{-\mathrm{H}}A_{12}^{\mathrm{T}} + B_1B_1^{\mathrm{T}} + B_1B_2^{\mathrm{T}}\Phi^{-\mathrm{H}}A_{12}^{\mathrm{T}} \\ &+ A_{12}\Phi^{-1}B_2B_1^{\mathrm{T}} + A_{12}\Phi^{-1}B_2B_2^{\mathrm{T}}\Phi^{-\mathrm{H}}A_{12}^{\mathrm{T}} \end{aligned}$$

Substituting $B_1B_1^{\mathrm{T}}$, $B_1B_2^{\mathrm{T}}$, $B_2B_1^{\mathrm{T}}$ and $B_2B_2^{\mathrm{T}}$ from (3.41) into the above equation, we get

$$\begin{aligned} J = &\ \Sigma_1 - A_{12}\Sigma_2A_{12}^{\mathrm{T}} - A_{12}\Sigma_2A_{22}^{\mathrm{T}}\Phi^{-\mathrm{H}}A_{12}^{\mathrm{T}} - A_{12}\Phi^{-1}A_{22}\Sigma_2A_{12}^{\mathrm{T}} \\ &- A_{12}\Phi^{-1}A_{22}\Sigma_2A_{22}^{\mathrm{T}}\Phi^{-\mathrm{H}}A_{12}^{\mathrm{T}} + A_{12}\Phi^{-1}\Sigma_2\Phi^{-\mathrm{H}}A_{12}^{\mathrm{T}} \end{aligned}$$

Collecting the terms involving Σ_2 yields

$$J - \Sigma_1 = A_{12}\Phi^{-1}(1 - |\alpha|^2)\Sigma_2\Phi^{-\mathrm{H}}A_{12}^{\mathrm{T}} = 0$$

since $|\alpha| = 1$. This proves the first Lyapunov equation $\Sigma_1 = A_r\Sigma_1A_r^{\mathrm{H}} + B_rB_r^{\mathrm{H}}$ of this lemma. In similar fashion, we can show $\Sigma_1 = A_r^{\mathrm{H}}\Sigma_1A_r + C_r^{\mathrm{H}}C_r$.

(ii) This can be proved similarly to that of (ii) of Lemma 3.7.

(iii) This part is omitted. See references [5, 71, 108, 185]. □

The reduced order model $G_r(z)$ obtained by Lemma 3.8 is called a balanced reduced order model for a discrete-time LTI system $G(z)$. The method of deriving $G_r(z)$ in Lemma 3.8 is called the singular perturbation approximation (SPA) method. It can easily be shown that $G(\alpha) = G_r(\alpha)$, and hence $G(1) = G_r(1)$. This implies that the reduced order model by the SPA method preserves the steady state gains of $G_r(z)$ and $G(z)$. However, this does not hold for the direct method of Lemma 3.7.

Before concluding this section, we provide a method of computing Gramians of unstable systems.

Definition 3.10. *[168, 186]. Suppose that $G(z) = (A, B, C, D)$, where A is possibly unstable, but has no eigenvalues on the unit circle. Then, the reachability and observability Gramians P and Q are respectively defined by*

$$P = \frac{1}{2\pi} \int_0^{2\pi} (e^{j\theta} I - A)^{-1} BB^{\mathrm{T}} (e^{-j\theta} I - A^{\mathrm{T}})^{-1} d\theta \tag{3.46}$$

and

$$Q = \frac{1}{2\pi} \int_0^{2\pi} (e^{-j\theta} I - A^{\mathrm{T}})^{-1} C^{\mathrm{T}} C (e^{j\theta} I - A)^{-1} d\theta \tag{3.47}$$

It should be noted that if A is stable, P and Q above reduce to standard Gramians of (3.34) and (3.35), respectively. □

Lemma 3.9. *[168, 186] Suppose that (A, B) is stabilizable, and (C, A) is detectable. Let X and Y respectively be the stabilizing solutions of the algebraic Riccati equations*

$$X = A^{\mathrm{T}}(X - XB[I_m + B^{\mathrm{T}}XB]^{-1}B^{\mathrm{T}}X)A$$

and

$$Y = A(Y - YC^{\mathrm{T}}[I_p + CYC^{\mathrm{T}}]^{-1}CY)A^{\mathrm{T}}$$

Also, define

$$F = -(I_m + B^{\mathrm{T}}XB)^{-1}B^{\mathrm{T}}XA, \qquad WW^{\mathrm{T}} = (I_m + B^{\mathrm{T}}XB)^{-1}$$

and

$$L = -AYC^{\mathrm{T}}(I_p + CYC^{\mathrm{T}})^{-1}, \qquad V^{\mathrm{T}}V = (I_p + CYC^{\mathrm{T}})^{-1}$$

Then, the Gramians P and Q respectively satisfy Lyapunov equations

$$P = (A + BF)P(A + BF)^{\mathrm{T}} + B(I_m + B^{\mathrm{T}}XB)^{-1}B^{\mathrm{T}} \tag{3.48}$$

and

$$Q = (A + LC)^{\mathrm{T}}Q(A + LC) + C^{\mathrm{T}}(I_p + CYC^{\mathrm{T}})^{-1}C \tag{3.49}$$

Proof. Consider the following right coprime factorization

$$(zI - A)^{-1}B = N(z)M^{-1}(z)$$

where $M(z)$ is an $m \times m$ inner matrix, satisfying $M^{\mathrm{T}}(z^{-1})M(z) = I_m$. Then, we have the following realization

$$\begin{bmatrix} N(z) \\ M(z) \end{bmatrix} = \left[\begin{array}{c|c} A + BF & BW \\ \hline I_n & 0 \\ F & W \end{array}\right], \quad F \in \mathbb{R}^{m \times n}, \quad W \in \mathbb{R}^{m \times m}$$

where $A_F := A + BF$ is stable. By using the above coprime factorization, we have

$$
\begin{aligned}
P &= \frac{1}{2\pi}\int_0^{2\pi} N(e^{j\theta})M^{-1}(e^{j\theta})M^{-\mathrm{T}}(e^{-j\theta})N^{\mathrm{T}}(e^{-j\theta})d\theta \\
&= \frac{1}{2\pi}\int_0^{2\pi} N(e^{j\theta})N^{\mathrm{T}}(e^{-j\theta})d\theta \\
&= \frac{1}{2\pi}\int_0^{2\pi} (e^{j\theta}I - A_F)^{-1}BWW^{\mathrm{T}}B^{\mathrm{T}}(e^{-j\theta}I - A_F^{\mathrm{T}})^{-1}d\theta \\
&= \frac{1}{2\pi}\int_0^{2\pi} (e^{j\theta}I - A_F)^{-1}B(I_m + B^{\mathrm{T}}XB)^{-1}B^{\mathrm{T}}(e^{-j\theta}I - A_F^{\mathrm{T}})^{-1}d\theta \\
&= \sum_{k=0}^{\infty} A_F^k B(I_m + B^{\mathrm{T}}XB)^{-1}B^{\mathrm{T}}(A_F^{\mathrm{T}})^k
\end{aligned}
$$

This shows that P satisfies (3.48). Similarly, let a left coprime factorization be given by

$$C(zI_n - A)^{-1} = \tilde{M}^{-1}(z)\tilde{N}(z),$$

where $\tilde{M}(z)$ is an $n \times n$ co-inner matrix with $\tilde{M}(z)\tilde{M}^{\mathrm{T}}(z^{-1}) = I_p$. A realization of $[\tilde{N}(z)\ \ \tilde{M}(z)]$ is then given by

$$[\tilde{N}(z)\ \ \tilde{M}(z)] = \left[\begin{array}{c|cc} A + LC & I_n & L \\ \hline VC & 0 & V \end{array}\right], \quad L \in \mathbb{R}^{n\times p}, \quad V \in \mathbb{R}^{p\times p}$$

Similarly, we can show that the observability Gramian Q satisfies (3.49). □

Example 3.4. Suppose that $(A,\ B,\ C)$ are given by

$$A = \begin{bmatrix} A_1 & 0 \\ 0 & A_2 \end{bmatrix}, \qquad B = \begin{bmatrix} B_1 \\ B_2 \end{bmatrix}, \qquad C = \begin{bmatrix} C_1 & C_2 \end{bmatrix} \tag{3.50}$$

where A_1 is stable, and A_2 is anti-stable. Define

$$
\begin{aligned}
P_1 &= A_1P_1A_1^{\mathrm{T}} + B_1B_1^{\mathrm{T}}, & P_2 &= A_2P_2A_2^{\mathrm{T}} + B_2B_2^{\mathrm{T}} \\
Q_1 &= A_1^{\mathrm{T}}Q_1A_1 + C_1^{\mathrm{T}}C_1, & Q_2 &= A_2^{\mathrm{T}}Q_2A_2 + C_2^{\mathrm{T}}C_2
\end{aligned}
$$

We wish to show that the reachability Gramian P and the observability Gramian Q of $(A,\ B,\ C)$ are given by $P = \begin{bmatrix} P_1 & 0 \\ 0 & P_2 \end{bmatrix}$ and $Q = \begin{bmatrix} Q_1 & 0 \\ 0 & Q_2 \end{bmatrix}$, respectively.

From (3.46), we have

$$
\begin{aligned}
P &= \frac{1}{2\pi}\int_0^{2\pi} (e^{j\theta}I - A)^{-1}BB^{\mathrm{T}}(e^{-j\theta}I - A^{\mathrm{T}})^{-1}d\theta \\
&= \frac{1}{2\pi j}\oint_{|z|=1} (zI - A)^{-1}BB^{\mathrm{T}}(z^{-1}I - A^{\mathrm{T}})^{-1}\frac{dz}{z}
\end{aligned}
$$

where $|z| = 1$ denotes the unit circle. According to (3.50), partition P as

$$P = \begin{bmatrix} P_{11} & P_{12} \\ P_{21} & P_{22} \end{bmatrix}$$

Noting that A_1 is stable and A_2 is anti-stable, we have

$$P_{11} = \frac{1}{2\pi j} \oint_{|z|=1} (zI - A_1)^{-1} B_1 B_1^{\mathrm{T}} (z^{-1} I - A_1^{\mathrm{T}})^{-1} \frac{dz}{z} = P_1$$

$$P_{22} = \frac{1}{2\pi j} \oint_{|z|=1} (zI - A_2)^{-1} B_2 B_2^{\mathrm{T}} (z^{-1} I - A_2^{\mathrm{T}})^{-1} \frac{dz}{z} = P_2$$

$$P_{12} = \frac{1}{2\pi j} \oint_{|z|=1} (zI - A_1)^{-1} B_1 B_2^{\mathrm{T}} (z^{-1} I - A_2^{\mathrm{T}})^{-1} \frac{dz}{z} = 0$$

We see that the third integral is zero since the integrand is analytic in $|z| > 1$, and similarly $P_{21} = 0$. Hence we have $P = \begin{bmatrix} P_1 & 0 \\ 0 & P_2 \end{bmatrix}$. That $Q = \begin{bmatrix} Q_1 & 0 \\ 0 & Q_2 \end{bmatrix}$ is proved in the same way. □

3.9 Realization Theory

In this section, we prove basic realization results, which will be used in Chapter 6 to discuss the deterministic realization theory.

Consider an infinite sequence $Y = (Y_1, Y_2, \cdots)$ with $Y_i \in \mathbb{R}^{p \times m}$. Let the infinite matrix formed by Y_i, $i = 1, 2, \cdots$ be given by

$$H = \begin{bmatrix} Y_1 & Y_2 & Y_3 & \cdots \\ Y_2 & Y_3 & Y_4 & \cdots \\ Y_3 & Y_4 & Y_5 & \cdots \\ \vdots & \vdots & \vdots & \ddots \end{bmatrix} \tag{3.51}$$

This is called an infinite block Hankel matrix. By using the shift operator σ, we define $\sigma^k Y = (Y_{k+1}, Y_{k+2}, \cdots)$ and the block Hankel matrix as

$$\sigma^k H = \begin{bmatrix} Y_{k+1} & Y_{k+2} & Y_{k+3} & \cdots \\ Y_{k+2} & Y_{k+3} & Y_{k+4} & \cdots \\ Y_{k+3} & Y_{k+4} & Y_{k+5} & \cdots \\ \vdots & \vdots & \vdots & \ddots \end{bmatrix}, \qquad k = 0, 1, \cdots \tag{3.52}$$

Definition 3.11. *If (A, B, C) satisfies*

$$Y_i = CA^{i-1}B, \qquad i = 1, 2, \cdots \tag{3.53}$$

then the triplet (A, B, C) is called a realization of H. □

Let the $k \times l$ block submatrix appearing in the upper-left corner of the infinite Hankel matrix H be given by

$$H_{k,l} = \begin{bmatrix} Y_1 & Y_2 & Y_3 & \cdots & Y_l \\ Y_2 & Y_3 & Y_4 & \cdots & Y_{l+1} \\ Y_3 & Y_4 & Y_5 & \cdots & Y_{l+2} \\ \vdots & \vdots & & \ddots & \vdots \\ Y_k & Y_{k+1} & \cdots & \cdots & Y_{k+l-1} \end{bmatrix} \tag{3.54}$$

Moreover, we define the k-observability matrix and l-reachability matrix as

$$\mathcal{O}_k = \begin{bmatrix} C \\ CA \\ \vdots \\ CA^{k-1} \end{bmatrix}, \qquad \mathcal{C}_l = [B \;\; AB \;\; \cdots \;\; A^{l-1}B] \tag{3.55}$$

where $A \in \mathbb{R}^{n\times n}$, $B \in \mathbb{R}^{n\times m}$, $C \in \mathbb{R}^{p\times n}$. If $k > n$ (or $l > n$), then $\mathcal{O}_k$ (or $\mathcal{C}_l$) is called an extended observability (or reachability) matrix, and the n-reachability (or n-observability) matrix is simply called the reachability (or observability) matrix.

Let α be the smallest positive integer such that $\text{rank}(\mathcal{C}_{\alpha+1}) = \text{rank}(\mathcal{C}_\alpha)$. Then this value of α is called the reachability index of $\Sigma = (A,\ B,\ C)$. Similarly, the smallest positive integer β such that $\text{rank}(\mathcal{O}_{\beta+1}) = \text{rank}(\mathcal{O}_\beta)$ is referred to as the observability index.

Lemma 3.10. *If $(A,\ B,\ C)$ is a realization of H, then*

$$H_{k,l} = \mathcal{O}_k \mathcal{C}_l, \qquad k,\ l = 1,\ 2, \cdots \tag{3.56}$$

holds, and vice versa. In this case, we have the following rank conditions.

$$\text{rank}(H_{k,l}) \le \max\{\,\text{rank}(\mathcal{O}_k),\ \text{rank}(\mathcal{C}_l)\,\} \le n$$

Proof. Equation (3.56) is clear from Definition 3.11 and (3.55). The inequalities above are also obvious from (3.56). □

From the canonical decomposition theorem (Theorem 3.11), if there exists a realization of H, then there exists a minimal realization. Thus, we have the following lemma.

Lemma 3.11. *If $(A,\ B,\ C)$ is a minimal realization, we have*

$$\text{rank}(H_{k,l}) = n, \qquad k,\ l = n,\ n+1, \cdots \tag{3.57}$$

Proof. For $k,\ l \ge n$, it follows that $\text{rank}(\mathcal{O}_k) = n$ and $\text{rank}(\mathcal{C}_l) = n$. Hence, we get

$$\mathcal{O}_k^\dagger \mathcal{O}_k = I_n, \qquad \mathcal{C}_l \mathcal{C}_l^\dagger = I_n$$

From (3.56), this implies that

$$I_n = \mathcal{O}_k^\dagger H_{k,l} \mathcal{C}_l^\dagger \quad \Rightarrow \quad \mathrm{rank}(\mathcal{O}_k^\dagger H_{k,l} \mathcal{C}_l^\dagger) = n$$

Thus $\mathrm{rank}(H_{k,l}) \geq n$ holds. But from Lemma 3.10, we have $\mathrm{rank}(H_{k,l}) \leq n$. This completes the proof. □

Now we define the rank of an infinite Hankel matrix. It may be, however, noted that the condition cannot be checked by a finite step procedure.

Definition 3.12. *The rank of the infinite block Hankel matrix H of (3.51) is defined by*

$$\mathrm{rank}(H) = \sup_{k,l} \mathrm{rank}(H_{k,l})$$ □

We consider the realizability condition of an infinite impulse response sequence in terms of the concept of recursive sequence. Suppose that there exist $\alpha_1, \cdots, \alpha_n \in \mathbb{R}$ such that

$$Y_{n+k+1} + \sum_{i=1}^{n} \alpha_{n-i+1} Y_{k+i} = 0, \qquad k = 0,\ 1,\ \cdots \tag{3.58}$$

holds. In this case, we say that $Y = (Y_1,\ Y_2, \cdots)$ is a recursive sequence of order n. The following theorem gives an important result for the realizability of the infinite block Hankel matrix, which is an extension of Lemma 2.14 to a matrix case.

Theorem 3.13. *An infinite block Hankel matrix H is realizable if and only if Y is recursive.*

Proof. To prove the necessity, let $Y_i = CA^{i-1}B,\ i = 1,\ 2,\ \cdots$ and let the characteristic polynomial of A be given by

$$\varphi_A(z) = z^n + \alpha_1 z^{n-1} + \cdots + \alpha_{n-1} z + \alpha_n$$

Then, from the Cayley-Hamilton theorem,

$$A^n + \alpha_1 A^{n-1} + \cdots + \alpha_{n-1} A + \alpha_n I = 0$$

Pre-multiplying this by CA^{k+1} and post-multiplying by B yield (3.58).

The sufficiency will be proved by constructing a realization and then a minimal realization. To this end, we consider the block Hankel matrix $\sigma^k H$ of (3.52). Let the $n \times n$ block submatrix appearing in the upper-left corner of $\sigma^k H$ be given by

$$(\sigma^k H)_{n,n} = \begin{bmatrix} Y_{k+1} & Y_{k+2} & \cdots & Y_{k+n} \\ Y_{k+2} & Y_{k+3} & \cdots & Y_{k+n+1} \\ \vdots & \vdots & \ddots & \vdots \\ Y_{k+n} & Y_{k+n+1} & \cdots & Y_{k+2n-1} \end{bmatrix} \in \mathbb{R}^{pn \times mn}$$

Also, let the block companion matrix be defined by

$$M = \begin{bmatrix} 0 & I_p & 0 & \cdots & 0 \\ 0 & 0 & I_p & \cdots & 0 \\ \vdots & \vdots & \vdots & \ddots & \vdots \\ 0 & 0 & 0 & \cdots & I_p \\ -\alpha_n I_p & -\alpha_{n-1} I_p & -\alpha_{n-2} I_p & \cdots & -\alpha_1 I_p \end{bmatrix} \in \mathbb{R}^{pn \times pn}$$

It follows from (3.58) that

$$M(\sigma^k H)_{n,n} = (\sigma^{k+1} H)_{n,n}, \qquad k = 0,\ 1,\ \cdots$$

Hence we have

$$M^k H_{n,n} = (\sigma^k H)_{n,n}, \qquad k = 0,\ 1,\ \cdots \tag{3.59}$$

Since the $(1, 1)$-block element of $(\sigma^k H)_{n,n}$ is just Y_{k+1}, we get

$$Y_{k+1} = [I_p\ 0\ \cdots\ 0] M^k H_{n,n} \begin{bmatrix} I_m \\ 0 \\ \vdots \\ 0 \end{bmatrix}, \qquad k = 0,\ 1,\ \cdots$$

For notational convenience, we define $E_p^{\mathrm{T}} = [I_p\ 0\ \cdots\ 0] \in \mathbb{R}^{p \times pn}$, and

$$E_m = \begin{bmatrix} I_m \\ 0 \\ \vdots \\ 0 \end{bmatrix} \in \mathbb{R}^{mn \times m}, \qquad \bar{B} = H_{n,n} E_m = \begin{bmatrix} Y_1 \\ Y_2 \\ \vdots \\ Y_n \end{bmatrix} \in \mathbb{R}^{pn \times m}$$

Also, define $\bar{A} = M$ and $\bar{C} = E_p$. Then, it follows that

$$Y_{k+1} = E_p^{\mathrm{T}} M^k H_{n,n} E_m = \bar{C} \bar{A}^k \bar{B}, \qquad k = 0,\ 1,\ \cdots \tag{3.60}$$

This concludes that $(\bar{A},\ \bar{B},\ \bar{C})$ is a (non-minimal) realization with $\bar{A} \in \mathbb{R}^{pn \times pn}$.

We derive a minimal realization from this non-minimal realization $(\bar{A},\ \bar{B},\ \bar{C})$. Define a block companion matrix

$$N = \begin{bmatrix} 0 & 0 & \cdots & 0 & -\alpha_n I_m \\ I_m & 0 & \cdots & 0 & -\alpha_{n-1} I_m \\ 0 & I_m & \cdots & 0 & -\alpha_{n-2} I_m \\ \vdots & \vdots & \ddots & \vdots & \vdots \\ 0 & 0 & \cdots & I_m & -\alpha_1 I_m \end{bmatrix} \in \mathbb{R}^{mn \times mn}$$

Then, similarly to the procedure of deriving (3.59),

$$H_{n,n} N^k = (\sigma^k H)_{n,n}, \qquad k = 0,\ 1,\ \cdots \tag{3.61}$$

Hence, from (3.59) and (3.61), we have $M^k H_{n,n} = H_{n,n} N^k$, $k = 0,\ 1,\ \cdots$.

Suppose that $\text{rank}(H_{n,n}) = r$. Let the SVD of $H_{n,n}$ be given by

$$H_{n,n} = U\Sigma V^{\mathrm{T}} = U \begin{bmatrix} \Sigma_r & 0 \\ 0 & 0 \end{bmatrix} V^{\mathrm{T}} = U_r \Sigma_r V_r^{\mathrm{T}} \in \mathbb{R}^{pn\times mn}$$

where $\Sigma_r > 0$, $\Sigma_r \in \mathbb{R}^{r\times r}$, and $U_r^{\mathrm{T}} U_r = I_r$, $V_r^{\mathrm{T}} V_r = I_r$, From Lemma 2.10, the pseudo-inverse of $H_{n,n}$ is given by $H_{n,n}^{\dagger} = V_r \Sigma_r^{-1} U_r^{\mathrm{T}}$, so that we have $H_{n,n} H_{n,n}^{\dagger} = U_r U_r^{\mathrm{T}}$ and $H_{n,n}^{\dagger} H_{n,n} = V_r V_r^{\mathrm{T}}$.

By using the SVD and pseudo-inverses, Y_{k+1} of (3.60) is computed as

$$\begin{aligned} Y_{k+1} &= E_p^{\mathrm{T}} M^k H_{n,n} E_m = E_p^{\mathrm{T}} H_{n,n} N^k E_m \\ &= E_p^{\mathrm{T}} H_{n,n} H_{n,n}^{\dagger} H_{n,n} N^k E_m \\ &= E_p^{\mathrm{T}} H_{n,n} H_{n,n}^{\dagger} M^k H_{n,n} E_m \\ &= E_p^{\mathrm{T}} H_{n,n} H_{n,n}^{\dagger} M^k H_{n,n} H_{n,n}^{\dagger} H_{n,n} E_m \\ &= E_p^{\mathrm{T}} U_r U_r^{\mathrm{T}} M^k H_{n,n} V_r V_r^{\mathrm{T}} E_m \\ &= (E_p^{\mathrm{T}} U_r \Sigma_r^{1/2})(\Sigma_r^{-1/2} U_r^{\mathrm{T}} M^k H_{n,n} V_r \Sigma_r^{-1/2})(\Sigma_r^{1/2} V_r^{\mathrm{T}} E_m) \end{aligned}$$

Define $A := \Sigma_r^{-1/2} U_r^{\mathrm{T}} M H_{n,n} V_r \Sigma_r^{-1/2} \in \mathbb{R}^{r\times r}$, $B := \Sigma_r^{1/2} V_r^{\mathrm{T}} E_m \in \mathbb{R}^{r\times m}$, and $C := E_p^{\mathrm{T}} U_r \Sigma_r^{1/2} \in \mathbb{R}^{p\times r}$. Then we see that

$$\begin{aligned} A^2 &= (\Sigma_r^{-1/2} U_r^{\mathrm{T}} M H_{n,n} V_r \Sigma_r^{-1/2})(\Sigma_r^{-1/2} U_r^{\mathrm{T}} M H_{n,n} V_r \Sigma_r^{-1/2}) \\ &= \Sigma_r^{-1/2} U_r^{\mathrm{T}} M H_{n,n} V_r \Sigma_r^{-1} U_r^{\mathrm{T}} M H_{n,n} V_r \Sigma_r^{-1/2} \\ &= \Sigma_r^{-1/2} U_r^{\mathrm{T}} M H_{n,n} H_{n,n}^{\dagger} H_{n,n} N V_r \Sigma_r^{-1/2} \\ &= \Sigma_r^{-1/2} U_r^{\mathrm{T}} M H_{n,n} N V_r \Sigma_r^{-1/2} = \Sigma_r^{-1/2} U_r^{\mathrm{T}} M^2 H_{n,n} V_r \Sigma_r^{-1/2} \end{aligned}$$

Thus, inductively, we can show that $Y_{k+1} = CA^k B$, implying that $(A,\ B,\ C)$ is a minimal realization with $A \in \mathbb{R}^{r\times r}$. □

It follows from this theorem that H is realizable if and only if the rank of H is finite. If $\text{rank}(H) = n$, then the rank of a minimal realization is also n. It may be, however, noted that this statement cannot be verified in an empirical way. But we have the following theorem in this connection.

Theorem 3.14. *Suppose that for $Y_i,\ i = 1, \cdots, 2n$, the following conditions are satisfied:*

$$\text{rank}(H_{n,n}) = \text{rank}(H_{n+1,n}) = \text{rank}(H_{n,n+1}) = n$$

Then, there exists a unique Markov sequence $Y = (Y_1, Y_2, \cdots)$ with rank n such that the first $2n$ parameters exactly equal the given Y_i, $i = 1, \cdots, 2n$.

Proof. From $\text{rank}(H_{n,n}) = \text{rank}(H_{n+1,n})$, the last p rows of $H_{n+1,n}$ must be linear combinations of the rows of $H_{n,n}$. Hence, there exist $p \times p$ matrices C_i, $i = 1, \cdots, n$ such that

$$Y_j = C_1 Y_{j-1} + \cdots + C_n Y_{j-n}, \qquad j = n+1, \cdots, 2n \tag{3.62}$$

Similarly, from $\mathrm{rank}(H_{n,n}) = \mathrm{rank}(H_{n,n+1})$, we see that the last m columns of $H_{n,n+1}$ must be linear combinations of the columns of $H_{n,n}$. Thus, there exist $m \times m$ matrices D_i, $i = 1, \cdots, n$ such that

$$Y_j = Y_{j-1} D_1 + \cdots + Y_{j-n} D_n, \qquad j = n+1, \cdots, 2n \tag{3.63}$$

Now we recursively define Y_j, $j = 2n+1, \cdots$ by means of (3.62), so that we have an infinite sequence $Y = (Y_1, Y_2, \cdots)$. By this construction, the rank of the infinite block Hankel matrix H has rank smaller than pn. We show that the rank is in fact n. To this end, we show that (3.63) also holds for $j = 2n+1, 2n+2, \cdots$. From (3.62) and (3.63), for $j > 2n$

$$\begin{aligned} Y_{j+1} &= \sum_{i=1}^{n} C_i Y_{j+1-i} = \sum_{i=1}^{n} C_i \sum_{k=1}^{n} Y_{j+1-i-k} D_k \\ &= \sum_{k=1}^{n} \left(\sum_{i=1}^{n} C_i Y_{j+1-i-k} \right) D_k = \sum_{k=1}^{n} Y_{j+1-k} D_k \end{aligned}$$

Thus the columns of H are linearly dependent on the first mn columns, and hence we have $\mathrm{rank}(H) = \mathrm{rank}(H_{n,n}) = n$.

Finally, the uniqueness is proved as follows. Suppose that we have two Markov sequences Y^1 and Y^2. Define $Y := Y^1 - Y^2$. Then we see that the rank of Y is at most $2n$ and that the first $2n$ parameters are zero. Therefore, from Theorem 3.13, applying (3.58) with $n := 2n$, we have $Y = 0$. This completes the proof. □

3.10 Notes and References

- After a brief review of z-transform in Section 3.1, we have introduced discrete-time systems and signals, together with their norms in Sections 3.2 and 3.3. Used are references [98, 121, 144] for systems and signals and [36] for complex function theory.
- In Sections 3.4 to 3.7, state-space methods are considered, including Lyapunov stability, reachability and observability of discrete-time LTI systems. In relation to the realization and system identification, the canonical decomposition theorem (Theorem 3.11) is theoretically most important because it tells us that the transfer matrix of an LTI system is related to the reachable and observable subsystem only. Thus the unreachable or unobservable parts of the system are irrelevant to system identification. The basic references are [27, 80, 185].
- In Section 3.8, we present balanced realization theory and model reduction techniques for discrete-time LTI systems by using [5, 108, 127, 168, 186]. It is well known that for continuous-time systems, a reduced-order model derived from a balanced realization by retaining principal modes is balanced, but this fact is no longer true for discrete-time systems.

- However, by using Lyapunov inequalities rather than Lyapunov equations, it can be shown in [71, 185] that reduced-order balanced models are obtained from higher-order balanced models. These model reduction theory and technique are employed in Chapter 8 to consider the theory of balanced stochastic realization and in Chapter 11 to compute reduced-order models in closed-loop identification algorithms.
- The basic results for the realization theory treated in Section 3.9 are found in [72,85,147]. The proof of Theorem 3.13 is based on [85] and the SVD technique due to [184], and this theorem is a basis of the classical deterministic realization theory to be developed in Chapter 6.

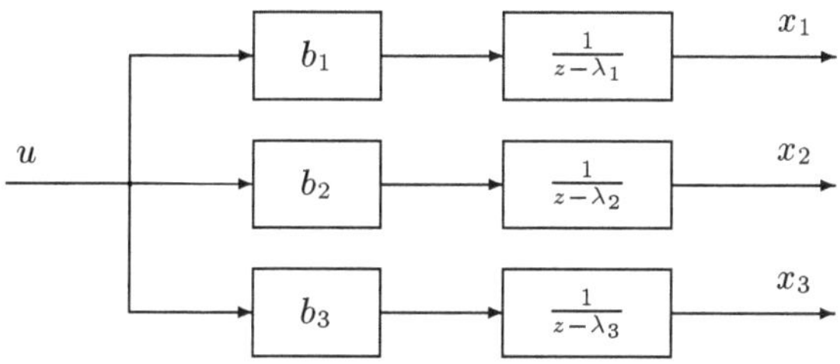

Figure 3.3. A diagonal system with $n = 3$

3.11 Problems

3.1 Suppose that the impulse response of $G(z)$ is given by

$$g_k = \frac{(-1)^{k-1}}{k}, \qquad k = 1, 2, \cdots$$

with $g_0 = 0$. Consider the stability of this system by means of Theorem 3.1.

3.2 Find a necessary and sufficient condition such that the second-order polynomial $f(z) := z^2 + a_1 z + a_2$ is stable. Note that a polynomial is called stable if all the roots are inside the unit circle.

3.3 Derive a state space model for the system shown in Figure 3.3, and obtain the reachability condition.

3.4 Consider a realization (A, b, c) of an SISO system with (A, b) reachable. Show that $\bar{A} = \mathcal{C}^{-1} A \mathcal{C}$ and $\bar{b} = \mathcal{C}^{-1} b$ are given by

$$\bar{A} = \begin{bmatrix} 0 & & & -\alpha_n \\ 1 & & & -\alpha_{n-1} \\ & \ddots & & \vdots \\ & & 1 & -\alpha_1 \end{bmatrix}, \qquad \bar{b} = \begin{bmatrix} 1 \\ 0 \\ \vdots \\ 0 \end{bmatrix}$$

where $\mathcal{C}$ is the reachability matrix.

3.5 Show that (A, B) is reachable (stabilizable) if and only if $(A + BK, B)$ is reachable (stabilizable). Also, (C, A) is observable (detectable) if and only if $(C, A + LC)$ is observable (detectable).

3.6 [73] Let $A \in \mathbb{R}^{n\times n}$. Show that for any $\varepsilon > 0$, there exists a constant $C > 0$ such that

$$\left|(A^k)_{ij}\right| \le C(\rho(A) + \varepsilon)^k, \qquad k = 1, 2, \cdots$$

where $i, j = 1, \cdots, n$. Recall that $\rho(A)$ is the spectral radius (see Lemma 2.1).

3.7 Consider a discrete-time LTI system of the form

$$x(t+1) = Ax(t) + f(t), \qquad t = 0,\ 1,\ \cdots$$

where $f(t) \in \mathbb{R}^n$ is an exogenous input, and $A \in \mathbb{R}^{n\times n}$ is stable. Show that if $\|f(t)\| \to 0$, then $x(t)$ converges to zero as $t \to \infty$.

3.8 Define the system matrix

$$S(z) = \begin{bmatrix} zI - A & B \\ -C & D \end{bmatrix}$$

Show that the following equality holds:

$$\text{rank}_z S(z) = n + \text{rank}_z G(z)$$

where rank_z denotes the maximal rank for $z \in \mathbb{C}$; note that this rank is called the normal rank.

3.9 [51] Consider the Hankel matrix H of (3.51) with scalar elements $Y_i = h_i$, $i = 1,\ 2, \cdots$. Then, H has finite rank if and only if the series

$$R(z) := \frac{h_1}{z} + \frac{h_2}{z^2} + \cdots$$

is a rational function of z. Moreover, the rank of H is equal to the number of poles of $R(z)$.

3.10 Show that the sequence $\{g_k,\ k = 1,\ 2,\ \cdots\}$ in Problem 3.1 cannot have a finite dimensional realization.

4

Stochastic Processes

This chapter is concerned with discrete-time stochastic processes and linear dynamic systems with random inputs. We introduce basic properties of stochastic processes, including stationarity, Markov and ergodic properties. We then study the spectral analysis of stationary stochastic processes. By defining Hilbert spaces generated by stationary processes, we discuss the optimal prediction problem for stationary processes. Finally, we turn to the study of linear stochastic systems driven by white noises, or Markov models, which play important roles in prediction, filtering and system identification. We also introduce backward Markov models for stationary processes.

4.1 Stochastic Processes

Consider a physical variable x that evolves in time in a manner governed by some probabilistic laws. There are many examples for these kinds of variables, including thermal noise in electrical circuits, radar signals, random fluctuation of ships due to ocean waves, temperature and pressure variations in chemical reactors, stock prices, waves observed in earthquakes, *etc*. The collection of all possible variations in time of any such variable is called a stochastic process, or a time series.

To be more precise, a stochastic process is a family of real valued (or complex valued) time functions, implying that a stochastic process is composed of a collection or ensemble of random variables over an index set, say, T. Let Ω be a sample space appropriately defined for the experiment under consideration. Then, a stochastic process is expressed as $\{x(t,\omega),\ t \in T\}$, where $\omega \in \Omega$. For a fixed $t = t_1$, we have a random variable $x(t_1, \cdot)$ on the sample space Ω. Also, if we fix $\omega = \omega_1$, then $x(\cdot, \omega_1)$ is a function of time called a sample function. This definition of stochastic process is very general, so that we usually assume a suitable statistical (or dynamic) model with a finite number of parameters for analyzing a random phenomenon (or system) of interest.

If the index set is $\mathbb{R}^1 = (-\infty, \infty)$, or the interval $[a, b] \subset \mathbb{R}^1$, then the process is called a continuous-time stochastic process. If, on the other hand, the index set is $\mathbb{Z} =$

$\{t = 0, \pm 1, \cdots\}$, we have a discrete-time stochastic process, or a time series. In this book, we consider discrete-time stochastic processes, so that they are expressed as $\{x(t), t = 0, \pm 1, \cdots\}$, $\{x(t)\}$, or simply x by suppressing the stochastic parameter $\omega \in \Omega$.

Consider the distribution of a stochastic process $\{x(t)\}$. Let $t_1, \cdots, t_k$ be k time instants. Then, for $a_1, \cdots, a_k \in \mathbb{R}$, the joint distribution of $x(t_1), \cdots, x(t_k)$ is defined by

$$\begin{aligned} &P\{x(t_1) \leq a_1, \cdots, x(t_k) \leq a_k\} \\ &\quad = \int_{-\infty}^{a_1} \cdots \int_{-\infty}^{a_k} p_{t_1, \cdots, t_k}(x_1, \cdots, x_k) dx_1 \cdots dx_k \end{aligned} \tag{4.1}$$

where $p_{t_1, \cdots, t_k}(x_1, \cdots, x_k)$ is the joint probability density function of $x(t_1), \cdots, x(t_k)$. The joint distribution of (4.1) is called a finite dimensional distribution of the stochastic process at $t_1, \cdots, t_k$. The distribution of a stochastic process can be determined by all the finite distributions of (4.1). In particular, if any finite distribution of x is Gaussian, then the distribution of x is called Gaussian.

Example 4.1. A stochastic process $\{v(t), t = 0, 1, \cdots\}$ is called a white noise, if $v(t)$ and $v(s)$ are independent for any $t \neq s$, *i.e.*,

$$p_{t,s}(v(t), v(s)) = p_t(v(t)) p_s(v(s)), \qquad t \neq s$$

The white noise is conveniently used for generating various processes with different stochastic properties. For example, a random walk $x(t)$ is expressed as a sum of white noises

$$x(t) = v(1) + v(2) + \cdots + v(t), \qquad t = 1, 2, \cdots \tag{4.2}$$

with $x(0) = 0$. It thus follows from (4.2) that

$$x(t) = x(t-1) + v(t), \qquad x(0) = 0 \tag{4.3}$$

Statistical property of the random walk is considered in Example 4.3. □

4.1.1 Markov Processes

Let $\{x(t), t = 0, \pm 1, \cdots\}$ be a stochastic process. We introduce the minimal σ-algebra that makes $\{x(s), s \leq t\}$ measurable, denoted by $\mathcal{F}_t = \sigma\{x(s), s \leq t\}$. The σ-algebra $\mathcal{F}_t$ satisfies $\mathcal{F}_{t_1} \subset \mathcal{F}_{t_2}$, $t_1 \leq t_2$, and is called a filtration. It involves all the information carried by $x(t), x(t-1), \cdots$.

Suppose that for $a \in \mathbb{R}$ and $t_{k+1} \geq t_k$, we have

$$\begin{aligned} P\{x(t_{k+1}) \leq a \mid \mathcal{F}_{t_k}\} &= P\{x(t_{k+1}) \leq a \mid \sigma(x(t_k))\} \\ &= P\{x(t_{k+1}) \leq a \mid x(t_k)\} \end{aligned} \tag{4.4}$$

Then we say that $\{x(t)\}$ has Markov property. In terms of the conditional probability density functions, the Markov property is written as

$$p(x(t_k) \mid x(t_{k-1}), \cdots, x(t_1)) = p(x(t_k) \mid x(t_{k-1})) \tag{4.5}$$

A stochastic process with the Markov property is called a Markov process. The random walk $x(t)$ in Example 4.1 is a Markov process, since for any $t > s \geq 0$,

$$p(x(t) \mid x(s-1), \cdots x(1)) = p(x(t) \mid x(s-1))$$

Let $t_1 < t_2 < \cdots < t_k < t_{k+1}$. Then, for a Markov process, the conditional probability of $x(t_{k+1})$ given $\mathcal{F}_{t_k}$ depends only on $x(t_k)$, and is independent of the information $\mathcal{F}_{t_{k-1}}$. Let $\mathcal{F}_{t_{k-1}}$ be the past, $x(t_k)$ present, and $x(t_{k+1})$ the future. Then, for Markov processes, the information for the present state $x(t_k)$ makes the past and the future independent. Also, by using Bayes' rule, the joint probability density function of a Markov process is expressed as

$$\begin{aligned} p(x(t_1), \cdots, x(t_k)) &= p(x(t_k) \mid x(t_1), \cdots, x(t_{k-1}))p(x(t_1), \cdots, x(t_{k-1})) \\ &= p(x(t_k) \mid x(t_{k-1}))p(x(t_1), \cdots, x(t_{k-1})) \end{aligned}$$

Continuing this procedure, we get

$$p(x(t_1), \cdots, x(t_k)) = p(x(t_1)) \prod_{i=2}^{k} p(x(t_i) \mid x(t_{i-1}))$$

We therefore see that the joint probability density function of a Markov process is determined by the first-order probability density functions $p(x(t_i))$ and the transition probability density functions $p(x(t_i) \mid x(t_{i-1}))$. Also, since

$$p(x(t_i) \mid x(t_{i-1})) = \frac{p(x(t_i), x(t_{i-1}))}{p(x(t_{i-1}))}$$

we can say that the distribution of a Markov process is determined by the first- and second-order probability density functions $p(x(t_i))$ and $p(x(t_i), x(t_{i-1}))$.

4.1.2 Means and Covariance Matrices

Let $\{x(t), t = 0, \pm 1, \cdots\}$ be a stochastic process. Given the distribution of $\{x(t)\}$ of (4.1), we can compute various expectations associated with the stochastic process. In particular, the expectation of the product $x(t_1) \cdots x(t_k)$ is called the kth-order moment function, which is given by

$$\begin{aligned} M(t_1, \cdots, t_k) &= E\{x(t_1) \cdots x(t_k)\} \\ &= \int_{-\infty}^{\infty} \cdots \int_{-\infty}^{\infty} x_1 \cdots x_k p_{t_1, \cdots, t_k}(x_1, \cdots, x_k) dx_1 \cdots dx_k \end{aligned}$$

where $E\{\cdot\}$ denotes the mathematical expectation.

In the following, we are mainly interested in the first- and second-order moment functions, which are respectively called the mean function and the (auto-) covariance function; they are written as

$$\mu_x(t) = E\{x(t)\}, \qquad \Lambda_{xx}(t, s) = E\{[x(t) - \mu_x(t)][x(s) - \mu_x(s)]\}$$

The covariance function is also written as $\text{cov}\{x(t), x(s)\}$, and in particular,

$$\sigma_x^2(t) = \text{cov}\{x(t), x(t)\} = \Lambda_{xx}(t, t)$$

is called the variance of $x(t)$. Stochastic processes with finite variances are called second-order processes.

Example 4.2. Let $\{x(t),\ t = 0,\ \pm 1,\ \cdots\}$ be a Gaussian stochastic process. Let the mean and covariance functions be given by $\mu_x(t)$ and $\sigma(t, s)$, respectively. Then, the joint probability density function of $x(t_1),\ \cdots,\ x(t_k)$ is expressed as

$$\begin{aligned} &p_{t_1,\cdots,t_k}(x_1,\ \cdots,\ x_k) \\ &\quad = \frac{1}{(2\pi)^{k/2}|\Sigma|^{1/2}} \exp\Big\{-\frac{1}{2} \sum_{i,j=1}^{k} \Sigma_{ij}^{-1}(x_i - \mu_x(t_i))(x_j - \mu_x(t_j))\Big\} \end{aligned}$$

where $\Sigma = (\sigma(t_i,\ t_j)) \in \mathbb{R}^{k\times k}$ is the covariance matrix, and Σ_{ij}^{-1} denotes the $(i,\ j)$-element of the inverse Σ^{-1}. If $x(t)$ is a white noise, we see that Σ becomes a diagonal matrix. □

Before concluding this section, we briefly discuss vector stochastic processes. Let $\{x(t),\ t = 0,\ \pm 1,\ \cdots\}$ be an n-dimensional vector process, *i.e.*,

$$x(t) = \begin{bmatrix} x_1(t) \\ \vdots \\ x_n(t) \end{bmatrix}$$

where $x_i(t)$ are scalar stochastic processes. Then we can respectively define the mean vector and covariance matrix as

$$\mu_x(t) = \begin{bmatrix} \mu_{x_1}(t) \\ \vdots \\ \mu_{x_n}(t) \end{bmatrix}$$

and

$$\begin{aligned} \Lambda_{xx}(t, s) &= E\{[x(t) - \mu_x(t)][x(s) - \mu_x(s)]^{\mathrm{T}}\} \\ &= \begin{bmatrix} E\{\tilde{x}_1(t)\tilde{x}_1(s)\} & E\{\tilde{x}_1(t)\tilde{x}_2(s)\} & \cdots & E\{\tilde{x}_1(t)\tilde{x}_n(s)\} \\ E\{\tilde{x}_2(t)\tilde{x}_1(s)\} & E\{\tilde{x}_2(t)\tilde{x}_2(s)\} & \cdots & E\{\tilde{x}_2(t)\tilde{x}_n(s)\} \\ \vdots & \vdots & \ddots & \vdots \\ E\{\tilde{x}_n(t)\tilde{x}_1(s)\} & E\{\tilde{x}_n(t)\tilde{x}_2(s)\} & \cdots & E\{\tilde{x}_n(t)\tilde{x}_n(s)\} \end{bmatrix} \end{aligned}$$

where $\tilde{x}(t) := x(t) - \mu_x(t)$. We see that the diagonal elements of $\Lambda_{xx}(t, s)$ are the covariance functions of $\{x_i(t),\ t = 0, \pm 1, \cdots\}$, and the non-diagonal elements are the cross-covariance functions of $x_i(t)$ and $x_j(t)$, $i \neq j$.

4.2 Stationary Stochastic Processes

Consider a stochastic process $\{x(t),\ t = 0, \pm 1, \cdots\}$ whose statistical properties do not change in time. This is roughly equivalent to saying that the future is statistically the same as the past, and can be expressed in terms of the joint probability density functions as

$$p_{t_1, \cdots, t_k}(x_1, \cdots, x_k) = p_{t_1+l, \cdots, t_k+l}(x_1, \cdots, x_k), \quad l = 0, \pm 1, \cdots \tag{4.6}$$

If (4.6) holds, $\{x(t)\}$ is called a strongly stationary process.

Let $\{x(t),\ t = 0, \pm 1, \cdots\}$ be a strongly stationary process with the finite kth-order moment function. It follows from (4.6) that

$$\begin{aligned} M(t_1, \cdots, t_k) &= M(t_1 + l, \cdots, t_k + l) \\ &= M(t_1 - t_k, \cdots, t_{k-1} - t_k, 0), \qquad l = 0, \pm 1, \cdots \end{aligned} \tag{4.7}$$

In particular, for the mean and covariance functions, we have

$$\mu_x(t) = E\{x(t)\} = \mu_x(0), \qquad \Lambda_{xx}(t,\ s) = \Lambda_{xx}(t - s,\ 0)$$

Thus, for a strongly stationary process with a finite second-order moment, we see that the mean function is constant and the covariance function depends only on the time difference. In this case, the covariance function is simply written as $\Lambda_{xx}(t - s)$ instead of $\Lambda_{xx}(t,\ s)$.

Let $\{x(t),\ t = 0, \pm 1, \cdots\}$ be a second-order stochastic process. If the mean function is constant, and if the covariance function is characterized by the time difference, then the process is called a weakly stationary process. Clearly, a strongly stationary process with a finite variance is weakly stationary; but the converse is not true. In fact, there are cases where the probability density function of (4.6) may not be a function of time difference for a second-order stationary process. However, note that a weakly stationary Gaussian process is strongly stationary.

Example 4.3. (Random walk) We compute the mean and variance of the random walk considered in Example 4.1. Since v is a zero mean Gaussian white noise with unit variance, we have $E\{v(t)\} = 0$ and $E\{v(t)v(s)\} = \delta_{ts}$. Hence, the mean of $x(t)$ becomes

$$\mu_x(t) = E\{v(1) + v(2) + \cdots + v(t)\} = 0$$

Since $x(t) - x(s) = \sum_{i=s+1}^{t} v(i)$ and $x(s) = \sum_{k=1}^{s} v(k)$ are independent for $t > s$, we get

$$\Lambda_{xx}(t,\ s) = E\{x(t)x(s)\} = E\{(x(t) - x(s))x(s)\} + E\{(x(s))^2\} = s$$

Similarly, for $t < s$, we get $\Lambda_{xx}(t, s) = t$. Thus the covariance function of the random walk is given by $\Lambda_{xx}(t, s) = \min(t, s)$, which is not a function of the difference $t - s$, so that the random walk is a non-stationary process. □

We now consider a second-order stationary process $\{x(t), t = 0, \pm 1, \cdots\}$. Since the mean function μ_x is constant, we put $x(t) := x(t) - \mu_x$. Then, without loss of generality, we can assume from the outset that a stationary process has zero mean. Moreover, being dependent only on the time difference $t - s$, the covariance function is written as

$$\Lambda_{xx}(l) = E\{x(t+l)x(t)\}, \qquad l = 0, \pm 1, \cdots \tag{4.8}$$

Lemma 4.1. *The covariance function $\Lambda_{xx}(l)$ has the following properties.*

(i) (Boundedness) $|\Lambda_{xx}(l)| \leq \Lambda_{xx}(0) = \sigma_x^2, \quad l = \pm 1, \pm 2, \cdots$

(ii) (Symmetry) $\Lambda_{xx}(l) = \Lambda_{xx}(-l), \quad l = 1, 2, \cdots$

(iii) (Nonnegativeness) For any $l_1, \cdots, l_n \in \mathbb{Z}; a_1, \cdots, a_n \in \mathbb{R}$*, we have*

$$\sum_{i,k=1}^{n} a_i a_k \Lambda_{xx}(l_i - l_k) \geq 0$$

Proof. Item (i) is proved by putting $\xi = x(l)$, $\eta = x(0)$ in the Schwartz inequality $|E\{\xi\eta\}|^2 \leq E\{\xi^2\}E\{\eta^2\}$. Item (ii) is obvious from stationarity, and (iii) is obtained from $E\{\left|\sum_{i=1}^{n} a_i x(l_i)\right|^2\} \geq 0$. □

Consider a joint process $\{x(t), y(t), t = 0, \pm 1, \cdots\}$ with means zero. If the vector process $w = \begin{bmatrix} x \\ y \end{bmatrix}$ is stationary, then we say that x and y are jointly stationary. Since the covariance matrix of w is given by

$$\begin{aligned} \begin{bmatrix} \Lambda_{xx}(t+l,t) & \Lambda_{xy}(t+l,t) \\ \Lambda_{yx}(t+l,t) & \Lambda_{yy}(t+l,t) \end{bmatrix} &= E\left\{ \begin{bmatrix} x(t+l) \\ y(t+l) \end{bmatrix} \begin{bmatrix} x^{\mathrm{T}}(t) & y^{\mathrm{T}}(t) \end{bmatrix} \right\} \\ &= \mathrm{cov}\{w(t+l)w(t)\} \end{aligned} \tag{4.9}$$

the stationarity of w implies that the four covariances of (4.9) are functions of the time difference l only. The expectation of the product $x(t+l)y(t)$, *i.e.*,

$$\Lambda_{xy}(l) = E\{x(t+l)y(t)\} \tag{4.10}$$

is called the cross-covariance function of x and y. If $\Lambda_{xy}(l) = 0$ for all l, then two processes x and y are mutually uncorrelated or orthogonal.

Lemma 4.2. *The cross-covariance function $\Lambda_{xy}(l)$ has the following properties.*

(i) (Anti-symmetry) $\Lambda_{xy}(l) = \Lambda_{yx}(-l), \qquad l = 1, 2, \cdots$

(ii) (Boundedness) $|\Lambda_{xy}(l)|^2 \leq \Lambda_{xx}(0)\Lambda_{yy}(0), \qquad l = \pm 1, \pm 2, \cdots$

Proof. (i) Obvious. (ii) This is easily proved by the Schwartz inequality. □

4.3 Ergodic Processes

A basic problem in analyses of stationary stochastic processes is to estimate statistical parameters from observed data. Since the parameters are related to expected values of some function of a stochastic process, we study the estimation problem of the mean of a stationary stochastic process.

Suppose that $\{x(t),\, t = 0,\, \pm 1,\, \cdots\}$ is a second-order stationary process with mean zero. Define a time average of the process by

$$r_{xx}(l) = \lim_{N \to \infty} \frac{1}{2N+1} \sum_{t=-N}^{N} x(t+l)x(t), \qquad l = 0, \pm 1, \cdots \tag{4.11}$$

This quantity is also called the (auto-) covariance function. The covariance function of (4.8) is defined as an ensemble average, but $r_{xx}(l)$ of (4.11) is defined as a time average for a sample process

$$x = (\,\cdots,\, x(-1),\, x(0),\, x(1),\, \cdots)$$

For data analysis, we deal with a time function, or a sample process, generated from a particular experiment rather than an ensemble. Hence, from practical points of view, the definition of moment functions in terms of the time average is preferable to the one defined by the ensemble average. But, we do not know whether the time average $r_{xx}(l)$ is equal to the ensemble average $\Lambda_{xx}(l)$ or not.

A stochastic process whose statistical properties are determined from its sample process is called an ergodic process. In other words, for an ergodic process, the time average equals the ensemble average. In the following, we state ergodic theorems for the mean and covariance functions.

We first consider the ergodic theorem for the mean. Let x be a second-order stationary stochastic process with mean μ_x, and consider the sample mean

$$m(N) = \frac{1}{2N+1} \sum_{t=-N}^{N} x(t) \tag{4.12}$$

Then, we see that $E\{m(N)\} = \mu_x$, which implies that the mathematical expectation of $m(N)$ is equal to the ensemble mean. Also, the variance of $m(N)$ is given by

$$\begin{aligned} E\{(m(N) - \mu_x)^2\} &= E\left\{ \left[\frac{1}{2N+1} \sum_{t=-N}^{N} (x(t) - \mu_x) \right]^2 \right\} \\ &= \frac{1}{(2N+1)^2} \sum_{t=-N}^{N} \sum_{s=-N}^{N} \Lambda_{xx}(t-s) \\ &= \frac{1}{2N+1} \sum_{k=-2N}^{2N} \left(1 - \frac{|k|}{2N+1}\right) \Lambda_{xx}(k) \end{aligned} \tag{4.13}$$

Thus we have the following theorem.

Theorem 4.1. *(Mean ergodic theorem) A necessary and sufficient condition that* $\lim_{N\to\infty} m(N) = \mu_x$ *holds in the quadratic mean is that*

$$\lim_{N\to\infty} \frac{1}{2N+1} \sum_{k=-2N}^{2N} \left(1 - \frac{|k|}{2N+1}\right) \Lambda_{xx}(k) = 0 \tag{4.14}$$

Proof. For a proof, see Problem 4.2. □

We see that if $\lim_{l\to\infty} \Lambda_{xx}(l) = 0$, then the Cesàro sum also does converge to zero, *i.e.*,

$$\lim_{N\to\infty} \frac{1}{N+1} \sum_{k=0}^{N} \Lambda_{xx}(k) = 0 \tag{4.15}$$

holds, and hence (4.14) follows (see Problem 4.3 (b)).

Next we consider an ergodic theorem for a covariance function. Suppose that x is a stationary process with mean zero. Let the sample average of the product $x(t+l)x(t)$ be defined by

$$r_{xx}(l;N) = \frac{1}{2N+1} \sum_{t=-N}^{N} x(t+l)x(t) \tag{4.16}$$

Obviously we have

$$E\{r_{xx}(l;N)\} = \frac{1}{2N+1} \sum_{t=-N}^{N} E\{x(t+l)x(t)\} = \Lambda_{xx}(l)$$

so that the expectation of $r_{xx}(l;N)$ equals the covariance function $\Lambda_{xx}(l)$.

To evaluate the variance of $r_{xx}(l;N)$, we define $\xi(t) = x(t+l)x(t)$, and apply the mean ergodic theorem to $\xi(t)$. We see that $\mu_\xi = E\{\xi(t)\} = \Lambda_{xx}(l)$ and that

$$\begin{aligned}\Lambda_{\xi\xi}(k) &= E\{[x(t+l+k)x(t+k) - \mu_\xi][x(t+l)x(t) - \mu_\xi]\} \\ &= E\{x(t+l+k)x(t+k)x(t+l)x(t)\} - \mu_\xi^2\end{aligned} \tag{4.17}$$

Also, similarly to the derivation of (4.13), it follows that

$$\begin{aligned}E\{[r_{xx}(l;N) - \Lambda_{xx}(l)]^2\} &= E\left\{\left[\frac{1}{2N+1} \sum_{t=-N}^{N} (\xi(t) - \mu_\xi)\right]^2\right\} \\ &= \frac{1}{2N+1} \sum_{k=-2N}^{2N} \left(1 - \frac{|k|}{2N+1}\right) \Lambda_{\xi\xi}(k)\end{aligned}$$

Thus, we have an ergodic theorem for the covariance function.

Theorem 4.2. *(Covariance ergodic theorem) A necessary and sufficient condition that*

$$\lim_{N\to\infty} r_{xx}(l;N) = \Lambda_{xx}(l) \tag{4.18}$$

holds in the quadratic mean is that

$$\lim_{N\to\infty} \frac{1}{2N+1} \sum_{k=-2N}^{2N} \left(1 - \frac{|k|}{2N+1}\right) \Lambda_{\xi\xi}(k) = 0 \tag{4.19}$$

Moreover, suppose that x is a Gaussian process with mean zero. If the condition

$$\lim_{N\to\infty} \frac{1}{N+1} \sum_{k=0}^{N} \Lambda_{xx}^2(k) = 0 \tag{4.20}$$

is satisfied, then (4.19) and hence (4.18) holds in the quadratic mean.

Proof. See Problem 4.4. □

Example 4.4. Consider a zero mean Gaussian process x with the covariance function $\Lambda_{xx}(l) = \sigma^2 a^{|l|}$, $l = 0, \pm 1, \cdots$ $(0 < |a| < 1, \sigma^2 > 0)$ (see Figure 4.2 below). Since (4.15) and (4.20) are satisfied, Theorems 4.1 and 4.2 indicate that a stochastic process with exponential covariance function is ergodic. □

4.4 Spectral Analysis

We consider a second-order stationary process $\{x(t),\, t = 0, \pm 1, \cdots\}$ with mean zero. Suppose that its covariance function $\{\Lambda_{xx}(l),\, l = 0, \pm 1, \cdots\}$ satisfies the summability condition

$$\sum_{l=-\infty}^{\infty} |\Lambda_{xx}(l)| < \infty \tag{4.21}$$

Definition 4.1. *Suppose that the covariance function satisfies the condition (4.21). Then, the Fourier transform (or two-sided z-transform) of $\Lambda_{xx}(l)$ is defined by*

$$\Phi_{xx}(z) = \sum_{l=-\infty}^{\infty} \Lambda_{xx}(l) z^{-l} \tag{4.22}$$

This is called the spectral density function of the stochastic process $\{x(t)\}$. □

Putting $z = e^{j\omega}$, $-\pi < \omega < \pi$, the spectral density function can be viewed as a function of ω (rad)

$$\Phi_{xx}(\omega) = \sum_{l=-\infty}^{\infty} e^{-j\omega l} \Lambda_{xx}(l), \qquad -\pi < \omega < \pi \tag{4.23}$$

We observe from the definition (4.23) that the spectral density function shows the distribution of power of the stationary process in the frequency domain.

It is well known that the covariance function is expressed as an inverse transform of the spectral density function as

$$\Lambda_{xx}(l) = \frac{1}{2\pi j}\int_{|z|=1} \Phi_{xx}(z)z^{l-1}dz \tag{4.24a}$$

$$= \frac{1}{2\pi}\int_{-\pi}^{\pi} e^{j\omega l}\Phi_{xx}(\omega)d\omega, \qquad l = 0, \pm 1, \cdots \tag{4.24b}$$

The relations in (4.24) are called the Wiener-Khinchine formula. We see from (4.22) and (4.24a) that the covariance function and spectral density function involve the same information about a stationary stochastic process since there exists a one-to-one correspondence between them.

If the sampling interval is given by Δt, then the spectral density function is defined by

$$\Phi_{xx}(\nu; \Delta t) = \Delta t \sum_{l=-\infty}^{\infty} e^{-j\nu\Delta t l}\Lambda_{xx}(l), \qquad -\frac{\pi}{\Delta t} < \nu < \frac{\pi}{\Delta t} \tag{4.25}$$

and its inverse is

$$\Lambda_{xx}(l) = \frac{1}{2\pi}\int_{-\pi/\Delta t}^{\pi/\Delta t} e^{j\nu\Delta t l}\Phi_{xx}(\nu; \Delta t)d\nu, \qquad l = 0, \pm 1, \cdots \tag{4.26}$$

It should be noted that ω in (4.23) and ν in (4.25) are related by $\omega = \nu\Delta t$, and hence ν has the dimension [rad/sec].

Lemma 4.3. *The spectral density function satisfies the following.*

(i) (Symmetry) $\quad \Phi_{xx}(\omega) = \Phi_{xx}(-\omega), \qquad -\pi < \omega \leq \pi$

(ii) (Nonnegativeness) $\quad \Phi_{xx}(\omega) \geq 0, \qquad -\pi < \omega \leq \pi$

Proof. (i) The symmetry is immediate from $\Lambda_{xx}(l) = \Lambda_{xx}(-l)$. (ii) This follows from the nonnegativeness of $\Lambda_{xx}(l)$ described in Lemma 4.1 (iii). For an alternate proof, see Problem 4.5. □

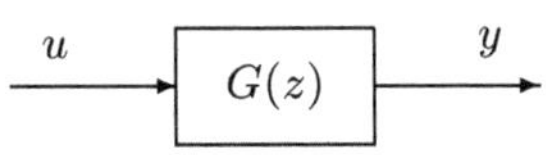

Figure 4.1. Discrete-time LTI system

Consider a discrete-time LTI system depicted in Figure 4.1, where u is the input and y is the output, and the impulse response of the system is given by $\{g(k),\ k = 0, 1, \cdots\}$ with $g(k) = 0,\ k < 0$. The transfer function is then expressed as

$$G(z) = \sum_{k=0}^{\infty} g(k) z^{-k} \tag{4.27}$$

As mentioned in Theorem 3.1, $G(z)$ is stable if and only if the impulse response sequence $\{g(k),\ k = 0,\ 1,\ \cdots\}$ is absolutely summable.

Lemma 4.4. *Consider an LTI system shown in Figure 4.1 with $G(z)$ stable. Suppose that the input u is a zero mean second-order stationary process with the covariance function $\Lambda_{uu}(l)$ satisfying*

$$\sum_{l=-\infty}^{\infty} |\Lambda_{uu}(l)| < \infty \tag{4.28}$$

Then, the output y is also a zero mean second-order process with the spectral density function of the form

$$\Phi_{yy}(z) = G(z)G(z^{-1})\Phi_{uu}(z) \tag{4.29}$$

or

$$\Phi_{yy}(\omega) = |G(e^{j\omega})|^2 \Phi_{uu}(\omega) \tag{4.30}$$

Further, the variance of y is given by

$$\sigma_y^2 = \frac{1}{2\pi} \int_{-\pi}^{\pi} |G(e^{j\omega})|^2 \Phi_{uu}(\omega) d\omega \tag{4.31}$$

Proof. Since $G(z)$ is stable, the output $y(t)$ is expressed as

$$y(t) = \sum_{i=0}^{\infty} g(i) u(t-i)$$

Hence, it follows that $\mu_y = E\{y(t)\} = 0$ and that the covariance function of y is given by

$$\begin{aligned} \Lambda_{yy}(l) &= E\{y(t+l)y(t)\} \\ &= \sum_{i=0}^{\infty} \sum_{k=0}^{\infty} g(i)g(k)E\{u(t+l-i)u(t-k)\} \\ &= \sum_{i=0}^{\infty} \sum_{k=0}^{\infty} g(i)g(k)\Lambda_{uu}(l+k-i) \end{aligned} \tag{4.32}$$

Since the right-hand side is a function of l, so is the left-hand side. Taking the sum of absolute values of the above equation, it follows from (4.28) and the stability of $G(z)$ that

$$\begin{aligned} \sum_{l=-\infty}^{\infty} |\Lambda_{yy}(l)| &\le \sum_{l=-\infty}^{\infty} \sum_{i=0}^{\infty} \sum_{k=0}^{\infty} |g(i)| \cdot |g(k)| \cdot |\Lambda_{uu}(l+k-i)| \\ &= \left(\sum_{i=0}^{\infty} |g(i)| \right)^2 \sum_{l=-\infty}^{\infty} |\Lambda_{uu}(l+k-i)| < \infty \end{aligned}$$

where the sum with respect to l should be taken first. From this equation, we see that $|\Lambda_{yy}(0)| < \infty$, implying that y is a second-order process. Also, we can take the Fourier transform of both sides of (4.32) to get

$$\begin{aligned}
\Phi_{yy}(z) &= \sum_{l=-\infty}^{\infty} \Lambda_{yy}(l) z^{-l} \\
&= \sum_{l=-\infty}^{\infty} z^{-l} \left(\sum_{i=0}^{\infty} \sum_{k=0}^{\infty} g(i) g(k) \Lambda_{uu}(l+k-i) \right) \\
&= \sum_{i=0}^{\infty} g(i) z^{-i} \sum_{k=0}^{\infty} g(k) z^{k} \sum_{l=-\infty}^{\infty} z^{-(l+k-i)} \Lambda_{uu}(l+k-i) \\
&= G(z) G(z^{-1}) \Phi_{uu}(z)
\end{aligned}$$

Equation (4.30) is trivial. Finally, putting $l = 0$ in (4.24b) gives (4.31). □

Example 4.5. Consider a system $G(z) = \sqrt{1-a^2}/(z-a)$, where the input is a Gaussian white noise e with mean zero and variance σ^2. We see that the output of the system is described by the first-order autoregressive (AR) model

$$y(t+1) = a y(t) + \sqrt{1-a^2}\, e(t), \qquad t = 0,\ 1,\ \cdots \tag{4.33}$$

The output process is called a first-order AR process. We observe that the future $y(t+1)$ depends partly on the present $y(t)$ and partly on the random noise $e(t)$, so that y is a Markov process. Since the spectral density function of e is $\Phi_{ee}(\omega) = \sigma^2$, it follows from (4.32) and (4.30) that the covariance function and the spectral density function of the output process y are respectively given by

$$\Lambda_{yy}(l) = \sigma^2 a^{|l|}, \qquad l = 0, \pm 1, \cdots$$

and

$$\Phi_{yy}(\omega) = \frac{\sigma^2(1-a^2)}{1 + a^2 - 2a\cos\omega}, \qquad |\omega| < \pi$$

The auto-covariance functions and spectral density functions for $a = 0.4, 0.8$ and $\sigma^2 = 1$ are displayed in Figures 4.2 and 4.3. For larger a, the value of the covariance function decreases slowly as $|l|$ gets larger, and the power is concentrated in the low frequency range. But, for smaller a, we see that the covariance function decreases rapidly and the power is distributed over the wide frequency range. □

The next example is concerned with an autoregressive moving average (ARMA) model.

Example 4.6. Let e be a zero mean white noise with variance σ^2. Suppose that the system is described by a difference equation

$$y(t) + \sum_{i=1}^{p} a_i y(t-i) = e(t) + \sum_{l=1}^{q} c_l e(t-l) \tag{4.34}$$

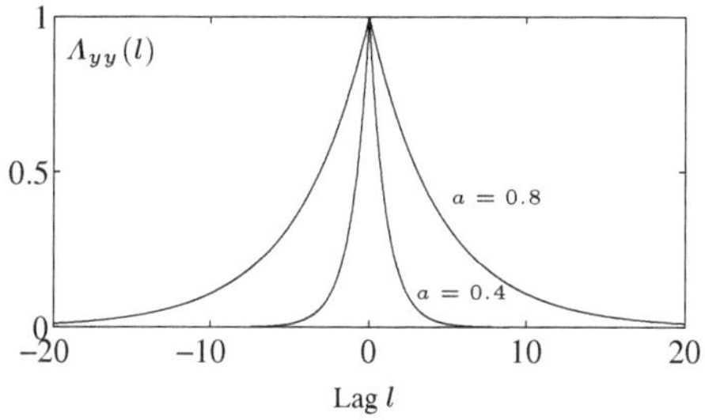

Figure 4.2. Auto-covariance functions

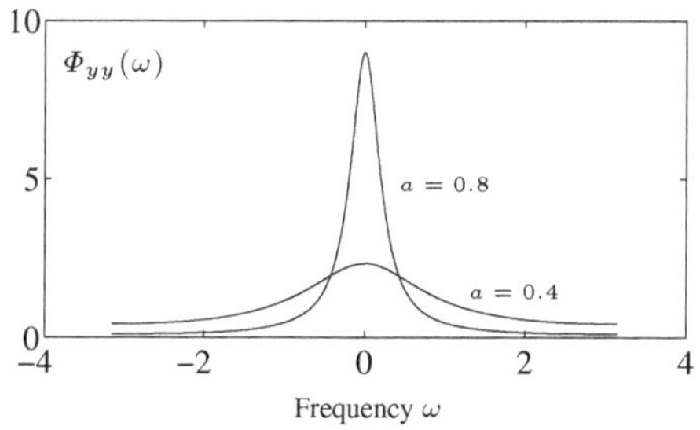

Figure 4.3. Spectral density functions

This equation is called an ARMA model of order $(p,\ q)$. From (4.34), we have

$$A(z)y(t) = C(z)e(t)$$

where $A(z)$ and $C(z)$ are given by

$$A(z) = 1 + a_1 z^{-1} + \cdots + a_p z^{-p}$$
$$C(z) = 1 + c_1 z^{-1} + \cdots + c_q z^{-q}$$

The process y generated by the ARMA model is called an ARMA process.

The condition that y is stationary is that all the zeros of $A(z)$ are within the unit circle ($|z| < 1$). Since the invertibility of the ARMA model requires that all the zeros of $C(z)$ are located within the unit circle, so that both

$$H(z) = \frac{C(z)}{A(z)}, \qquad \frac{1}{H(z)} = \frac{A(z)}{C(z)}$$

are stable[1]. Since both $H(z)$ and $1/H(z)$ are stable, we can generate the noise process e by feeding the output y to the inverse filter $1/H(z)$ as shown in Figure 4.4. Thus $1/H(z)$ is called an whitening filter. Also, by feeding the white noise to the

[1]We say that a transfer function with this property is of minimal phase.

filter $H(z)$, we have the output y. Hence, the spectral density function of y is given by

$$\Phi_{yy}(\omega) = \sigma^2 \left| \frac{1 + c_1 e^{-j\omega} + \cdots + c_q e^{-j\omega q}}{1 + a_1 e^{-j\omega} + \cdots + a_p e^{-j\omega p}} \right|^2 \tag{4.35}$$

We see that the output spectral has a certain distribution over the range $(-\pi, \pi)$ corresponding to the filter used. Therefore, $H(z)$ is often called a shaping filter.

It should be noted that the design of a shaping filter is closely related to the spectral factorization and the stochastic realization problem to be discussed in later chapters. □

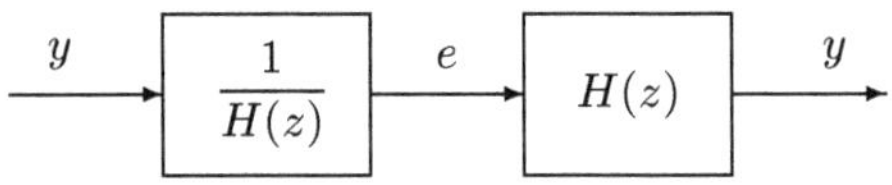

Figure 4.4. Whitening filter and shaping filter

The rest of this section is devoted to the spectral analysis for an n-dimensional vector stochastic process $\{x(t),\ t = 0,\ \pm 1,\ \cdots\}$. For simplicity, we assume that x has zero mean. Then the covariance matrix is given by

$$\Lambda_{xx}(l) = E\{x(t+l)x^{\mathrm{T}}(t)\}, \qquad l = 0,\ 1,\ \cdots \tag{4.36}$$

Obviously, we have $\Lambda_{xx}(l) = \Lambda_{xx}^{\mathrm{T}}(-l)$. Let the diagonal elements of $\Lambda_{xx}(l)$ be $\Lambda_{ii}(l),\ i = 1, \cdots, n$. Suppose that

$$\sum_{l=-\infty}^{\infty} |\Lambda_{ii}(l)| < \infty, \qquad i = 1,\ \cdots,\ n$$

hold. Then, we can define the spectral density matrix by means of the Fourier transform of the covariance matrix as

$$\Phi_{xx}(z) = \sum_{l=-\infty}^{\infty} \Lambda_{xx}(l) z^{-l} \tag{4.37}$$

or

$$\Phi_{xx}(\omega) = \sum_{l=-\infty}^{\infty} e^{-j\omega l} \Lambda_{xx}(l), \qquad -\pi < \omega < \pi \tag{4.38}$$

where $\Phi_{xx}(z)$ and $\Phi_{xx}(\omega)$ are $n \times n$ matrices. In the matrix case, we have also Wiener-Khinchine formula

$$\Lambda_{xx}(l) = \frac{1}{2\pi j} \int_{|z|=1} \Phi_{xx}(z) z^{l-1} dz \tag{4.39a}$$

$$= \frac{1}{2\pi} \int_{-\pi}^{\pi} e^{j\omega l} \Phi_{xx}(\omega) d\omega, \qquad l = 0, \pm 1, \cdots \tag{4.39b}$$

The $(i,\ k)$-elements of $\Phi_{xx}(z)$ and $\Phi_{xx}(\omega)$ are respectively expressed as $\Phi_{ik}(z)$ and $\Phi_{ik}(\omega)$, which are called the cross-spectral density function of $x_i(t)$ and $x_k(t)$. We see that $\Phi_{ii}(\omega)$ are real functions with respect to the angular frequency ω, but $\Phi_{ik}(\omega)$, $i \neq k$ are complex functions.

Lemma 4.5. *The spectral density matrix $\Phi_{xx}(\omega)$ has the following properties.*

(i) (Hermite) $\quad \Phi_{xx}(\omega) = \Phi_{xx}^{\mathrm{H}}(-\omega), \qquad -\pi < \omega \leq \pi$

(ii) (Nonnegativeness) $\quad \Phi_{xx}(\omega) \geq 0, \qquad -\pi < \omega \leq \pi$

Proof. Noting that $\Lambda_{xx}(l) = \Lambda_{xx}^{\mathrm{T}}(-l)$, we get

$$\begin{aligned}\Phi_{xx}(\omega) &= \sum_{l=-\infty}^{\infty} e^{-j\omega l} \Lambda_{xx}(l) = \sum_{l=-\infty}^{\infty} e^{-j\omega l} \Lambda_{xx}^{\mathrm{T}}(-l) \\ &= \sum_{l=-\infty}^{\infty} e^{j\omega l} \Lambda_{xx}^{\mathrm{T}}(l) = \Phi_{xx}^{\mathrm{H}}(-\omega)\end{aligned}$$

which proves (i). Now we prove (ii). Let the element of $x(t)$ be $x_i(t),\ i = 1, \cdots, n$, and let $\xi(t) = \sum_{i=1}^{n} a_i x_i(t)$ with $a_i \in \mathbb{C}$. It is easy to see that ξ is a second-order stationary process with

$$\Lambda_{\xi\xi}(l) = \sum_{i,k=1}^{n} a_i \bar{a}_k \Lambda_{ik}(l)$$

where $\Lambda_{ik}(l) = E\{x_i(t+l) x_k(t)\}$. Taking the Fourier transform of $\Lambda_{\xi\xi}(l)$ yields

$$\Phi_{\xi\xi}(\omega) = \sum_{i,k=1}^{n} a_i \bar{a}_k \Phi_{ik}(\omega) \tag{4.40}$$

From Lemma 4.2, we have $\Phi_{\xi\xi}(\omega) \geq 0$, so that the right-hand side of (4.40) becomes nonnegative for any $a_1, \cdots, a_n \in \mathbb{C}$. This indicates that $\Phi_{xx}(\omega)$ is nonnegative definite. □

4.5 Hilbert Space and Prediction Theory

We consider a Hilbert space generated by a stationary stochastic process and a related problem of prediction. Let $\{y(t),\ t = 0, \pm 1, \cdots\}$ be a zero mean second-order stochastic process. Let the space generated by all finite linear combinations of y be given by

$$\mathcal{H} = \left\{ \xi = \sum_{k=k_1}^{k_2} a_k y(k) \,\middle|\, a_k \in \mathbb{R} \right\}, \quad -\infty < k_1 \leq k_2 < \infty$$

Define $\xi,\ \eta \in \mathcal{H}$ as

$$\xi = \sum_{i=i_1}^{i_2} a_i y(i), \qquad \eta = \sum_{j=j_1}^{j_2} b_j y(j)$$

where $i_1 \leq i_2$ and $j_1 \leq j_2$. Then we define the inner product of ξ and η as $(\xi, \eta)_{\mathcal{H}} = E\{\xi\eta\}$. Hence we have

$$(\xi, \eta)_{\mathcal{H}} = E\left\{\left(\sum_{i=i_1}^{i_2} a_i y(i)\right)\left(\sum_{j=j_1}^{j_2} b_j y(j)\right)\right\}$$
$$= \sum_{(i,j)\in D} E\{y(i)y(j)\}a_i b_j = \sum_{(i,j)\in D} \Lambda_{yy}(i-j)a_i b_j \tag{4.41}$$

where $D = \{(i, j) \mid i_1 \leq i \leq i_2;\ j_1 \leq j \leq j_2\}$ is a finite set of indices.

Now suppose that the covariance matrix $\{\Lambda_{yy}(i-j)\}$ is positive definite. Then we can define

$$\|\xi\|_{\mathcal{H}}^2 = (\xi, \xi)_{\mathcal{H}} = \sum_{(i,j)\in D} \Lambda_{yy}(i-j)a_i a_j$$

Then $\|\cdot\|_{\mathcal{H}}$ becomes a norm in $\mathcal{H}$ [106]. Hence the space $\mathcal{H}$ becomes a Hilbert space by completing it with respect to the norm $\|\cdot\|_{\mathcal{H}}$. The Hilbert space so obtained is written as

$$\mathcal{H} = \overline{\text{span}}\{y(t) \,\big|\, -\infty < t < \infty\}$$

where $\overline{\text{span}}$ denotes the closure of the vector space spanned by linear combinations of its elements. The Hilbert space generated by y is a subspace of the Hilbert space $L_2(\Omega)$ of square integrable random variables.

Example 4.7. Let $\{e(t),\ t = 0, 1, \cdots\}$ be a white noise with zero mean and unit variance. The Hilbert space

$$\mathcal{H} = \overline{\text{span}}\{e(t) \mid t = 0, 1, \cdots\}$$

generated by the white noise $\{e(t)\}$ is defined as follows. For $a = (a_1, a_2, \cdots) \in l_2[0, \infty)$, we define the set consisting of partial sums of e as

$$\mathcal{H} = \left\{\xi_n = \sum_{k=0}^{n} a_k e(k) \,\middle|\, \sum_{k=0}^{\infty} |a_k|^2 < \infty,\ a_k \in \mathbb{R}\right\}$$

Taking the limit $m > n \to \infty$ yields

$$\|\xi_m - \xi_n\|_{\mathcal{H}}^2 = \sum_{k=n+1}^{m} |a_k|^2 \to 0$$

Thus $\{\xi_n\}$ becomes a Cauchy sequence, so that there exists a quadratic mean limit

$$\xi = q\text{-}\lim_{n\to\infty} \xi_n = q\text{-}\lim_{n\to\infty} \sum_{k=0}^{n} a_k e(k)$$

Thus, by adjoining all possible quadratic mean limits, $\mathcal{H}$ becomes a Hilbert space. Hence $\mathcal{H}$ may be written as

$$\mathcal{H} = \left\{ \xi = q\text{-}\lim_{n\to\infty} \sum_{k=0}^{n} a_k e(k) \,\middle|\, \sum_{k=0}^{\infty} |a_k|^2 < \infty,\ a_k \in \mathbb{R} \right\}$$

The norm of $\xi \in \mathcal{H}$ is given by $\|\xi\|_{\mathcal{H}}^2 = \sum_{k=0}^{\infty} |a_k|^2 < \infty$. In this sense, the Hilbert space is also written as $\mathcal{H} = L_2(\Omega)$, where Ω is a set of stochastic parameters. □

For $\xi,\ \eta \in \mathcal{H}$, if $(\xi,\ \eta)_{\mathcal{H}} = 0$ holds, then we say that ξ and η are orthogonal, and the orthogonality is written as $\xi \perp \eta$. Let $\mathcal{W}$ be a subspace of the Hilbert space $\mathcal{H}$. If $(\xi,\ w)_{\mathcal{H}} = 0$ holds for any $w \in \mathcal{W}$, then ξ is orthogonal to $\mathcal{W}$, which is written as $\xi \perp \mathcal{W}$, and the orthogonal complement is written as $\mathcal{W}^{\perp}$.

Lemma 4.6. *Let $\mathcal{W}$ be a closed subspace of a Hilbert space $\mathcal{H}$. For any element $\xi \in \mathcal{H}$, there exists a unique $w_0 \in \mathcal{W}$ such that*

$$\|\xi - w_0\|_{\mathcal{H}} \le \|\xi - w\|_{\mathcal{H}}, \qquad \forall w \in \mathcal{W} \tag{4.42}$$

Moreover, w_0 is a minimizing vector if and only if $\xi - w_0 \perp \mathcal{W}$. The element w_0 satisfying (4.42) is the orthogonal projection of ξ onto the subspace $\mathcal{W}$, so that we write $w_0 = \hat{E}\{\xi \mid \mathcal{W}\}$.

Proof. See [111, 183]. □

Let $\{y(t),\ t = 0,\ \pm 1,\ \cdots\}$ be a second-order stationary stochastic process with mean zero. We consider the problem of predicting the future $y(t+m),\ m = 1,\ 2,\ \cdots$ in terms of a linear combination of the present and past $y(t),\ y(t-1),\ \cdots$ in the least-squares sense. To this end, we define a Hilbert subspace generated by the present and past $y(t),\ y(t-1),\ \cdots$ as

$$\mathcal{Y}_t = \left\{ \xi(t) = \sum_{k=0}^{\infty} a_k y(t-k) \,\middle|\, \sum_{k=0}^{\infty} |a_k| < \infty,\ a_k \in \mathbb{R} \right\}$$

By definition, $A(z) = \sum_{k=0}^{\infty} a_k z^{-k}$ is a stable filter. Thus $\mathcal{Y}_t$ is a linear space generated by the outputs of stable LTI systems subjected to the inputs $y(\tau),\ \tau \le t$, so that it is a subspace of the Hilbert space $\mathcal{H} = \overline{\text{span}}\{y(t) \mid -\infty < t < \infty\}$. In fact, for $\xi,\ \eta \in \mathcal{Y}_t$, we have

$$\xi = \sum_{k=0}^{\infty} a_k y(t-k), \qquad \eta = \sum_{k=0}^{\infty} b_k y(t-k)$$

where $\sum_{k=0}^{\infty} |a_k| < \infty,\ \sum_{k=0}^{\infty} |b_k| < \infty$. It follows that

$$\xi + \eta = \sum_{k=0}^{\infty} (a_k + b_k) y(t-k)$$

Since $|a_k + b_k| \le |a_k| + |b_k|$, we see that $\sum_{k=0}^{\infty} |a_k + b_k| < \infty$. Hence, it follows that $\xi + \eta \in \mathcal{Y}_t$ holds. Moreover, for any $\alpha \in \mathbb{R}$, we have $\alpha\xi \in \mathcal{Y}_t$, implying that $\mathcal{Y}_t$ is a linear space.

Since, in general, $y(t+m) \in \mathcal{H}$, $m > 0$ does not belong to $\mathcal{Y}_t$, the linear prediction problem is reduced to a problem of finding the nearest element $\hat{y}(t+m)$ in $\mathcal{Y}_t$ to $y(t+m)$. It therefore follows from Lemma 4.6 that the optimal predictor is given by the orthogonal projection

$$\hat{y}(t+m) = \hat{E}\{y(t+m) \mid \mathcal{Y}_t\}$$

Define the variance of the prediction error by

$$\sigma_m^2 = E\{[y(t+m) - \hat{y}(t+m)]^2\}, \qquad m = 1, 2, \cdots$$

Then, we see that the variance is independent of time t due to the stationarity of y. Also, since $\mathcal{Y}_s \subset \mathcal{Y}_t$, $s \le t$, the variance σ_m^2 is a non-decreasing function with respect to m, i.e.,

$$0 \le \sigma_1^2 \le \sigma_2^2 \le \cdots$$

Definition 4.2. *Consider the linear prediction problem for a second-order stationary stochastic process y with mean zero. If $\sigma_1^2 > 0$, we say that y is regular, or non-deterministic. On the other hand, if $\sigma_1^2 = 0$, then y is called singular, or deterministic.* □

If $\sigma_1^2 > 0$, we have $\sigma_m^2 > 0$ for all $m = 1, 2, \cdots$. Also, if $\sigma_1^2 = 0$, then it follows that

$$\hat{y}(t+1) = \hat{E}\{y(t+1) \mid \mathcal{Y}_t\} = y(t+1) \in \mathcal{Y}_t$$

holds for any $y(t+1)$. Thus, $\mathcal{Y}_{t+1} = \mathcal{Y}_t$ holds for t, so that $\mathcal{Y}_t$ is equal for all t. Hence, we get

$$0 = \sigma_1^2 = \sigma_2^2 = \cdots = \sigma_m^2 = \cdots$$

Therefore, the variances σ_m^2 of prediction errors are either positive, or zero. For the latter case, y is completely predictable by means of its past values.

The following theorem is stated without proof.

Theorem 4.3. *(Wold decomposition theorem) Let y be a second-order stationary stochastic process with mean zero. Then, y is uniquely decomposed as*

$$y(t) = u(t) + v(t) \tag{4.43}$$

where u and v have the following properties.

(i) The processes u and v are mutually uncorrelated.

(ii) The process u has a moving average (MA) representation

$$u(t) = \sum_{i=0}^{\infty} h(i)\varepsilon(t-i) \tag{4.44}$$

where ε is a white noise, and is uncorrelated with v. Further, $\{h(i)\}$ satisfy

$$\sum_{i=0}^{\infty} |h(i)|^2 < \infty, \qquad h(0) = 1 \tag{4.45}$$

(iii) Define $\mathcal{U}_t = \overline{\text{span}}\{u(t),\, u(t-1),\, \cdots\}$ and $\mathcal{E}_t = \overline{\text{span}}\{\varepsilon(t),\, \varepsilon(t-1),\, \cdots\}$. Then we have $\mathcal{U}_t = \mathcal{E}_t$ and the process u is regular.

(iv) Define $\mathcal{V}_t = \overline{\text{span}}\{v(t),\, v(t-1),\, \cdots\}$. Then $\mathcal{V}_t = \mathcal{V}_s$ holds for all t, s, so that the process v is called singular in the sense that it can be completely determined by linear functions of its past values.

Proof. For proofs, see Anderson [13], Koopmans [95], and Doob [44] (pp. 159–164 and pp. 569–577). □

Example 4.8. (Singular process) From Theorem 4.3 (iv), it follows that $\mathcal{V}_{t+l} = \mathcal{V}_t$ for all t, l. Hence $v(t+l) \in \mathcal{V}_{t+l}$ is also in $\mathcal{V}_t$. Thus $\hat{E}\{v(t+l) \mid \mathcal{V}_t\} = v(t+l)$, implying that if $v(s), s \le t$ are observed for some t, the future $v(t+1), v(t+2), \cdots$ are determined as linear functions of past observed values, like a sinusoid. Thus such a process is called deterministic. □

Theorem 4.4. *Let y be a zero mean, stationary process with the spectral density function $\Phi_{yy}(\omega)$. Then, y is regular if and only if*

$$c_0 = \frac{1}{2\pi} \int_{-\pi}^{\pi} \log \Phi_{yy}(\omega) d\omega > -\infty \tag{4.46}$$

This is called the regularity condition due to Szegö [65]. Under the regularity condition, there exists a unique sequence $\{h(i),\, i = 0, 1, \cdots\}$ such that (4.45) holds, and and the transfer function

$$H(z) = \sum_{i=0}^{\infty} h(i) z^{-i}, \qquad h(0) = 1 \tag{4.47}$$

has no zeros in $|z| > 1$, and provides a spectral factorization of the form

$$\Phi_{yy}(z) = \sigma^2 H(z) H(z^{-1}) \tag{4.48}$$

where the spectral factor $H(z)$ is analytic outside the unit circle ($|z| > 1$), satisfying (4.45) and $\sigma^2 = e^{c_0}$.

Proof. For a complete proof, see Doob [44] (pp. 159–164 and pp. 569–577). But, we follow [134, 178] to prove (4.47) and (4.48) under a stronger assumption that $\log \Phi_{yy}(z)$ is analytic in an annulus $\rho < |z| < 1/\rho$ with $0 < \rho < 1$.

Under this assumption, $\log \Phi_{yy}(z)$ has a Laurent expansion

$$\log \Phi_{yy}(z) = \sum_{l=-\infty}^{\infty} c_l z^{-l}, \qquad \rho < |z| < 1/\rho \tag{4.49}$$

By using the inversion formula [see (4.24)], we have

$$c_l = \frac{1}{2\pi}\int_{-\pi}^{\pi} e^{j\omega l}\log \Phi_{yy}(\omega)d\omega, \qquad l = 0, \pm 1, \cdots \tag{4.50}$$

For $l = 0$, we have the equality in (4.46). Since c_l are the Fourier coefficients of an even, real-valued function $\log \Phi_{yy}(\omega)$, they satisfy $c_{-l} = c_l, l = 0, \pm 1, \cdots$. Thus, for $\rho < |z| < 1/\rho$,

$$\Phi_{yy}(z) = \exp\Big\{\sum_{l=-\infty}^{\infty} c_l z^{-l}\Big\} = e^{c_0}\exp\Big\{\sum_{l=1}^{\infty} c_l z^{l}\Big\}\exp\Big\{\sum_{l=1}^{\infty} c_l z^{-l}\Big\}$$

Now we define

$$H(z) = \exp\Big\{\sum_{l=1}^{\infty} c_l z^{-l}\Big\} \tag{4.51}$$

Since the power series in the bracket $\{\cdots\}$ of (4.51) converges in $|z| > \rho$, we see that $H(z)$ is analytic in $|z| > \rho$, and $H(\infty) = 1$. Thus, $H(z)$ of (4.51) has a Taylor series expansion

$$H(z) = \sum_{i=0}^{\infty} h(i)z^{-i}, \qquad |z| > \rho$$

with $h(0) = 1$. This shows that (4.47) and (4.48) hold. This power series converges in $|z| > \rho$, so that $H(z)$ has no poles in $|z| \geq 1$. Also, it follows that $|h(l)| \leq M\rho_1^l$ for any $l \geq 0$, where $M > 0$ and $0 < \rho_1 \leq \rho < 1$. Hence,

$$\sum_{l=0}^{\infty} |h(l)| < \infty \quad \Rightarrow \quad \sum_{l=0}^{\infty} |h(l)|^2 < \infty$$

Moreover, from (4.51), we see that

$$[H(z)]^{-1} = \exp\Big\{-\sum_{l=1}^{\infty} c_l z^{-l}\Big\}$$

is analytic in $|z| > \rho$, and hence $H(z)$ has no zeros in $|z| \geq 1$. This completes the proof that $H(z)$ is of minimal phase. □

Since $\Phi_{yy}(\omega) \geq 0$, it follows that $\log \Phi_{yy}(\omega) \leq \Phi_{yy}(\omega)$ holds. Thus we get

$$c_0 \leq \frac{1}{2\pi}\int_{-\pi}^{\pi} \Phi_{yy}(\omega)d\omega < \infty$$

This implies that c_0 is always bounded above. Thus, if c_0 is bounded below, the process is regular; on the other hand, if $c_0 = -\infty$, we have $\sigma^2 = 0$, so that the process becomes singular (or deterministic).

It should be noted that under the assumption that $c_0 > -\infty$ of (4.46), there is a possibility that $\Phi_{yy}(z)$ has zeros on the unit circle, and hence the assumption of (4.46) is weaker than the analyticity of $\log \Phi_{yy}(z)$ in the neighborhood of the unit circle $|z| = 1$, as shown in the following example.

Example 4.9. Consider a simple MA process

$$y(t) = \varepsilon(t) - \varepsilon(t-1)$$

where ε is a zero mean white noise process with variance σ^2. Thus, we have

$$H(z) = 1 - z^{-1} \;\Rightarrow\; \Phi_{yy}(z) = 2 - (z + z^{-1}), \quad \Phi_{yy}(\omega) = 2 - 2\cos\omega$$

It is easy to see that $\log \Phi_{yy}(z)|_{z=1} = -\infty$, so that $\log \Phi_{yy}(z)$ is not analytic in the neighborhood of $|z| = 1$. But, we can show that (see Problem 4.7)

$$\int_{-\pi}^{\pi} \log \Phi_{yy}(\omega) d\omega = \int_{-\pi}^{\pi} \log(2 - 2\cos\omega) d\omega = 0 > -\infty \tag{4.52}$$

Thus, the condition of (4.46) is satisfied. But, in this case, it is impossible to have the inverse representation such that

$$\varepsilon(t) = \sum_{i=0}^{\infty} a_i y(t-i), \qquad \sum_{i=0}^{\infty} a_i^2 < \infty$$

In fact, the inverse $1/H(z)$ shows that $a_i = 1, i = 0, 1, \cdots$; but the sequence $a = (1, 1, \cdots)$ is not square summable. □

Example 4.10. Consider a regular stationary process y. It follows from Theorems 4.3 and 4.4 that y can be expressed as

$$y(t) = \sum_{i=0}^{\infty} h(i)\varepsilon(t-i), \qquad H(z) = \sum_{i=0}^{\infty} h(i) z^{-i}$$

where we assume that $H(z)$ is of minimal phase. For $m > 0$, we consider the m-step prediction problem of the stationary process y. Since $\mathcal{Y}_t = \mathcal{E}_t$, the m-step predictor is expressed as the following two different expressions:

$$\hat{y}(t+m \mid t) = \hat{E}\{y(t+m) \mid \mathcal{Y}_t\} = \sum_{i=0}^{\infty} g_i y(t-i) \tag{4.53}$$

$$= \hat{E}\{y(t+m) \mid \mathcal{E}_t\} = \sum_{i=0}^{\infty} f_i \varepsilon(t-i) \tag{4.54}$$

In terms of coefficients $\{g_i\}$ and $\{f_i\}$, define the transfer functions

$$G(z) = \sum_{i=0}^{\infty} g(i) z^{-i}, \qquad F(z) = \sum_{i=0}^{\infty} f(i) z^{-i}$$

We see that feeding y into the inverse filter $1/H(z)$ yields the innovation process ε as shown in Figure 4.5. Thus, by using the filter $F(z)$, the optimal filter $G(z)$ is expressed as

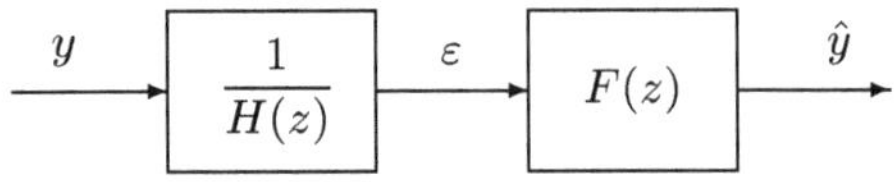

Figure 4.5. Optimal prediction: the innovation approach

$$G(z) = \frac{F(z)}{H(z)}$$

Hence, it suffices to obtain the optimal filter $F(z)$ acting on the innovation process.

We derive the optimal transfer function $F(z)$. Define the prediction error

$$\tilde{y}(t+m) = y(t+m) - \hat{y}(t+m \mid t)$$

Then, by using (4.54), the prediction error is expressed as

$$\begin{aligned}\tilde{y}(t+m) &= \sum_{i=0}^{\infty} h(i)\varepsilon(t+m-i) - \sum_{i=0}^{\infty} f(i)\varepsilon(t-i) \\ &= \sum_{i=0}^{m-1} h(i)\varepsilon(t+m-i) + \sum_{i=0}^{\infty} [h(i+m) - f(i)]\varepsilon(t-i)\end{aligned}$$

Since ε is a white noise, the variance of $\tilde{y}(t+m)$ is written as

$$E\{\tilde{y}^2(t+m)\} = \sigma_e^2 \sum_{i=0}^{m-1} h^2(i) + \sigma_e^2 \sum_{i=0}^{\infty} [h(i+m) - f(i)]^2 \tag{4.55}$$

Hence, the coefficients of the filter minimizing the variance of estimation error are given by

$$f(i) = h(i+m), \qquad i = 0,\, 1,\, \cdots \tag{4.56}$$

This indicates that the optimal predictor has the form

$$\hat{y}(t+m \mid t) = \sum_{i=0}^{\infty} h(m+i)\varepsilon(t-i) = \sum_{i=-\infty}^{t} h(t+m-i)\varepsilon(i)$$

We compute the transfer function of the optimal predictor. It follows from (4.56) that

$$F(z) = \sum_{i=0}^{\infty} h(i+m)z^{-i} = h(m) + h(m+1)z^{-1} + \cdots$$

Multiplying $H(z)$ by z^m yields

$$z^m H(z) = h(0)z^m + \cdots + h(m-1)z + h(m) + h(m+1)z^{-1} + \cdots$$

We see that $F(z)$ is equal to the causal part of $z^m H(z)$; the causal part is obtained by deleting the polynomial part. Let $[\,\cdot\,]_+$ be the operation to retrieve the causal part. Then, we have $F(z) = [z^m H(z)]_+$, so that the optimal transfer function is given by

$$G(z) = \frac{F(z)}{H(z)} = \frac{[z^m H(z)]_+}{H(z)} \tag{4.57}$$

From (4.55), the associated minimal variance of prediction error is given by

$$\sigma_m^2 = \sigma_e^2 \sum_{i=0}^{m-1} h^2(i), \qquad m = 1, 2, \cdots$$

Since $y(t+m) = \hat{y}(t+m \mid t) + \tilde{y}(t+m)$ with $\hat{y}(t+m \mid t) \perp \tilde{y}(t+m)$,

$$\sigma_m^2 = \sigma_y^2 - \sigma_{\hat{y}}^2$$

where $\sigma_{\hat{y}}^2 = E\{\hat{y}^2(t+m \mid t)\}$. Noting that $y = H(z)\varepsilon$ and $\hat{y} = F(z)\varepsilon$, the minimum variance is expressed as

$$\sigma_m^2 = \frac{\sigma_e^2}{2\pi} \int_{-\pi}^{\pi} \left(|H(e^{j\omega})|^2 - |F(e^{j\omega})|^2 \right) d\omega$$

by using the formula (4.31). □

4.6 Stochastic Linear Systems

We consider a stochastic linear system described by the state space model

$$x(t+1) = A(t)x(t) + w(t) \tag{4.58a}$$

$$y(t) = C(t)x(t) + v(t), \qquad t = 0, 1, \cdots \tag{4.58b}$$

where $x \in \mathbb{R}^n$ is the state vector, $y \in \mathbb{R}^p$ the observation vector, $w \in \mathbb{R}^n$ the plant noise vector, and $v \in \mathbb{R}^p$ the observation noise vector. Also, $A(t) \in \mathbb{R}^{n \times n}$, $C(t) \in \mathbb{R}^{p \times n}$ are deterministic functions of time t. Moreover, w and v are zero mean Gaussian white noise vectors with covariance matrices

$$E\left\{ \begin{bmatrix} w(t) \\ v(t) \end{bmatrix} \begin{bmatrix} w^{\mathrm{T}}(s) & v^{\mathrm{T}}(s) \end{bmatrix} \right\} = \begin{bmatrix} Q(t) & S(t) \\ S^{\mathrm{T}}(t) & R(t) \end{bmatrix} \delta_{ts} \tag{4.59}$$

where $Q(t) \in \mathbb{R}^{n \times n}$ is nonnegative definite, and $R(t) \in \mathbb{R}^{p \times p}$ is positive definite for all $t = 0, 1, \cdots$. The initial state $x(0)$ is Gaussian with mean $E\{x(0)\} = \mu_x(0)$ and covariance matrix

$$E\{[x(0) - \mu_x(0)][x(0) - \mu_x(0)]^{\mathrm{T}}\} = \Pi(0)$$

and is uncorrelated with the noises $w(t)$, $v(t)$, $t = 0, 1, \cdots$. The system described by (4.58) is schematically shown in Figure 4.6. This model is also called a Markov model for the process y.

In order to study the statistical properties of the state vector $x(t)$ of (4.58), we define the state transition matrix

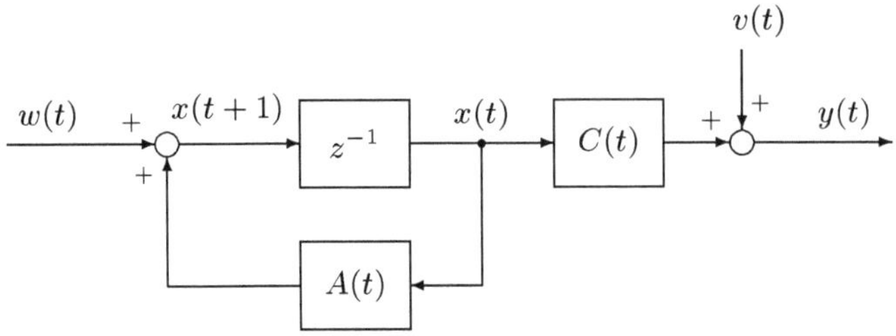

Figure 4.6. Stochastic linear state space system

$$\Phi(t,s) = \begin{cases} A(t-1)A(t-2)\cdots A(s), & t > s \\ I, & t = s \end{cases} \tag{4.60}$$

For any $k \le s \le t$, it follows that

$$\Phi(t,k) = \Phi(t,s)\Phi(s,k) \tag{4.61}$$

In terms of the transition matrix, the solution of (4.58a) is written as

$$x(t) = \Phi(t,s)x(s) + \sum_{k=s}^{t-1} \Phi(t,\, k+1)w(k) \tag{4.62}$$

Then, we can easily prove the lemma that characterizes the process $x(t)$.

Lemma 4.7. *The process x of (4.58a) is a Gauss-Markov process.*

Proof. Putting $s = 0$ in (4.62),

$$x(t) = \Phi(t,0)x(0) + \sum_{k=0}^{t-1} \Phi(t,k+1)w(k) \tag{4.63}$$

This shows that $x(t)$ is a linear combination of a Gaussian random vector $x(0)$ and the noises $\{w(0), \cdots, w(t-1)\}$, so that $x(t)$ is a Gaussian random vector. Thus x is a Gaussian process. Suppose that $s < t$. Then, we see from (4.62) that $x(t)$ is also a linear combination of $x(s)$, $w(s)$, $\cdots$, $w(t-1)$, and that $\{w(s), \cdots, w(t-1)\}$ are Gaussian white noises independent of $x(s)$. Hence, we have

$$p(x(t) \mid x(s),\, x(s-1),\, \cdots,\, x(0)) = p(x(t) \mid x(s)), \qquad t > s$$

This implies that $\{x(t),\, t = 0,\, 1,\, \cdots\}$ is a Markov process. □

It should be noted that a Gaussian process can be characterized by the mean and covariance matrix

$$\mu_x(t) = E\{x(t)\}, \qquad \Lambda_{xx}(t,s) = E\{[x(t) - \mu_x(t)][x(s) - \mu_x(s)]^{\mathrm{T}}\}$$

Lemma 4.8. *The mean vector and the covariance matrix of the state process x of (4.58a) are respectively given by*

$$\mu_x(t) = \Phi(t,0)\mu_x(0) \tag{4.64}$$

and

$$\Lambda_{xx}(t,s) = \begin{cases} \Phi(t,s)\Pi(s), & t \geq s \\ \Pi(t)\Phi^{\mathrm{T}}(s,t), & t < s \end{cases} \tag{4.65}$$

where $\Pi(t) := \Lambda_{xx}(t,t) = \operatorname{cov}\{x(t) - \mu_x(t)\}$ is the state covariance matrix that satisfies

$$\Pi(t) = \Phi(t,0)\Pi(0)\Phi^{\mathrm{T}}(t,0) + \sum_{k=0}^{t-1} \Phi(t,k+1)Q(k)\Phi^{\mathrm{T}}(t,k+1) \tag{4.66}$$

Proof. Taking the expectation of both sides of (4.63) immediately yields (4.64). We prove (4.65). Suppose that $t \geq s$. Then it follows from (4.63) that

$$\begin{aligned} \Lambda_{xx}(t,s) = E\Big\{ &\Big[\Phi(t,0)[x(0) - \mu_x(0)] + \sum_{l=0}^{t-1} \Phi(t,l+1)w(l)\Big] \\ &\times \Big[\Phi(s,0)[x(0) - \mu_x(0)] + \sum_{k=0}^{s-1} \Phi(s,k+1)w(k)\Big]^{\mathrm{T}} \Big\} \end{aligned}$$

Expanding the right-hand side of the above equation and using (4.59) yield

$$\Lambda_{xx}(t,s) = \Phi(t,0)\Pi(0)\Phi^{\mathrm{T}}(s,0) + \sum_{k=0}^{s-1} \Phi(t,k+1)Q(k)\Phi^{\mathrm{T}}(s,k+1)$$

Putting $s = t$ gives (4.66). Since $\Phi(t,0) = \Phi(t,s)\Phi(s,0)$, we see from the above equation that (4.65) holds for $t \geq s$. Similarly, we can prove (4.65) for $t < s$. □

It can be shown that the state covariance matrix $\Pi(t)$ satisfies

$$\Pi(t+1) = A(t)\Pi(t)A^{\mathrm{T}}(t) + Q(t), \qquad t = 0, 1, \cdots \tag{4.67}$$

Thus, for a given initial condition $\Pi(0) = \operatorname{cov}\{x(0)\}$, we can recursively compute the covariance matrix $\Pi(t)$ for $t = 1, 2, \cdots$.

Lemma 4.9. *The process y defined by (4.58) is a Gaussian process, whose mean vector $\mu_y(t)$ and covariance matrix $\Lambda_{yy}(t,s)$ are respectively given by*

$$\mu_y(t) = C(t)\mu_x(t) = C(t)\Phi(t,0)\mu_x(0) \tag{4.68}$$

and

$$\Lambda_{yy}(t,s) = \begin{cases} C(t)\Phi(t,s)\Pi(s)C^{\mathrm{T}}(s) + C(t)\Phi(t,s+1)S(s), & t > s \\ C(t)\Pi(t)C^{\mathrm{T}}(t) + R(t), & t = s \\ \Lambda_{yy}^{\mathrm{T}}(s,t), & t < s \end{cases} \tag{4.69}$$

Proof. Equation (4.68) is obvious from (4.58b) and (4.64). To prove (4.69), we assume that $t \geq s$. It follows from (4.58b), (4.62) and (4.68) that

$$\begin{aligned} y(t) - \mu_y(t) &= C(t)[x(t) - \mu_x(t)] + v(t) \\ &= C(t)\Phi(t,s)[x(s) - \mu_x(s)] + C(t)\sum_{k=s}^{t-1} \Phi(t,k+1)w(k) + v(t) \end{aligned}$$

Thus we have

$$\begin{aligned} \Lambda_{yy}(t,s) &= E\{[y(t) - \mu_y(t)][y(s) - \mu_y(s)]^{\mathrm{T}}\} \\ &= E\bigg\{\Big[C(t)\Phi(t,s)\tilde{x}(s) + C(t)\sum_{k=s}^{t-1}\Phi(t,k+1)w(k) + v(t)\Big] \\ &\qquad \times \Big[C(s)\tilde{x}(s) + v(s)\Big]^{\mathrm{T}}\bigg\} \end{aligned}$$

where $\tilde{x}(t) := x(t) - \mu_x(t)$. From (4.59) and the fact that $E\{w(k)\tilde{x}^{\mathrm{T}}(s)\} = 0$ and $E\{v(k)\tilde{x}^{\mathrm{T}}(s)\} = 0,\ k \geq s$, we have the first and second equations of (4.69). Similarly, for $t < s$. □

4.7 Stochastic Linear Time-Invariant Systems

In this section, we consider a stochastic LTI system, where $A(t)$, $C(t)$, $Q(t)$, $R(t)$, $S(t)$ in (4.58) and (4.59) are independent of time t.

Consider a stochastic LTI system described by

$$x(t+1) = Ax(t) + w(t) \tag{4.70a}$$

$$y(t) = Cx(t) + v(t), \qquad t = t_0,\ t_0 + 1,\ \cdots \tag{4.70b}$$

where t_0 is the initial time, and $x(t_0)$ is a Gaussian random vector with mean $\mu_x(t_0)$ and the covariance matrix $\Pi(t_0)$.

We see from (4.60) that the state transition matrix becomes $\Phi(t,s) = A^{t-s},\ t \geq s$. It thus follows from (4.64) and (4.66) that the mean vector is given by

$$\mu_x(t) = A^{t-t_0}\mu_x(t_0) \tag{4.71}$$

and the state covariance matrix becomes

$$\begin{aligned}\Pi(t) &= A^{t-t_0}\Pi(t_0)(A^{\mathrm{T}})^{t-t_0} + \sum_{k=t_0}^{t-1} A^{t-k-1}Q(A^{\mathrm{T}})^{t-k-1} \\ &= A^{t-t_0}\Pi(t_0)(A^{\mathrm{T}})^{t-t_0} + \sum_{k=0}^{t-t_0-1} A^{k}Q(A^{\mathrm{T}})^{k}\end{aligned} \tag{4.72}$$

Also, from (4.72) and (4.67), $\Pi(t)$ satisfies

$$\Pi(t+1) = A\Pi(t)A^{\mathrm{T}} + Q, \qquad t = t_0,\ t_0+1, \cdots \tag{4.73}$$

Lemma 4.10. *Suppose that A in (4.70a) is stable, i.e., $\rho(A) < 1$. Letting $t_0 \to -\infty$, the process x becomes a stationary Gauss-Markov process with mean zero and covariance matrix*

$$\Lambda_{xx}(l) = \begin{cases} A^{l}\Pi, & l \geq 0 \\ \Pi(A^{\mathrm{T}})^{-l}, & l < 0 \end{cases} \tag{4.74}$$

where Π is a unique solution of the Lyapunov equation

$$\Pi = A\Pi A^{\mathrm{T}} + Q \tag{4.75}$$

Proof. Since A is stable, we get $\lim_{t_0 \to -\infty} A^{t-t_0} = 0$. Thus, from (4.71),

$$\lim_{t_0 \to -\infty} \mu_x(t) = 0$$

Also taking $t_0 \to -\infty$ in (4.72),

$$\lim_{t_0 \to -\infty} \Pi(t) = \sum_{k=0}^{\infty} A^{k}Q(A^{\mathrm{T}})^{k} =: \Pi$$

It can be shown that Π satisfies (4.75), whose uniqueness is proved in Theorem 3.3. Since the right-hand side of (4.74) is a function of the time difference, the process x is stationary. The Gauss-Markov property of x follows from Lemma 4.7. □

Lemma 4.11. *Suppose that A is stable. For $t_0 \to -\infty$, the process y of (4.70b) becomes a stationary Gaussian process with mean zero and covariance matrix*

$$\Lambda_{yy}(l) = \begin{cases} CA^{l-1}\bar{C}^{\mathrm{T}}, & l > 0 \\ C\Pi C^{\mathrm{T}} + R, & l = 0 \\ \bar{C}(A^{\mathrm{T}})^{-l-1}C^{\mathrm{T}}, & l < 0 \end{cases} \tag{4.76}$$

where $\bar{C}^{\mathrm{T}}$ is defined by

$$\bar{C}^{\mathrm{T}} = A\Pi C^{\mathrm{T}} + S \tag{4.77}$$

Proof. As in Lemma 4.10, it can easily be shown that since A is stable, for $t_0 \to -\infty$, y of (4.70b) becomes a stationary Gaussian process with mean zero. From Lemmas 4.9 and 4.10, the covariance matrix of y becomes

$$\Lambda_{yy}(l) = \begin{cases} CA^l \Pi C^{\mathrm{T}} + CA^{l-1}S, & l > 0 \\ C\Pi C^{\mathrm{T}} + R, & l = 0 \\ C\Pi(A^{\mathrm{T}})^{-l}C^{\mathrm{T}} + S^{\mathrm{T}}(A^{\mathrm{T}})^{-l-1}C^{\mathrm{T}}, & l < 0 \end{cases}$$

This reduces to (4.76) by using $\bar{C}$ of (4.77). □

Example 4.11. For the Markov model (4.70), the matrices A, C, $\bar{C}$ are expressed as

$$\begin{aligned} A &= E\{x(t+1)x^{\mathrm{T}}(t)\}\Pi^{-1} \\ C &= E\{y(t)x^{\mathrm{T}}(t)\}\Pi^{-1} \\ \bar{C} &= E\{y(t)x^{\mathrm{T}}(t+1)\} \end{aligned}$$

In fact, post-multiplying (4.70a) by $x^{\mathrm{T}}(t)$, and noting that $E\{w(t)x^{\mathrm{T}}(t)\} = 0$, we have

$$E\{x(t+1)x^{\mathrm{T}}(t)\} = AE\{x(t)x^{\mathrm{T}}(t)\} = A\Pi$$

showing that the first relation holds. The second relation is proved similarly by using (4.70b). Finally, from (4.70),

$$\begin{aligned} E\{y(t)x^{\mathrm{T}}(t+1)\} &= E\{[Cx(t) + v(t)][x^{\mathrm{T}}(t)A^{\mathrm{T}} + w^{\mathrm{T}}(t)]\} \\ &= C\Pi A^{\mathrm{T}} + S^{\mathrm{T}} = \bar{C} \end{aligned}$$

This completes the proof. □

Example 4.12. We compute the spectral density matrix $\Phi_{yy}(z)$ of y with covariance matrix (4.76). We assume for simplicity that $S = 0$, so that $\bar{C}^{\mathrm{T}} = A\Pi C^{\mathrm{T}}$. Thus, we have

$$\Lambda_{yy}(l) = \begin{cases} C\Lambda_{xx}(l)C^{\mathrm{T}}, & l \neq 0 \\ C\Lambda_{xx}(0)C^{\mathrm{T}} + R, & l = 0 \end{cases}$$

so that the spectral density matrix is given by

$$\Phi_{yy}(z) = C\Phi_{xx}(z)C^{\mathrm{T}} + R \tag{4.78}$$

where $\Phi_{xx}(z)$ is the spectral density matrix of x. It follows from (4.37) and (4.74) that

$$\begin{aligned} \Phi_{xx}(z) &= \sum_{l=-\infty}^{\infty} \Lambda_{xx}(l)z^{-l} = \sum_{l=-\infty}^{-1} \Pi(A^{\mathrm{T}})^{-l}z^{-l} + \Pi + \sum_{l=1}^{\infty} A^k \Pi\, z^{-l} \\ &= \Pi + \Pi\left(\sum_{l=1}^{\infty} z^k (A^{\mathrm{T}})^l\right) + \left(\sum_{l=1}^{\infty} z^{-l} A^l\right)\Pi \end{aligned} \tag{4.79}$$

Let $\rho := \rho(A)$. Since A is stable, we have $0 < \rho < 1$, so that

$$\sum_{l=1}^{\infty} z^{-l} A^l = (zI - A)^{-1} A, \qquad |z| > \rho$$

and

$$\sum_{l=1}^{\infty} z^l (A^{\mathrm{T}})^l = A^{\mathrm{T}} (z^{-1} I - A^{\mathrm{T}})^{-1}, \qquad |z| < \rho^{-1}$$

This shows that the right-hand side of (4.79) is absolutely convergent for $\rho < |z| < \rho^{-1}$. Thus, by using the Lyapunov equation (4.75),

$$\begin{aligned} \Phi_{xx}(z) &= \Pi + \Pi A^{\mathrm{T}} (z^{-1} I - A^{\mathrm{T}})^{-1} + (zI - A)^{-1} A \Pi \\ &= (zI - A)^{-1} (\Pi - A \Pi A^{\mathrm{T}}) (z^{-1} I - A^{\mathrm{T}})^{-1} \\ &= (zI - A)^{-1} Q (z^{-1} I - A^{\mathrm{T}})^{-1} \end{aligned}$$

Also, let $W(z) = C(zI - A)^{-1}$. Then $\Phi_{yy}(z)$ is expressed as

$$\Phi_{yy}(z) = R + W(z) Q W^{\mathrm{T}}(z^{-1}) \tag{4.80}$$

This is an extended version of (4.29) to a multivariable LTI system. If $S \neq 0$, (4.80) becomes

$$\Phi_{yy}(z) = R + W(z) S + S^{\mathrm{T}} W^{\mathrm{T}}(z^{-1}) + W(z) Q W^{\mathrm{T}}(z^{-1}) \tag{4.81}$$

For a proof of (4.81), see Problem 4.12. □

4.8 Backward Markov Models

In the previous section, we have shown that the stochastic LTI system defined by (4.70) generates a stationary process y, so that the system of (4.70) is often called a Markov model for the stationary process y. In this section, we introduce a dual Markov model for the stationary process y; the dual model is also called a backward Markov model corresponding to the forward Markov model.

For the Markov model of (4.70), we assume that the state covariance matrix $\Pi = E\{x(t) x^{\mathrm{T}}(t)\}$ of (4.75) is positive definite. Then, we define

$$w_b(t) := \Pi^{-1} x(t) - A^{\mathrm{T}} \Pi^{-1} x(t+1) \tag{4.82}$$

It follows from (4.74) that

$$\begin{aligned} E\{w_b(t) x^{\mathrm{T}}(t+l)\} &= \Pi^{-1} E\{x(t) x^{\mathrm{T}}(t+l)\} - A^{\mathrm{T}} \Pi^{-1} E\{x(t+1) x^{\mathrm{T}}(t+l)\} \\ &= \Pi^{-1} \Lambda_{xx}^{\mathrm{T}}(l) - A^{\mathrm{T}} \Pi^{-1} \Lambda_{xx}^{\mathrm{T}}(l-1) \\ &= \Pi^{-1} \Pi (A^{\mathrm{T}})^l - A^{\mathrm{T}} \Pi^{-1} \Pi (A^{\mathrm{T}})^{l-1} = 0, \quad l = 1, 2, \cdots \end{aligned}$$

Hence, $w_b(t)$ defined above is orthogonal to the future $x(t+l)$, $l = 1, 2, \cdots$, so that it is a backward white noise. In fact, by definition, since

$$w_b(t+l) \in \text{span}\{x(t+l), x(t+l+1)\}$$

we see that $w_b(t)$ is orthogonal to $w_b(t+l)$. This implies that

$$E\{w_b(t+l)w_b^{\mathrm{T}}(t)\} = 0, \qquad l \neq 0$$

Motivated by the above observation, we prove a lemma that gives a backward Markov model.

Lemma 4.12. *Define $x_b(t) = \bar{\Pi}x(t+1)$ with $\bar{\Pi} = \Pi^{-1}$. Then, the model with x_b as the state vector*

$$x_b(t-1) = A^{\mathrm{T}}x_b(t) + w_b(t) \tag{4.83a}$$

$$y(t) = \bar{C}x_b(t) + v_b(t) \tag{4.83b}$$

is a backward Markov model for the stationary process y, where $\bar{C} \in \mathbb{R}^{p\times n}$ is called the backward output matrix, and w_b and v_b are zero mean white noises with covariance matrices

$$E\left\{\begin{bmatrix} w_b(t) \\ v_b(t)\end{bmatrix}\begin{bmatrix} w_b^{\mathrm{T}}(s) & v_b^{\mathrm{T}}(s)\end{bmatrix}\right\} = \begin{bmatrix} \bar{Q} & \bar{S} \\ \bar{S}^{\mathrm{T}} & \bar{R}\end{bmatrix}\delta_{ts} \tag{4.84}$$

Moreover, we have $\text{cov}\{x_b(t)\} = \bar{\Pi}$ *and*

$$\bar{Q} = \bar{\Pi} - A^{\mathrm{T}}\bar{\Pi}A, \quad \bar{S} = C^{\mathrm{T}} - A^{\mathrm{T}}\bar{\Pi}\bar{C}^{\mathrm{T}}, \quad \bar{R} = \Lambda_{yy}(0) - \bar{C}\bar{\Pi}\bar{C}^{\mathrm{T}} \tag{4.85}$$

Proof. Equation (4.83a) is immediate from (4.82). We show that the following relations hold.

$$E\{w_b(t)x_b^{\mathrm{T}}(t+l-1)\} = 0, \quad E\{w_b(t)y^{\mathrm{T}}(t+l)\} = 0, \quad l = 1, 2, \cdots \tag{4.86}$$

$$E\{v_b(t)x_b^{\mathrm{T}}(t+l-1)\} = 0, \quad E\{v_b(t)y^{\mathrm{T}}(t+l)\} = 0, \quad l = 1, 2, \cdots \tag{4.87}$$

(i) The first relation of (4.86) follows from the fact that $w_b(t) \perp x(t+l)$, $l = 1, 2, \cdots$ and $x(t+l) = \Pi x_b(t+l-1)$. We show the second relation in (4.86). From (4.82),

$$\begin{aligned}
E\{w_b(t)y^{\mathrm{T}}(t+l)\} &= \Pi^{-1}E\{x(t)[Cx(t+l) + v(t+l)]^{\mathrm{T}}\} \\
&\quad - A^{\mathrm{T}}\Pi^{-1}E\{x(t+1)[Cx(t+l) + v(t+l)]^{\mathrm{T}}\} \\
&= \Pi^{-1}E\{x(t)x^{\mathrm{T}}(t+l)\}C^{\mathrm{T}} + \Pi^{-1}E\{x(t)v^{\mathrm{T}}(t+l)\} \\
&\quad - A^{\mathrm{T}}\Pi^{-1}E\{x(t+1)x^{\mathrm{T}}(t+l)\}C^{\mathrm{T}} \\
&\quad - A^{\mathrm{T}}\Pi^{-1}E\{x(t+1)v^{\mathrm{T}}(t+l)\}
\end{aligned}$$

Since $v(t+l) \perp \{x(t+1), x(t)\}$, $l = 1, 2, \cdots$, the second and the fourth terms in the above equation vanish. Thus, it follows from (4.74) that for $l = 1, 2, \cdots$,

$$E\{w_b(t)y^{\mathrm{T}}(t+l)\} = \Pi^{-1}\Pi(A^{\mathrm{T}})^l C^{\mathrm{T}} - A^{\mathrm{T}}\Pi^{-1}\Pi(A^{\mathrm{T}})^{l-1}C^{\mathrm{T}} = 0$$

as was to be proved.

(ii) From (4.83b) and $x_b(t-1) = \Pi^{-1}x(t)$,

$$\begin{aligned} E\{v_b(t)x_b^{\mathrm{T}}(t+l-1)\} &= E\{[y(t) - \bar{C}x_b(t)]x^{\mathrm{T}}(t+l)\}\Pi^{-1} \\ &= E\{y(t)x^{\mathrm{T}}(t+l)\}\Pi^{-1} \\ &\quad - \bar{C}\Pi^{-1}E\{x(t+1)x^{\mathrm{T}}(t+l)\}\Pi^{-1} \end{aligned} \tag{4.88}$$

The first term in the right-hand side of the above equation is reduced to

$$\begin{aligned} E\{y(t)x^{\mathrm{T}}(t+l)\}\Pi^{-1} &= E\{y(t)[Ax(t+l-1) + w(t+l-1)]^{\mathrm{T}}\}\Pi^{-1} \\ &= E\{y(t)x^{\mathrm{T}}(t+l-1)\}A^{\mathrm{T}}\Pi^{-1} \\ &\quad + E\{y(t)w^{\mathrm{T}}(t+l-1)\}\Pi^{-1} \\ &= E\{y(t)x^{\mathrm{T}}(t+l-1)\}A^{\mathrm{T}}\Pi^{-1} \end{aligned}$$

where we have used the fact that $E\{y(t)w^{\mathrm{T}}(t+l-1)\} = 0,\ l = 2,\ 3,\ \cdots$. Repeating this procedure gives

$$\begin{aligned} E\{y(t)x^{\mathrm{T}}(t+l)\}\Pi^{-1} &= E\{y(t)x^{\mathrm{T}}(t+l-2)\}(A^{\mathrm{T}})^2\Pi^{-1} \\ &= E\{y(t)x^{\mathrm{T}}(t+1)\}(A^{\mathrm{T}})^{l-1}\Pi^{-1} \\ &= \bar{C}(A^{\mathrm{T}})^{l-1}\Pi^{-1} \end{aligned} \tag{4.89}$$

Also, from (4.74), the second term of the right-hand side of (4.88) becomes

$$\begin{aligned} \bar{C}\Pi^{-1}E\{x(t+1)x^{\mathrm{T}}(t+l)\}\Pi^{-1} &= \bar{C}\Pi^{-1}\Pi(A^{\mathrm{T}})^{l-1}\Pi^{-1} \\ &= \bar{C}(A^{\mathrm{T}})^{l-1}\Pi^{-1} \end{aligned}$$

Thus, it follows that the right-hand side of (4.88) vanishes, implying that the first equation in (4.87) is proved. Similarly, we can prove the second equation. In fact, we see from (4.83b) and (4.70b) that

$$\begin{aligned} E\{v_b(t)y^{\mathrm{T}}(t+l)\} &= E\{[y(t) - \bar{C}x_b(t)]y^{\mathrm{T}}(t+l)\} \\ &= E\{y(t)y^{\mathrm{T}}(t+l)\} \\ &\quad - \bar{C}\Pi^{-1}E\{x(t+1)[Cx(t+l) + v(t+l)]^{\mathrm{T}}\} \\ &= E\{y(t)y^{\mathrm{T}}(t+l)\} - \bar{C}\Pi^{-1}E\{x(t+1)x^{\mathrm{T}}(t+l)\}C^{\mathrm{T}} \end{aligned}$$

By using (4.74) and (4.76), we see that the right-hand side of the above equation vanishes for $l = 1, 2, \cdots$.

Having proved (4.86) and (4.87), we can easily show that w_b, v_b are white noises. By using (4.87),

$$E\{v_b(t)v_b^{\mathrm{T}}(t+l)\} = E\{v_b(t)[y(t+l) - \bar{C}x_b(t+l)]^{\mathrm{T}}\} = 0, \quad l = 1, 2, \cdots$$

so that $v_b(t)$ is a white noise. Similarly, we see from (4.86) and (4.87) that for $l = 1, 2, \cdots$,

$$E\{v_b(t)w_b^{\mathrm{T}}(t+l)\} = E\{v_b(t)[x_b(t+l-1) - A^{\mathrm{T}}x_b(t+l)]^{\mathrm{T}}\} = 0$$

$$E\{w_b(t)v_b^{\mathrm{T}}(t+l)\} = E\{w_b(t)[y(t+l) - \bar{C}x_b(t+l)]^{\mathrm{T}}\} = 0$$

Hence, $w_b(t+l) \perp v_b(t)$ holds for $l \neq 0$. Thus we have shown that $(w_b(t),\, v_b(t))$ are jointly white noises. Finally, it can easily be shown that $\mathrm{cov}\{x_b(t)\} = \bar{\Pi}$ and (4.85) hold. □

The backward Markov models introduced above, together with forward Markov models, play important roles in the analysis and modeling of stationary stochastic processes. Especially, the backward Markov model is utilized for proving the positive real lemma in Chapter 7, and it is also instrumental for deriving a balanced (reduced) stochastic realization for a stationary stochastic process in Chapter 8.

4.9 Notes and References

- A large number of books and papers are available for stochastic processes and systems. Sections 4.1 and 4.2, dealing with an introduction to stochastic processes, are based on [13, 44, 68, 134, 142], where the last one contains many practical examples.
- Section 4.3 has discussed ergodic properties of stochastic processes based on [123]. Also, for spectral analysis of stationary stochastic processes in Section 4.4, see [123, 134, 150].
- In Section 4.5, we have introduced Hilbert spaces generated by stochastic processes, and then stated the Wold decomposition theorem; see [44, 95]. This theorem is needed in developing a stochastic realization theory in the presence of exogenous inputs in Chapter 9. The regularity condition of stationary processes due to Szegö [65] is proved under a certain restricted condition [134, 178], and the linear prediction theory is briefly discussed. Some advanced results on prediction theory are found in [115, 179, 180]. Other related references in this section are books [13, 33, 138].
- Sections 4.6 and 4.7 have dealt with the stochastic linear dynamical systems, or the Markov models, based on [11, 68, 144]. Moreover, Section 4.8 derives dual or backward Markov models for stationary stochastic processes based on [39, 42, 106].

4.10 Problems

4.1 Prove the following identity for double sums.

$$\sum_{i=1}^{N}\sum_{j=1}^{N}\phi(i-j) = \sum_{k=-N+1}^{N-1}(N-|k|)\phi(k)$$

4.2 Prove (4.13).

4.3 Prove the formulas for Cèsaro sums.

(a) $$\lim_{n\to\infty} a_n = 0 \ \Rightarrow\ \lim_{n\to\infty}\frac{1}{n}\sum_{i=1}^{n} a_i = 0$$

(b) $$\lim_{n\to\infty}\frac{1}{n}\sum_{k=1}^{n} a_k = 0 \ \Rightarrow\ \lim_{n\to\infty}\frac{1}{n}\sum_{k=1}^{n}\left(1-\frac{k}{n}\right)a_k = 0$$

(c) $$\lim_{n\to\infty}\sum_{k=1}^{\infty} a_k = S \ \Rightarrow\ \lim_{n\to\infty}\sum_{k=1}^{n}\left(1-\frac{k}{n}\right)a_k = S$$

4.4 Prove Theorem 4.2.

4.5 Prove Lemma 4.3 (ii) by means of the relation $\Phi(\omega) = \lim_{N\to\infty} I_N(\omega)$, where $I_N(\omega)$ is given by

$$I_N(\omega) = \frac{1}{2N+1}E\left\{\left|\sum_{l=-N}^{N} e^{-j\omega l}x(l)\right|^2\right\} \geq 0$$

4.6 For the linear system shown in Figure 4.1, show that the following relation

$$\Phi_{yu}(\omega) = G(e^{j\omega})\Phi_{uu}(\omega)$$

holds, where the cross-spectral density function $\Phi_{yu}(\omega)$ is the Fourier transform of $\Lambda_{yu}(l)$. (Hint: See Lemma 4.4.)

4.7 Prove (4.52).

4.8 Suppose that the spectral density function of an ARMA process y is given by

$$\Phi_y(\omega) = \frac{1.25+\cos\omega}{1.81-1.8\cos\omega}$$

Obtain the difference equation satisfied by y.

4.9 By using the result of Example 4.10, solve the m-step prediction problem for the ARMA process

$$y(t) + ay(t-1) = e(t) + ce(t-1), \qquad |a| < 1, \qquad |c| < 1$$

where $m > 0$ and $a \neq c$.

4.10 Show that y in (4.58) is not a Markov process, but the joint process (x, y) is a Markov process.

4.11 Let $\Pi(t)$, $t = 1, 2, \cdots$ be the solution of the Lyapunov equation (4.73) with the initial condition $\Pi(0) = 0$. Let $M_0 := A$, $N_0 := Q$. For $k = 1, 2, \cdots$, we iterate

$$N_k := M_{k-1} N_{k-1} M_{k-1}^{\mathrm{T}} + N_{k-1}$$
$$M_k := M_{k-1}^2$$

Show that $\Pi(2^k) = N_k$, $k = 1, 2, \cdots$ holds. This scheme is called a doubling algorithm for solving a stationary Lyapunov equation.

4.12 Prove (4.81).

5

Kalman Filter

This chapter is concerned with the discrete-time Kalman filter for stochastic dynamic systems. First, we review a multi-dimensional Gaussian probability density function, and the least-squares (or minimum variance) estimation problem. We then derive the Kalman filter algorithm for discrete-time stochastic linear systems by using the orthogonal projection. The filter algorithm is extended so that it can be applied to stochastic systems with exogenous inputs. Moreover, we derive stationary forward and backward Kalman filter algorithms, which are useful for the study of stochastic realization problem.

5.1 Multivariate Gaussian Distribution

We consider a multivariate Gaussian distribution and the minimum variance estimation problem. Let $x \in \mathbb{R}^n$ and $y \in \mathbb{R}^p$ be jointly Gaussian random vectors. Let the mean vectors be given by $\mu_x = E\{x\}$ and $\mu_y = E\{y\}$ and the covariance matrices by

$$\Sigma = \begin{bmatrix} \Sigma_{xx} & \Sigma_{xy} \\ \Sigma_{yx} & \Sigma_{yy} \end{bmatrix} := \begin{bmatrix} \operatorname{cov}\{x, x\} & \operatorname{cov}\{x, y\} \\ \operatorname{cov}\{y, x\} & \operatorname{cov}\{y, y\} \end{bmatrix}$$

where we assume that the covariance matrix $\Sigma \in \mathbb{R}^{(n+p)\times(n+p)}$ is positive definite. For convenience, we define a quadratic form

$$Q(x, y) = [(x - \mu_x)^{\mathrm{T}} \ \ (y - \mu_y)^{\mathrm{T}}]\Sigma^{-1} \begin{bmatrix} x - \mu_x \\ y - \mu_y \end{bmatrix} \tag{5.1}$$

Then the joint probability density function of $(x,\ y)$ can be written as

$$p(x,\ y) = \frac{1}{C} \exp\left\{-\frac{1}{2} Q(x, y)\right\}, \tag{5.2}$$

where $C = (2\pi)^{(n+p)/2} |\Sigma|^{1/2}$ is the normalization factor.

Lemma 5.1. *Let the probability density function $p(x, y)$ be given by (5.2). Then, the conditional distribution of x given y is also Gaussian with mean*

$$E\{x \mid y\} = \mu_x + \Sigma_{xy}\Sigma_{yy}^{-1}(y - \mu_y) \tag{5.3}$$

and covariance matrix

$$E\{[x - E\{x \mid y\}][x - E\{x \mid y\}]^{\mathrm{T}}\} = \Sigma_{xx} - \Sigma_{xy}\Sigma_{yy}^{-1}\Sigma_{yx} \tag{5.4}$$

Moreover, the vector $x - E\{x \mid y\}$ is independent of y, i.e., the orthogonality condition $x - E\{x \mid y\} \perp y$ holds.

Proof. First we compute the joint probability density function $p(x, y)$. Define

$$\Sigma^{-1} = \begin{bmatrix} \Sigma_{xx} & \Sigma_{xy} \\ \Sigma_{yx} & \Sigma_{yy} \end{bmatrix}^{-1} =: \begin{bmatrix} V_{xx} & V_{xy} \\ V_{yx} & V_{yy} \end{bmatrix},$$

Also, define $\Upsilon := \Sigma_{xx} - \Sigma_{xy}\Sigma_{yy}^{-1}\Sigma_{yx}$. Then, it follows from Problem 2.3 (c) that

$$V_{xx} = \Upsilon^{-1}, \quad V_{xy} = -\Upsilon^{-1}\Sigma_{xy}\Sigma_{yy}^{-1}, \quad V_{yx} = -\Sigma_{yy}^{-1}\Sigma_{yx}\Upsilon^{-1}$$
$$V_{yy} = \Sigma_{yy}^{-1} + \Sigma_{yy}^{-1}\Sigma_{yx}\Upsilon^{-1}\Sigma_{xy}\Sigma_{yy}^{-1}$$

Thus $Q(x, y)$ defined by (5.1) becomes

$$\begin{aligned} Q(x, y) &= (x - \mu_x)^{\mathrm{T}} V_{xx}(x - \mu_x) + (x - \mu_x)^{\mathrm{T}} V_{xy}(y - \mu_y) \\ &\quad + (y - \mu_y)^{\mathrm{T}} V_{yx}(x - \mu_x) + (y - \mu_y)^{\mathrm{T}} V_{yy}(y - \mu_y) \\ &= [x - \mu_x + V_{xx}^{-1}V_{xy}(y - \mu_y)]^{\mathrm{T}}\Upsilon^{-1}[x - \mu_x + V_{xx}^{-1}V_{xy}(y - \mu_y)] \\ &\quad + (y - \mu_y)^{\mathrm{T}}[V_{yy} - V_{yx}V_{xx}^{-1}V_{xy}](y - \mu_y) \\ &= (x - \alpha)^{\mathrm{T}}\Upsilon^{-1}(x - \alpha) + (y - \mu_y)^{\mathrm{T}}\Sigma_{yy}^{-1}(y - \mu_y) \end{aligned}$$

where $\alpha := \mu_x + \Sigma_{xy}\Sigma_{yy}^{-1}(y - \mu_y)$. Therefore, the joint probability density function $p(x, y)$ is given by

$$\begin{aligned} p(x, y) &= \frac{1}{C'} \exp\left\{-\frac{1}{2}(x - \alpha)^{\mathrm{T}}\Upsilon^{-1}(x - \alpha)\right\} \\ &\quad \times \frac{1}{C''} \exp\left\{-\frac{1}{2}(y - \mu_y)^{\mathrm{T}}\Sigma_{yy}^{-1}(y - \mu_y)\right\} \end{aligned} \tag{5.5}$$

where $C' = (2\pi)^{n/2}|\Upsilon|^{1/2}$ and $C'' = (2\pi)^{p/2}|\Sigma_{yy}|^{1/2}$. Thus integrating $p(x, y)$ with respect to x yields the marginal probability density function

$$p(y) = \frac{1}{C''} \exp\left\{-\frac{1}{2}(y - \mu_y)^{\mathrm{T}}\Sigma_{yy}^{-1}(y - \mu_y)\right\} \tag{5.6}$$

It also follows from (5.5) and (5.6) that the conditional probability density function is given by

$$p(x \mid y) = \frac{1}{C'} \exp\left\{-\frac{1}{2}(x-\alpha)^{\mathrm{T}} \Upsilon^{-1}(x-\alpha)\right\}$$

From this, (5.3) and (5.4) hold. Also, we see from (5.5) that $x - \alpha = x - E\{x \mid y\}$ and y are independent. □

In the above proof, it is assumed that Σ is nonsingular. For the case where Σ is singular, the results still hold if we replace the inverse Σ^{-1} by the pseudo-inverse $\Sigma^{\dagger}$ introduced in Lemma 2.10.

Lemma 5.2. *Suppose that (x, y) are jointly Gaussian random vectors. Then, the minimum variance estimate of x based on y is given by the conditional mean*

$$\hat{x} := E\{x \mid y\} = \mu_x + \Sigma_{xy}\Sigma_{yy}^{-1}(y - \mu_y) \tag{5.7}$$

Proof. It may be noted that the minimum variance estimate $\hat{x}$ is a y-measurable function $f(y)$ that minimizes $E\{\|x - f(y)\|^2\}$. It can be shown that

$$\begin{aligned} E\{\|x - f(y)\|^2\} &= E\{\|x - \alpha + \alpha - f(y)\|^2\} \\ &= E\{\|x - \alpha\|^2\} + 2E\{[x-\alpha]^{\mathrm{T}}[\alpha - f(y)]\} \\ &\quad + E\{\|\alpha - f(y)\|^2\} \end{aligned}$$

Since $\alpha - f(y)$ is y-measurable and since $E\{x \mid y\} = \alpha$, the second term in the right-hand side becomes

$$\begin{aligned} E\{[x-\alpha]^{\mathrm{T}}[\alpha - f(y)]\} &= E\{E\{[x-\alpha]^{\mathrm{T}}[\alpha - f(y)] \mid y\}\} \\ &= E\{E\{[x-\alpha]^{\mathrm{T}} \mid y\}[\alpha - f(y)]\} = 0 \end{aligned}$$

Thus we have

$$E\{\|x - f(y)\|^2\} = E\{\|x - \alpha\|^2\} + E\{\|\alpha - f(y)\|^2\} \geq E\{\|x - \alpha\|^2\}$$

where the equality holds if and only if $f(y) = \alpha$. Hence, the minimum variance estimate is given by the conditional mean $\alpha = E\{x \mid y\}$. □

Suppose that x, y are jointly Gaussian random vectors. Then, from Lemma 5.2, the conditional expectation $E\{x \mid y\}$ is a linear function in y, so that for Gaussian case, the minimum variance estimate is obtained by the orthogonal projection of x onto the linear space generated by y (see Section 5.2).

Example 5.1. Consider a linear regression model given by

$$y = Hx + v$$

where $x \in \mathbb{R}^n$ is the input Gaussian random vector with $\mathcal{N}(\mu_x, P)$, $y \in \mathbb{R}^p$ the output vector, $v \in \mathbb{R}^p$ a Gaussian white noise vector with $\mathcal{N}(0, R)$, and $H \in \mathbb{R}^{p \times n}$ a constant matrix. We compute the minimum variance estimate of x based on the observation y, together with the error covariance matrix. From the regression model,

$$\mu_y = E\{Hx + v\} = H\mu_x$$
$$\Sigma_{xy} = E\{(x - \mu_x)(y - \mu_y)^{\mathrm{T}}\} = PH^{\mathrm{T}}$$
$$\Sigma_{yy} = E\{(y - \mu_y)(y - \mu_y)^{\mathrm{T}}\} = HPH^{\mathrm{T}} + R$$

Therefore, from Lemma 5.2, the minimum variance estimate is given by

$$\hat{x} = \mu_x + PH^{\mathrm{T}}[HPH^{\mathrm{T}} + R]^{-1}(y - H\mu_x)$$

Also, from (5.4), the error covariance matrix $\hat{P} := E\{[x - \hat{x}][x - \hat{x}]^{\mathrm{T}}\}$ is

$$\hat{P} = P - PH^{\mathrm{T}}[HPH^{\mathrm{T}} + R]^{-1}HP \tag{5.8}$$

where $HPH^{\mathrm{T}} + R$ is assumed to be nonsingular. □

Lemma 5.3. *For $P \in \mathbb{R}^{n\times n}$, $H \in \mathbb{R}^{p\times n}$, $R \in \mathbb{R}^{p\times p}$, we have*

$$PH^{\mathrm{T}}[R + HPH^{\mathrm{T}}]^{-1} = [P^{-1} + H^{\mathrm{T}}R^{-1}H]^{-1}H^{\mathrm{T}}R^{-1} \tag{5.9}$$

where it is assumed that P and R are nonsingular.

Proof. The following identity is immediate:

$$[P^{-1} + H^{\mathrm{T}}R^{-1}H]PH^{\mathrm{T}} = H^{\mathrm{T}}R^{-1}[R + HPH^{\mathrm{T}}]$$

Pre-multiplying the above equation by $[P^{-1} + H^{\mathrm{T}}R^{-1}H]^{-1}$ and post-multiplying by $[R + HPH^{\mathrm{T}}]^{-1}$ yield (5.9). □

It follows from (5.9) that the right-hand side of (5.8) becomes

$$P - PH^{\mathrm{T}}[HPH^{\mathrm{T}} + R]^{-1}HP = [P^{-1} + H^{\mathrm{T}}R^{-1}H]^{-1} \tag{5.10}$$

Equations (5.9) and (5.10) are usually called the matrix inversion lemmas.

Lemma 5.4. *Let (x, y, z) be jointly Gaussian random vectors. If y and z are mutually uncorrelated, we have*

$$E\{x \mid y,\, z\} = E\{x \mid y\} + E\{x \mid z\} - \mu_x \tag{5.11}$$

Proof. Define $w^{\mathrm{T}} := (y^{\mathrm{T}},\, z^{\mathrm{T}})$ and $\mu_w^{\mathrm{T}} := (\mu_y^{\mathrm{T}},\, \mu_z^{\mathrm{T}})$. Then we have $E\{x \mid w\} = E\{x \mid y,\, z\}$. Since y and z are uncorrelated,

$$\Sigma_{xw} = [\Sigma_{xy}\;\; \Sigma_{xz}], \qquad \Sigma_{ww}^{-1} = \begin{bmatrix} \Sigma_{yy}^{-1} & 0 \\ 0 & \Sigma_{zz}^{-1} \end{bmatrix}$$

Thus, from Lemma 5.1,

$$\begin{aligned} E\{x \mid w\} &= \mu_x + \Sigma_{xw}\Sigma_{ww}^{-1}(w - \mu_w) \\ &= \mu_x + \Sigma_{xy}\Sigma_{yy}^{-1}(y - \mu_y) + \Sigma_{xz}\Sigma_{zz}^{-1}(z - \mu_z) \end{aligned}$$

Since $E\{x \mid y\} = \mu_x + \Sigma_{xy}\Sigma_{yy}^{-1}(y - \mu_y)$ and $E\{x \mid z\} = \mu_x + \Sigma_{xz}\Sigma_{zz}^{-1}(z - \mu_z)$, we see that (5.11) holds. □

We consider the minimum variance estimation problem by using the orthogonal projection in a Hilbert space of random vectors with finite variances. Let $x \in \mathbb{R}^n$ be a random vector with the finite second-order moment

$$E\{\|x\|^2\} = \sum_{i=1}^{n} E\{x_i^2\} < \infty$$

Let a set of random vectors with finite second-order moments be

$$\mathcal{H} = \left\{ x \;\middle|\; E\{\|x\|^2\} < \infty \right\}$$

Then, it is easy to show that $\mathcal{H}$ is a linear space.

For $x,\ y \in \mathcal{H}$, we define the inner product by

$$(x, y)_{\mathcal{H}} = E\{x^{\mathrm{T}} y\} = \mathrm{trace} E\{x y^{\mathrm{T}}\}$$

and the norm by

$$\|x\|_{\mathcal{H}} = \sqrt{(x, x)_{\mathcal{H}}} = \sqrt{E\{\|x\|^2\}}$$

By completing $\mathcal{H}$ by means of this norm, we have a Hilbert space of n-dimensional random vectors with finite variances, which is again written as $\mathcal{H} = L_2(\Omega)$.

Let $x,\ y \in \mathcal{H}$. If $(x, y)_{\mathcal{H}} = 0$ holds, then we say that x and y are orthogonal, and write $x \perp y$. Suppose that $\mathcal{Y}$ is a subspace of $\mathcal{H}$. If $(x, y)_{\mathcal{H}} = 0$ holds for any $y \in \mathcal{Y}$, then we say that x is orthogonal to $\mathcal{Y}$, and write $x \perp \mathcal{Y}$. Let $x \in \mathcal{H}$. Then, from Lemma 4.6, there exists a unique $y_0 \in \mathcal{Y}$ such that

$$\|x - y_0\|_{\mathcal{H}} \le \|x - y\|_{\mathcal{H}}, \qquad \forall y \in \mathcal{Y}$$

Thus y_0 is a minimizing vector, and the optimality condition is that $x - y_0 \perp \mathcal{Y}$.

Suppose that $y_1, \cdots, y_N$ be p-dimensional random vectors with finite second-order moments. Let $\mathcal{Y}$ be the subspace generated by $y_1, \cdots, y_N$, *i.e.*,

$$\mathcal{Y} = \left\{ a + \sum_{i=1}^{N} A_i y_i \;\middle|\; a \in \mathbb{R}^n,\ A_i \in \mathbb{R}^{n \times p} \right\} \tag{5.12}$$

Any element $\hat{x} \in \mathcal{Y}$ is an n-dimensional random vector with finite second-order moment. By completing the linear space $\mathcal{Y}$ by the norm $\|\cdot\|_{\mathcal{H}}$ defined above, we see that $\mathcal{Y}$ becomes a Hilbert subspace of $\mathcal{H}$, *i.e.*, $\mathcal{Y} \subset \mathcal{H}$.

Lemma 5.5. *Let $\tilde{x}$ be an element in $\mathcal{H}$, and $\mathcal{Y}$ be the subspace defined by (5.12). Then, $\tilde{x}$ is orthogonal to $\mathcal{Y}$ if and only if the following conditions hold.*

$$E\{\tilde{x}\} = 0, \qquad E\{\tilde{x} y_i^{\mathrm{T}}\} = 0, \qquad i = 1, \cdots, N \tag{5.13}$$

Proof. Since any element $\hat{x} \in \mathcal{Y}$ is expressed as in (5.12), we see that if (5.13) holds,

$$(\tilde{x}, \hat{x})_{\mathcal{H}} = \left(\tilde{x}, a + \sum_{i=1}^{N} A_i y_i\right) = E\{\tilde{x}^{\mathrm{T}}\}a + \sum_{i=1}^{N} \mathrm{trace} E\{\tilde{x} y_i^{\mathrm{T}}\} A_i^{\mathrm{T}} = 0$$

Conversely, suppose that $\tilde{x} \perp \mathcal{Y}$ holds, *i.e.*, $(\tilde{x}, \hat{x})_{\mathcal{H}} = 0$ for any $\hat{x} \in \mathcal{Y}$. Putting $\hat{x} = a$, we have $(\tilde{x}, a)_{\mathcal{H}} = \mathrm{trace}(E\{\tilde{x}\}a^{\mathrm{T}}) = 0$. Taking $a = E\{\tilde{x}\}$ yields $\|E\{\tilde{x}\}\|^2 = 0$, implying that $E\{\tilde{x}\} = 0$. Next, let $\hat{x} = A_i y_i$. It follows that $(\tilde{x}, A_i y_i)_{\mathcal{H}} = \mathrm{trace}\left(E\{\tilde{x} y_i^{\mathrm{T}}\} A_i^{\mathrm{T}}\right) = 0$. Similarly, taking $A_i = E\{\tilde{x} y_i^{\mathrm{T}}\} \in R^{n \times p}$ yields

$$\mathrm{trace}\left(E\{\tilde{x} y_i^{\mathrm{T}}\} E\{\tilde{x} y_i^{\mathrm{T}}\}^{\mathrm{T}}\right) = \|E\{\tilde{x} y_i^{\mathrm{T}}\}\|_F^2 = 0 \quad \Rightarrow \quad E\{\tilde{x} y_i^{\mathrm{T}}\} = 0$$

This completes the proof of lemma. □

Example 5.2. Consider the random vectors x, y with probability density function of (5.2). Let the data space be given by $\mathcal{Y} = \{b + Ay \mid b \in \mathbb{R}^n,\ A \in \mathbb{R}^{n \times p}\}$. Then, the orthogonal projection of x onto the space $\mathcal{Y}$ is given by

$$\hat{E}\{x \mid \mathcal{Y}\} = \mu_x + \Sigma_{xy} \Sigma_{yy}^{-1} (y - \mu_y)$$

In fact, let $\tilde{x} = x - (b + Ay)$. Then, from the conditions of Lemma 5.5, we have

$$0 = E\{\tilde{x}\} = E\{x - (b + Ay)\} = \mu_x - b - A\mu_y$$
$$0 = E\{\tilde{x} y^{\mathrm{T}}\} = E\{[x - (b + Ay)] y^{\mathrm{T}}\}$$

From the first condition, we have $b = \mu_x - A\mu_y$. Substituting this condition into the second relation gives $E\{[x - \mu_x - A(y - \mu_y)] y^{\mathrm{T}}\} = 0$, so that

$$E\{[x - \mu_x - A(y - \mu_y)][y - \mu_y]^{\mathrm{T}}\} = 0$$

Thus we obtain

$$\Sigma_{xy} - A\Sigma_{yy} = 0 \quad \Rightarrow \quad A = \Sigma_{xy} \Sigma_{yy}^{-1}$$

and hence

$$\hat{x} = \mu_x + A(y - \mu_y) = \mu_x + \Sigma_{xy} \Sigma_{yy}^{-1} (y - \mu_y)$$

Thus we have shown that the orthogonal projection is equivalent to the conditional expectation (5.3), and hence to the minimum variance estimate (5.7). □

Suppose that $y_1, \cdots, y_N$ be p-dimensional random vectors, and that there exist a set of p-dimensional independent random vectors $\tilde{y}_1, \cdots, \tilde{y}_N$ such that

$$\sigma\{\tilde{y}_i,\ i = 1, \cdots, k\} = \sigma\{y_i,\ i = 1, \cdots, k\}, \qquad k \le N \tag{5.14}$$

where $\sigma\{y_i,\ i = 1, \cdots, k\}$ is the σ-algebra generated by $\{y_i,\ i = 1, \cdots, k\}$, which is roughly the information contained in $\{y_i,\ i = 1, \cdots, k\}$. In this case, the random vectors $\tilde{y}_1, \cdots, \tilde{y}_N$ are called the innovations of $y_1, \cdots, y_N$.

Example 5.3. We derive the innovations for p-dimensional Gaussian random vectors $y_1, \cdots, y_N$. Let $\mathfrak{F}_k = \sigma\{y_i, i = 1, \cdots, k\}$, and define $\tilde{y}_1, \cdots, \tilde{y}_N$ as

$$\begin{aligned} \tilde{y}_1 &= y_1 - E\{y_1 \mid \mathfrak{F}_0\} = y_1 - E\{y_1\} \\ \tilde{y}_2 &= y_2 - E\{y_2 \mid \mathfrak{F}_1\} \\ &\vdots \\ \tilde{y}_N &= y_N - E\{y_N \mid \mathfrak{F}_{N-1}\} \end{aligned} \tag{5.15}$$

Since, for Gaussian random vectors, the conditional expectation coincides with the orthogonal projection onto the data space, we have

$$E\{y_k \mid \mathfrak{F}_{k-1}\} = \hat{E}\{y_k \mid y_1, \cdots, y_{k-1}\} = a_k + \sum_{i=1}^{k-1} A_{ki} y_i, \quad A_{ki} \in \mathbb{R}^{p \times p}$$

We see from (5.15) that

$$\begin{bmatrix} \tilde{y}_1 \\ \tilde{y}_2 \\ \vdots \\ \tilde{y}_k \end{bmatrix} = \begin{bmatrix} I_p & & & 0 \\ -A_{21} & I_p & & \\ \vdots & & \ddots & \\ -A_{k1} & \cdots & -A_{k,k-1} & I_p \end{bmatrix} \begin{bmatrix} y_1 \\ y_2 \\ \vdots \\ y_k \end{bmatrix} - \begin{bmatrix} a_1 \\ a_2 \\ \vdots \\ a_k \end{bmatrix} \tag{5.16}$$

This shows that $\tilde{y}_k$ is a Gaussian random vector, since it is a linear combination of $a_1, \cdots, a_k$ and $y_1, \cdots, y_k$. Since the $pk \times pk$ lower triangular matrix in (5.16) is nonsingular, we see that y_k is also expressed as a linear combination of $\tilde{y}_1, \cdots, \tilde{y}_k$, $a_1, \cdots, a_k$. Hence (5.14) holds.

We show that $\tilde{y}_1, \cdots, \tilde{y}_N$ are independent. From Lemma 5.1, $\tilde{y}_k$ and $\mathfrak{F}_{k-1}$ are independent, so that we get $E\{\tilde{y}_k \mid \mathfrak{F}_{k-1}\} = 0$ and $E\{\tilde{y}_k\} = 0$. Since for $k > l$, $\tilde{y}_l$ is $\mathfrak{F}_{k-1}$-measurable,

$$E\{\tilde{y}_k \tilde{y}_l^{\mathrm{T}}\} = E\{E\{\tilde{y}_k \tilde{y}_l^{\mathrm{T}} \mid \mathfrak{F}_{k-1}\}\} = E\{E\{\tilde{y}_k \mid \mathfrak{F}_{k-1}\} \tilde{y}_l^{\mathrm{T}}\} = 0$$

It can be shown that the above relation also holds for $k < l$, so that $E\{\tilde{y}_k \tilde{y}_l^{\mathrm{T}}\} = 0,\ k \neq l$. Since the uncorrelated two Gaussian random vectors are independent, $\tilde{y}_k$ are $\tilde{y}_l$ $(k \neq l)$ are independent. Hence, we see that $\tilde{y}_1, \cdots, \tilde{y}_N$ are the innovations for the Gaussian random vectors $y_1, \cdots, y_N$. □

5.2 Optimal Estimation by Orthogonal Projection

We consider a state estimation problem for discrete-time stochastic linear dynamic systems. This is the celebrated Kalman filtering problem.

We deal with a discrete-time stochastic linear system described by

$$x(t+1) = A(t)x(t) + w(t) \tag{5.17a}$$

$$y(t) = C(t)x(t) + v(t), \qquad t = 0, 1, \cdots \tag{5.17b}$$

where $x \in \mathbb{R}^n$ is the state vector, $y \in \mathbb{R}^p$ the observation vector, $w \in \mathbb{R}^n$ the plant noise vector, and $v \in \mathbb{R}^p$ the observation noise vector. Also, $A(t) \in \mathbb{R}^{n\times n}$, $C(t) \in \mathbb{R}^{p\times n}$ are deterministic functions of time t. Moreover, w and v are zero mean Gaussian white noise vectors with covariance matrices

$$E\left\{\begin{bmatrix} w(t) \\ v(t) \end{bmatrix} \begin{bmatrix} w^{\mathrm{T}}(s) & v^{\mathrm{T}}(s) \end{bmatrix}\right\} = \begin{bmatrix} Q(t) & S(t) \\ S^{\mathrm{T}}(t) & R(t) \end{bmatrix} \delta_{ts} \tag{5.18}$$

where $Q(t) \in \mathbb{R}^{n\times n}$ is nonnegative definite, and $R(t) \in \mathbb{R}^{p\times p}$ is positive definite for all $t = 0, 1, \cdots$. The initial state $x(0)$ is Gaussian with mean $E\{x(0)\} = \mu_x(0)$ and covariance matrix

$$E\{[x(0) - \mu_x(0)][x(0) - \mu_x(0)]^{\mathrm{T}}\} = \Pi(0)$$

and is uncorrelated with the noises $w(t)$, $v(t)$, $t = 0, 1, \cdots$. A block diagram of the Markov model is depicted in Figure 5.1.

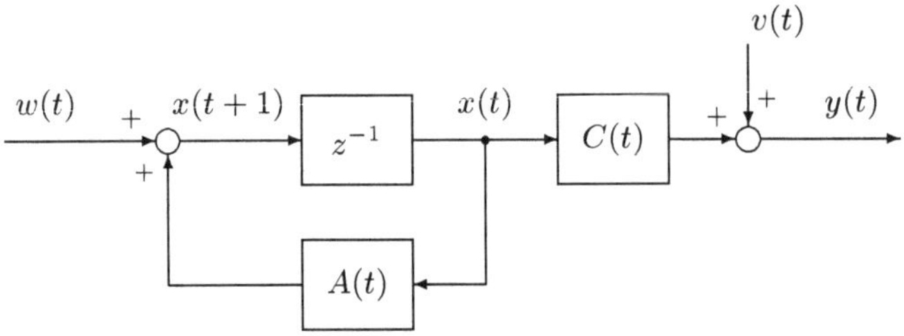

Figure 5.1. Stochastic linear dynamic system

Let $\mathcal{F}_t = \sigma\{y(0), y(1), \cdots, y(t)\}$ be the σ-algebra generated by the observations up to the present time t. We see that $\mathcal{F}_t$ is the information carried by the output observations, satisfying $\mathcal{F}_s \subset \mathcal{F}_t$, $s \le t$. Thus $\mathcal{F}_t$ is called an increasing family of σ-algebras, or a filtration. We now formulate the state estimation problem.

State Estimation Problem

The problem is to find the minimum variance estimate $\hat{x}(t+m \mid t)$ of the state vector $x(t+m)$ based on the observations up to time t. This is equivalent to designing a filter that produces $\hat{x}(t+m \mid t)$ minimizing the performance index

$$J = E\{\|x(t+m) - \hat{x}(t+m \mid t)\|^2\} \tag{5.19}$$

where $\hat{x}(t+m \mid t)$ is $\mathcal{F}_t$-measurable. The estimation problem is called the prediction, filtering or smoothing according as $m > 0$, $m = 0$ or $m < 0$. □

We see from Lemma 5.2 that the optimal estimate $\hat{x}(t+m \mid t)$ that minimizes the performance index of (5.19) is expressed in terms of the conditional expectation of $x(t+m)$ given $\mathcal{F}_t$ as

$$\hat{x}(t+m \mid t) = E\{x(t+m) \mid \mathcal{F}_t\}$$

Let the estimation error be defined by $\tilde{x}(t+m \mid t) := x(t+m) - \hat{x}(t+m \mid t)$ and the error covariance matrix be

$$P(t+m \mid t) := E\{[x(t+m) - \hat{x}(t+m \mid t)][x(t+m) - \hat{x}(t+m \mid t)]^{\mathrm{T}}\}$$

From Lemmas 4.7, 4.8 and 4.9, we see that (x, y) of (5.17) are jointly Gaussian processes. For Gaussian processes, the conditional expectation $\hat{x}(t+m \mid t)$ is a linear function of observations $y(0), y(1), \cdots, y(t)$, so that the optimal estimate coincides with the linear minimum variance estimate of $x(t+m)$ given observations up to time t. More precisely, we define a linear space generated by the observations as

$$\mathcal{Y}_t = \left\{ c + \sum_{i=0}^{t} A_i y(i) \;\middle|\; c \in \mathbb{R}^n, \;\; A_i \in \mathbb{R}^{n \times p} \right\} \tag{5.20}$$

The space $\mathcal{Y}_t$ is called the data space at time t. Then, from Lemma 5.5, we have the following results.

Lemma 5.6. *The minimum variance estimate $\hat{x}(t+m \mid t)$ is given by the orthogonal projection of $x(t+m)$ onto $\mathcal{Y}_t$, i.e.,*

$$\hat{x}(t+m \mid t) = \hat{E}\{x(t+m) \mid \mathcal{Y}_t\} \tag{5.21}$$

The optimality of $\hat{x}(t+m \mid t)$ is that the estimation error $\tilde{x}(t+m \mid t)$ is orthogonal to the data space (see Figure 5.2):

$$\tilde{x}(t+m \mid t) = x(t+m) - \hat{x}(t+m \mid t) \perp \mathcal{Y}_t \tag{5.22}$$

Moreover, the minimum variance estimate is unbiased.

Proof. Equations (5.21) and (5.22) are obvious from Lemma 5.5. Since the data space $\mathcal{Y}_t$ contains constant vectors, it also follows that $E\{\tilde{x}(t+m)\} = 0, \; t = 0, 1, \cdots$. Thus the minimum variance estimate is unbiased. □

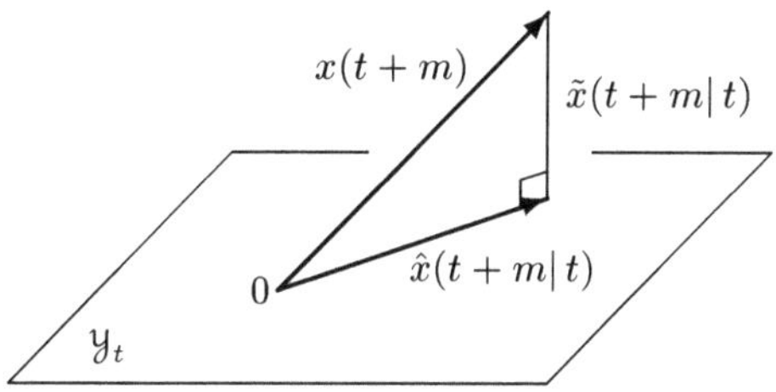

Figure 5.2. Orthogonal projection onto data space $\mathcal{Y}_t$

5.3 Prediction and Filtering Algorithms

Now we define

$$e(t) = y(t) - E\{y(t) \mid \mathcal{F}_{t-1}\}, \qquad t = 1, 2, \cdots \tag{5.23}$$

where $e(0) = y(0) - E\{y(0) \mid \mathcal{F}_{-1}\} = y(0) - \mu_y(0)$. Then, as in Example 5.3, it can be shown that e is the innovation process for y.

Lemma 5.7. *The innovation process $e \in \mathbb{R}^p$ is a Gaussian process with mean zero and covariance matrix*

$$E\{e(t)e^{\mathrm{T}}(s)\} = [C(t)P(t \mid t-1)C^{\mathrm{T}}(t) + R(t)]\delta_{ts} \tag{5.24}$$

where $P(t \mid t-1)$ is the error covariance matrix defined by

$$P(t \mid t-1) = E\{[x(t) - \hat{x}(t \mid t-1)][x(t) - \hat{x}(t \mid t-1)]^{\mathrm{T}}\}$$

Proof. Since y is Gaussian, the conditional expectation $E\{y(t) \mid \mathcal{F}_{t-1}\}$ is Gaussian, and hence e is Gaussian. By the definition (5.23), we see that

$$E\{e(t) \mid \mathcal{F}_{t-1}\} = 0, \qquad E\{e(t)\} = 0$$

Since $e(s)$ is a function of $y(0), y(1), \cdots, y(s)$, it is $\mathcal{F}_{t-1}$-measurable if $t > s$. Therefore, by the property of conditional expectation,

$$\begin{aligned} E\{e(t)e^{\mathrm{T}}(s)\} &= E\{E\{e(t)e^{\mathrm{T}}(s) \mid \mathcal{F}_{t-1}\}\} \\ &= E\{E\{e(t) \mid \mathcal{F}_{t-1}\}e^{\mathrm{T}}(s)\} = 0 \end{aligned}$$

Similarly, we can prove that the above equality holds for $t < s$. Thus $e(t)$ and $e(s)$, $t \neq s$ are uncorrelated.

We show that (5.24) holds for $t = s$. It follows from (5.17b) that

$$\begin{aligned} e(t) &= y(t) - E\{C(t)x(t) + v(t) \mid \mathcal{Y}_{t-1}\} \\ &= y(t) - C(t)\hat{x}(t \mid t-1) = C(t)\tilde{x}(t \mid t-1) + v(t) \end{aligned}$$

so that

$$\begin{aligned} E\{e(t)e^{\mathrm{T}}(t)\} &= E\{[C(t)\tilde{x}(t \mid t-1) + v(t)][C(t)\tilde{x}(t \mid t-1) + v(t)]^{\mathrm{T}}\} \\ &= C(t)E\{\tilde{x}(t \mid t-1)\tilde{x}^{\mathrm{T}}(t \mid t-1)\}C^{\mathrm{T}}(t) \\ &\quad + C(t)E\{\tilde{x}(t \mid t-1)v^{\mathrm{T}}(t)\} \\ &\quad + E\{v(t)\tilde{x}^{\mathrm{T}}(t \mid t-1)\}C^{\mathrm{T}}(t) + E\{v(t)v^{\mathrm{T}}(t)\} \end{aligned} \tag{5.25}$$

Since $v(t)$ is uncorrelated with $x(t)$ and $\hat{x}(t \mid t-1)$, we have

$$E\{\tilde{x}(t \mid t-1)v^{\mathrm{T}}(t)\} = E\{[x(t) - \hat{x}(t \mid t-1)]v^{\mathrm{T}}(t)\} = 0$$

Thus we see that the second and third terms of the right-hand side of (5.25) vanish; thus (5.24) holds from the definitions of $R(t)$ and $P(t \mid t-1)$. □

In the following, we derive a recursive algorithm that produces the one-step predicted estimates $\hat{x}(t+1 \mid t)$ and $\hat{x}(t \mid t-1)$ by using the orthogonal projection. We employ (5.21) as the definition of the optimal estimate.

From the definition of $e(t)$ and $\mathcal{Y}_t$, the innovation process is also expressed as

$$e(t) = y(t) - \hat{E}\{y(t) \mid \mathcal{Y}_{t-1}\}$$

Thus, we have $\mathcal{Y}_t = \mathcal{Y}_{t-1} \oplus \text{span}\,\{e(t)\}$, where $\oplus$ denotes the orthogonal sum. It therefore follows that

$$\begin{aligned}\hat{x}(t+1 \mid t) &= \hat{E}\{x(t+1) \mid \mathcal{Y}_t\} = \hat{E}\{x(t+1) \mid \mathcal{Y}_{t-1} \oplus e(t)\} \\ &= \hat{E}\{x(t+1) \mid \mathcal{Y}_{t-1}\} + \hat{E}\{x(t+1) \mid e(t)\} \end{aligned} \tag{5.26}$$

The first term in the right-hand side is expressed as

$$\begin{aligned}\hat{E}\{x(t+1) \mid \mathcal{Y}_{t-1}\} &= \hat{E}\{A(t)x(t) + w(t) \mid \mathcal{Y}_{t-1}\} \\ &= A(t)\hat{x}(t \mid t-1) \end{aligned} \tag{5.27}$$

and the second term is given by

$$\hat{E}\{x(t+1) \mid e(t)\} = K(t)e(t) \tag{5.28}$$

where $K(t) \in \mathbb{R}^{n \times p}$ is to be determined below.

Recall that the optimality condition for $K(t)$ is $x(t+1) - K(t)e(t) \perp e(t)$, *i.e.*

$$E\{[x(t+1) - K(t)e(t)]e^{\mathrm{T}}(t)\} = 0$$

so that

$$K(t) = E\{x(t+1)e^{\mathrm{T}}(t)\}(E\{e(t)e^{\mathrm{T}}(t)\})^{-1} \tag{5.29}$$

We see from the definition of $e(t)$ that

$$\begin{aligned}E\{x(t+1)e^{\mathrm{T}}(t)\} &= E\{[A(t)x(t) + w(t)][C(t)\tilde{x}(t \mid t-1) + v(t)]^{\mathrm{T}}\} \\ &= A(t)E\{x(t)\tilde{x}^{\mathrm{T}}(t \mid t-1)\}C^{\mathrm{T}}(t) \\ &\quad + A(t)E\{x(t)v^{\mathrm{T}}(t)\} \\ &\quad + E\{w(t)\tilde{x}^{\mathrm{T}}(t \mid t-1)\}C^{\mathrm{T}}(t) \\ &\quad + E\{w(t)v^{\mathrm{T}}(t)\} \end{aligned} \tag{5.30}$$

Noting that $w(t)$, $v(t)$ are white noises, the second and the third terms in the above equation vanish, and the fourth term becomes $S(t)$. Also, we have

$$x(t) = \hat{x}(t \mid t-1) + \tilde{x}(t \mid t-1), \qquad \hat{x}(t \mid t-1) \perp \tilde{x}(t \mid t-1)$$

It therefore follows that

$$E\{x(t)\tilde{x}^{\mathrm{T}}(t \mid t-1)\} = E\{\tilde{x}(t \mid t-1)\tilde{x}^{\mathrm{T}}(t \mid t-1)\} = P(t \mid t-1)$$

Thus from (5.30), we get

$$E\{x(t+1)e^{\mathrm{T}}(t)\} = A(t)P(t \mid t-1)C^{\mathrm{T}}(t) + S(t)$$

Since $R(t) > 0$, we see that $E\{e(t)e^{\mathrm{T}}(t)\} = C(t)P(t \mid t-1)C^{\mathrm{T}}(t) + R(t) > 0$. Thus, from (5.29)

$$K(t) = [A(t)P(t)C^{\mathrm{T}}(t) + S(t)][C(t)P(t)C^{\mathrm{T}}(t) + R(t)]^{-1} \tag{5.31}$$

where $K(t) \in \mathbb{R}^{n \times p}$ is called the Kalman gain.

For simplicity, we write the one-step predicted estimate as $\hat{x}(t)$. Accordingly, the corresponding estimation error and error covariance matrix are respectively written as $\tilde{x}(t)$ and $P(t)$. But, the filtered estimate and filtered error covariance matrix are respectively written as $\hat{x}(t \mid t)$ and $P(t \mid t)$ without abbreviation.

Lemma 5.8. *The one-step predicted estimate satisfies*

$$\hat{x}(t+1) = A(t)\hat{x}(t) + K(t)[y(t) - C(t)\hat{x}(t)] \tag{5.32}$$

with $\hat{x}(0) = \mu_x(0)$*, and the error covariance matrix is given by*

$$\begin{aligned} P(t+1) = A(t)P(t)A^{\mathrm{T}}(t) - K(t)[C(t)P(t)C^{\mathrm{T}}(t) + R(t)]K^{\mathrm{T}}(t) \\ + Q(t), \qquad P(0) = \Pi(0) \end{aligned} \tag{5.33}$$

Also, the predicted estimate $\hat{x}(t+1 \mid t)$ *is unbiased, i.e.*

$$E\{x(t+1) - \hat{x}(t+1)\} = 0, \qquad t = 0, 1, \cdots \tag{5.34}$$

Proof. Equation (5.32) is immediate from (5.26), (5.27) and (5.28). Now, it follows from (5.17a) and (5.32) that the prediction error satisfies

$$\tilde{x}(t+1) = [A(t) - K(t)C(t)]\tilde{x}(t) + w(t) - K(t)v(t) \tag{5.35}$$

Since $w(t)$ and $v(t)$ are white noises with mean zero, the expectation of both sides of (5.35) yields

$$E\{\tilde{x}(t+1)\} = [A(t) - K(t)C(t)]E\{\tilde{x}(t)\}$$

From the initial condition $\hat{x}(0) = \mu_x(0)$, we have $E\{\tilde{x}(0)\} = 0$, so that

$$E\{\tilde{x}(t+1)\} = (A(t) - K(t)C(t)) \cdots (A(0) - K(0)C(0))E\{\tilde{x}(0)\} = 0$$

This proves (5.34). Also, $w(t)$ and $v(t)$ are independent of $\tilde{x}(t)$, so that from (5.35),

$$\begin{aligned} E\{\tilde{x}(t+1)\tilde{x}^{\mathrm{T}}(t+1)\} = [A(t) - K(t)C(t)]E\{\tilde{x}(t)\tilde{x}^{\mathrm{T}}(t)\}[A(t) - K(t)C(t)]^{\mathrm{T}} \\ + [I \quad -K(t)] \begin{bmatrix} Q(t) & S(t) \\ S^{\mathrm{T}}(t) & R(t) \end{bmatrix} \begin{bmatrix} I \\ -K^{\mathrm{T}}(t) \end{bmatrix} \end{aligned}$$

Hence, we see that

$$P(t+1) = [A(t) - K(t)C(t)]P(t)[A(t) - K(t)C(t)]^{\mathrm{T}} \\ + Q(t) + K(t)R(t)K^{\mathrm{T}}(t) - S(t)K^{\mathrm{T}}(t) - K(t)S^{\mathrm{T}}(t)$$

By using $K(t)$ of (5.31), we have (5.33). □

If the matrix $C(t)P(t)C^{\mathrm{T}}(t) + R(t)$ is singular, the inverse in (5.31) is to be replaced by the pseudo-inverse. Given the one-step prediction $\hat{x}(t) := \hat{x}(t \mid t-1)$ and the new observation $y(t)$, we can compute the new one-step prediction $\hat{x}(t+1) := \hat{x}(t+1 \mid t)$ from (5.32), in which we observe that the Kalman gain $K(t)$ represents the relative weight of the information about the state vector $x(t+1)$ contained in the innovation $e(t)$.

Lemma 5.9. *Given the one-step predicted estimate $\hat{x}(t)$, the filtered estimate $\hat{x}(t \mid t)$ and its error covariance matrix $P(t \mid t)$ are respectively given by*

$$\hat{x}(t \mid t) = \hat{x}(t) + K_f(t)e(t) \tag{5.36}$$

$$K_f(t) = P(t)C^{\mathrm{T}}(t)[C(t)P(t)C^{\mathrm{T}}(t) + R(t)]^{-1} \tag{5.37}$$

and

$$P(t \mid t) = P(t) - P(t)C^{\mathrm{T}}(t)[C(t)P(t)C^{\mathrm{T}}(t) + R(t)]^{-1}C(t)P(t) \tag{5.38}$$

Proof. By definition, the filtered estimate is given by

$$\hat{x}(t \mid t) = \hat{E}\{x(t) \mid \mathcal{Y}_t\} = \hat{E}\{x(t) \mid \mathcal{Y}_{t-1} \oplus e(t)\} \\ = \hat{E}\{x(t) \mid \mathcal{Y}_{t-1}\} + \hat{E}\{x(t) \mid e(t)\} = \hat{x}(t) + \hat{E}\{x(t) \mid e(t)\}$$

where we have

$$\hat{E}\{x(t) \mid e(t)\} = E\{x(t)e^{\mathrm{T}}(t)\}(E\{e(t)e^{\mathrm{T}}(t)\})^{-1}e(t) \\ = E\{x(t)[\tilde{x}^{\mathrm{T}}(t)C^{\mathrm{T}}(t) + v^{\mathrm{T}}(t)]\}(E\{e(t)e^{\mathrm{T}}(t)\})^{-1}e(t) \\ = P(t)C^{\mathrm{T}}(t)[C(t)P(t)C^{\mathrm{T}}(t) + R(t)]^{-1}e(t) =: K_f(t)e(t)$$

This proves (5.36) and (5.37). Moreover, the estimation error is given by

$$\tilde{x}(t \mid t) = \tilde{x}(t) - P(t)C^{\mathrm{T}}(t)[R(t) + C(t)P(t)C^{\mathrm{T}}(t)]^{-1}e(t)$$

Noting that $E\{\tilde{x}(t)e^{\mathrm{T}}(t)\} = P(t)C^{\mathrm{T}}(t)$ and taking the covariance matrices of both sides of the above equation yields (5.38). □

Using the algorithm (5.36) ~ (5.38), the filtered estimate $\hat{x}(t \mid t)$ and associated error covariance matrix $P(t \mid t)$ can be computed from the predicted estimate $\hat{x}(t)$ and the associated error covariance matrix $P(t)$. Summarizing the above results, we have the following filter algorithm.

Theorem 5.1. *(Kalman filter) The algorithm of Kalman filter for the discrete-time stochastic system described by (5.17) and (5.18) is given by the following (i) ∼ (v).*

(i) Filter equations

$$\hat{x}(t+1) = A(t)\hat{x}(t) + K(t)[y(t) - C(t)\hat{x}(t)] \tag{5.39a}$$

$$\hat{x}(t \mid t) = \hat{x}(t) + K_f(t)[y(t) - C(t)\hat{x}(t)] \tag{5.39b}$$

(ii) The innovation process

$$e(t) = y(t) - C(t)\hat{x}(t) \tag{5.40}$$

(iii) Kalman gains

$$K(t) = [A(t)P(t)C^{\mathrm{T}}(t) + S(t)][C(t)P(t)C^{\mathrm{T}}(t) + R(t)]^{-1} \tag{5.41a}$$

$$K_f(t) = P(t)C^{\mathrm{T}}(t)[C(t)P(t)C^{\mathrm{T}}(t) + R(t)]^{-1} \tag{5.41b}$$

(iv) Error covariance matrices

$$P(t+1) = A(t)P(t)A^{\mathrm{T}}(t) - K(t)[C(t)P(t)C^{\mathrm{T}}(t) + R(t)]K^{\mathrm{T}}(t) + Q(t) \tag{5.42a}$$

$$P(t \mid t) = P(t) - P(t)C^{\mathrm{T}}(t)[C(t)P(t)C^{\mathrm{T}}(t) + R(t)]^{-1}C(t)P(t) \tag{5.42b}$$

(v) Initial conditions

$$\hat{x}(0) = \mu_x(0), \qquad P(0) = \Pi(0) \tag{5.43}$$

Figure 5.3 displays a block diagram of Kalman filter that produces the one-step predicted estimates $\hat{x}(t)$ and $\hat{x}(t+1)$ with the input $y(t)$. □

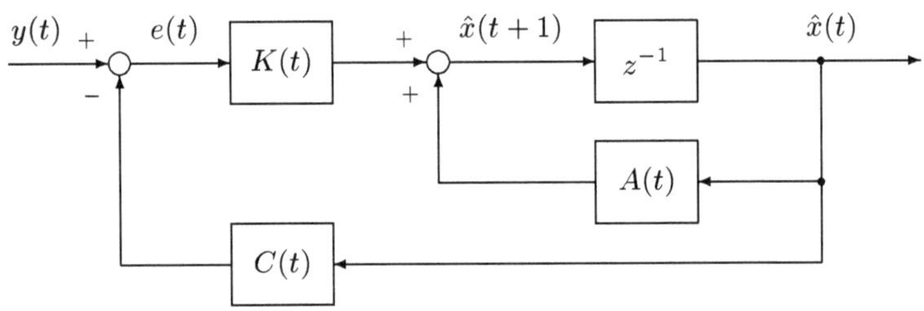

Figure 5.3. Block diagram of Kalman filter

The structure of Kalman filter shown above is quite similar to that of the discrete-time stochastic system of Figure 5.1 except that Kalman filter has a feedback-loop with a time-varying gain $K(t)$. We see that the Kalman filter is a dynamic system that recursively produces the state estimates $\hat{x}(t+1)$ and $\hat{x}(t \mid t)$ by updating the old estimates based on the received output data $y(t)$. The Kalman filter is, therefore, an algorithm suitable for the on-line state estimation.

Equation (5.42a) is a discrete-time Riccati equation satisfied by $P(t) \in \mathbb{R}^{n\times n}$. Being symmetric, $P(t)$ consists of $n(n+1)/2$ nonlinear difference equations. We see that the Riccati equation is determined by the model and the statistics of noise processes, and is independent of the observations. Thus, given the initial condition $P(0) = \Pi(0)$, we can recursively compute $P(t)$, $t = 1, 2, \cdots$, and hence $K(t)$, $t = 1, 2, \cdots$ off-line.

It follows from the definition of the innovation process e that the Kalman filter equation is also written as

$$\hat{x}(t+1) = A(t)\hat{x}(t) + K(t)e(t) \tag{5.44a}$$

$$y(t) = C(t)\hat{x}(t) + e(t) \tag{5.44b}$$

Equation (5.44) as a model of the process y has a different state vector and a noise process than those of the state space model of (5.17), but the two models are equivalent state space representations that simulate the same output process y. The model of (5.44) is called the innovation representation, or innovation model. The innovation model is less redundant in the noise models, and is often used in the stochastic realization, or the state space system identification.

Example 5.4. Consider an AR model described by

$$y(t) = \theta y(t-1) + v(t), \qquad t = 0, 1, \cdots; \qquad |\theta| < 1$$

where θ is an unknown parameter, and v is a white noise with $\mathcal{N}(0, r)$. The problem is to estimate the unknown parameter θ based on the observations $\mathcal{Y}_t$. Define $x(t) = \theta$ and $w = 0$. The AR model is rewritten as a state space model

$$x(t+1) = x(t), \qquad y(t) = c(t)x(t) + v(t) \tag{5.45}$$

where $c(t) = y(t-1)$. The state estimate based on the observations gives the least-squares estimate of the unknown parameter, *i.e.*,

$$\hat{x}(t+1) := \hat{E}\{x(t+1) \mid \mathcal{Y}_t\} = \hat{\theta}(t+1) = \hat{E}\{\theta \mid \mathcal{Y}_t\}$$

Applying Kalman filter algorithm of Theorem 5.1 to (5.45) yields

$$\hat{\theta}(t+1) = \hat{\theta}(t) + \frac{c(t)p(t)}{c^2(t)p(t) + r}[y(t) - c(t)\hat{\theta}(t)], \qquad \hat{\theta}(0) = 0$$

$$p(t+1) = \frac{rp(t)}{c^2(t)p(t) + r}, \qquad p(0) = p_0 > 0$$

Since $p_0 > 0$, we have $p(t) > 0$ for all $t > 0$. Thus the inverse $p^{-1}(t)$ satisfies

$$p^{-1}(t+1) = p^{-1}(t) + \frac{c^2(t)}{r}, \qquad p^{-1}(0) = p_0^{-1}$$

so that

$$p^{-1}(t) = p_0^{-1} + \frac{1}{r}\sum_{i=1}^{t} y^2(i-1)$$

Since, from Example 4.4, the process y is ergodic, we have

$$\lim_{t\to\infty} \frac{1}{t}\sum_{i=1}^{t} y^2(i-1) = \sigma_y^2$$

in the quadratic mean. Hence for large t,

$$p^{-1}(t) \sim p_0^{-1} + \frac{\sigma_y^2}{r}t \quad \Rightarrow \quad p(t) \sim \left(\frac{r}{\sigma_y^2}\right)\frac{1}{t}$$

showing that the estimate $\hat{\theta}(t)$ converges to the true θ in the mean square sense with the asymptotic variance of the order $1/t$. □

Remark 5.1. Recall that it is assumed that the coefficients matrices in the state space model of (5.17) are deterministic functions of time t. However, the state space model of (5.45) does not satisfy this basic assumption, because $c(t)$ is a function of the observation $y(t-1)$. Thus strictly speaking the algorithm of Theorem 5.1 cannot be applied to the state space model with random coefficients. □

In this regard, we have the following result [28]. Recall that the σ-algebra $\mathcal{F}_t$ is defined by $\mathcal{F}_t = \sigma\{y(0), y(1), \cdots, y(t)\}$.

Lemma 5.10. *Suppose that for the state space system of (5.17), the conditions (i) ~ (iv) are satisfied.*

(i) The noise vectors w and v are Gaussian white noises.

(ii) The a priori distribution of the initial state $x(0)$ is Gaussian.

(iii) The matrices $A(t)$, $Q(t)$ are $\mathcal{F}_t$-measurable, and $C(t)$, $S(t)$, $R(t)$ are $\mathcal{F}_{t-1}$-measurable.

(iv) The elements of $A(t)$, $C(t)$, $Q(t)$, $S(t)$, $R(t)$ are bounded.

Then the conditional probability density function $p(x(t) \mid \mathcal{F}_t)$ of the state vector given the observations is Gaussian.

Proof. For a proof, see [28]. □

This lemma implies that if the random coefficient matrices satisfy the conditions above, then the algorithm of Theorem 5.1 is valid, and so is the algorithm of Example 5.4. In this case, however, the estimates $\hat{x}(t \mid t)$ and $\hat{x}(t)$ are to be understood as the conditional mean estimates.

In the next section, we consider stochastic systems with exogenous inputs, which may be control inputs, reference inputs or some probing signals for identification. A version of Kalman filter will be derived under the assumption that the inputs are $\mathcal{F}_t$-measurable.

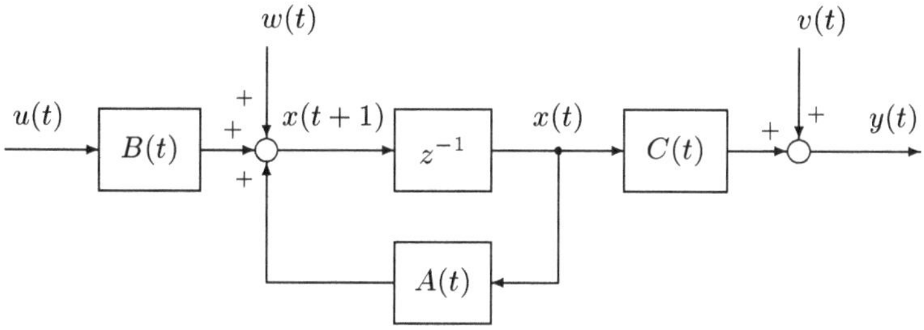

Figure 5.4. Stochastic system with inputs

5.4 Kalman Filter with Inputs

Since there are no external inputs in the state space model of (5.17), the Kalman filter algorithm in Theorem 5.1 cannot be applied to the system subjected to exogenous or control inputs. In this subsection, we modify the Kalman filter algorithm so that it can be applied to state space models with inputs.

Consider a discrete-time stochastic linear system

$$x(t+1) = A(t)x(t) + Bu(t) + w(t) \tag{5.46a}$$

$$y(t) = C(t)x(t) + v(t) \tag{5.46b}$$

where $u(t) \in \mathbb{R}^m$ is the input vector, and $B(t) \in \mathbb{R}^{n \times m}$ is a matrix connecting the input vector to the system as shown in Figure 5.4. We assume that $u(t)$ is $\mathcal{F}_t$-measurable, *i.e.*, $u(t)$ is a function of the outputs $y(0), y(1), \cdots, y(t)$, including deterministic time functions. We say that $\mathcal{F}_t$-measurable inputs are admissible inputs. Since the class of admissible inputs includes $\mathcal{F}_t$-measurable nonlinear functions, the process x generated by (5.46a) may not be Gaussian nor Markov. Of course, if $u(t)$ is a linear output feedback such that

$$u(t) = L(t)y(t) = L(t)C(t)x(t) + L(t)v(t), \qquad L(t) \in \mathbb{R}^{m \times p}$$

then $\{x(t),\ t = 0, 1, \cdots\}$ becomes a Gauss-Markov process.

In the following, we derive a filtering algorithm for (5.46) that produces the one-step predicted estimates $\hat{x}(t)$ and $\hat{x}(t+1)$.

Lemma 5.11. *Suppose that $x_w(t)$ and $x_u(t)$ are the solutions of*

$$x_w(t+1) = A(t)x_w(t) + w(t), \qquad x_w(0) = x(0) \tag{5.47}$$

and

$$x_u(t+1) = A(t)x_u(t) + B(t)u(t), \qquad x_u(0) = 0 \tag{5.48}$$

respectively. Then, the solution $x(t)$ of (5.46a) is expressed as

$$x(t) = x_w(t) + x_u(t), \qquad t = 0, 1, \cdots \tag{5.49}$$

Proof. A proof is immediate from the linearity of the system. □

By using the state transition matrix of (4.60), the solution of (5.48) is given by

$$x_u(t) = \sum_{k=0}^{t-1} \Phi(t, k+1)B(k)u(k), \qquad t = 0,\ 1,\ \cdots \tag{5.50}$$

Thus it follows that $x_u(t)$ is a function of $u(0),\ u(1),\ \cdots,\ u(t-1)$, so that $x_u(t)$ is $\mathcal{F}_{t-1}$-measurable, and hence $\mathcal{F}_t$-measurable. From the property of conditional expectation,

$$\hat{x}(t \mid t) = E\{x_w(t) \mid \mathcal{F}_t\} + x_u(t) \tag{5.51a}$$

$$\hat{x}(t) = E\{x_w(t) \mid \mathcal{F}_{t-1}\} + x_u(t) \tag{5.51b}$$

Since $x_u(t)$ is known, it suffices to derive an algorithm for computing the estimates of the vector $x_w(t)$ of (5.47) based on observations.

Lemma 5.12. *By using (5.46b) and (5.49), we define*

$$h(t) := y(t) - C(t)x_u(t) = C(t)x_w(t) + v(t) \tag{5.52}$$

Let $\mathcal{F}_t^h$ be the σ-algebra generated by $\{h(i),\ i = 0,\ 1,\ \cdots,\ t\}$. Then, $\mathcal{F}_t^h = \mathcal{F}_t$ holds, implying that the process h of (5.52) contains the same information carried by the output process y.

Proof. Since $x_u(t)$ is $\mathcal{F}_t$-measurable, we see from (5.52) that $h(t)$ is $\mathcal{F}_t$-measurable. Thus, we get $\mathcal{F}_t^h = \sigma\{h(0),\ h(1),\ \cdots,\ h(t)\} \subset \mathcal{F}_t$. Now, we show that $\mathcal{F}_t \subset \mathcal{F}_t^h$. From (5.50) and (5.52),

$$y(t) = h(t) + C(t) \sum_{k=0}^{t-1} \Phi(t, k+1)B(k)u(k)$$

For $t = 0$, we have $y(0) = h(0)$, so that $\mathcal{F}_0 = \mathcal{F}_0^h$ holds. For $t = 1$, $y(1) = h(1) + C(1)B(0)u(0)$. Since $u(0)$ is $\mathcal{F}_0$-measurable, and $\mathcal{F}_0 = \mathcal{F}_0^h$ holds, $y(1)$ is the sum of $h(1)$ and $\mathcal{F}_0^h$-measurable $C(1)B(0)u(0)$, implying that $y(1)$ is $\mathcal{F}_1^h$-measurable. Thus, we get $\mathcal{F}_1 \subset \mathcal{F}_1^h$. Similarly, we can show that $\mathcal{F}_t \subset \mathcal{F}_t^h$ holds. Hence, we have $\mathcal{F}_t^h = \mathcal{F}_t$, $t = 0, 1, \cdots$. □

Let the predicted estimates of the state vector $x_w(t)$ of (5.47) be given by

$$\hat{x}_w(t+1) = E\{x_w(t+1) \mid \mathcal{F}_t^h\}, \qquad \hat{x}_w(t) = E\{x_w(t) \mid \mathcal{F}_{t-1}^h\}$$

It follows from (5.51) that

$$\hat{x}(t+1) = \hat{x}_w(t+1) + x_u(t+1) \tag{5.53a}$$

$$\hat{x}(t) = \hat{x}_w(t) + x_u(t) \tag{5.53b}$$

Since the state vector $x_u(t)$ is given by (5.50), the algorithm is completed if we can compute $\hat{x}_w(t)$ and $\hat{x}_w(t+1)$. From (5.47) and (5.52), we have

$$x_w(t+1) = A(t)x_w(t) + w(t) \tag{5.54a}$$

$$h(t) = C(t)x_w(t) + v(t) \tag{5.54b}$$

This is a stochastic linear system with the state vector $x_w(t)$ and with the observation vector $h(t)$. Moreover, this state space model does not contain external inputs; hence we can apply Theorem 5.1 to derive the Kalman filter algorithm for (5.54).

The innovation process for h of (5.54) is given by

$$\begin{aligned} e_h(t) &= h(t) - C(t)\hat{x}_w(t) \\ &= y(t) - C(t)x_u(t) - C(t)\hat{x}_w(t) \\ &= y(t) - C(t)\hat{x}(t) = e(t) \end{aligned}$$

so that e_h coincides with the innovation process e for the observation y. Also, from (5.49) and (5.53), we have

$$\begin{aligned} x(t) - \hat{x}(t \mid t) &= x_w(t) - \hat{x}_w(t \mid t) \\ x(t+1) - \hat{x}(t+1) &= x_w(t+1) - \hat{x}_w(t+1) \end{aligned}$$

Thus the error covariance matrices are given by

$$P(t \mid t) = E\{[x_w(t) - \hat{x}_w(t \mid t)][x_w(t) - \hat{x}_w(t \mid t)]^{\mathrm{T}}\} \tag{5.55a}$$

$$P(t+1) = E\{[x_w(t+1) - \hat{x}_w(t+1)][x_w(t+1) - \hat{x}_w(t+1)]^{\mathrm{T}}\} \tag{5.55b}$$

This implies that the error covariance matrices are independent of the admissible input u, so that they coincides with the error covariance matrices of the system defined by (5.54). Hence, the prediction error $\tilde{x}(t) = x(t) - \hat{x}(t)$, $t = 0, 1, \cdots$ is a Gauss-Markov process with mean zero and covariance matrix $P(t)$.

Theorem 5.2. *Suppose that $u(t)$ in (5.46) be $\mathcal{F}_t$-measurable. Then, the Kalman filter algorithm for the stochastic system with admissible inputs is given by (i) ~ (iv).*

(i) Filter equations

$$\hat{x}(t+1) = A(t)\hat{x}(t) + B(t)u(t) + K(t)e(t) \tag{5.56a}$$

$$\hat{x}(t \mid t) = \hat{x}(t) + K_f(t)e(t) \tag{5.56b}$$

$$e(t) = y(t) - C(t)\hat{x}(t) \tag{5.56c}$$

(ii) Filter gains

$$K(t) = [A(t)P(t)C^{\mathrm{T}}(t) + S(t)][C(t)P(t)C^{\mathrm{T}}(t) + R(t)]^{-1} \tag{5.57a}$$

$$K_f(t) = P(t)C^{\mathrm{T}}(t)[C(t)P(t)C^{\mathrm{T}}(t) + R(t)]^{-1} \tag{5.57b}$$

(iii) Error covariance matrices

$$P(t+1) = A(t)P(t)A^{\mathrm{T}}(t) - K(t)[C(t)P(t)C^{\mathrm{T}}(t) + R(t)]K^{\mathrm{T}}(t) + Q(t) \tag{5.58a}$$

$$P(t \mid t) = P(t) - P(t)C^{\mathrm{T}}(t)[C(t)P(t)C^{\mathrm{T}}(t) + R(t)]^{-1}C(t)P(t) \tag{5.58b}$$

(iv) Initial conditions

$$\hat{x}(0) = \mu_x(0), \qquad P(0) = \Pi(0) \tag{5.59}$$

Proof. It follows from Theorem 5.1 that the Kalman filter for the system described by (5.54) is given by

$$\hat{x}_w(t+1) = A(t)\hat{x}_w(t) + K(t)e_h(t) \tag{5.60a}$$

$$\hat{x}_w(t \mid t) = \hat{x}_w(t) + K_f(t)e_h(t) \tag{5.60b}$$

From (5.48), (5.53), (5.60) and the fact that $e_h(t) = e(t)$, we get

$$\begin{aligned}\hat{x}(t+1) &= \hat{x}_w(t+1) + x_u(t+1) \\ &= A(t)\hat{x}_w(t) + K(t)e(t) + A(t)x_u(t) + B(t)u(t) \\ &= A(t)\hat{x}(t) + B(t)u(t) + K(t)e(t)\end{aligned}$$

Thus we have (5.56a). From (5.53) and (5.60),

$$\begin{aligned}\hat{x}(t \mid t) &= \hat{x}_w(t \mid t) + x_u(t) = \hat{x}_w(t) + K_f(t)e(t) + x_u(t) \\ &= \hat{x}(t) + K_f(t)e(t)\end{aligned}$$

This proves (5.56b). Equations (5.57) ~ (5.59) are obvious from (5.55). □

Figure 5.5 shows a block diagram of the optimal filter. It seems that the form of optimal filter is quite obvious in view of Figure 5.4, but $\mathcal{F}_t$-measurability of the inputs is needed for the filter in Figure 5.5 to be optimal in the sense of least-squares.

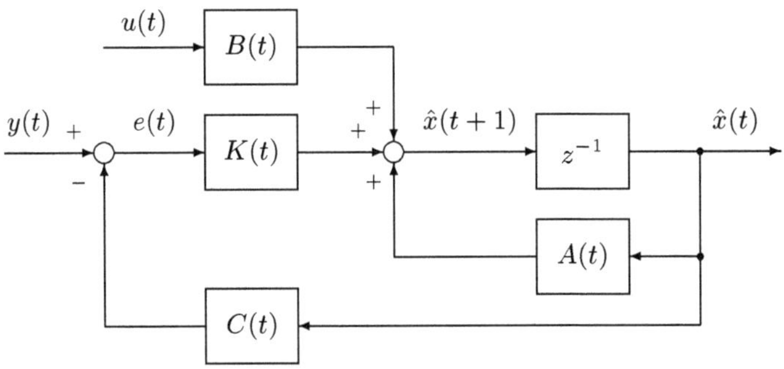

Figure 5.5. Block diagram of Kalman filter with inputs

5.5 Covariance Equation of Predicted Estimate

Recall from (5.17a) that the covariance matrix $\Pi(t) = \mathrm{cov}\{x(t)\}$ of the state vector $x(t)$ satisfies (4.67). As in (4.77), we define

$$\bar{C}^{\mathrm{T}}(t) = A(t)\Pi(t)C^{\mathrm{T}}(t) + S(t)$$

It then follows from Lemma 4.9 that the covariance matrix of y is expressed as

$$\Lambda_{yy}(t,s) = \begin{cases} C(t)A(t)\cdots A(s+1)\bar{C}^{\mathrm{T}}(s), & t > s \\ C(t)\Pi(t)C^{\mathrm{T}}(t) + R(t), & t = s \\ \bar{C}(s)A^{\mathrm{T}}(s+1)\cdots A^{\mathrm{T}}(t)C^{\mathrm{T}}(t), & t < s \end{cases} \tag{5.61}$$

For simplicity, we define $\Lambda(t) := \Lambda_{yy}(t,t)$. Then, in terms of $A(t)$, $C(t)$, $\bar{C}(t)$, $\Lambda(t)$, we define a new Riccati equation

$$\begin{aligned} \Sigma(t+1) = A(t)\Sigma(t)A^{\mathrm{T}}(t) &+ (\bar{C}^{\mathrm{T}}(t) - A(t)\Sigma(t)C^{\mathrm{T}}(t)) \\ &\times [\Lambda(t) - C(t)\Sigma(t)C^{\mathrm{T}}(t)]^{-1}(\bar{C}(t) - C(t)\Sigma(t)A^{\mathrm{T}}(t)) \end{aligned} \tag{5.62}$$

with $\Sigma(0) = 0$. The following theorem gives a relation between the new Riccati equation (5.62) and the Riccati equation (5.42a) satisfied by $P(t)$.

Theorem 5.3. *The solution $\Sigma(t)$ of Riccati equation (5.62) is the covariance matrix of the predicted estimate $\hat{x}(t)$, and the relation*

$$P(t) = \Pi(t) - \Sigma(t) \tag{5.63}$$

holds. Moreover, the Kalman gain of (5.41a) is equivalently expressed as

$$K(t) = [\bar{C}^{\mathrm{T}}(t) - A(t)\Sigma(t)C^{\mathrm{T}}(t)][\Lambda(t) - C(t)\Sigma(t)C^{\mathrm{T}}(t)]^{-1} \tag{5.64}$$

Proof. From $\Sigma(0) = 0$, (5.63) is obvious for $t = 0$. Since

$$\Lambda(0) = C(0)\Pi(0)C^{\mathrm{T}}(0) + R(0), \qquad \bar{C}^{\mathrm{T}}(0) = A(0)\Pi(0)C^{\mathrm{T}}(0) + S(0)$$

we see from (5.64) that

$$K(0) = [A(0)\Pi(0)C^{\mathrm{T}}(0) + S(0)][C(0)\Pi(0)C^{\mathrm{T}}(0) + R(0)]^{-1}$$

The right-hand side of the above equation equals the Kalman gain at $t = 0$ [see (5.41a)]. Suppose that (5.63) and (5.64) are valid up to time t. Then, from (4.67) and the definition of $\Lambda(t)$,

$$\begin{aligned} \Pi(t+1) - \Sigma(t+1) &= A(t)\Pi(t)A^{\mathrm{T}}(t) + Q(t) - A(t)\Sigma(t)A^{\mathrm{T}}(t) \\ &\quad - K(t)[\Lambda(t) - C(t)\Sigma(t)C^{\mathrm{T}}(t)]K^{\mathrm{T}}(t) \\ &= A(t)P(t)A^{\mathrm{T}}(t) + Q(t) \\ &\quad - K(t)[C(t)P(t)C^{\mathrm{T}}(t) + R(t)]K^{\mathrm{T}}(t) = P(t+1) \end{aligned}$$

This implies that (5.63) holds for time $t+1$. Further, we have

$$\begin{aligned} \bar{C}^{\mathrm{T}}(t+1) &- A(t+1)\Sigma(t+1)C^{\mathrm{T}}(t+1) \\ &= A(t+1)P(t+1)C^{\mathrm{T}}(t+1) + S(t+1) \\ \Lambda(t+1) &- C(t+1)\Sigma(t+1)C^{\mathrm{T}}(t+1) \\ &= C(t+1)P(t+1)C^{\mathrm{T}}(t+1) + R(t+1) \end{aligned}$$

so that (5.64) also holds for time $t+1$.

We show $\Sigma(t) = \mathrm{cov}\{\hat{x}(t)\}$. By the property of conditional expectation,

$$E\{\hat{x}(t)\} = E\{E\{x(t) \mid \mathcal{F}_{t-1}\}\} = E\{x(t)\} = \mu_x(t)$$

Hence, we have

$$x(t) - \mu_x(t) = \hat{x}(t) - \mu_x(t) + \tilde{x}(t)$$

where $\hat{x}(t) - \mu_x(t) \perp \tilde{x}(t)$. Computing the covariance matrix of the above equation yields

$$\Pi(t) = E\{[\hat{x}(t) - \mu_x(t)][\hat{x}(t) - \mu_x(t)]^{\mathrm{T}}\} + E\{\tilde{x}(t)\tilde{x}^{\mathrm{T}}(t)\}$$

where $E\{\tilde{x}(t)\tilde{x}^{\mathrm{T}}(t)\} = P(t)$, so that

$$E\{[\hat{x}(t) - \mu_x(t)][\hat{x}(t) - \mu_x(t)]^{\mathrm{T}}\} = \Sigma(t)$$

as was to be proved. □

It should be noted that the Riccati equation of (5.62) is defined by using only the covariance data of the output signal y [see (5.61)], so that no information about noise covariance matrices $Q(t)$, $S(t)$, $R(t)$ is used. Thus, if the statistical property of y is the same, even if the state space realizations are different, the Kalman gains are the same [146]. The Riccati equation (5.62) satisfied by the covariance matrix $\Sigma(t)$ of the predicted estimate plays an important role in stochastic realization theory to be developed in Chapter 7.

5.6 Stationary Kalman Filter

Consider the Kalman filter for the stochastic LTI system of (4.70). Since all the system parameters are time-invariant, it follows from (5.39a) in Theorem 5.1 that the Kalman filter is expressed as

$$\hat{x}(t+1) = A\hat{x}(t) + K(t)[y(t) - C\hat{x}(t)] \tag{5.65}$$

where $\hat{x}(t) := \hat{x}(t \mid t-1)$ with the initial condition $\hat{x}(0) = \mu_x(0)$, and where the Kalman gain is given by

$$K(t) = [AP(t)C^{\mathrm{T}} + S][CP(t)C^{\mathrm{T}} + R]^{-1}$$

Also, the error covariance matrix $P(t) := P(t \mid t-1)$ satisfies the Riccati equation

$$P(t+1) = AP(t)A^{\mathrm{T}} - K(t)[CP(t)C^{\mathrm{T}} + R]K^{\mathrm{T}}(t) + Q \tag{5.66}$$

with $P(0) = \Pi(0)$.

Suppose that a solution $P(t)$ of the Riccati equation (5.66) converges to a constant matrix as $t \to \infty$. Put $P(t) = P(t+1) = P$ in (5.66) to get an algebraic Riccati equation (ARE)

$$P = APA^{\mathrm{T}} - (APC^{\mathrm{T}} + S)(CPC^{\mathrm{T}} + R)^{-1}(APC^{\mathrm{T}} + S)^{\mathrm{T}} + Q \tag{5.67}$$

In this case, $K(t)$ converges to the stationary Kalman gain

$$K = (APC^{\mathrm{T}} + S)(CPC^{\mathrm{T}} + R)^{-1} \tag{5.68}$$

Hence, the filter equation (5.65) becomes

$$\hat{x}(t+1) = (A - KC)\hat{x}(t) + Ky(t) \tag{5.69}$$

This filter is called a stationary Kalman filter that produces the one-step predicted estimates of the state vector.

In the following, we define $\Phi := A - SR^{-1}C$ and $M := Q - SR^{-1}S^{\mathrm{T}}$. Then, it can be shown that the ARE of (5.67) reduces to

$$P = \Phi(P - PC^{\mathrm{T}}[CPC^{\mathrm{T}} + R]^{-1}CP)\Phi^{\mathrm{T}} + M \tag{5.70}$$

Theorem 5.4. *The following statements are equivalent.*

(i) The pair $(\Phi,\ M^{1/2})$ is stabilizable and $(C,\ \Phi)$ is detectable.

(ii) There exists a unique nonnegative definite solution P of the ARE (5.70); moreover, P is stabilizing, i.e., $\Phi - \Gamma C$ is stable, where

$$\Gamma = \Phi PC^{\mathrm{T}}(CPC^{\mathrm{T}} + R)^{-1}$$

Under the above condition (i), the solution $P(t)$ of the Riccati equation (5.66) for any $P(0) \geq 0$ converges to a unique nonnegative definite solution P.

Proof. For proof, see [11, 20, 97]. □

Example 5.5. Consider a scalar system

$$x(t+1) = ax(t) + w(t), \qquad y(t) = x(t) + v(t)$$

where $A = a$, $C = 1$, $Q = q > 0$, $S = 0$, $R = r > 0$. From (5.39a) ~ (5.42b) of Theorem 5.1, the Kalman filter and Riccati equation are given by

$$\hat{x}(t+1) = \frac{ar}{p(t) + r}\hat{x}(t) + \frac{ap(t)}{p(t) + r}y(t), \qquad \hat{x}(1) = 0 \tag{5.71}$$

$$p(t+1) = \frac{a^2 rp(t)}{p(t) + r} + q, \qquad p(1) = p_1 \tag{5.72}$$

Thus the ARE reduces to $p^2 + [(1 - a^2)r - q]p - qr = 0$, so that the ARE has two solutions

$$p_+ = \frac{1}{2}\left[(a^2 - 1)r + q + \sqrt{[(a^2 - 1)r + q]^2 + 4rq}\right] > 0$$

$$p_- = \frac{1}{2}\left[(a^2 - 1)r + q - \sqrt{[(a^2 - 1)r + q]^2 + 4rq}\right] < 0$$

Putting $a = 0.8$, $r = 1$, $q = 2$, we have $p_+ = 2.4547$ and $p_- = -0.8147$. In Figure 5.6, the solutions of (5.72) for ten random initial values $p_1 \sim \mathcal{N}(0, 4)$ are shown. In this case, all the solutions have converged to $p_+ = 2.4547$. □

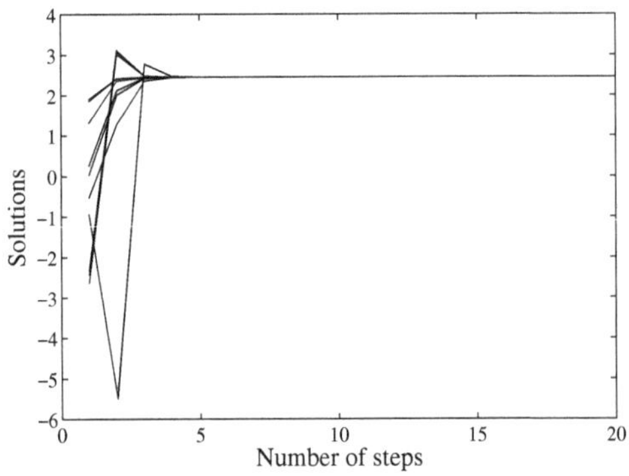

Figure 5.6. Solutions of the Riccati equation (5.72) for random initial values, where the initial time is taken as $t = 1$ for convenience

We see that the stationary Kalman filter is expressed as

$$\hat{x}(t+1) = A\hat{x}(t) + Ke(t) \tag{5.73a}$$

$$y(t) = C\hat{x}(t) + e(t) \tag{5.73b}$$

By means of Theorem 5.3, the stationary Kalman gain is also expressed as

$$K = (\bar{C}^{\mathrm{T}} - A\Sigma C^{\mathrm{T}})(\Lambda(0) - C\Sigma C^{\mathrm{T}})^{-1} \tag{5.74}$$

where the covariance matrix $\Sigma = \mathrm{cov}\{\hat{x}(t)\}$ satisfies the ARE

$$\Sigma = A\Sigma A^{\mathrm{T}} + (\bar{C}^{\mathrm{T}} - A\Sigma C^{\mathrm{T}})(\Lambda(0) - C\Sigma C^{\mathrm{T}})^{-1}(\bar{C} - C\Sigma A^{\mathrm{T}}) \tag{5.75}$$

The state space equation (5.73) is called a stationary forward innovation model for the stationary process y, where the noise model is less redundant than that of (4.70).

5.7 Stationary Backward Kalman Filter

In this section, we derive the Kalman filter for the backward Markov model for the stationary process y, which is useful for modeling stationary processes.

Consider the backward Markov model of (4.83), *i.e.*,

$$x_b(t-1) = A^{\mathrm{T}} x_b(t) + w_b(t) \tag{5.76a}$$

$$y(t) = \bar{C} x_b(t) + v_b(t) \tag{5.76b}$$

where w_b and v_b are white noises with covariance matrices

$$E\left\{ \begin{bmatrix} w_b(t) \\ v_b(t) \end{bmatrix} \begin{bmatrix} w_b^{\mathrm{T}}(s) & v_b^{\mathrm{T}}(s) \end{bmatrix} \right\} = \begin{bmatrix} \bar{Q} & \bar{S} \\ \bar{S}^{\mathrm{T}} & \bar{R} \end{bmatrix} \delta_{ts} \tag{5.77}$$

Moreover, we have $\mathrm{cov}\{x_b(t)\} = \bar{\Pi} = \Pi^{-1}$ and

$$\bar{Q} = \bar{\Pi} - A^{\mathrm{T}} \bar{\Pi} A, \quad \bar{S} = C^{\mathrm{T}} - A^{\mathrm{T}} \bar{\Pi} \bar{C}^{\mathrm{T}}, \quad \bar{R} = \Lambda(0) - \bar{C} \bar{\Pi} \bar{C}^{\mathrm{T}} \tag{5.78}$$

In order to derive the backward Kalman filter, we define the future space

$$\mathcal{Y}_t^+ := \overline{\mathrm{span}}\{y(t),\, y(t+1),\, \cdots\} \tag{5.79}$$

Since we deal with stationary processes with mean zero, no constant vectors are included in the right-hand side of (5.79), unlike the past space defined by (5.20). Then, the one-step backward predicted estimate is defined by

$$\hat{x}_b(t) = \hat{E}\{x_b(t) \mid \mathcal{Y}_{t+1}^+\} \tag{5.80}$$

Also, we define the backward innovation process by

$$e_b(t) = y(t) - \hat{E}\{y(t) \mid \mathcal{Y}_{t+1}^+\} \tag{5.81}$$

Lemma 5.13. *The backward process e_b is a backward white noise with mean zero and covariance matrix*

$$\mathrm{cov}\{e_b(t)\} = \Lambda(0) - \bar{C} \bar{\Sigma} \bar{C}^{\mathrm{T}} \tag{5.82}$$

where $\bar{\Sigma} = \mathrm{cov}\{\hat{x}_b(t)\}$.

Proof. By the definition of orthogonal projection, we see that $\hat{E}\{e_b(t) \mid \mathcal{Y}_{t+1}^+\} = 0$, $E\{e_b(t)\} = 0$. For $s < t$, it follows that $e_b(t) \in \mathcal{Y}_t^+ \subset \mathcal{Y}_{s+1}^+$, so that

$$\begin{aligned} E\{e_b(t) e_b^{\mathrm{T}}(s)\} &= E\{E\{e_b(t) e_b^{\mathrm{T}}(s) \mid \mathcal{Y}_{s+1}^+\}\} \\ &= E\{e_b(t) E\{e_b^{\mathrm{T}}(s) \mid \mathcal{Y}_{s+1}^+\}\} = 0 \end{aligned}$$

Similarly, one can show that the above equality holds for $s > t$, implying that $e_b(t)$ and $e_b(s)$ are uncorrelated for $s \neq t$. This shows that e_b is a zero mean white noise.

We compute the covariance matrix of e_b. It follows from (5.76b) and (5.81) that

$$\begin{aligned} e_b(t) &= y(t) - \hat{E}\{\bar{C}x_b(t) + v_b(t) \mid \mathcal{Y}_{t+1}\} \\ &= y(t) - \bar{C}\hat{x}_b(t) = \bar{C}[x_b(t) - \hat{x}_b(t)] + v_b(t) \end{aligned}$$

Hence, noting that $v_b(t)$ is uncorrelated with $x_b(t)$ and $\hat{x}_b(t) \in \mathcal{Y}_{t+1}$, we have

$$\begin{aligned} \mathrm{cov}\{e_b(t)\} &= \bar{C}E\{[x_b(t) - \hat{x}_b(t)][x_b(t) - \hat{x}_b(t)]^{\mathrm{T}}\}\bar{C}^{\mathrm{T}} + E\{v_b(t)v_b^{\mathrm{T}}(t)\} \\ &= \bar{C}E\{x_b(t)x_b^{\mathrm{T}}(t)\}\bar{C}^{\mathrm{T}} - \bar{C}E\{x_b(t)\hat{x}_b^{\mathrm{T}}(t)\}\bar{C}^{\mathrm{T}} \\ &\quad - \bar{C}E\{\hat{x}_b(t)x_b^{\mathrm{T}}(t)\}\bar{C}^{\mathrm{T}} + \bar{C}E\{\hat{x}_b(t)\hat{x}_b^{\mathrm{T}}(t)\}\bar{C}^{\mathrm{T}} + \bar{R} \end{aligned} \tag{5.83}$$

Since, from (5.80), $x_b(t) - \hat{x}_b(t) \perp \hat{x}_b(t)$, we have

$$E\{x_b(t)\hat{x}_b^{\mathrm{T}}(t)\} = E\{\hat{x}_b(t)\hat{x}_b^{\mathrm{T}}(t)\} = \bar{\Sigma} \tag{5.84}$$

Applying this relation to (5.83) yields

$$\begin{aligned} \mathrm{cov}\{e_b(t)\} &= \bar{C}\bar{\Pi}\bar{C}^{\mathrm{T}} - \bar{C}\bar{\Sigma}\bar{C}^{\mathrm{T}} + \Lambda(0) - \bar{C}\bar{\Pi}\bar{C}^{\mathrm{T}} \\ &= \Lambda(0) - \bar{C}\bar{\Sigma}\bar{C}^{\mathrm{T}} \end{aligned}$$

This completes the proof. □

The backward Kalman filter is given by the next theorem.

Theorem 5.5. *(Backward Kalman filter) The backward Kalman filter equations for the backward Markov model are given by (i) ~ (iv).*

(i) The filter equation

$$\hat{x}_b(t-1) = A^{\mathrm{T}}\hat{x}_b(t) + \bar{K}^{\mathrm{T}}[y(t) - \bar{C}\hat{x}(t)] \tag{5.85}$$

where $A^{\mathrm{T}} - \bar{K}^{\mathrm{T}}\bar{C}$ is stable.

(ii) The innovation process

$$e_b(t) = y(t) - \bar{C}\hat{x}(t) \tag{5.86}$$

(iii) The backward Kalman gain

$$\bar{K}^{\mathrm{T}} = (C^{\mathrm{T}} - A^{\mathrm{T}}\bar{\Sigma}\bar{C}^{\mathrm{T}})(\Lambda(0) - \bar{C}\bar{\Sigma}\bar{C}^{\mathrm{T}})^{-1} \tag{5.87}$$

(iv) The ARE satisfied by the covariance matrix of the backward predicted estimate $\hat{x}_b(t)$ is given by

$$\bar{\Sigma} = A^{\mathrm{T}}\bar{\Sigma}A + (C^{\mathrm{T}} - A^{\mathrm{T}}\bar{\Sigma}\bar{C}^{\mathrm{T}})(\Lambda(0) - \bar{C}\bar{\Sigma}\bar{C}^{\mathrm{T}})^{-1}(C - \bar{C}\bar{\Sigma}A) \tag{5.88}$$

This is the dual ARE of (5.75), satisfied by the covariance matrix $\Sigma = \mathrm{cov}\{\hat{x}(t)\}$.

Proof. It follows from (5.76a) and Lemma 5.13 that

$$\begin{aligned}\hat{E}\{x_b(t-1) \mid \mathcal{Y}_t\} &= \hat{E}\{x_b(t-1) \mid \mathcal{Y}_{t+1} \oplus e_b(t)\} \\ &= \hat{E}\{A^{\mathrm{T}} x_b(t) + w_b(t) \mid \mathcal{Y}_{t+1}\} + \hat{E}\{x_b(t-1) \mid e_b(t)\} \\ &= A^{\mathrm{T}} \hat{E}\{x_b(t) \mid \mathcal{Y}_{t+1}\} + \hat{E}\{x_b(t-1) \mid e_b(t)\}\end{aligned}$$

Thus from (5.80), we get (5.85), where the backward Kalman gain is determined by

$$\bar{K}^{\mathrm{T}} = \mathrm{cov}\{x_b(t-1)e_b^{\mathrm{T}}(t)\}(\mathrm{cov}\{e_b(t)\})^{-1}$$

It follows from (5.84) that

$$\begin{aligned}E\{x_b(t-1)e_b^{\mathrm{T}}(t)\} &= E\{[A^{\mathrm{T}} x_b(t) + w_b(t)][\bar{C}[x_b(t) - \hat{x}_b(t)] + v_b(t)]^{\mathrm{T}}\} \\ &= A^{\mathrm{T}} \bar{\Pi} \bar{C}^{\mathrm{T}} - A^{\mathrm{T}} \bar{\Sigma} \bar{C}^{\mathrm{T}} + \bar{S} \\ &= C^{\mathrm{T}} - A^{\mathrm{T}} \bar{\Sigma} \bar{C}^{\mathrm{T}}\end{aligned}$$

Thus the backward Kalman gain is given by (5.87). Finally, the dual ARE (5.88) can be easily derived by computing the covariance matrix of (5.85). □

In view of Theorem 5.5, the backward innovation model is given by

$$\hat{x}_b(t-1) = A^{\mathrm{T}} \hat{x}_b(t) + \bar{K}^{\mathrm{T}} e_b(t) \tag{5.89a}$$

$$y(t) = \bar{C} \hat{x}_b(t) + e_b(t) \tag{5.89b}$$

This should be compared with the forward innovation model of (5.73).

We are now in a position to summarize the different Markov models for a stationary process y, including forward and backward Markov models defined in Sections 4.7 and 4.8, and the forward and backward innovation models obtained in Sections 5.3 and 5.7 through the stationary Kalman filters.

Table 5.1. Schematic diagram of different Markov models

Forward model	Kalman filter	Forward innovation model
$(\Pi, A, C, \bar{C}, \Lambda(0))$	$\longrightarrow$	$(\Sigma, A, K, C, \Lambda(0))$
$\downarrow$		
$(\Pi^{-1}, A^{\mathrm{T}}, \bar{C}, C, \Lambda(0))$	$\longrightarrow$	$(\bar{\Sigma}, A^{\mathrm{T}}, \bar{K}^{\mathrm{T}}, \bar{C}, \Lambda(0))$
Backward model	Backward Kalman filter	Backward innovation model

Table 5.1 displays a schematic diagram of different Markov models for the same stationary process y with the covariance matrix $\Lambda_{yy}(l)$. From Lemmas 4.10 and 4.11, we see that (A, C, Q, S, R) determines $(\Pi, A, C, \bar{C}, \Lambda(0))$, and *vice versa*.

Hence, we say that the stationary forward model (4.70) with $t_0 \to -\infty$ is characterized by the quintuple $(\Pi, A, C, \bar{C}, \Lambda(0))$, where $\Lambda(0) := \Lambda_{yy}(0)$. Also, by the similar argument, we see that the backward Markov model (4.83) is specified by $(\Pi^{-1}, A^{\mathrm{T}}, \bar{C}, C, \Lambda(0))$. On the other hand, the forward innovation model of (5.73) is characterized by the quintuple $(\Sigma, A, K, C, \Lambda(0))$, and the backward innovation model by $(\bar{\Sigma}, A^{\mathrm{T}}, \bar{K}^{\mathrm{T}}, \bar{C}, \Lambda(0))$; however, note that $\bar{\Sigma} \neq \Sigma^{-1}$.

5.8 Numerical Solution of ARE

The stabilizing solution Σ of the ARE (5.75) can be obtained by using a solution of the generalized eigenvalue problem (GEP) associated with the ARE. Consider the ARE given by (5.75), *i.e.*

$$\Sigma = A\Sigma A^{\mathrm{T}} + (\bar{C}^{\mathrm{T}} - A\Sigma C^{\mathrm{T}})(\Lambda(0) - C\Sigma C^{\mathrm{T}})^{-1}(\bar{C} - C\Sigma A^{\mathrm{T}}) \qquad (5.90)$$

where the Kalman gain is given by (5.74).

Define $F := A - \bar{C}^{\mathrm{T}}\Lambda^{-1}(0)C$. Then, (5.90) is rewritten as (see Problem 5.7)

$$\Sigma = F\Sigma F^{\mathrm{T}} + F\Sigma C^{\mathrm{T}}(\Lambda(0) - C\Sigma C^{\mathrm{T}})^{-1}C\Sigma F^{\mathrm{T}} + \bar{C}^{\mathrm{T}}\Lambda^{-1}(0)\bar{C} \qquad (5.91)$$

Associated with (5.91), we define the GEP

$$\begin{bmatrix} F^{\mathrm{T}} & 0 \\ -\bar{C}^{\mathrm{T}}\Lambda^{-1}(0)\bar{C} & I_n \end{bmatrix} \begin{bmatrix} z_1 \\ z_2 \end{bmatrix} = \lambda \begin{bmatrix} I_n & -C^{\mathrm{T}}\Lambda^{-1}(0)C \\ 0 & F \end{bmatrix} \begin{bmatrix} z_1 \\ z_2 \end{bmatrix} \qquad (5.92)$$

Suppose that there are no eigenvalues on the unit circle ($|z| = 1$). Then, we can show that if $\lambda \neq 0$ is an eigenvalue, then the inverse $1/\lambda$ is also an eigenvalue (see Problem 5.8). Hence, (5.92) has $2n$ eigenvalues, and n of them are stable and other n are unstable.

Let $U = \begin{bmatrix} U_1 \\ U_2 \end{bmatrix} \in \mathbb{C}^{2n \times n}$ be the matrix formed by the eigenvectors corresponding to the n stable eigenvalues of (5.92). Thus, we have

$$\begin{bmatrix} F^{\mathrm{T}} & 0 \\ -\bar{C}^{\mathrm{T}}\Lambda^{-1}(0)\bar{C} & I_n \end{bmatrix} \begin{bmatrix} U_1 \\ U_2 \end{bmatrix} = \begin{bmatrix} I_n & -C^{\mathrm{T}}\Lambda^{-1}(0)C \\ 0 & F \end{bmatrix} \begin{bmatrix} U_1 \\ U_2 \end{bmatrix} J_0 \qquad (5.93)$$

where all the eigenvalues of $J_0 \in \mathbb{C}^{n \times n}$ are stable.

Lemma 5.14. *Suppose that the GEP of (5.92) has no eigenvalues on the unit circle. Also, suppose that* $\det(U_1) \neq 0$ *and that* $R(\Sigma) := \Lambda(0) - C\Sigma C^{\mathrm{T}} > 0$*. Then, the stabilizing solution of the ARE (5.90) is given by the formula*

$$\Sigma = U_2 U_1^{-1} \qquad (5.94)$$

Proof. [124] We show that $\Sigma = U_2 U_1^{-1}$ is a solution of the ARE of (5.91). From (5.93), we get

$$F^{\mathrm{T}}U_1 = U_1 J_0 - C^{\mathrm{T}} \Lambda^{-1}(0) C U_2 J_0$$
$$F U_2 J_0 = -\bar{C}^{\mathrm{T}} \Lambda^{-1}(0) \bar{C} U_1 + U_2$$

Post-multiplying the above equations by U_1^{-1} yields

$$F^{\mathrm{T}} = U_1 J_0 U_1^{-1} - C^{\mathrm{T}} \Lambda^{-1}(0) C U_2 J_0 U_1^{-1} \tag{5.95a}$$
$$U_2 U_1^{-1} = F U_2 J_0 U_1^{-1} + \bar{C}^{\mathrm{T}} \Lambda^{-1}(0) \bar{C} \tag{5.95b}$$

From (5.94) and (5.95b),

$$\begin{aligned}\mathrm{Ric}(\Sigma) :&= F\Sigma F^{\mathrm{T}} - \Sigma + F\Sigma C^{\mathrm{T}}(\Lambda(0) - C\Sigma C^{\mathrm{T}})^{-1} C\Sigma F^{\mathrm{T}} + \bar{C}^{\mathrm{T}} \Lambda^{-1}(0)\bar{C} \\ &= F\Sigma F^{\mathrm{T}} - F U_2 J_0 U_1^{-1} + F\Sigma C^{\mathrm{T}}(\Lambda(0) - C\Sigma C^{\mathrm{T}})^{-1} C\Sigma F^{\mathrm{T}}\end{aligned}$$

Also, from (5.95a), we have

$$\begin{aligned}\mathrm{Ric}(\Sigma) &= F U_2 U_1^{-1}[U_1 J_0 U_1^{-1} - C^{\mathrm{T}} \Lambda^{-1}(0) C U_2 J_0 U_1^{-1}] - F U_2 J_0 U_1^{-1} \\ &\quad + F U_2 U_1^{-1} C^{\mathrm{T}}(\Lambda(0) - C U_2 U_1^{-1} C^{\mathrm{T}})^{-1} C U_2 U_1^{-1} \\ &\qquad \times [U_1 J_0 U_1^{-1} - C^{\mathrm{T}} \Lambda^{-1}(0) C U_2 J_0 U_1^{-1}] \\ &= -F U_2 U_1^{-1} C^{\mathrm{T}} \Lambda^{-1}(0) C U_2 J_0 U_1^{-1} \\ &\quad + F U_2 U_1^{-1} C^{\mathrm{T}}(\Lambda(0) - C U_2 U_1^{-1} C^{\mathrm{T}})^{-1} C U_2 J_0 U_1^{-1} \\ &\quad - F U_2 U_1^{-1} C^{\mathrm{T}}(\Lambda(0) - C U_2 U_1^{-1} C^{\mathrm{T}})^{-1} \\ &\qquad \times C U_2 U_1^{-1} C^{\mathrm{T}} \Lambda^{-1}(0) C U_2 J_0 U_1^{-1}\end{aligned}$$

Define $\alpha := F U_2 U_1^{-1} C^{\mathrm{T}}$, $\beta := C U_2 J_0 U_1^{-1}$ and $\gamma := C U_2 U_1^{-1} C^{\mathrm{T}}$. Then, it follows that

$$\begin{aligned}\mathrm{Ric}(\Sigma) &= -\alpha \Lambda^{-1}(0)\beta + \alpha(\Lambda(0) - \gamma)^{-1}\beta - \alpha(\Lambda(0) - \gamma)^{-1}\gamma\Lambda^{-1}(0)\beta \\ &= -\alpha[\Lambda^{-1}(0) - (\Lambda(0) - \gamma)^{-1} + (\Lambda(0) - \gamma)^{-1}\gamma\Lambda^{-1}(0)]\beta \\ &= -\alpha(\Lambda(0) - \gamma)^{-1}[(\Lambda(0) - \gamma)\Lambda^{-1}(0) - I + \gamma\Lambda^{-1}(0)]\beta = 0\end{aligned}$$

as was to be proved.

Finally we show that the closed loop matrix $A_K := A - KC$ is stable. Recall that $K = (\bar{C}^{\mathrm{T}} - A\Sigma C^{\mathrm{T}})(\Lambda(0) - C\Sigma C^{\mathrm{T}})^{-1}$. Since $A = F + \bar{C}^{\mathrm{T}} \Lambda^{-1}(0) C$, we see that

$$\begin{aligned}A_K^{\mathrm{T}} &= F^{\mathrm{T}} + C^{\mathrm{T}} \Lambda^{-1}(0)\bar{C} \\ &\quad - C^{\mathrm{T}}(\Lambda(0) - C\Sigma C^{\mathrm{T}})^{-1}(\bar{C} - C\Sigma[F^{\mathrm{T}} + C^{\mathrm{T}} \Lambda^{-1}(0)\bar{C}]) \\ &= F^{\mathrm{T}} + C^{\mathrm{T}}(\Lambda(0) - C\Sigma C^{\mathrm{T}})^{-1} C\Sigma F^{\mathrm{T}} + C^{\mathrm{T}} \Lambda^{-1}(0)\bar{C} \\ &\quad - C^{\mathrm{T}}(\Lambda(0) - C\Sigma C^{\mathrm{T}})^{-1}(I - C\Sigma C^{\mathrm{T}} \Lambda^{-1}(0))\bar{C}\end{aligned}$$

It is easy to see that the second and third terms in the right-hand side of the above equation vanish, so that

$$A_K^{\mathrm{T}} = F^{\mathrm{T}} + C^{\mathrm{T}}(\Lambda(0) - C\Sigma C^{\mathrm{T}})^{-1}C\Sigma F^{\mathrm{T}} \tag{5.96}$$

Substituting F^{T} of (5.95a) into (5.96) yields

$$\begin{aligned}
A_K^{\mathrm{T}} &= F^{\mathrm{T}} + C^{\mathrm{T}}(\Lambda(0) - C\Sigma C^{\mathrm{T}})^{-1}CU_2U_1^{-1}F^{\mathrm{T}} \\
&= U_1J_0U_1^{-1} - C^{\mathrm{T}}\Lambda^{-1}(0)CU_2J_0U_1^{-1} + C^{\mathrm{T}}(\Lambda(0) - C\Sigma C^{\mathrm{T}})^{-1} \\
&\qquad\qquad \times CU_2U_1^{-1}[U_1J_0U_1^{-1} - C^{\mathrm{T}}\Lambda^{-1}(0)CU_2J_0U_1^{-1}] \\
&= U_1J_0U_1^{-1} + C^{\mathrm{T}}(\Lambda(0) - C\Sigma C^{\mathrm{T}})^{-1}CU_2J_0U_1^{-1} \\
&\qquad - C^{\mathrm{T}}(\Lambda(0) - C\Sigma C^{\mathrm{T}})^{-1}[(\Lambda(0) - C\Sigma C^{\mathrm{T}}) + C\Sigma C^{\mathrm{T}}] \\
&\qquad\qquad \times \Lambda^{-1}(0)CU_2J_0U_1^{-1} \\
&= U_1J_0U_1^{-1}
\end{aligned}$$

Thus the eigenvalues of A_K are equal to those of J_0. This completes the proof. □

5.9 Notes and References

- There is a vast literature on the Kalman filter; but readers should consult basic papers due to Kalman [81], Kalman and Bucy [84], and then proceed to a survey paper [78], books [11, 23, 66, 79], and a recent monograph for adaptive filtering [139], *etc.*
- Section 5.1 reviews a multivariate Gaussian probability density function based on [14]; see also books [79, 111, 136] for the least-squares estimation (minimum variance estimation). The state estimation problem for a Markov model is defined in Section 5.2, and the Kalman filter algorithm is derived in Section 5.3 based on the technique of orthogonal projection; see [11, 23, 66]. Also, in Section 5.4, the Kalman filter in the presence of external inputs is developed by extending the result of [182] to a discrete-time system.
- Section 5.5 derives the Riccati equation satisfied by the covariance matrix $\Sigma(t)$ of the one-step predicted estimate, which is a companion Riccati equation satisfied by the error covariance matrix. It should be noted that the Riccati equation for $\Sigma(t)$ is defined by using only the covariance data for the output process y. Thus, if the covariance information of y is the same, the Kalman gain is the same even if state space realizations are different [11, 146]. This fact is called invariance of the Kalman filter with respect to the signal model.
- The stationary Kalman filter and the associated ARE are derived in Section 5.6. The existence of a stabilizing solution of the discrete-time ARE has been discussed. For proofs, see [97, 117] and monographs [20, 99].

- In Section 5.7, the backward Kalman filter is introduced based on a backward Markov model for a stationary process, and relation among four different Markov models for a stationary process is briefly discussed. These Markov models will play important roles in stochastic realization problems to be studied in Chapters 7 and 8.
- Section 5.8 is devoted to a direct solution method of the ARE (5.75) due to [103, 124], in which numerical methods for the Kalman filter ARE of (5.70) are developed in terms of the solution of GEP. See also a monograph [125] in which various numerical algorithms arising in control area are included.

5.10 Problems

5.1 Suppose that the probability density function of (x, y) is given by the Gaussian density function

$$p(x,y) = \frac{1}{2\pi(1-\rho^2)^{1/2}\sigma_x\sigma_y} \times e^{-\frac{1}{2(1-\rho^2)}\left[\frac{(x-\mu_x)^2}{\sigma_x^2} - \frac{2\rho(x-\mu_x)(y-\mu_y)}{\sigma_x\sigma_y} + \frac{(y-\mu_y)^2}{\sigma_y^2}\right]}$$

where $\sigma_x,\ \sigma_y > 0$ and $|\rho| < 1$. Show that the following relations hold:

$$E\{x \mid y\} = \mu_x + \rho\frac{\sigma_x}{\sigma_y}(y - \mu_y), \quad E\{(x - E\{x \mid y\})^2\} = \sigma_x^2(1-\rho^2)$$

5.2 Let $K_\alpha(t)$ be the Kalman gain for the covariance matrices $\alpha Q(t)$, $\alpha S(t)$, $\alpha R(t)$, $\alpha P(0)$ with $\alpha > 0$. By using (5.41a) and (5.42a), show that $K_\alpha(t)$ is the same as $K(t)$ of (5.41a).

5.3 Define the state covariance matrix $\varPi(t) = E\{[x(t) - \mu_x(t)][x(t) - \mu_x(t)]^{\mathrm{T}}\}$. Show that the following inequalities hold:

$$\varPi(t) \geq P(t) \geq P(t \mid t) \geq 0, \qquad \varPi(t) \geq \varSigma(t)$$

5.4 Consider an AR model

$$y(t) = a_1 y(t-1) + \cdots + a_n y(t-n) + w(t)$$

Then a state space model for y is given by

$$x(t+1) = \begin{bmatrix} 0 & \cdots & 0 & a_n \\ 1 & & & a_{n-1} \\ & \ddots & & \vdots \\ & & 1 & a_1 \end{bmatrix} x(t) + \begin{bmatrix} a_n \\ a_{n-1} \\ \vdots \\ a_1 \end{bmatrix} w(t) \tag{5.97a}$$

$$y(t) = [0 \cdots 0\ 1]x(t) + w(t) \tag{5.97b}$$

Derive the Kalman filter for the above state space model, and show that the Kalman filter is the same as the state space model. (Hint: The state variables of (5.97a) are expressed as

$$x_1(t) = a_n y(t-1), \quad x_2(t) = x_1(t-1) + a_{n-1}y(t-1), \cdots,$$
$$x_n(t) = x_{n-1}(t-1) + a_1 y(t-1)$$

Thus we see that the state vector $x(t) := (x_1(t), \cdots, x_n(t))^{\mathrm{T}}$, $t > n$ can be determined from $y(k)$, $k = t-1, \cdots, t-n$, so that we have $P(t) = 0$ for $t > n$, and hence $\Pi(t) = \Sigma(t)$. It also follows from $Q(t) = R(t) = q$ and $S(t) = Bq$ that $K(t) = B$ for $t > n$.)

5.5 By using (5.44) with $A(t) = A$, $C(t) = C$, show that

$$y(t) = e(t) + CK(t-1)e(t-1) + CAK(t-2)e(t-2)$$
$$+ \cdots + CA^{t-1}K(0)e(0) + CA^t\hat{x}(0)$$

and that

$$\begin{bmatrix} y(0) \\ y(1) \\ \vdots \\ y(t-1) \end{bmatrix} = \begin{bmatrix} C \\ CA \\ \vdots \\ CA^{t-1} \end{bmatrix} \hat{x}(0) + \begin{bmatrix} I & & & \\ CK(0) & I & & \\ \vdots & \ddots & \ddots & \\ CA^{t-2}K(0) & \cdots & CK(t-2) & I \end{bmatrix} \begin{bmatrix} e(0) \\ e(1) \\ \vdots \\ e(t-1) \end{bmatrix}$$

This is useful for giving a triangular factorization of the covariance matrix of the stacked output vector.

5.6 In Section 5.6, we defined $\Phi = A - SR^{-1}C$ and $\Gamma = \Phi PC^{\mathrm{T}}(CPC^{\mathrm{T}} + R)^{-1}$. Show that $\Phi - \Gamma C = A - KC$ holds. Also, derive (5.70) from (5.67).

5.7 Derive (5.91) from (5.90).

5.8 Consider the GEP of (5.92):

$$Nz = \lambda Lz, \quad z \in \mathbb{C}^{2n}, \quad \lambda \in \mathbb{C}$$

Let $\hat{J} = \begin{bmatrix} 0 & I_n \\ -I_n & 0 \end{bmatrix}$. Show that $L\hat{J}L^{\mathrm{T}} = N\hat{J}N^{\mathrm{T}}$ holds. By using this fact, show that if $\lambda \neq 0$ is an eigenvalue of (5.92), so is $1/\lambda$.

Part II

Realization Theory

6

Realization of Deterministic Systems

We introduce the basic idea of deterministic subspace identification methods for a discrete-time LTI system from the point of view of classical realization theory. We first present the realization algorithm due to Ho-Kalman [72] based on the SVD of the block Hankel matrix formed by impulse responses. Then we define a data matrix generated by the observed input-output data for the system, and explain the relation between the data matrix and the block Hankel matrix by means of zero-input responses. Based on the LQ decomposition of data matrices, we develop two subspace identification methods, *i.e.*, the MOESP method [172] and N4SID method [164, 165]. Finally, we consider the effect of additive white noise on the SVD of a wide rectangular matrix. Some numerical results are included.

6.1 Realization Problems

Consider a discrete-time LTI system described by

$$x(t+1) = Ax(t) + Bu(t) \tag{6.1a}$$

$$y(t) = Cx(t) + Du(t), \qquad t = 0, 1, \cdots \tag{6.1b}$$

where $x \in \mathbb{R}^n$ is the state vector, $u \in \mathbb{R}^m$ the control input, $y \in \mathbb{R}^p$ the output vector, and $A \in \mathbb{R}^{n\times n}$, $B \in \mathbb{R}^{n\times m}$, $C \in \mathbb{R}^{p\times n}$, $D \in \mathbb{R}^{p\times m}$ are constant matrices. In the following, we assume that $(A,\ B)$ is reachable and $(C,\ A)$ is observable; in this case, we say that (A, B, C) is minimal.

From (6.1), the transfer matrix and the impulse response matrices of the system are respectively given by

$$G(z) = D + C(zI - A)^{-1}B \tag{6.2}$$

and

$$G_t = \begin{cases} D, & t = 0 \\ CA^{t-1}B, & t = 1, 2, \cdots \end{cases} \tag{6.3}$$

where $\{G_t, t = 0, 1, \cdots\}$ are also called the Markov parameters. We see that given (A, B, C, D), the transfer matrix and impulse response matrices can uniquely be determined by (6.2) and (6.3), respectively (see also Section 3.4).

This chapter considers the inverse problems called realization problems [72].

Problem A Suppose that a sequence of impulse responses $\{G_t, t = 0, 1, \cdots\}$, or a transfer matrix $G(z)$, of a discrete-time LTI system is given. The realization problem is to find the dimension n and the system matrices (A, B, C, D), up to similarity transforms.

Problem B Suppose that input-output data $\{u(t), y(t), t = 0, 1, \cdots, N-1\}$ are given. The problem is to identify the dimension n and the system matrices (A, B, C, D), up to similarity transforms. This is exactly a subspace identification problem for a deterministic LTI system.

6.2 Ho-Kalman's Method

In this section, we present the realization method of Ho-Kalman based on the results stated in Section 3.9, providing a complete solution to Problem A. Let the impulse response of the system be given by $(G_0, G_1, G_2, \cdots)$. Then, since $D = G_0$, we must identify three matrices (A, B, C).

Consider the input u that assumes non-zero values up to time $t = -1$ and zero for $t = 0, 1, \cdots$, *i.e.*,

$$u = (\cdots, u(-3), u(-2), u(-1), 0, 0, 0, \cdots) \tag{6.4}$$

Applying this input to a discrete-time LTI system, we observe the output for $t = 0, 1, \cdots$ as shown in Figure 6.1. For the input sequence of (6.4), the output is expressed as

$$y(t) = \sum_{i=-\infty}^{-1} G_{t-i}u(i), \qquad t = 0, 1, \cdots \tag{6.5}$$

This is a zero-input response with the initial state $x(0)$, which is determined by the past inputs. It should be noted that the responses $y(t)$ for $t = -1, -2, \cdots$ are shown by dotted lines.

We define the block Hankel operator with infinite dimension as

$$H = \begin{bmatrix} G_1 & G_2 & G_3 & G_4 & \cdots \\ G_2 & G_3 & G_4 & G_5 & \cdots \\ G_3 & G_4 & G_5 & G_6 & \cdots \\ G_4 & G_5 & G_6 & G_7 & \cdots \\ \vdots & \vdots & \vdots & \vdots & \ddots \end{bmatrix} \tag{6.6}$$

Then the input-output relation is expressed as

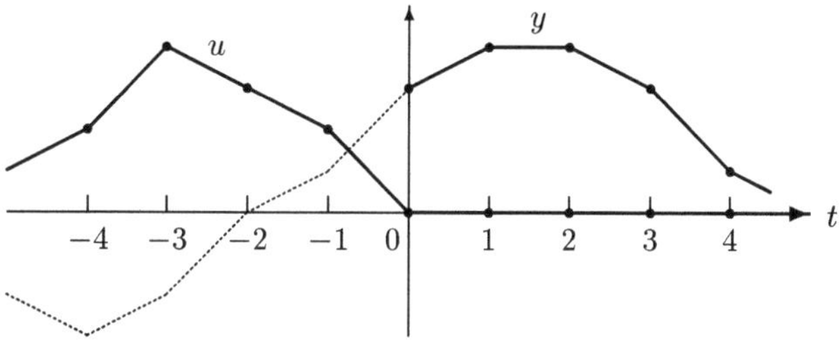

Figure 6.1. Zero-input response of an LTI system

$$\boldsymbol{y}_+ = H\boldsymbol{u}_- \tag{6.7}$$

where $\boldsymbol{y}_+$ and $\boldsymbol{u}_-$ are infinite dimensional vectors defined by

$$\boldsymbol{y}_+ = \begin{bmatrix} y(0) \\ y(1) \\ \vdots \end{bmatrix}, \qquad \boldsymbol{u}_- = \begin{bmatrix} u(-1) \\ u(-2) \\ \vdots \end{bmatrix}$$

Moreover, the observability and reachability operators are defined by

$$\mathcal{O} = \begin{bmatrix} C \\ CA \\ CA^2 \\ \vdots \end{bmatrix}, \qquad \mathcal{C} = \begin{bmatrix} B & AB & A^2B & \cdots \end{bmatrix}$$

We present the basic theorem for the properties of block Hankel matrix, which plays an important role in the later developments.

Theorem 6.1. *(Properties of Hankel matrix) Suppose that (A, B, C) is minimal. Then, the following (i) ~ (iv) hold.*

(i) The block Hankel matrix H of (6.6) has finite rank if and only if the impulse response has a factorization like (6.3).

(ii) The block Hankel matrix has rank n, i.e., $\mathrm{rank}(H) = n$. Moreover, H has the factorization of the form

$$H = \mathcal{O}\mathcal{C} = \mathcal{O}TT^{-1}\mathcal{C}, \qquad |T| \neq 0$$

(iii) Let the state vector at $t = 0$ be given by $x(0) = \mathcal{C}\boldsymbol{u}_-$. Then (6.7) is written as

$$\boldsymbol{y}_+ = \mathcal{O}x(0) \tag{6.8}$$

(iv) The block Hankel matrix is shift invariant, i.e.,

$$H^{\uparrow} = \mathcal{O}^{\uparrow}\mathcal{C} = \mathcal{O}A \cdot \mathcal{C} = \mathcal{O} \cdot A\mathcal{C} = \mathcal{O}\mathcal{C}^{\leftarrow} = H^{\leftarrow}$$

where $(\cdot)^{\uparrow}$ denotes the upward shift that removes the first block row, and $(\cdot)^{\leftarrow}$ the left shift that removes the first block column.

Proof. Item (i) follows from Theorem 3.13, and items (ii), (iii), (iv) are obvious from the definition. See also [162]. □

Item (iv) has the following physical meaning. From (6.7), we see that $\mathrm{Im}(H)$ contains all the outputs after $t = 0$ due to the past inputs up to $t = -1$. Define

$$\boldsymbol{y}_+^{\uparrow} = \begin{bmatrix} y(1) \\ y(2) \\ \vdots \end{bmatrix}$$

Then, we have

$$\boldsymbol{y}_+^{\uparrow} = H^{\uparrow}\boldsymbol{u}_- \tag{6.9}$$

Hence, it follows from (6.9) that $\mathrm{Im}(H^{\uparrow})$ contains all possible outputs after $t = 1$ due to the past inputs up to $t = -1$. Since the system is time-invariant, this is equivalent to saying that $\mathrm{Im}(H^{\uparrow})$ contains the output after $t = 0$ due to the past input up to $t = -2$. Since the set of all inputs up to $t = -2$ is a subspace of the space of all the past inputs, we see that all resulting outputs $\mathrm{Im}(H^{\uparrow})$ should be included in $\mathrm{Im}(H)$.

The above properties of the block Hankel operator are extensively used for deriving realization algorithms. In fact, the celebrated Ho-Kalman algorithm described below is entirely based on Theorem 6.1.

For the actual computation using finite number of impulse response matrices, however, we must use the truncated block Hankel matrix of the form

$$H_{k,l} = \begin{bmatrix} G_1 & G_2 & G_3 & \cdots & G_l \\ G_2 & G_3 & G_4 & \cdots & G_{l+1} \\ G_3 & G_4 & G_5 & \cdots & G_{l+2} \\ \vdots & \vdots & \vdots & \ddots & \vdots \\ G_k & G_{k+1} & G_{k+2} & \cdots & G_{k+l-1} \end{bmatrix} \in \mathbb{R}^{kp \times lm} \tag{6.10}$$

Also, the extended observability matrix $\mathcal{O}_k$ and the extended reachability matrix $\mathcal{C}_l$ are defined by

$$\mathcal{O}_k = \begin{bmatrix} C \\ CA \\ \vdots \\ CA^{k-1} \end{bmatrix}, \qquad \mathcal{C}_l = \begin{bmatrix} B & AB & A^2B & \cdots & A^{l-1}B \end{bmatrix}$$

where k and l should be greater than n[1]. Usually, we take $n < k \le l$.

In the finite dimensional case, if $\mathrm{rank}(H_{k,l}) = n$, we see from Theorem 6.1 (ii) that

$$H_{k,l} = \mathcal{O}_k\mathcal{C}_l = \mathcal{O}_kTT^{-1}\mathcal{C}_l, \qquad |T| \ne 0 \tag{6.11}$$

---

[1] In practice, the dimension n is not known. Since it is impossible to find an upper bound of n by a finite procedure, it is necessary to assume an upper bound *a priori*.

where $\operatorname{rank}(\mathcal{O}_k) = n$, $\operatorname{rank}(\mathcal{C}_l) = n$. Also, concerning the extended observability matrix, we have the following identity

$$\begin{bmatrix} C \\ CA \\ \vdots \\ CA^{k-2} \end{bmatrix} A = \begin{bmatrix} CA \\ CA^2 \\ \vdots \\ CA^{k-1} \end{bmatrix} \quad \Rightarrow \quad \mathcal{O}_{k-1}A = \mathcal{O}_k(p+1:kp,:) \tag{6.12}$$

To get a unique least-squares solution of A from (6.12), we see that $\mathcal{O}_{k-1}$ should be full column rank, so that $p(k-1) \geq n$. Thus, for a single output case ($p = 1$), the relation turns out to be $k \geq n+1$, so that k should be strictly greater than n.

Similarly, from the extended reachability matrix, we have

$$A\mathcal{C}_{l-1} = \mathcal{C}_l(:, m+1 : lm) \tag{6.13}$$

Lemma 6.1. *(Deterministic realization algorithm [72, 184])*

Step 1: Compute the SVD of $H_{k,l}$ as

$$H_{k,l} = [U_s \ \ U_n] \begin{bmatrix} \Sigma_s & 0 \\ 0 & 0 \end{bmatrix} \begin{bmatrix} V_s^{\mathrm{T}} \\ V_n^{\mathrm{T}} \end{bmatrix} = U_s \Sigma_s V_s^{\mathrm{T}} \tag{6.14}$$

where Σ_s is a diagonal matrix with the first n non-zero singular values of $H_{k,l}$, so that we have

$$\sigma_1 \geq \sigma_2 \geq \cdots \geq \sigma_n > 0 = \sigma_{n+1} = \sigma_{n+2} = \cdots$$

Step 2: Compute the extended observability and reachability matrices as

$$\mathcal{O}_k = U_s \Sigma_s^{1/2} T, \qquad \mathcal{C}_l = T^{-1} \Sigma_s^{1/2} V_s^{\mathrm{T}} \tag{6.15}$$

where $T \in \mathbb{R}^{n \times n}$ is an arbitrary nonsingular matrix.

Step 3: Compute the matrices A, B, C as

$$A = \mathcal{O}_{k-1}^{\dagger} \overline{\mathcal{O}}_k, \quad B = \mathcal{C}_l(1:n, 1:m), \quad C = \mathcal{O}_k(1:p, 1:n) \tag{6.16}$$

where $\overline{\mathcal{O}}_k = \mathcal{O}_k(p+1:kp, 1:n)$ $(= \mathcal{O}_k^{\uparrow})$. □

For computation of A, the identity of (6.12) is used. It follows from (6.13) that the matrix A in *Step 3* is also given by

$$A = \mathcal{C}_l^{\leftarrow} \mathcal{C}_{l-1}^{\dagger} \tag{6.17}$$

Example 6.1. Suppose that for the impulse input $u = (1,\ 0,\ 0, \cdots)$, we observe the outputs

$$y = (0,\ 1,\ 1,\ 2,\ 3,\ 5,\ 8,\ 13,\ 21,\ 34,\ \cdots)$$

This output sequence is the well-known Fibonacci sequence generated by

$$G_t = G_{t-1} + G_{t-2}, \qquad t = 2, 3, \cdots$$

with the initial conditions $G_0 = 0$, $G_1 = 1$. A realization of this impulse response sequence is given, *e.g.*, by [143]

$$A = \begin{bmatrix} 0 & 1 \\ 1 & 1 \end{bmatrix}, \qquad B = \begin{bmatrix} 1 \\ 1 \end{bmatrix}, \qquad C = [1 \ \ 0], \qquad D = 0 \tag{6.18}$$

and the transfer function is given by

$$G(z) = \frac{z}{z^2 - z - 1} \tag{6.19}$$

Now we use the algorithm of Lemma 6.1 to compute a realization. Recall that it is necessary to take the number of rows k should be greater than n. Thus, taking $k = l = 5$, we have the following Hankel matrix

$$H_{5,5} = \begin{bmatrix} 1 & 1 & 2 & 3 & 5 \\ 1 & 2 & 3 & 5 & 8 \\ 2 & 3 & 5 & 8 & 13 \\ 3 & 5 & 8 & 13 & 21 \\ 5 & 8 & 13 & 21 & 34 \end{bmatrix} \in \mathbb{R}^{5\times 5}$$

By the SVD of $H_{5,5}$, we get

$$\sigma_1 = 54.5601, \qquad \sigma_2 = 0.4399, \qquad \sigma_i = 0, \quad i \geq 3$$

so that we have $n = 2$. By putting $T = I_5$,

$$A = \begin{bmatrix} 1.6179 & 0.0185 \\ 0.0185 & -0.6179 \end{bmatrix}, \quad B = \begin{bmatrix} 0.8550 \\ -0.5187 \end{bmatrix}, \quad C = [0.8550 \ \ -0.5187]$$

where the transfer function $G(z) = (A, B, C)$ is also given by (6.19). □

We see that the matrices obtained above satisfy an interesting property that $A = A^{\mathrm{T}}$ and $B = C^{\mathrm{T}}$. If we use an asymmetric Hankel matrix, say $H_{4,6}$, then this property does not hold; however, the correct transfer function is obtained as long as $k,\ l > 2$.

Lemma 6.2. *Suppose that $H_{k,l}$ is symmetric in Lemma 6.1 and that $T = I_n$ in Step 2. If all the elements of Σ are different, i.e., $\sigma_1 > \sigma_2 > \cdots > \sigma_n$, there exists a matrix $S = \mathrm{diag}(\pm 1, \cdots, \pm 1)$ such that $A = SA^{\mathrm{T}}S$ and $C = B^{\mathrm{T}}S$ hold.*

Proof. The fact that $H_{k,l}$ is symmetric implies that $k = l, m = p$ and $G_t^{\mathrm{T}} = G_t$, $t = 1, 2, \cdots$. Let $H := H_{k,k}$ in (6.14). Since $H^{\mathrm{T}} = H$, it follows that $H = U\Sigma V^{\mathrm{T}} = V\Sigma U^{\mathrm{T}}$, so that $\mathrm{Im}(H) = \mathrm{Im}(U) = \mathrm{Im}(V)$. Thus there exists a nonsingular matrix $S \in \mathbb{R}^{n\times n}$ such that $U = VS$. Since $I_n = U^{\mathrm{T}}U = S^{\mathrm{T}}V^{\mathrm{T}}VS = S^{\mathrm{T}}S$, we see that $S = V^{\mathrm{T}}U \in \mathbb{R}^{n\times n}$ is an orthogonal matrix. Let

$$S = \begin{bmatrix} S_{n-1} & a \\ b^{\mathrm{T}} & c \end{bmatrix}, \qquad S_{n-1} \in \mathbb{R}^{(n-1)\times(n-1)} \tag{6.20}$$

where $a,\ b \in \mathbb{R}^{n-1}$ and $c \in \mathbb{R}$. Comparing the $(2,2)$-block elements of the identities $S^{\mathrm{T}}S = I_n$ and $SS^{\mathrm{T}} = I_n$, we see from (6.20) that

$$\|a\|^2 + c^2 = \|b\|^2 + c^2 = 1 \quad \Rightarrow \quad \|a\| = \|b\|$$

Also, $U\Sigma V^{\mathrm{T}} = V\Sigma U^{\mathrm{T}}$ implies $\Sigma S^{\mathrm{T}} = S\Sigma$, so that from (6.20),

$$\Sigma_{n-1}S_{n-1}^{\mathrm{T}} = S_{n-1}\Sigma_{n-1}, \qquad \Sigma_{n-1}b = a\sigma_n$$

From the second relation in the above equation,

$$a = \frac{1}{\sigma_n}\Sigma_{n-1}b = \begin{bmatrix} \alpha_1 & & & \\ & \alpha_2 & & \\ & & \ddots & \\ & & & \alpha_{n-1} \end{bmatrix} b$$

where $\alpha_i = \sigma_i/\sigma_n,\ i = 1, \cdots, n-1$. Since $\|a\| = \|b\|$, we have

$$a_1^2 + \cdots + a_{n-1}^2 = \alpha_1^2 b_1^2 + \cdots + \alpha_{n-1}^2 b_{n-1}^2 = b_1^2 + \cdots + b_{n-1}^2$$

so that $(\alpha_1^2 - 1)b_1^2 + \cdots + (\alpha_{n-1}^2 - 1)b_{n-1}^2 = 0$. But, since $\alpha_i > 1,\ i = 1, \cdots, n-1$, we have $b_i = 0,\ i = 1, \cdots, n-1$, implying that $a = b = 0$ and $c^2 = 1$. By means of these relations, we have $S = \begin{bmatrix} S_{n-1} & 0 \\ 0 & \pm 1 \end{bmatrix}$ and

$$\Sigma_{n-1}S_{n-1}^{\mathrm{T}} = S_{n-1}\Sigma_{n-1}, \quad S_{n-1}S_{n-1}^{\mathrm{T}} = I_{n-1}, \quad S_{n-1}^{\mathrm{T}}S_{n-1} = I_{n-1}$$

Applying the above procedure to $S_{n-1} \in \mathbb{R}^{(n-1)\times(n-1)}$, we can inductively prove that $S = \mathrm{diag}(\pm 1, \cdots, \pm 1)$, a signature matrix.

Since $T = I_n$ in (6.15), it follows that

$$\mathcal{O}_k = U\Sigma^{1/2}, \qquad \mathcal{C}_k = \Sigma^{1/2}V^{\mathrm{T}}$$

Also, S and $\Sigma^{1/2}$ are diagonal, so that $S\Sigma^{1/2} = \Sigma^{1/2}S$. Thus, by using $V = US$,

$$\mathcal{C}_k = \Sigma^{1/2}V^{\mathrm{T}} = \Sigma^{1/2}SU^{\mathrm{T}} = S\Sigma^{1/2}U^{\mathrm{T}} = S\mathcal{O}_k^{\mathrm{T}}$$

Hence we get

$$B = \mathcal{C}_k(:, 1:p) = S\mathcal{O}_k^{\mathrm{T}}(1:p, :) = SC^{\mathrm{T}}$$

and also

$$\mathcal{C}_k^{\leftarrow} = S(\mathcal{O}_k^{\uparrow})^{\mathrm{T}}, \qquad \mathcal{C}_{k-1}^{\dagger} = (\mathcal{O}_{k-1}^{\mathrm{T}})^{\dagger}S$$

Thus from (6.13), we have

$$A = \mathcal{C}_k^{\leftarrow}\mathcal{C}_{k-1}^{\dagger} = S(\mathcal{O}_k^{\uparrow})^{\mathrm{T}}(\mathcal{O}_{k-1}^{\dagger})^{\mathrm{T}}S = S(\mathcal{O}_{k-1}^{\dagger}\mathcal{O}_k^{\uparrow})^{\mathrm{T}}S = SA^{\mathrm{T}}S$$

as was to be proved. □

It should be noted that $(A,\ B,\ C)$ derived in Lemma 6.2 is not balanced (see Problem 6.2).

Example 6.2. Consider a scalar transfer function

$$G(z) = \frac{\beta_1 z^2 + \beta_2 z + \beta_3}{z^3 + \alpha_1 z^2 + \alpha_2 z + \alpha_3} = g_1 z^{-1} + g_2 z^{-2} + \cdots$$

where it is assumed that the transfer function is coprime and that the dimension is *a priori* known ($n = 3$). Suppose that we are given an impulse response sequence $(g_1,\ g_2,\ \cdots)$. Since $\operatorname{rank}(H) = 3$, there exists a vector $\xi^{\mathrm{T}} = [a_3\ a_2\ a_1\ 1]$ such that

$$H_{3,4}\xi = \begin{bmatrix} g_1 & g_2 & g_3 & g_4 \\ g_2 & g_3 & g_4 & g_5 \\ g_3 & g_4 & g_5 & g_6 \end{bmatrix} \begin{bmatrix} a_3 \\ a_2 \\ a_1 \\ 1 \end{bmatrix} = \begin{bmatrix} 0 \\ 0 \\ 0 \end{bmatrix}$$

so that $\xi \in \operatorname{Ker}(H_{3,4})$. Let the SVD of $H_{3,4}$ be given by

$$H_{3,4} = U \begin{bmatrix} \sigma_1 & 0 & 0 & 0 \\ 0 & \sigma_2 & 0 & 0 \\ 0 & 0 & \sigma_3 & 0 \end{bmatrix} \begin{bmatrix} v_1^{\mathrm{T}} \\ v_2^{\mathrm{T}} \\ v_3^{\mathrm{T}} \\ v_4^{\mathrm{T}} \end{bmatrix}, \qquad \sigma_1 \geq \sigma_2 \geq \sigma_3 > 0$$

Hence we have

$$H_{3,4} v_4 = 0 \quad \Rightarrow \quad v_4 \in \operatorname{Ker}(H_{3,4})$$

Since both ξ and v_4 belong to the one-dimensional subspace $\operatorname{Ker}(H_{3,4})$, we get ξ by normalizing v_4 so that $v_4(4) = 1$.

In view of (6.11), we have

$$H_{3,4} = \begin{bmatrix} c \\ cA \\ cA^2 \end{bmatrix} \begin{bmatrix} b & Ab & A^2 b & A^3 b \end{bmatrix}$$

Since $(c,\ A)$ is observable, it follows from $H_{3,4}\xi = 0$ that

$$(A^3 + a_1 A^2 + a_2 A + a_3 I)b = 0$$

Pre-multiplying the above equation by A and A^2, and re-arranging the terms yield

$$(A^3 + a_1 A^2 + a_2 A + a_3 I)[b\ \ Ab\ \ A^2 b] = 0$$

By the reachability of $(A,\ b)$, we have $\det[b\ \ Ab\ \ A^2 b] \neq 0$, and hence

$$A^3 + a_1 A^2 + a_2 A + a_3 I = 0$$

Since $G(z)$ is coprime, the characteristic polynomial of A coincides with the denominator polynomial of $G(z)$, so that $\alpha_i = a_i,\ i = 1,\ 2,\ 3$. Moreover, from the identity

$$\begin{bmatrix} \beta_1 \\ \beta_2 \\ \beta_3 \end{bmatrix} = \begin{bmatrix} g_1 & 0 & 0 \\ g_2 & g_1 & 0 \\ g_3 & g_2 & g_1 \end{bmatrix} \begin{bmatrix} 1 \\ \alpha_1 \\ \alpha_2 \end{bmatrix}$$

we get the coefficients of the numerator polynomial. □

We see from above examples that state space models and transfer functions of LTI systems can be obtained by utilizing the SVD of block Hankel matrices formed by the impulse response matrices.

6.3 Data Matrices

Consider an discrete-time LTI system, for which we assume that the system is at rest for $t < 0$, *i.e.*, $u(t) = 0,\ y(t) = 0,\ t = -1, -2, \cdots$. Suppose that the input-output data $u = (u(0), u(1), \cdots u(N-1))$ and $y = (y(0), y(1), \cdots y(N-1))$ are given, where N is sufficiently large. Then, for $k > 0$, we get

$$\begin{aligned} &[y(0)\ y(1)\ \cdots\ y(N-1)] \\ &= [g_{k-1}\ \cdots\ g_0] \begin{bmatrix} 0 & \cdots & 0 & u(0) & \cdots & u(N-k) \\ \vdots & \iddots & \iddots & u(1) & \cdots & u(N-k+1) \\ 0 & \iddots & \iddots & \vdots & & \vdots \\ u(0) & u(1) & \cdots & u(k-1) & \cdots & u(N-1) \end{bmatrix} \end{aligned}$$

Suppose that the wide matrix in the right-hand side formed by the inputs has full rank. Then, the impulse responses $(g_{k-1}, \cdots, g_1, g_0)$ can be obtained by solving the above equation by the least-squares method[2]. This indicates that under certain assumptions, we can compute a minimal realization of an LTI system by using an input-output data, without using impulse responses.

Suppose that the inputs and outputs

$$(u(0)\ \ u(1)\ \ \cdots\ \ u(k+N-2))$$

and

$$(y(0)\ \ y(1)\ \ \cdots\ \ y(k+N-2))$$

are given, where k is strictly greater than n, the dimension of state vector. For these data, we form block Hankel matrices

$$U_{0|k-1} = \begin{bmatrix} u(0) & u(1) & \cdots & u(N-1) \\ u(1) & u(2) & \cdots & u(N) \\ \vdots & \vdots & & \vdots \\ u(k-1) & u(k) & \cdots & u(k+N-2) \end{bmatrix} \in \mathbb{R}^{km \times N}$$

[2]For the exact computation of impulse responses from the input-output data, see (6.67) in Section 6.6.

and

$$Y_{0|k-1} = \begin{bmatrix} y(0) & y(1) & \cdots & y(N-1) \\ y(1) & y(2) & \cdots & y(N) \\ \vdots & \vdots & & \vdots \\ y(k-1) & y(k) & \cdots & y(k+N-2) \end{bmatrix} \in \mathbb{R}^{kp \times N}$$

where the indices 0 and $k-1$ denote the arguments of the upper-left and lower-left element, respectively, and the number of columns of block Hankel matrices is usually fixed as N, which is sufficiently large.

We now derive matrix input-output equations, which play a fundamental role in subspace identification. By the repeated use of (6.1), we get[3]

$$\begin{bmatrix} y(t) \\ y(t+1) \\ \vdots \\ y(t+k-1) \end{bmatrix} = \begin{bmatrix} C \\ CA \\ \vdots \\ CA^{k-1} \end{bmatrix} x(t) + \begin{bmatrix} D & & & \\ CB & D & & \\ \vdots & \ddots & \ddots & \\ CA^{k-2}B & \cdots & CB & D \end{bmatrix} \begin{bmatrix} u(t) \\ u(t+1) \\ \vdots \\ u(t+k-1) \end{bmatrix}$$

For notational simplicity, we define

$$\boldsymbol{y}_k(t) = \begin{bmatrix} y(t) \\ y(t+1) \\ \vdots \\ y(t+k-1) \end{bmatrix} \in \mathbb{R}^{kp}, \qquad \boldsymbol{u}_k(t) = \begin{bmatrix} u(t) \\ u(t+1) \\ \vdots \\ u(t+k-1) \end{bmatrix} \in \mathbb{R}^{km}$$

and the block Toeplitz matrix

$$\Psi_k = \begin{bmatrix} D & & & \\ CB & D & & \\ \vdots & \ddots & \ddots & \\ CA^{k-2}B & \cdots & CB & D \end{bmatrix} \in \mathbb{R}^{kp \times km}$$

Then we have

$$\boldsymbol{y}_k(t) = \mathcal{O}_k x(t) + \Psi_k \boldsymbol{u}_k(t), \qquad t = 0, 1, \cdots \tag{6.21}$$

We see that in terms of $\boldsymbol{u}_k(t)$ and $\boldsymbol{y}_k(t)$, the block Hankel matrices $U_{0|k-1}$ and $Y_{0|k-1}$ are expressed as

$$U_{0|k-1} = [\boldsymbol{u}_k(0) \ \ \boldsymbol{u}_k(1) \ \ \cdots \ \ \boldsymbol{u}_k(N-1)]$$

and

$$Y_{0|k-1} = [\boldsymbol{y}_k(0) \ \ \boldsymbol{y}_k(1) \ \ \cdots \ \ \boldsymbol{y}_k(N-1)]$$

It thus follows from (6.21) that

[3]This type of equations have been employed in state space identification problems in earlier papers [62, 155]; see also Problem 5.5.

$$Y_{0|k-1} = \mathcal{O}_k X_0 + \Psi_k U_{0|k-1} \tag{6.22}$$

where $X_0 = [x(0)\ x(1)\ \cdots\ x(N-1)] \in \mathbb{R}^{n \times N}$ is the initial state matrix.

Similarly, we define

$$U_{k|2k-1} = [\boldsymbol{u}_k(k)\ \ \boldsymbol{u}_k(k+1)\ \cdots\ \ \boldsymbol{u}_k(N+k-1)]$$

$$Y_{k|2k-1} = [\boldsymbol{y}_k(k)\ \ \boldsymbol{y}_k(k+1)\ \cdots\ \ \boldsymbol{y}_k(N+k-1)]$$

Then, using (6.21) for $t = k,\ k+1,\ \cdots, k+N-1$, we get

$$Y_{k|2k-1} = \mathcal{O}_k X_k + \Psi_k U_{k|2k-1} \tag{6.23}$$

where $X_k = [x(k)\ x(k+1)\ \cdots\ x(k+N-1)] \in \mathbb{R}^{n \times N}$.

Equations (6.22) and (6.23) are the matrix input-output equations with initial states X_0 and X_k, respectively. The block Hankel matrices $U_{0|k-1}$ and $Y_{0|k-1}$ are usually called the past inputs and outputs, respectively, whereas the block Hankel matrices $U_{k|2k-1}$ and $Y_{k|2k-1}$ are called the future inputs and outputs, respectively.

We assume that the following conditions are satisfied for the exogenous input and the initial state matrix.

Assumption 6.1. *A1)* $\operatorname{rank}(X_0) = n$.

A2) $\operatorname{rank}(U_{0|k-1}) = km$, *where* $k > n$.

A3) $\operatorname{span}(X_0) \cap \operatorname{span}(U_{0|k-1}) = \{0\}$, *where* $\operatorname{span}(\cdot)$ *denotes the space spanned by the row vectors of a matrix.* □

Assumption 6.1 A1) implies that the state vector is sufficiently excited, or the system is reachable. Indeed, if A1) is not satisfied, there exists a non-zero vector $\eta \in \mathbb{R}^n$ such that $\eta^{\mathrm{T}} X_0 = 0$, which implies that $X_0 \in \mathbb{R}^{n \times N}$ does not span the n-dimensional state space. Assumption 6.1 A2) shows that the input sequence $u \in \mathbb{R}^m$ should satisfy the persistently exciting (PE) condition of order k. For the PE condition, see Definition B.1 of Appendix B for more details. Also, A3) means that the row vectors of X_0 and $U_{0|k-1}$ are linearly independent, or there is no linear feedback from the states to the inputs. This implies that the input-output data are obtained from an open-loop experiment.

Lemma 6.3. *[118, 119] Suppose that A1)* $\sim$ *A3) and rank*$(\mathcal{O}_k) = n$ *are satisfied. Then, the following rank condition holds:*

$$\operatorname{rank} \begin{bmatrix} U_{0|k-1} \\ Y_{0|k-1} \end{bmatrix} = km + n \tag{6.24}$$

Proof. [86] It follows from (6.22) that

$$\begin{bmatrix} U_{0|k-1} \\ Y_{0|k-1} \end{bmatrix} = \begin{bmatrix} I_{km} & 0_{km \times n} \\ \Psi_k & \mathcal{O}_k \end{bmatrix} \begin{bmatrix} U_{0|k-1} \\ X_0 \end{bmatrix} \tag{6.25}$$

where $k > n$. From Assumption 6.1, we see that the two block matrices in the right-hand side of the above equation have rank $km + n$. This proves (6.24). □

Lemma 6.3 implies that for the LTI system (6.1), if we delete row vectors in $Y_{0|k-1}$ that are dependent on the row vectors in $U_{0|k-1}$, there remain exactly n independent row vectors in $Y_{0|k-1}$, where n is the dimension of the state space.

In the following, the matrix

$$W_{0|k-1} := \begin{bmatrix} U_{0|k-1} \\ Y_{0|k-1} \end{bmatrix}, \qquad k > n$$

is referred to as data matrix. In order to study the relation between the block Hankel matrix $H_{k,l}$ formed by the impulse responses and the data matrix $W_{0|k-1}$ defined above, we begin with a simple example.

Example 6.3. Suppose that $y(t) = 0,\ t < 0$. Let $u = (0,\ 0,\ 0,\ 1,\ 0,\ 0,\ \cdots)$ be the unit impulse at $t = 3$. We apply the input u to an LTI system, and observe the impulse response with three steps delay

$$y = (0,\ 0,\ 0,\ g_0,\ g_1,\ g_2,\ g_3,\ \cdots)$$

Let $k = 4,\ N = 8$. Then, we have

$$\begin{bmatrix} U_{0|3} \\ Y_{0|3} \end{bmatrix} = \left[\begin{array}{cccc|cccc} 0 & 0 & 0 & 1 & 0 & 0 & 0 & 0 \\ 0 & 0 & 1 & 0 & 0 & 0 & 0 & 0 \\ 0 & 1 & 0 & 0 & 0 & 0 & 0 & 0 \\ 1 & 0 & 0 & 0 & 0 & 0 & 0 & 0 \\ \hline 0 & 0 & 0 & g_0 & g_1 & g_2 & g_3 & g_4 \\ 0 & 0 & g_0 & g_1 & g_2 & g_3 & g_4 & g_5 \\ 0 & g_0 & g_1 & g_2 & g_3 & g_4 & g_5 & g_6 \\ g_0 & g_1 & g_2 & g_3 & g_4 & g_5 & g_6 & g_7 \end{array}\right] \tag{6.26}$$

This data matrix has a particular block structure, in which the upper-right block is a zero matrix, and the lower-right block is exactly the Hankel matrix $H_{4,4}$. Also, if we post-multiply (6.26) by a nonsingular matrix $\begin{bmatrix} J_4 & 0 \\ 0 & I_4 \end{bmatrix}$, where J_4 is a permutation matrix with 1 along the principal anti-diagonal [see (2.39)], then the right-hand side of (6.26) has the form $\left[\begin{array}{c|c} I_4 & 0 \\ \hline \Psi_4 & \mathcal{O}_4\mathcal{C}_4 \end{array}\right]$, which is similar to the block matrix with the upper-right block zero appearing in the right-hand side of (6.25). □

Data matrices formed by generic input-output data do not have a nice structure like (6.26). However, by exploiting the linearity of the system, we can transform the data matrices into block matrices with zeros in the upper-right block. This fact is indeed guaranteed by the following lemma.

Lemma 6.4. *[181] Suppose that the input-output data*

$$W_{0|k-1} = \begin{bmatrix} U_{0|k-1} \\ Y_{0|k-1} \end{bmatrix}, \qquad k > n \tag{6.27}$$

are given. Then, under the assumption of Lemma 6.3, any input-output pair

$$\tilde{\boldsymbol{u}}_k(0) = \begin{bmatrix} \tilde{u}(0) \\ \tilde{u}(1) \\ \vdots \\ \tilde{u}(k-1) \end{bmatrix}, \qquad \tilde{\boldsymbol{y}}_k(0) = \begin{bmatrix} \tilde{y}(0) \\ \tilde{y}(1) \\ \vdots \\ \tilde{y}(k-1) \end{bmatrix}$$

of length k can be expressed as a linear combination of the column vectors of $W_{0|k-1}$. In other words, there exists a vector $\zeta \in \mathbb{R}^N$ such that

$$\begin{bmatrix} \tilde{\boldsymbol{u}}_k(0) \\ \tilde{\boldsymbol{y}}_k(0) \end{bmatrix} = \begin{bmatrix} U_{0|k-1} \\ Y_{0|k-1} \end{bmatrix} \zeta \tag{6.28}$$

Proof. Post-multiplying (6.22) by a vector $\zeta \in \mathbb{R}^N$ yields

$$Y_{0|k-1}\zeta = \mathcal{O}_k X_0 \zeta + \Psi_k U_{0|k-1}\zeta$$

Thus it should be noted that $\tilde{\boldsymbol{u}}_k(0) := U_{0|k-1}\zeta$ and $\tilde{\boldsymbol{y}}_k(0) := Y_{0|k-1}\zeta$ are an input-output pair with the initial state vector $\tilde{x}(0) := X_0\zeta$. This is a version of the well-known principle of superposition for an LTI system.

To prove the lemma, let $(\tilde{\boldsymbol{u}}_k(0), \tilde{\boldsymbol{y}}_k(0))$ be an input-output pair. Then, it follows from (6.21) that there exists an initial state $\tilde{x}(0) \in \mathbb{R}^n$ such that

$$\tilde{\boldsymbol{y}}_k(0) = \mathcal{O}_k \tilde{x}(0) + \Psi_k \tilde{\boldsymbol{u}}_k(0) \tag{6.29}$$

By assumption, $\begin{bmatrix} U_{0|k-1} \\ X_0 \end{bmatrix} \in \mathbb{R}^{(km+n)\times N}$ has full row rank, so that there exists a vector $\zeta \in \mathbb{R}^N$ such that

$$\begin{bmatrix} \tilde{\mathbf{u}}_k(0) \\ \tilde{x}(0) \end{bmatrix} = \begin{bmatrix} U_{0|k-1} \\ X_0 \end{bmatrix} \zeta$$

Thus from (6.29) and (6.25), we have

$$\begin{aligned} \begin{bmatrix} \tilde{\boldsymbol{u}}_k(0) \\ \tilde{\boldsymbol{y}}_k(0) \end{bmatrix} &= \begin{bmatrix} I_{km} & 0_{km\times n} \\ \Psi_k & \mathcal{O}_k \end{bmatrix} \begin{bmatrix} \tilde{\boldsymbol{u}}_k(0) \\ \tilde{x}(0) \end{bmatrix} \\ &= \begin{bmatrix} I_{km} & 0_{km\times n} \\ \Psi_k & \mathcal{O}_k \end{bmatrix} \begin{bmatrix} U_{0|k-1} \\ X_0 \end{bmatrix} \zeta = \begin{bmatrix} U_{0|k-1} \\ Y_{0|k-1} \end{bmatrix} \zeta \end{aligned}$$

as was to be proved. □

The above lemma ensures that any input-output pair can be generated by using a sufficiently long input-output data, if the input has a certain PE condition.

Example 6.4. Consider a scalar system described by

$$y(t) = ay(t-1) + u(t), \qquad t = 0, 1, \cdots; \; y(-1) = 0$$

Then, the output is expressed as

$$y(t) = a^t u(0) + a^{t-1}u(1) + \cdots + au(t-1) + u(t), \qquad t = 0,\ 1,\ \cdots$$

Suppose that we have the following set of input-output data.

t	u	y
0	1	$y_0 = 1$
1	2	$y_1 = a + 2$
2	1	$y_2 = a^2 + 2a + 1$
3	-1	$y_3 = a^3 + 2a^2 + a - 1$
4	-1	$y_4 = a^4 + 2a^3 + a^2 - a - 1$
5	1	$y_5 = a^5 + 2a^4 + a^3 - a^2 - a + 1$
6	1	$y_6 = a^6 + 2a^5 + a^4 - a^3 - a^2 + a + 1$
7	-1	$y_7 = a^7 + 2a^6 + a^5 - a^4 - a^3 + a^2 + a - 1$

Let $k = 3$, $N = 6$. Then, the data matrix is given by

$$\begin{bmatrix} U_{0|2} \\ Y_{0|2} \end{bmatrix} = \left[\begin{array}{ccc|ccc} 1 & 2 & 1 & -1 & -1 & 1 \\ 2 & 1 & -1 & -1 & 1 & 1 \\ 1 & -1 & -1 & 1 & 1 & -1 \\ \hline y_0 & y_1 & y_2 & y_3 & y_4 & y_5 \\ y_1 & y_2 & y_3 & y_4 & y_5 & y_6 \\ y_2 & y_3 & y_4 & y_5 & y_6 & y_7 \end{array}\right] \tag{6.30}$$

We observe that three vectors in the upper-left 3×3 block are linearly independent. By applying the column operation using these three column vectors, we make the upper-right 3×3 block a zero matrix. By this column operation, the lower-right block is also affected by the column vectors in the lower-left block, so that

$$\begin{bmatrix} U_{0|2} \\ Y_{0|2} \end{bmatrix}' = \left[\begin{array}{ccc|ccc} 1 & 2 & 1 & 0 & 0 & 0 \\ 2 & 1 & -1 & 0 & 0 & 0 \\ 1 & -1 & -1 & 0 & 0 & 0 \\ \hline y_0 & y_1 & y_2 & y_3' & y_4' & y_5' \\ y_1 & y_2 & y_3 & ay_3' & ay_4' & ay_5' \\ y_2 & y_3 & y_4 & a^2y_3' & a^2y_4' & a^2y_5' \end{array}\right] \tag{6.31}$$

In fact, by taking $\zeta = (-2/3\ \ 4/3\ \ -1\ \ 1\ \ 0\ \ 0)^{\mathrm{T}} \in \mathbb{R}^6$, it follows that

$$\left[\begin{array}{ccc|ccc} 1 & 2 & 1 & -1 & -1 & 1 \\ 2 & 1 & -1 & -1 & 1 & 1 \\ 1 & -1 & -1 & 1 & 1 & -1 \\ \hline y_0 & y_1 & y_2 & y_3 & y_4 & y_5 \\ y_1 & y_2 & y_3 & y_4 & y_5 & y_6 \\ y_2 & y_3 & y_4 & y_5 & y_6 & y_7 \end{array}\right] \left[\begin{array}{c} -\frac{2}{3} \\ \frac{4}{3} \\ -1 \\ \hline 1 \\ 0 \\ 0 \end{array}\right] = \begin{bmatrix} 0 \\ 0 \\ 0 \\ y_3' \\ ay_3' \\ a^2y_3' \end{bmatrix}$$

where $y_3' = a^3 + a^2 + a/3$. We see that y_3' is the output at $t = 3$ due to the input $u = (1\ \ 1\ \ 1/3\ \ 0)$, so that $(y_3'\ \ ay_3'\ \ a^2y_3')$ is a zero-input response with the initial condition $\tilde{y}(0) = y'_3$. Similarly, we can show that $y'_4 = a^4 + 2a^3 + 2a^2 + a$ and $y'_5 = a^5 + 2a^4 + a^3 + 2a/3$, so that y'_4 and y'_5 are the outputs at $t = 4$ and $t = 5$ with inputs $u = (1\ 2\ 2\ 1\ 0)$ and $u = (1\ 2\ 1\ 0\ 2/3\ 0)$, respectively. It should be noted that these inputs are fictitious inputs, which are not actually fed into the system.

We can write the lower-right block of (6.31) as

$$\begin{bmatrix} 1 \\ a \\ a^2 \end{bmatrix} [y'_3\ y'_4\ y'_5] = \begin{bmatrix} C \\ CA \\ CA^2 \end{bmatrix} [y'_3\ y'_4\ y'_5] = \mathcal{O}_3 [y'_3\ y'_4\ y'_5]$$

Clearly, the image of the above matrix is equal to the image of the extended observability matrix $\mathcal{O}_3 \in \mathbb{R}^{3\times 1}$. Thus, we have $C = 1$, $A = a$. □

We need a column operation to make the upper-right block of the data matrix a zero matrix as shown in (6.31). However, this is easily performed by means of the LQ decomposition, which is the dual of the QR decomposition.

6.4 LQ Decomposition

We usually consider rectangular data matrices with a large number of columns. Thus if we apply the LQ decomposition to rectangular matrices, then we get block lower triangular matrices with a zero block at the upper-right corner.

Let the LQ decomposition of a data matrix be given by

$$\begin{bmatrix} U_{0|k-1} \\ Y_{0|k-1} \end{bmatrix} = \begin{bmatrix} L_{11} & 0 \\ L_{21} & L_{22} \end{bmatrix} \begin{bmatrix} Q_1^{\mathrm{T}} \\ Q_2^{\mathrm{T}} \end{bmatrix} \tag{6.32}$$

where $L_{11} \in \mathbb{R}^{km\times km}$, $L_{21} \in \mathbb{R}^{kp\times km}$, $L_{22} \in \mathbb{R}^{kp\times kp}$ with L_{11}, L_{22} lower triangular, and $Q_1 \in \mathbb{R}^{N\times km}$, $Q_2 \in \mathbb{R}^{N\times kp}$ are orthogonal. The actual computation of LQ decomposition is performed by taking the transpose of the QR decomposition of the tall matrix

$$[U_{0|k-1}^{\mathrm{T}}\ Y_{0|k-1}^{\mathrm{T}}] \in \mathbb{R}^{N\times k(m+p)}$$

A MATLAB® program for the LQ decomposition is displayed in Table 6.1.

Example 6.5. Let $a = 0.9$ in Example 6.4. Then, from (6.30) and (6.31), it follows that

$$\begin{bmatrix} U_{0|2} \\ Y_{0|2} \end{bmatrix} = \left[\begin{array}{ccc|ccc} 1 & 2 & 1 & -1 & -1 & 1 \\ 2 & 1 & -1 & -1 & 1 & 1 \\ 1 & -1 & -1 & 1 & 1 & -1 \\ \hline 1 & 2.9 & 3.61 & 2.249 & 1.0241 & 1.92169 \\ 2.9 & 3.61 & 2.249 & 1.0241 & 1.92169 & 2.729521 \\ 3.61 & 2.249 & 1.0241 & 1.92169 & 2.729521 & 1.4565689 \end{array}\right] \tag{6.33}$$

and

Table 6.1. LQ decomposition

```
% Program
% LQ decomposition
function [L11,L21,L22]=lq(U,Y)
km=size(U,1); kp=size(Y,1);
[Q,L]=qr([U;Y]',0);
Q=Q'; L=L';
L11=L(1:km,1:km);
L21=L(km+1:km+kp,1:km);
L22=L(km+1:km+kp,km+1:km+kp);
```

$$\begin{bmatrix} U_{0|2} \\ Y_{0|2} \end{bmatrix}' = \left[\begin{array}{ccc|ccc} 1 & 2 & 1 & 0 & 0 & 0 \\ 2 & 1 & -1 & 0 & 0 & 0 \\ 1 & -1 & -1 & 0 & 0 & 0 \\ \hline 1 & 2.9 & 3.61 & 1.839 & 4.6341 & 4.17069 \\ 2.9 & 3.61 & 2.249 & 1.6551 & 4.17069 & 3.753621 \\ 3.61 & 2.249 & 1.0241 & 1.48959 & 3.753621 & 3.3782589 \end{array}\right] \tag{6.34}$$

Also, the LQ decomposition of (6.33) gives

$$L = \left[\begin{array}{ccc|ccc} -3.0000 & & & 0 & 0 & 0 \\ -1.3333 & 2.6874 & & 0 & 0 & 0 \\ 1.6667 & 1.1990 & -1.3359 & 0 & 0 & 0 \\ \hline -3.0159 & -0.7588 & -1.3353 & -4.5569 & 0 & 0 \\ -4.0509 & 2.0045 & -1.2017 & -4.1012 & 0 & 0 \\ -1.9792 & 3.0030 & -2.4175 & -3.6911 & 0 & 0 \end{array}\right] \tag{6.35}$$

We see that in (6.34) and (6.35), multiplying the first row of the lower-right block by 0.9 yields the second row, and multiplying the second row by 0.9 yields the third row, so that the rank of these matrices is one, which is the same as the dimension of the system treated in Example 6.4. □

From (6.32), we obtain

$$\begin{bmatrix} L_{11} & 0 \\ L_{21} & L_{22} \end{bmatrix} = \begin{bmatrix} U_{0|k-1} \\ Y_{0|k-1} \end{bmatrix} \begin{bmatrix} Q_1 & Q_2 \end{bmatrix} \tag{6.36}$$

The following lemma provides a system theoretic meaning of the L-matrix in terms of zero-input responses of the system.

Lemma 6.5. *Under Assumption 6.1, each column of the L-matrix is an input-output pair; in particular, each column of L_{22} contains a zero-input response of the system. Moreover, we have* $\operatorname{rank}(L_{22}) = n$*, i.e. the dimension of the system.*

Proof. Since Q_1, Q_2 are formed by N-dimensional column vectors, it follows from (6.28) of Lemma 6.4 that each column of L-matrix of (6.36) is an input-output pair.

Since $L_{12} = 0$, we see that L_{22} consists of zero-input responses. Recall from (6.8) and (6.21) that the zero-input response is expressed as $\boldsymbol{y}_k(0) = \mathcal{O}_k x(0)$. We see that the number of independent zero-input responses is $n = \dim x(0)$, so that we have the desired result. □

The scenario of the realization procedure based on the LQ decomposition is to compute the SVD of L_{22} in order to recover the information about the extended observability matrix, and then to estimate the matrices A and C by using the relation of (6.12). On the other hand, the information about matrices B and D is included in the matrices L_{11} and L_{21} of (6.32). To retrieve this information, however, the matrix input-output equation of (6.22) [and/or (6.23)] should be employed together with L_{11} and L_{21}, as explained in the next section.

Thus, in the next section, we shall present a solution to Problem B, stated in Section 6.1, based on a subspace identification method, called the MOESP method, in which the LQ decomposition technique and the SVD are employed. Another solution to Problem B is provided by the N4SID subspace identification method, which will be discussed in Section 6.6.

6.5 MOESP Method

In this section, we discuss the basic subspace identification method called MOESP method[4] due to Verhaegen and Dewilde [172, 173]. In the following, the orthogonal projection is expressed as $\hat{E}\{\cdot \mid \cdot\}$.

We see from (6.32) that

$$U_{0|k-1} = L_{11} Q_1^{\mathrm{T}} \tag{6.37a}$$

$$Y_{0|k-1} = L_{21} Q_1^{\mathrm{T}} + L_{22} Q_2^{\mathrm{T}} \tag{6.37b}$$

where $L_{11} \in \mathbb{R}^{km \times km}$, $L_{22} \in \mathbb{R}^{kp \times kp}$ are lower triangular, and $Q_1 \in \mathbb{R}^{N \times km}$, $Q_2 \in \mathbb{R}^{N \times kp}$ are orthogonal. Under Assumption 6.1, we see that L_{11} is nonsingular, so that $Q_1^{\mathrm{T}} = L_{11}^{-1} U_{0|k-1}$. Thus, it follows that (6.37b) is written as

$$Y_{0|k-1} = L_{21} L_{11}^{-1} U_{0|k-1} + L_{22} Q_2^{\mathrm{T}}$$

Since Q_1, Q_2 are orthogonal, the first term in the right-hand side of the above equation is spanned by the row vectors in $U_{0|k-1}$, and the second term is orthogonal to it. Hence, the orthogonal projection of the row space of $Y_{0|k-1}$ onto the row space of $U_{0|k-1}$ is given by

$$\hat{E}\{Y_{0|k-1} \mid U_{0|k-1}\} = L_{21} Q_1^{\mathrm{T}} = L_{21} L_{11}^{-1} U_{0|k-1}$$

Also, the orthogonal projection of the row space of $Y_{0|k-1}$ onto the complement $U_{0|k-1}^{\perp}$ of the row space of $U_{0|k-1}$ is given by

[4] MOESP=Multivariable Output Error State Space

$$\hat{E}\{Y_{0|k-1} \mid U_{0|k-1}^{\perp}\} = L_{22}Q_2^{\mathrm{T}}$$

In summary, the right-hand side of $Y_{0|k-1}$ in (6.37b) is the orthogonal sum decomposition of the output matrix $Y_{0|k-1}$ onto the row space of the input matrix $U_{0|k-1}$ and its complement.

Also, it follows from (6.22) and (6.37b) that

$$\mathcal{O}_k X_0 + \Psi_k L_{11} Q_1^{\mathrm{T}} = L_{21}Q_1^{\mathrm{T}} + L_{22}Q_2^{\mathrm{T}} \tag{6.38}$$

where it should be noted that though the right-hand side is an orthogonal sum, the left-hand side is a direct sum, so that two quantities therein are not necessarily orthogonal. This implies that $\mathcal{O}_k X_0 \neq L_{22}Q_2^{\mathrm{T}}$ and $\Psi_k L_{11}Q_1^{\mathrm{T}} \neq L_{21}Q_1^{\mathrm{T}}$.

Post-multiplying (6.38) by Q_2 yields

$$\mathcal{O}_k X_0 Q_2 = L_{22}$$

where $Q_1^{\mathrm{T}}Q_2 = 0,\ Q_2^{\mathrm{T}}Q_2 = I_{kp}$ are used. Under the assumptions of Lemma 6.3, the product X_0Q_2 has full row rank n and $\mathrm{rank}(\mathcal{O}_k) = n$, which is equal to $\mathrm{rank}(L_{22})$. Thus we can obtain the image of the extended observability matrix $\mathcal{O}_k$ and hence the dimension n from the SVD of $L_{22} \in \mathbb{R}^{kp\times kp}$.

Let the SVD of L_{22} be given by

$$L_{22} = [U_1 \ \ U_2]\begin{bmatrix} \Sigma_1 & 0 \\ 0 & 0 \end{bmatrix}\begin{bmatrix} V_1^{\mathrm{T}} \\ V_2^{\mathrm{T}} \end{bmatrix} = U_1\Sigma_1V_1^{\mathrm{T}} \tag{6.39}$$

where $U_1 \in \mathbb{R}^{kp\times n}$ and $U_2 \in \mathbb{R}^{kp\times(kp-n)}$. Then, we have

$$\mathcal{O}_k X_0 Q_2 = U_1\Sigma_1V_1^{\mathrm{T}}$$

so that we define the extended observability matrix as

$$\mathcal{O}_k = U_1\Sigma_1^{1/2} \tag{6.40}$$

and $n = \dim \Sigma_1$. The matrix C is readily given by

$$C = \mathcal{O}_k(1:p, 1:n) \tag{6.41}$$

and A is obtained by solving the linear equation (see Lemma 6.1)

$$\mathcal{O}_k(1:p(k-1), 1:n)A = \mathcal{O}_k(p+1:kp, 1:n) \tag{6.42}$$

Now we consider the estimation of matrices B and D. Since $U_2^{\mathrm{T}}L_{22} = 0$ and $U_2^{\mathrm{T}}\mathcal{O}_k = 0$, pre-multiplying (6.38) by $U_2^{\mathrm{T}} \in \mathbb{R}^{(kp-n)\times kp}$ yields

$$U_2^{\mathrm{T}}\Psi_k L_{11}Q_1^{\mathrm{T}} = U_2^{\mathrm{T}}L_{21}Q_1^{\mathrm{T}}$$

Further post-multiplying this equation by Q_1 yields

$$U_2^{\mathrm{T}} \begin{bmatrix} D & 0 & \cdots & 0 \\ CB & D & \cdots & 0 \\ \vdots & \vdots & \ddots & \vdots \\ CA^{k-2}B & CA^{k-3}B & \cdots & D \end{bmatrix} = U_2^{\mathrm{T}} L_{21} L_{11}^{-1} \tag{6.43}$$

This is a linear equation with respect to B and D, so that we can use the least-squares method to find them. In fact, define

$$U_2^{\mathrm{T}} := [\mathcal{L}_1 \ \mathcal{L}_2 \ \cdots \ \mathcal{L}_k], \qquad U_2^{\mathrm{T}} L_{21} L_{11}^{-1} := [\mathcal{M}_1 \ \mathcal{M}_2 \ \cdots \ \mathcal{M}_k]$$

where $\mathcal{L}_i \in \mathbb{R}^{(kp-n)\times p}$, $i = 1, \cdots, k$ and $\mathcal{M}_i \in \mathbb{R}^{(kp-n)\times m}$. Thus, from (6.43),

$$\begin{aligned} \mathcal{L}_1 D + \mathcal{L}_2 CB + \cdots + \mathcal{L}_{k-1} CA^{k-3}B + \mathcal{L}_k CA^{k-2}B &= \mathcal{M}_1 \\ \mathcal{L}_2 D + \mathcal{L}_3 CB + \cdots + \mathcal{L}_k CA^{k-3}B &= \mathcal{M}_2 \\ &\vdots \\ \mathcal{L}_{k-1} D + \mathcal{L}_k CB &= \mathcal{M}_{k-1} \\ \mathcal{L}_k D &= \mathcal{M}_k \end{aligned}$$

Defining $\bar{\mathcal{L}}_i = [\mathcal{L}_i \ \cdots \ \mathcal{L}_k] \in \mathbb{R}^{(kp-n)\times(k+1-i)p}$, $i = 2, \cdots, k$, we get the following overdetermined linear equations:

$$\begin{bmatrix} \mathcal{L}_1 & \bar{\mathcal{L}}_2 \mathcal{O}_{k-1} \\ \mathcal{L}_2 & \bar{\mathcal{L}}_3 \mathcal{O}_{k-2} \\ \vdots & \vdots \\ \mathcal{L}_{k-1} & \bar{\mathcal{L}}_k \mathcal{O}_1 \\ \mathcal{L}_k & 0 \end{bmatrix} \begin{bmatrix} D \\ B \end{bmatrix} = \begin{bmatrix} \mathcal{M}_1 \\ \mathcal{M}_2 \\ \vdots \\ \mathcal{M}_{k-1} \\ \mathcal{M}_k \end{bmatrix} \tag{6.44}$$

where the block coefficient matrix in the left-hand side is $k(kp-n) \times (p+n)$ - dimensional. To obtain a unique least-squares solution $(D,\ B)$ of (6.44), the block matrix has full column rank, so that $k(kp-n) \geq (p+n)$ should be satisfied. It can be shown that if $k > n$, this condition is satisfied.

Summarizing the above, we can provide a subspace identification method that solves Problem B. Suppose that we have the input and output data $U_{0|k-1}$ and $Y_{0|k-1}$. Then, we have the following lemma.

Lemma 6.6. *(MOESP algorithm)*

Step 1: Compute the LQ decomposition of (6.32).

Step 2: Compute the SVD of (6.39), and let $n := \dim \Sigma_1$, *and define the extended observability matrix as*

$$\mathcal{O}_k = U_1 \Sigma_1^{1/2}$$

Step 3: Obtain C and A from (6.41) and (6.42), respectively.

Step 4: Solve (6.44) by the least-squares method to estimate B and D. □

In [172, 173], this algorithm is called the ordinary MOESP, and a program of the above algorithm is given in Table D.2 of Appendix D.

Example 6.6. Consider a simple 2nd-order system with

$$A = \begin{bmatrix} 0.6 & 0.4 \\ -0.4 & 0.6 \end{bmatrix}, \quad B = \begin{bmatrix} 0 \\ 1 \end{bmatrix}, \quad C = [1 \;\; 0.5], \quad D = 0$$

The transfer function is then given by

$$G(z) = \frac{0.5z + 0.1}{z^2 - 1.2z + 0.52} = \frac{b_0 z^2 + b_1 z + b_2}{z^2 + a_1 z + a_2}$$

We have performed simulation studies by using 100 sets of input-output data with the length $N = 100$, where the input is a white noise with mean zero and unit variance, and then a white noise v is added to the output y so that the S/N ratio in the output becomes approximately $\sigma_y^2/\sigma_v^2 \simeq 100$.

By using the MOESP algorithm of Lemma 6.6 with $k = 8$, we have identified the dimension n and parameters of the transfer function. The means and standard deviations of the first five singular values of L_{22} are displayed in Table 6.2, where s.d. denotes the standard deviation.

Table 6.2. Singular values of L_{22}

	σ_1	σ_2	σ_3	σ_4	σ_5
mean	15.4313	5.7956	1.1010	1.0354	0.9664
s.d.	1.8690	0.5360	0.1004	0.0887	0.0790

The singular values σ_i, $i = 3, 4, \cdots$ are relatively small compared with the first two σ_1 and σ_2, so that the dimension is correctly identified as $n = 2$. It should be noted that for the noise free case ($v = 0$), we observed that σ_i, $i = 3, 4, \cdots$ are nearly zero (order of 10^{-14}). Also, the identification result of the transfer function is displayed in Table 6.3. Thus we see that for this simple system, the identification result is quite satisfactory. □

Table 6.3. Simulation result

	a_1	a_2	b_0	b_1	b_2
True	−1.2000	0.5200	0.0000	0.5000	0.1000
mean	−1.1999	0.5204	−0.0016	0.5019	0.1002
s.d.	0.0147	0.0110	0.0102	0.0156	0.0185

6.6 N4SID Method

In this section, we show another method of solving Problem B. This is fulfilled by introducing the basic idea of the subspace identification method called N4SID method[5] developed by Van Overschee and De Moor [164,165]. Prior to describing the method, we briefly review the oblique projection in subspaces (see Section 2.5).

Let $\mathcal{A}$, $\mathcal{B}$, $\mathcal{C}$ be the row spaces generated by the row vectors of matrices A, B, C, respectively. We assume that $\mathcal{B} \cap \mathcal{C} = \{0\}$, which corresponds to Condition A3) of Assumption 6.1. For $\alpha \in \mathcal{A}$, we have the following decomposition

$$\hat{E}\{\alpha \mid \mathcal{B} \vee \mathcal{C}\} = \hat{E}_{\|\mathcal{C}}\{\alpha \mid \mathcal{B}\} + \hat{E}_{\|\mathcal{B}}\{\alpha \mid \mathcal{C}\} \tag{6.45}$$

where the left-hand side is the orthogonal projection, while the right-hand side is a direct sum decomposition; $\hat{E}_{\|\mathcal{C}}\{\alpha \mid \mathcal{B}\}$ is the oblique projection of α onto $\mathcal{B}$ along $\mathcal{C}$, and $\hat{E}_{\|\mathcal{B}}\{\alpha \mid \mathcal{C}\}$ is the oblique projection of α onto $\mathcal{C}$ along $\mathcal{B}$.

Let $k > n$ be the present time. Define $U_p := U_{0|k-1}$, $Y_p := Y_{0|k-1}$, $X_p := X_0$ and $U_f := U_{k|2k-1}$, $Y_f := Y_{k|2k-1}$, $X_f := X_k$, where the subscripts p and f denote the past and future, respectively. In order to explain the N4SID method, we recall two matrix input-output equations derived in Section 6.3, *i.e.*,

$$Y_p = \mathcal{O}_k X_p + \Psi_k U_p \tag{6.46}$$

$$Y_f = \mathcal{O}_k X_f + \Psi_k U_f \tag{6.47}$$

Further, define $W_p, W_f \in \mathbb{R}^{k(m+p)\times N}$ as

$$W_p := \begin{bmatrix} U_p \\ Y_p \end{bmatrix} = \begin{bmatrix} U_{0|k-1} \\ Y_{0|k-1} \end{bmatrix}, \qquad W_f := \begin{bmatrix} U_f \\ Y_f \end{bmatrix} = \begin{bmatrix} U_{k|2k-1} \\ Y_{k|2k-1} \end{bmatrix}$$

The following lemma explains a role of the state vector for an LTI system.

Lemma 6.7. *[41,118] Suppose that* $\text{rank}(\mathcal{O}_k) = \text{rank}(\mathcal{C}_k) = n$ *with* $k > n$. *Under A1) ∼A3) of Assumption 6.1 with* k *replaced by* $2k$*, the following relation holds.*

$$\text{span}\,(X_f) = \text{span}\,(W_p) \cap \text{span}\,(W_f) \tag{6.48}$$

Proof. First we show that $\text{rank}(X_f) = n$. It follows from (6.1a) that

$$x(k+i) = A^k x(i) + [A^{k-1}B \;\; A^{k-2}B \;\; \cdots \;\; B] \begin{bmatrix} u(i) \\ u(i+1) \\ \vdots \\ u(i+k-1) \end{bmatrix}$$

so that

$$X_f = A^k X_p + \bar{\mathcal{C}}_k U_p \tag{6.49}$$

[5]N4SID= Numerical Algorithms for Subspace State Space System Identification.

where $\bar{\mathcal{C}}_k = [A^{k-1}B \;\; A^{k-2}B \;\; \cdots \;\; B]$ is the reversed extended reachability matrix. Since $\mathrm{rank}(\bar{\mathcal{C}}_k U_p) = n$ and $\mathrm{span}(X_p) \cap \mathrm{span}(U_p) = \{0\}$ by Assumption 6.1, we see from (6.49) that $\mathrm{rank}(X_f) \geq n$. But, by definition, $\mathrm{rank}(X_f) \leq n$, so that we have $\mathrm{rank}(X_f) = n$. Moreover, from (6.46) and (6.47),

$$X_p = \mathcal{O}_k^\dagger Y_p - \mathcal{O}_k^\dagger \Psi_k U_p \in \mathrm{span}\,(W_p) \tag{6.50}$$

$$X_f = \mathcal{O}_k^\dagger Y_f - \mathcal{O}_k^\dagger \Psi_k U_f \in \mathrm{span}\,(W_f) \tag{6.51}$$

where $\mathcal{O}_k^\dagger$ is the pseudo-inverse of $\mathcal{O}_k$. Thus, from (6.50) and (6.49), $\mathrm{span}\,(X_f) \subset \mathrm{span}\,(W_p)$. It therefore follows from (6.51) that

$$\mathrm{span}\,(X_f) \subset \mathrm{span}\,(W_p) \cap \mathrm{span}\,(W_f)$$

We show that the dimension of the space in the right-hand side of the above relation is equal to n. From Lemma 6.3, we have $\dim(W_p) = \dim(W_f) = km + n$ and $\dim\,(W_p \vee W_f) = 2km + n$, where $\dim(\,\cdot\,)$ denotes the dimension of the row space. On the other hand, for dimensions of subspaces, the following identity holds:

$$\dim(W_p \cap W_f) = \dim(W_p) + \dim(W_f) - \dim(W_p \vee W_f)$$

Thus we have $\dim(W_p \cap W_f) = n$. This completes the proof. □

Since W_p and W_f are the past and future data matrices, respectively, Lemma 6.7 means that the state vector X_f is a basis of the intersection of the past and future subspaces. Hence, we observe that the state vector plays a role of memory for exchanging information between the past and the future, where the state vector X_f can be computed by the SVD of $\begin{bmatrix} W_p \\ W_f \end{bmatrix} \in \mathbb{R}^{2k(p+m)\times N}$; see [118].

Consider the LQ decomposition

$$\begin{bmatrix} U_f \\ U_p \\ Y_p \\ Y_f \end{bmatrix} = \begin{bmatrix} L_{11} & 0 & 0 & 0 \\ L_{21} & L_{22} & 0 & 0 \\ L_{31} & L_{32} & L_{33} & 0 \\ L_{41} & L_{42} & L_{43} & L_{44} \end{bmatrix} \begin{bmatrix} Q_1^{\mathrm{T}} \\ Q_2^{\mathrm{T}} \\ Q_3^{\mathrm{T}} \\ Q_4^{\mathrm{T}} \end{bmatrix} \tag{6.52}$$

where $L_{11}, L_{22} \in \mathbb{R}^{km\times km}$, $L_{33}, L_{44} \in \mathbb{R}^{kp\times kp}$ are lower triangular, and where $Q_1, Q_2 \in \mathbb{R}^{N\times km}$, $Q_3, Q_4 \in \mathrm{R}^{N\times kp}$ are orthogonal. Then, we have the following theorem, which is an extended version of Lemma 6.5.

Theorem 6.2. *Suppose that A1) ∼ A3) of Assumption 6.1 hold with k replaced by $2k$, so that the PE condition of order $2k$ holds. Then, for the LQ decomposition of (6.52), we have*

$$\mathrm{rank}(L_{42}) = n, \quad \mathrm{rank}(L_{43}) = n, \quad \mathrm{rank}\begin{bmatrix} L_{33} \\ L_{43} \end{bmatrix} = n, \quad \mathrm{rank}[L_{42} \;\; L_{43}] = n$$

Moreover, it follows that $L_{44} = 0$ and hence $\mathrm{rank}(L_{33}) = n$.

Proof. Recall from Lemma 6.5 that each column vector in the L-matrix of (6.52) is an input-output pair, though the blocks U_f and U_p are interchanged. Also, we see that three block matrices L_{12}, L_{13}, L_{14} corresponding to future inputs are zero. This implies that each column of the last three block columns of the L-matrix is an input-output pair with zero future inputs. In particular, columns of L_{42}, L_{43}, $\begin{bmatrix} L_{33} \\ L_{43} \end{bmatrix}$, L_{44}, $[L_{42} \;\; L_{43}]$ consist of zero-input responses. Since the number of independent zero-input responses equals the dimension of the system, we have all rank conditions stated in this theorem. Also, we see that $L_{44} = 0$, since past inputs and outputs together with the future inputs generating it are zero ($L_{14} = 0$, $L_{24} = 0$, $L_{34} = 0$).

From (6.24) with $k := 2k$, the rank of the left-hand side of (6.52) is $2km + n$, so is the rank of the L-matrix. Since $\text{rank}(L_{11}) = \text{rank}(L_{22}) = km$ and $L_{44} = 0$, we see that $\text{rank}(L_{33}) = n$. This completes the proof. □

We are now in a position to present a theorem that provides a basis of the N4SID method.

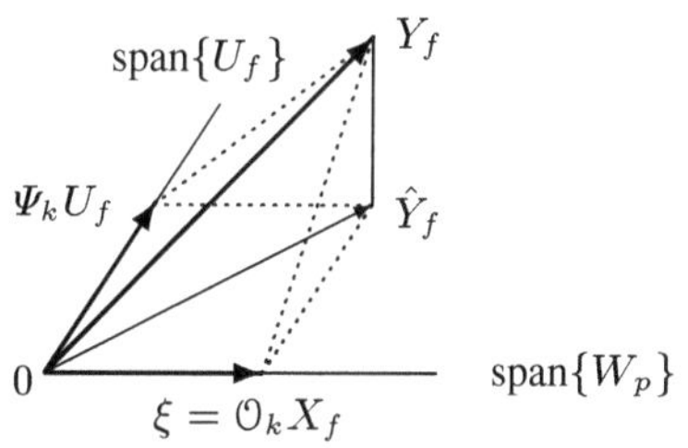

Figure 6.2. Oblique projection

Theorem 6.3. *[165] Suppose that A1) $\sim$ A3) of Assumption 6.1 hold with k replaced by $2k$. Let the oblique projection of Y_f onto W_p along U_f be given by*

$$\xi = \hat{E}_{\|U_f}\{Y_f \mid W_p\} \tag{6.53}$$

(see Figure 6.2). Also, let the SVD of ξ be given by

$$\xi = [U_1 \;\; U_2] \begin{bmatrix} \Sigma_1 & 0 \\ 0 & 0 \end{bmatrix} \begin{bmatrix} V_1^{\mathrm{T}} \\ V_2^{\mathrm{T}} \end{bmatrix} = U_1 \Sigma_1 V_1^{\mathrm{T}} \tag{6.54}$$

Then, we have the following results.

$$n = \dim \Sigma_1 \tag{6.55}$$

$$\xi = \mathcal{O}_k X_f \in \mathbb{R}^{kp \times N} \tag{6.56}$$

$$\mathcal{O}_k = U_1 \Sigma_1^{1/2} T \in \mathbb{R}^{kp \times n}, \qquad |T| \neq 0 \tag{6.57}$$

$$X_f = T^{-1} \Sigma_1^{1/2} V_1^{\mathrm{T}} \in \mathbb{R}^{n \times N} \tag{6.58}$$

Proof. Since $L_{44} = 0$, the future Y_f is completely determined by the past W_p and the future inputs U_f. Thus, $Y_f = \hat{Y}_f$ in Figure 6.2[6].

For convenience, we rewrite (6.52) as

$$\begin{bmatrix} U_f \\ W_p \\ Y_f \end{bmatrix} = \begin{bmatrix} R_{11} & 0 & 0 \\ R_{21} & R_{22} & 0 \\ R_{31} & R_{32} & 0 \end{bmatrix} \begin{bmatrix} \bar{Q}_1^{\mathrm{T}} \\ \bar{Q}_2^{\mathrm{T}} \\ \bar{Q}_3^{\mathrm{T}} \end{bmatrix} \tag{6.59}$$

From Theorem 6.2, we see that $\mathrm{rank}(R_{22}) = \mathrm{rank} \begin{bmatrix} L_{22} & 0 \\ L_{32} & L_{33} \end{bmatrix} = km + n$, which is less than $k(m+p)$, so that R_{22} is rank deficient. Also, it follows from $\mathrm{rank}(L_{22}) = km$ and the third condition in Theorem 6.2 that (see Problem 6.6)

$$\mathrm{Ker}(R_{22}) \subset \mathrm{Ker}(R_{32}) \tag{6.60}$$

Now from (6.59), we have

$$R_{22}\bar{Q}_2^{\mathrm{T}} = W_p - R_{21}\bar{Q}_1^{\mathrm{T}}$$

Thus, there exists a $\Xi \in \mathbb{R}^{k(p+m)\times N}$ such that

$$\bar{Q}_2^{\mathrm{T}} = R_{22}^{\dagger}(W_p - R_{21}\bar{Q}_1^{\mathrm{T}}) + [I_{k(p+m)} - R_{22}^{\dagger}R_{22}]\Xi \tag{6.61}$$

where $R_{22}^{\dagger}$ is the pseudo-inverse defined in Lemma 2.10; see also Lemma 2.11.

From the third relation of (6.59), we have $Y_f = R_{31}\bar{Q}_1^{\mathrm{T}} + R_{32}\bar{Q}_2^{\mathrm{T}}$, so that by using (6.61) and $\bar{Q}_1^{\mathrm{T}} = R_{11}^{-1}U_f$,

$$\begin{aligned} Y_f = {} & (R_{31} - R_{32}R_{22}^{\dagger}R_{21})R_{11}^{-1}U_f + R_{32}R_{22}^{\dagger}W_p \\ & + R_{32}\left[I_{k(p+m)} - R_{22}^{\dagger}R_{22}\right]\Xi \end{aligned} \tag{6.62}$$

But, from (6.60), $R_{32}\left[I_{k(p+m)} - R_{22}^{\dagger}R_{22}\right] = 0$, since $\Pi := I_{k(p+m)} - R_{22}^{\dagger}R_{22}$ is the orthogonal projection onto $\mathrm{Ker}(R_{22})$. Thus, (6.62) reduces to

$$Y_f = (R_{31} - R_{32}R_{22}^{\dagger}R_{21})R_{11}^{-1}U_f + R_{32}R_{22}^{\dagger}W_p \tag{6.63}$$

where $\mathrm{span}(U_f) \cap \mathrm{span}(W_p) = \{0\}$ from A3) of Assumption 6.1. It thus follows that the right-hand side of (6.63) is a direct sum of the oblique projections of Y_f onto $\mathrm{span}(U_f)$ along $\mathrm{span}(W_p)$ and of Y_f onto $\mathrm{span}(W_p)$ along $\mathrm{span}(U_f)$.

On the other hand, from (6.47),

$$Y_f = \Psi_k U_f + \mathcal{O}_k X_f \tag{6.64}$$

Again, from A3) of Assumption 6.1, $\mathrm{span}(U_f) \cap \mathrm{span}(X_f) = \{0\}$, so that the right-hand side of (6.64) is the direct sum of the oblique projections of Y_f onto $\mathrm{span}(U_f)$

[6]Note that if the output y is disturbed by a noise, then we have $L_{44} \neq 0$, implying that $Y_f \neq \hat{Y}_f$.

along $\mathrm{span}(X_f)$ and of Y_f onto $\mathrm{span}(X_f)$ along $\mathrm{span}(U_f)$. Thus, comparing (6.63) and (6.64), we have the desired result, and hence

$$\xi = \hat{E}_{\|U_f}\{Y_f \mid W_p\} = R_{32}R_{22}^{\dagger}W_p = \mathcal{O}_k X_f \tag{6.65}$$

This proves (6.56). Equations (6.55), (6.57) and (6.58) are obvious from (6.54). □

It follows from Theorem 6.3 that the estimates of A and C can be obtained from the extended observability matrix of (6.57). Also, we see from (6.63) and (6.64) that

$$\Psi_k = (R_{31} - R_{32}R_{22}^{\dagger}R_{21})R_{11}^{-1} \tag{6.66}$$

holds. Hence, by the definition of Ψ_k,

$$\begin{bmatrix} D & 0 & \cdots & 0 \\ CB & D & \cdots & 0 \\ \vdots & \vdots & \ddots & \vdots \\ CA^{k-2}B & CA^{k-3}B & \cdots & D \end{bmatrix} = (R_{31} - R_{32}R_{22}^{\dagger}R_{21})R_{11}^{-1} \tag{6.67}$$

which is similar to the expression (6.43). Thus we can apply the same method used in the MOESP algorithm of Lemma 6.6 to compute the estimates of B and D.

Van Overschee and De Moor [165] have developed a subspace method of identifying state space models by using the state vector given by (6.58). In fact, from (6.58), we have the estimate of the state vector

$$X_k = [x(k)\ \ x(k+1)\ \cdots\ x(k+N-2)\ \ x(k+N-1)] \tag{6.68}$$

We define the following matrices with $N-1$ columns as

$$\bar{X}_{k+1} := [x(k+1)\ \ \cdots\ \ x(k+N-1)] \tag{6.69a}$$

$$\bar{X}_k := [x(k)\ \ \cdots\ \ x(k+N-2)] \tag{6.69b}$$

$$\bar{U}_{k|k} := [u(k)\ \ \cdots\ \ u(k+N-2)] \tag{6.69c}$$

$$\bar{Y}_{k|k} := [y(k)\ \ \cdots\ \ y(k+N-2)] \tag{6.69d}$$

Then, it follows that

$$\begin{bmatrix} \bar{X}_{k+1} \\ \bar{Y}_{k|k} \end{bmatrix} = \begin{bmatrix} A & B \\ C & D \end{bmatrix} \begin{bmatrix} \bar{X}_k \\ \bar{U}_{k|k} \end{bmatrix} \tag{6.70}$$

This is a system of linear equations for the system matrices, so that they can be estimated by applying the least-squares method:

$$\begin{bmatrix} \hat{A} & \hat{B} \\ \hat{C} & \hat{D} \end{bmatrix} = \left(\begin{bmatrix} \bar{X}_{k+1} \\ \bar{Y}_{k|k} \end{bmatrix} \begin{bmatrix} \bar{X}_k \\ \bar{U}_{k|k} \end{bmatrix}^{\mathrm{T}} \right) \left(\begin{bmatrix} \bar{X}_k \\ \bar{U}_{k|k} \end{bmatrix} \begin{bmatrix} \bar{X}_k \\ \bar{U}_{k|k} \end{bmatrix}^{\mathrm{T}} \right)^{-1}$$

Summarizing the above, we have the following lemma.

Lemma 6.8. *(N4SID algorithm)*

Step 1: Compute ξ *by using (6.65) and the LQ decomposition of (6.59).*

Step 2: Compute the state vector X_k *from (6.58), and define* $\bar{X}_{k+1}$, $\bar{X}_k$, $\bar{Y}_{k|k}$, $\bar{U}_{k|k}$ *as in (6.69).*

Step 3: Compute the matrices A, B, C, D *by solving the regression equation (6.70) by using the least-squares technique.* □

Remark 6.1. This is a slightly modified version of Algorithm 1 in Chapter 3 of [165]. The LQ decomposition of (6.52) has been employed by Verhaegen [171] to develop the PO-MOESP algorithm. Here we have used it to compute the oblique projections. The LQ decomposition is frequently used in Chapters 9 and 10 in order to compute orthogonal and oblique projections. □

6.7 SVD and Additive Noises

Up to now, we have presented deterministic realization results under the assumption that complete noise-free input-output data are available; but it is well known that real data are corrupted by noises. Thus, in this section, we consider the SVD of a data matrix corrupted by a uniform white noise. We show that the left singular vectors of a wide rectangular matrix are not very sensitive to additive white noise.

Consider a real rectangular matrix $X \in \mathbb{R}^{M \times N}$ with $M \ll N$. Let $\operatorname{rank}(X) = r \le M$, and let the SVD of $X/\sqrt{N}$ be given by

$$\frac{1}{\sqrt{N}} X = U \Sigma V^{\mathrm{T}} = [U_s \ \ U_n] \begin{bmatrix} \Sigma_s & 0 \\ 0 & 0 \end{bmatrix} \begin{bmatrix} V_s^{\mathrm{T}} \\ V_n^{\mathrm{T}} \end{bmatrix} = U_s \Sigma_s V_s^{\mathrm{T}} \tag{6.71}$$

where $\Sigma_s = \operatorname{diag}(\sigma_1, \cdots, \sigma_r)$, and where

$$\sigma_1 \ge \sigma_2 \ge \cdots \ge \sigma_r > \sigma_{r+1} = \cdots = \sigma_M = 0$$

The matrices $U \in \mathbb{R}^{M \times M}$, $V \in \mathbb{R}^{N \times N}$ are orthogonal, and $U_s := U(1:M, 1:r)$, $V_s := V(1:N, 1:r)$. From (6.71),

$$\frac{1}{N} X X^{\mathrm{T}} U = U \Sigma \Sigma^{\mathrm{T}} = U \begin{bmatrix} \Sigma_s^2 & 0 \\ 0 & 0 \end{bmatrix} \in \mathbb{R}^{M \times M} \tag{6.72}$$

and hence,

$$\frac{1}{N} X X^{\mathrm{T}} u_i = \sigma_i^2 u_i, \qquad i = 1, \cdots, r$$

We see that the left singular vectors u_i of $X/\sqrt{N}$ are the eigenvectors of XX^{T}/N, so that we have

$$\sigma_i^2 = \lambda_i(XX^{\mathrm{T}}/N), \qquad i = 1, \cdots, r$$

We consider the effect of white noise on the SVD of X. Let X be perturbed by white noise Ξ. Then, the observed data Y is expressed as

$$Y = X + \Xi \tag{6.73}$$

where the element ξ_{ij} of Ξ is white noise with mean 0 and variance σ_ξ^2. Thus, we have

$$\lim_{N\to\infty} \frac{1}{N} E(YY^{\mathrm{T}}) = \lim_{N\to\infty} \frac{1}{N} XX^{\mathrm{T}} + \sigma_\xi^2 I_M$$

Hence, from (6.72), YY^{T}/N is approximated as

$$\begin{aligned} \frac{1}{N} YY^{\mathrm{T}} &\simeq \frac{1}{N} XX^{\mathrm{T}} + \sigma_\xi^2 I_M \\ &= U \left(\begin{bmatrix} \Sigma_s^2 & 0 \\ 0 & 0 \end{bmatrix} \right) U^{\mathrm{T}} + \sigma_\xi^2 \, I_M \end{aligned} \tag{6.74}$$

where N is sufficiently large and where $U(\sigma_\xi^2 I_M)U^{\mathrm{T}} = U^{\mathrm{T}}(\sigma_\xi^2 I_M)U = \sigma_\xi^2 I_M$ is used. Defining $S_s^2 = \Sigma_s^2 + \sigma_\xi^2 I_r$, (6.74) becomes

$$\frac{1}{N} YY^{\mathrm{T}} \simeq [U_s \;\; U_n] \begin{bmatrix} S_s^2 & 0 \\ 0 & \sigma_\xi^2 I_{M-r} \end{bmatrix} \begin{bmatrix} U_s^{\mathrm{T}} \\ U_n^{\mathrm{T}} \end{bmatrix} = US^2U^{\mathrm{T}} \tag{6.75}$$

where $S = \mathrm{diag}(s_1, \cdots, s_M)$, $s_i > 0$ with

$$s_i = \begin{cases} \sqrt{\sigma_i^2 + \sigma_\xi^2}, & i = 1, \cdots, r \\ \sigma_\xi, & i = r+1, \cdots, M \end{cases} \tag{6.76}$$

as shown in Figure 6.3. It should be noted that the right-hand side of (6.75) is the eigenvalue decomposition of the sample covariance matrix of Y.

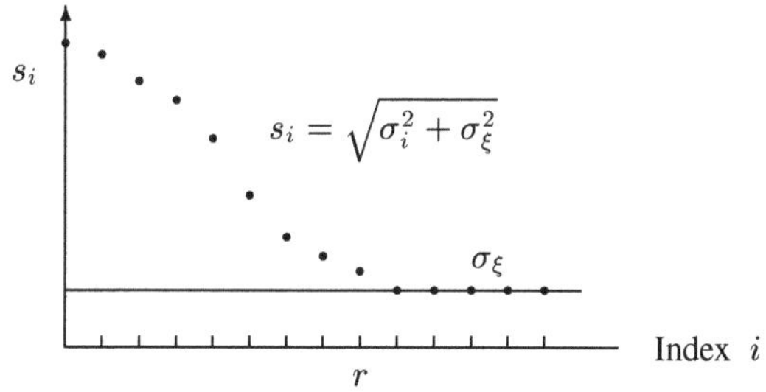

Figure 6.3. Singular values of $Y/\sqrt{N}$ and $X/\sqrt{N}$

Lemma 6.9. *Suppose that the variance σ_ξ^2 of the white noise ξ_{ij} is relatively small, and that N is sufficiently large. Then, for the SVDs of $X/\sqrt{N}$ and $Y/\sqrt{N}$, the following (i) $\sim$ (iii) hold.*

(i) The singular values of $Y/\sqrt{N}$ *are given by* $(s_1, s_2, \cdots, s_M)$, *where*

$$s_1 \geq s_2 \geq \cdots \geq s_r > s_{r+1} = \cdots = s_M = \sigma_\xi$$

Thus, the $M - r$ *minimum singular values of* $Y/\sqrt{N}$ *are nearly equal to* σ_ξ.

(ii) The eigenvalues of XX^{T}/N *are obtained by* $\sigma_i^2 = s_i^2 - \sigma_\xi^2$, $i = 1, \cdots, r$.

(iii) The left singular vectors of $Y/\sqrt{N}$ *are close to those of* $X/\sqrt{N}$. *Hence, we know the left singular vectors* $u_1, \cdots, u_r$ *of* $X/\sqrt{N}$ *from the left singular vectors corresponding to* r *singular values* $s_1 \geq s_2 \geq \cdots \geq s_r$ *of* $Y/\sqrt{N}$. *This fact implies that information about* X *can be extracted from the noise corrupted observation* Y *by using the SVD.*

Proof. See [40, 162]. □

Remark 6.2. We see from Lemma 6.9 that the information about X is contained in the r left singular vectors corresponding to the r largest singular values of Y, and no information is in the singular vectors corresponding to smaller singular values. Hence the subspace spanned by the left singular vectors $U_s = [u_1, \cdots, u_r]$ corresponding to the first r singular values is called the signal subspace. Also, the subspace spanned by $U_n = [u_{r+1}, \cdots, u_M]$ is called the noise subspace associated with the space spanned by Y. It should be noted here that noise subspaces are no less important than signal subspaces in applications. In fact, noise subspaces are successfully used for solving blind identification problems [156, 162]. □

We give a numerical example which is related to a frequency estimation problem based on noisy data. This clearly shows the meaning of Lemma 6.9.

Example 6.7. Consider a simple sinusoidal model

$$x(t) = a_1 \sin(\omega_1 t + \varphi_1) + a_2 \sin(\omega_2 t + \varphi_2)$$

where $a_1 = 10$, $a_2 = 5$ denote the amplitudes, $\omega_1 = 0.24\pi$, $\omega_2 = 0.26\pi$ two adjacent angular frequencies, φ_1, φ_2 random variables with uniform distribution on the interval $(-\pi, \pi)$. We assume that the observation is given by

$$y(t) = x(t) + e(t)$$

where e is a Gaussian white noise with mean zero and variance σ^2.

For random initial phases φ_1, φ_2, and a Gaussian white noise e, we generated the data $y(t)$, $t = 1, \cdots, 1024$. Assuming that $k = 16$, we formed X and Y, and computed the singular values σ_i and s_i of $X/\sqrt{N}$ and $Y/\sqrt{N}$, respectively; the results are shown in Table 6.4.

We observe that the singular values s_i, $i \geq 5$ of $Y/\sqrt{N}$ decrease very slowly as the index i. In this case, it is not difficult to determine $\operatorname{rank}(X/\sqrt{N}) = 4$ based on the distribution of singular values of $Y/\sqrt{N}$. If the noise variance of e increases, the decision becomes difficult. Also, if the difference $|\omega_1 - \omega_2|$ of the two angular frequencies get larger, the singular values s_3, s_4 become larger, while s_i, $i \geq 5$ remain almost the same. Thus, if the difference of two angular frequencies is expanded, the rank determination of $X/\sqrt{N}$ becomes easier. □

Table 6.4. Singular values of $X/\sqrt{N}$ and $Y/\sqrt{N}$

i	1	2	3	4	5	6	$\cdots$	16
σ_i	22.6323	22.0416	2.5283	2.5128	0	0	$\cdots$	0
s_i	22.6903	22.1013	2.7308	2.7276	1.1132	1.0769	$\cdots$	0.9046

6.8 Notes and References

- The germ of realization theory is found in the papers by Gilbert [56] and Kalman [82]. Then Ho and Kalman [72] has first solved the deterministic realization problem and derived a method of constructing a discrete-time state space model based on a given impulse response sequence; see also Kalman, Falb and Arbib [85]. Zeiger and McEwen [184] have further studied this algorithm based on the SVD to make its numerical implementation easier. Other references cited in this chapter are review articles [162, 175] and books [59, 147, 157].
- Two realization problems are stated in Section 6.1; one is the classical realization problem to recover state space models from given impulse responses, and the other is the deterministic identification problem to construct state space models from observed input and output data. The classical solution based on the SVD of the Hankel matrix formed by the given impulse responses is described in Section 6.2. A program of Ho-Kalman algorithm of Lemma 6.1 is provided in Table D.1 of Appendix D.
- A crucial problem of the realization method based on the infinite Hankel matrix is that it is necessary to assume that the Hankel matrix has finite rank *a priori*. In fact, it is impossible to determine the rank of infinite dimensional matrix in finite steps, and also it is not practical to assume that infinite impulse responses are available. The realization problem based on finite impulse response matrices is related to the partial realization problem; see Theorem 3.14.
- In Section 6.3, we have defined the data matrix generated by the input-output observations, and basic assumptions for the inputs and system are introduced. It is shown by using some examples that information about the image of extended observability matrix can be retrieved from the data matrix [162]. Lemma 6.3 is due to Moonen *et al.* [118, 119], but essentially the same result is proved in Gopinath [62], of which relation to subspace methods has been explored in [177]. The proof of Lemma 6.3 is based on the author's review paper [86], and Lemma 6.4 is adapted from the technical report [181].
- In Section 6.4, we have shown that the image of extended observability matrix can be extracted by using the LQ decomposition of the data matrix, followed by the SVD. Lemma 6.5 provides a system theoretic interpretation of the L-matrix, each column of which is an input-output pair of the system.
- Two subspace identification methods, the MOESP method [172, 173] and N4SID method [164, 165], are introduced in Sections 6.5 and 6.6, respectively. A proof

of the N4SID method is based on Lemma 6.5 and a new Theorem 6.2. Some numerical results using the MOESP method are also included.

- In Section 6.7, based on [40, 150, 162], we considered the SVD of wide rectangular matrices and the influence of white noise on the SVD, and defined signal and noise subspaces. It is shown that, since the SVD is robust to a white noise, the column space of unknown signal is recovered from the column space of noise corrupted observed signal. Lemma 6.9 gives a basis of MUSIC method; for more details, see [150, 162].

6.9 Problems

6.1 Find the realizations of the following sequences.

(a) Natural numbers: $(0,\ 1,\ 2,\ \cdots)$

(b) A periodic signal: $(0,\ 1,\ 0,\ -1,\ 0,\ 0,\ 1,\ 0,\ -1,\ 0,\ 0\ \cdots)$

6.2 Compute the reachability and observability Gramians for $(A,\ B,\ C)$ obtained by the algorithm of Lemma 6.1 under the conditions of Lemma 6.2.

6.3 Suppose that $A \in \mathbb{R}^{p\times N}$, $B \in \mathbb{R}^{q\times N}$, where $p,\ q < N$. Let the orthogonal projection of the row vectors of A onto the space spanned by the row vectors of B be defined by $\hat{E}\{A \mid B\}$. Prove the following.

$$\hat{E}\{A \mid B\} = AB^{\mathrm{T}}(BB^{\mathrm{T}})^{\dagger}B$$

If B has full row rank, the pseudo-inverse is replaced by the inverse.

6.4 Let A and B be defined in Problem 6.3. Consider the LQ decomposition of (6.32):

$$\begin{bmatrix} B \\ A \end{bmatrix} = \begin{bmatrix} L_{11} & 0 \\ L_{21} & L_{22} \end{bmatrix} \begin{bmatrix} Q_1^{\mathrm{T}} \\ Q_2^{\mathrm{T}} \end{bmatrix}$$

Suppose that B has full row rank. Show that the orthogonal projection is given by

$$\hat{E}\{A \mid B\} = L_{21}Q_1^{\mathrm{T}} = L_{21}L_{11}^{-1}B = A(Q_1Q_1^{\mathrm{T}})$$

Let $B^{\perp}$ be the orthogonal complement of the space spanned by the row vectors of B. Then, the orthogonal projection of the row vectors of A onto $B^{\perp}$ is expressed as

$$\hat{E}\{A \mid B^{\perp}\} = L_{22}Q_2^{\mathrm{T}} = A(Q_2Q_2^{\mathrm{T}})$$

6.5 Suppose that $A \in \mathbb{R}^{p\times N}$, $B \in \mathbb{R}^{q\times N}$, $C \in \mathbb{R}^{r\times N}$, where $p,\ q,\ r < N$. Suppose that B and C have row full rank, and $\mathrm{span}\{B\} \cap \mathrm{span}\{C\} = \{0\}$. (Note that this condition corresponds to A3) in Assumption 6.1.) Then, $\hat{E}_{\|C}\{A \mid B\}$, the oblique projection of the row vectors of A onto B along C, is expressed as

$$\hat{E}_{\|C}\{A \mid B\} = A[B^{\mathrm{T}}\ \ C^{\mathrm{T}}] \begin{bmatrix} BB^{\mathrm{T}} & BC^{\mathrm{T}} \\ CB^{\mathrm{T}} & CC^{\mathrm{T}} \end{bmatrix}^{-1} \begin{bmatrix} B \\ 0 \end{bmatrix}$$

6.6 Prove (6.60). (Hint: Use Lemmas 6.4 and 6.5.)

7

Stochastic Realization Theory (1)

Stochastic realization theory provides a method of constructing Markov models that simulate a stationary stochastic process with a prescribed covariance matrix, and serves as a basis for the subspace identification methods. In this chapter, we present a method of stochastic realization by using the deterministic realization theory and a linear matrix inequality (LMI) satisfied by the state covariance matrix. We show that all solutions to the stochastic realization problem are derived from solutions of the LMI. Using the approach due to Faurre [45–47], we show that the positive realness of covariance matrices and the existence of Markov models are equivalent, and then derive matrix Riccati equations that compute the boundary solutions of the LMI. Moreover, we discuss results for strictly positive real conditions and present a stochastic realization algorithm based on a finite covariance data.

7.1 Preliminaries

Consider a second-order vector stationary process $y \in \mathbb{R}^p$ with zero mean and covariance matrices

$$\Lambda(l) = E\{y(t+l)y^{\mathrm{T}}(t)\}, \qquad l = 0, \pm 1, \cdots \tag{7.1}$$

where the covariance matrices satisfy the condition

$$\sum_{l=-\infty}^{\infty} \|\Lambda(l)\| < \infty \tag{7.2}$$

It therefore follows that the spectral density matrix of y is given by

$$\Phi(z) = \sum_{l=-\infty}^{\infty} \Lambda(l) z^{-l} \qquad (p \times p \text{ matrix}) \tag{7.3}$$

In the following, we assume that y is regular and of full rank, in the sense that the spectral density matrix $\Phi(z)$ has full rank [68, 138].

The stochastic realization problem is to find all Markov models whose outputs simulate given covariance data of (7.1), or spectral density matrix of (7.3). In this chapter, we assume that an infinite data $\{y(t),\, t = 0, \pm 1, \cdots\}$, or a complete sequence of covariances $\{\Lambda(l)\}$, is available, so that here we develop a theoretical foundation for the existence of stochastic realizations, thereby leaving a more realistic problem of identifying state space models based on finite input-output data for later chapters.

Let t be the present time. Let the stacked infinite dimensional vectors of the future and past be given by[1]

$$f(t) := \begin{bmatrix} y(t) \\ y(t+1) \\ \vdots \\ \end{bmatrix}, \qquad p(t) := \begin{bmatrix} y(t-1) \\ y(t-2) \\ \vdots \\ \end{bmatrix}$$

Then, the covariance matrix of the future and past is defined by

$$H = E\{f(t)p^{\mathrm{T}}(t)\} = \begin{bmatrix} \Lambda(1) & \Lambda(2) & \Lambda(3) & \cdots \\ \Lambda(2) & \Lambda(3) & \Lambda(4) & \cdots \\ \Lambda(3) & \Lambda(4) & \Lambda(5) & \cdots \\ \vdots & \vdots & \vdots & \ddots \end{bmatrix} \tag{7.4}$$

and the auto-covariance matrices of the future and the past are respectively given by

$$T_+ = E\{f(t)f^{\mathrm{T}}(t)\} = \begin{bmatrix} \Lambda(0) & \Lambda^{\mathrm{T}}(1) & \Lambda^{\mathrm{T}}(2) & \cdots \\ \Lambda(1) & \Lambda(0) & \Lambda^{\mathrm{T}}(1) & \cdots \\ \Lambda(2) & \Lambda(1) & \Lambda(0) & \cdots \\ \vdots & \vdots & \vdots & \ddots \end{bmatrix} \tag{7.5}$$

and

$$T_- = E\{p(t)p^{\mathrm{T}}(t)\} = \begin{bmatrix} \Lambda(0) & \Lambda(1) & \Lambda(2) & \cdots \\ \Lambda^{\mathrm{T}}(1) & \Lambda(0) & \Lambda(1) & \cdots \\ \Lambda^{\mathrm{T}}(2) & \Lambda^{\mathrm{T}}(1) & \Lambda(0) & \cdots \\ \vdots & \vdots & \vdots & \ddots \end{bmatrix} \tag{7.6}$$

where H is an infinite dimensional block Hankel matrix, and $T_\pm$ are infinite dimensional block Toeplitz matrices.

As in the deterministic case, it is assumed that $\mathrm{rank}(H) = n < \infty$. Then, from the deterministic realization theory described in Section 6.2, there exists a minimal realization $(A,\, C,\, \bar{C},\, \Lambda(0))$ satisfying

$$\Lambda(l) = \begin{cases} \Lambda(0), & l = 0 \\ CA^{l-1}\bar{C}^{\mathrm{T}}, & l = 1,\, 2,\, \cdots \end{cases} \tag{7.7}$$

[1]Though the present time t is included in the future, we could include it in the past as well. Then, by definition, we obtain a model without a noise term in the output equation [2].

where $A \in \mathbb{R}^{n\times n}$ and $C,\ \bar{C} \in \mathbb{R}^{p\times n}$ are constant matrices. It should be noted from (7.2) and (7.7) that A is stable. Thus, the deterministic state space realization

$$G(z) = \left[\begin{array}{c|c} A & \bar{C}^{\mathrm{T}} \\ \hline C & \Lambda(0) \end{array}\right]$$

is observable and reachable, and that the impulse response matrices are given by $\{\Lambda(t),\ t = 0,\ 1,\ \cdots\}$. In the following, we say that $(A,\ C,\ \bar{C}^{\mathrm{T}})$ is minimal, if $(C,\ A)$ is observable and $(A,\ \bar{C}^{\mathrm{T}})$ is reachable.

Define the infinite dimensional observability and reachability matrices as

$$\mathcal{O} = \begin{bmatrix} C \\ CA \\ \vdots \\ \end{bmatrix}, \qquad \mathcal{C} = [\bar{C}^{\mathrm{T}}\ \ A\bar{C}^{\mathrm{T}}\ \ \ A^2\bar{C}^{\mathrm{T}}\ \ \cdots]$$

Then, we see from (7.7) that the block Hankel matrix of (7.4) has a factorization

$$H = \mathcal{O}\mathcal{C} \tag{7.8}$$

This is exactly the same factorization we have seen for the deterministic factorization in Section 6.2; see Theorem 6.1.

It can be shown from (7.7) that

$$\Lambda(l) = CA^{l-1}\bar{C}^{\mathrm{T}}\,1(l-1) + \Lambda(0)\delta_{l0} + \bar{C}(A^{\mathrm{T}})^{-l-1}C^{\mathrm{T}}1(-l-1) \tag{7.9}$$

where

$$1(l) = \begin{cases} 1, & l = 0,\ 1,\ \cdots \\ 0, & l = -1,\ -2,\ \cdots \end{cases}$$

Hence, from (7.3), the spectral density matrix is expressed as

$$\begin{aligned} \Phi(z) &= \sum_{l=1}^{\infty} CA^{l-1}\bar{C}^{\mathrm{T}}z^{-l} + \Lambda(0) + \sum_{l=-\infty}^{-1} \bar{C}(A^{\mathrm{T}})^{-l-1}C^{\mathrm{T}}z^{-l} \\ &= \sum_{l=1}^{\infty} CA^{l-1}\bar{C}^{\mathrm{T}}z^{-l} + \Lambda(0) + \sum_{l=1}^{\infty} \bar{C}(A^{\mathrm{T}})^{l-1}C^{\mathrm{T}}z^{l} \\ &= C(zI-A)^{-1}\bar{C}^{\mathrm{T}} + \frac{1}{2}\Lambda(0) + \bar{C}(z^{-1}I - A^{\mathrm{T}})^{-1}C^{\mathrm{T}} + \frac{1}{2}\Lambda(0) \end{aligned} \tag{7.10}$$

If we define

$$Z(z) = C(zI-A)^{-1}\bar{C}^{\mathrm{T}} + \frac{1}{2}\Lambda(0) \tag{7.11}$$

then the spectral density matrix satisfies

$$\Phi(z) = Z(z) + Z^{\mathrm{T}}(z^{-1}) \tag{7.12}$$

This is a well-known additive decomposition of the spectral density matrix.

Let $\rho := \rho(A)$, the spectral radius. Since A is stable, we get $0 < \rho < 1$. Hence the right-hand side of (7.10) is absolutely convergent for $\rho < |z| < \rho^{-1}$, implying that the spectral density matrix is analytic in the annular domain $\rho < |z| < \rho^{-1}$ that includes the unit circle ($|z| = 1$) (see Example 4.12). Let $\Phi(\omega) := \Phi(z)|_{z=e^{j\omega}}$. Then, we have

$$\begin{aligned}\Phi(\omega) &= Z(e^{j\omega}) + Z^{\mathrm{T}}(e^{-j\omega}) \\ &= Z(e^{j\omega}) + Z^{\mathrm{H}}(e^{j\omega}) \geq 0, \quad -\pi < \omega \leq \pi \end{aligned} \tag{7.13}$$

For scalar systems, this is equivalently written as $\mathfrak{Re} Z(e^{j\omega}) \geq 0$, where $\mathfrak{Re}$ denotes the real part.

Definition 7.1. *(Positive real matrix) A square matrix $Z(z)$ is positive real if the conditions (i) and (ii) are satisfied:*

(i) $Z(z)$ is analytic in $|z| \geq 1$.
(ii) $Z(z)$ satisfies (7.13).

If, together with item (i), a stronger condition

(ii') $Z(e^{j\omega}) + Z^{\mathrm{H}}(e^{j\omega}) > 0, \quad -\pi < \omega \leq \pi$

holds, then $Z(z)$ is called strictly positive real. In this case, it follows that $\Phi(\omega) > 0$ for $-\pi < \omega \leq \pi$, and such $\Phi(z)$ is called coercive. □

7.2 Stochastic Realization Problem

In this section we introduce a forward Markov model for a stationary stochastic process, and define the stochastic realization problem due to Faurre [45].

Consider a state space model of the form

$$x(t+1) = A_0 x(t) + w(t) \tag{7.14a}$$

$$y(t) = C_0 x(t) + v(t) \tag{7.14b}$$

where $y \in \mathbb{R}^p$ is the output vector, $x \in \mathbb{R}^n$ the state vector, and where $w \in \mathbb{R}^n$ and $v \in \mathbb{R}^p$ are white noises with mean zero and covariance matrices

$$E\left\{ \begin{bmatrix} w(t) \\ v(t) \end{bmatrix} \begin{bmatrix} w^{\mathrm{T}}(s) & v^{\mathrm{T}}(s) \end{bmatrix} \right\} = \begin{bmatrix} Q & S \\ S^{\mathrm{T}} & R \end{bmatrix} \delta_{ts} \tag{7.15}$$

It is assumed that A_0 is stable, (C_0, A_0) is observable, and $(A_0, Q^{1/2})$ is reachable. In this case, the model of (7.14) is called a stationary Markov model as discussed in Section 4.7 (see Figure 7.1).

Let $\Pi = E\{x(t)x^{\mathrm{T}}(t)\}$. Then, we have

$$\Pi = \sum_{i=0}^{\infty} A_0^i Q (A_0^{\mathrm{T}})^i > 0$$

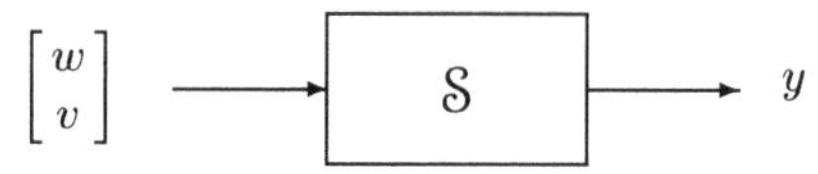

Figure 7.1. Markov model

and hence Π satisfies the Lyapunov equation (see Section 4.7)

$$\Pi = A_0 \Pi A_0^{\mathrm{T}} + Q \tag{7.16}$$

From Lemma 4.10, the covariance matrix of x satisfies

$$\Lambda_{xx}(l) = \begin{cases} A_0^l \Pi, & l = 0,\, 1,\, \cdots \\ \Pi (A_0^{\mathrm{T}})^{-l}, & l = -1,\, -2,\, \cdots \end{cases} \tag{7.17}$$

Moreover, from Lemma 4.11, the covariance matrix of y is given by

$$\Lambda(l) = \begin{cases} C_0 A_0^{l-1}(A_0 \Pi C_0^{\mathrm{T}} + S), & l = 1,\, 2,\, \cdots \\ C_0 \Pi C_0^{\mathrm{T}} + R, & l = 0 \\ \Lambda^{\mathrm{T}}(-l), & l = -1,\, -2,\, \cdots \end{cases} \tag{7.18}$$

Thus, comparing (7.7) and (7.18), we conclude that

$$A_0 = A \tag{7.19a}$$

$$C_0 = C \tag{7.19b}$$

$$A_0 \Pi C_0^{\mathrm{T}} + S = \bar{C}^{\mathrm{T}} \tag{7.19c}$$

$$C_0 \Pi C_0^{\mathrm{T}} + R = \Lambda(0) \tag{7.19d}$$

It may be noted that A, C, $\bar{C}^{\mathrm{T}}$ in the right-hand side of (7.19) can respectively be replaced by $T^{-1}AT$, CT, $T^{-1}\bar{C}^{\mathrm{T}}$ for an arbitrary nonsingular matrix T; but for simplicity, it is assumed that $T = I_n$.

Since A, C, $\bar{C}$ in (7.19) are given by the factorization (7.7), they are regarded as given data. Recall from Example 4.11 that these matrices A, C, $\bar{C}$ are expressed as

$$\begin{aligned} A &= E\{x(t+1)x^{\mathrm{T}}(t)\}\Pi^{-1} \\ C &= E\{y(t)x^{\mathrm{T}}(t)\}\Pi^{-1} \\ \bar{C} &= E\{y(t)x^{\mathrm{T}}(t+1)\} \end{aligned} \tag{7.20}$$

It follows from (7.16) and (7.19) that

$$\Pi - A\Pi A^{\mathrm{T}} = Q \tag{7.21a}$$

$$\bar{C}^{\mathrm{T}} - A\Pi C^{\mathrm{T}} = S \tag{7.21b}$$

$$\Lambda(0) - C\Pi C^{\mathrm{T}} = R \tag{7.21c}$$

where

$$\begin{bmatrix} Q & S \\ S^{\mathrm{T}} & R \end{bmatrix} \geq 0, \qquad \Pi > 0 \tag{7.22}$$

For given data $(A,\ C,\ \bar{C},\ \Lambda(0))$, the stochastic realization problem considered by Faurre [46,47] is to find four covariance matrices $(\Pi,\ Q,\ R,\ S)$ satisfying (7.21) and (7.22). This problem can be solved by using the techniques of linear matrix inequality (LMI) and spectral factorization as shown in Section 7.3.

From Lemma 4.12, recall that a backward Markov model for a stationary process is given by the following lemma .

Lemma 7.1. *Define* $x_b(t-1) = \bar{\Pi}x(t)$ *with* $\bar{\Pi} = \Pi^{-1}$. *Then, the backward Markov model is given by*

$$x_b(t-1) = A^{\mathrm{T}}x_b(t) + w_b(t) \tag{7.23a}$$

$$y(t) = \bar{C}x_b(t) + v_b(t) \tag{7.23b}$$

where A *and* $\bar{C}$, *called the backward output matrix, are given by (7.20), and where* w_b *and* v_b *are zero mean white noises with covariance matrices*

$$E\left\{\begin{bmatrix} w_b(t) \\ v_b(t) \end{bmatrix} \begin{bmatrix} w_b^{\mathrm{T}}(s) & v_b^{\mathrm{T}}(s) \end{bmatrix}\right\} = \begin{bmatrix} \bar{Q} & \bar{S} \\ \bar{S}^{\mathrm{T}} & \bar{R} \end{bmatrix} \delta_{ts} \tag{7.24}$$

Moreover, we have $\mathrm{cov}\{x_b(t)\} = \bar{\Pi}$ *and*

$$\bar{Q} = \bar{\Pi} - A^{\mathrm{T}}\bar{\Pi}A, \quad \bar{S} = C^{\mathrm{T}} - A^{\mathrm{T}}\bar{\Pi}\bar{C}^{\mathrm{T}}, \quad \bar{R} = \Lambda(0) - \bar{C}\bar{\Pi}\bar{C}^{\mathrm{T}} \tag{7.25}$$

Proof. See the proof of Lemma 4.12. □

We see that the forward Markov model is characterized by $(\Pi,\ A,\ C, \bar{C}, \Lambda(0))$, whereas the backward model is characterized by $(\bar{\Pi},\ A^T,\ \bar{C},\ C,\ \Lambda(0))$. Thus there exists a one-to-one correspondence between the forward and backward models of the form

$$\Pi \leftrightarrow \Pi^{-1}, \quad A \leftrightarrow A^{\mathrm{T}}, \quad C \leftrightarrow \bar{C}, \quad Q \leftrightarrow \bar{Q}, \quad S \leftrightarrow \bar{S}, \quad R \leftrightarrow \bar{R}$$

This fact is employed to prove the boundedness of the solution set of the ARE satisfied by Π of Theorem 7.4 (see Section 7.4).

7.3 Solution of Stochastic Realization Problem

Theory of stochastic realization provides a technique of computing all the Markov models that generate a stationary stochastic process with a prescribed covariance matrix. Besides, it is very important from practical points of view since it serves a theoretical foundation for subspace methods for identifying state space models from given input-output data.

7.3.1 Linear Matrix Inequality

Suppose that the data $(A,\ C,\ \bar{C},\ \Lambda(0))$ are given. Substituting $(Q,\ R,\ S)$ of (7.21) into (7.22), we see that the stochastic realization problem is reduced to finding solutions $\Pi > 0$ satisfying the LMI such that

$$M(\Pi) := \begin{bmatrix} \Pi - A\Pi A^{\mathrm{T}} & \bar{C}^{\mathrm{T}} - A\Pi C^{\mathrm{T}} \\ \bar{C} - C\Pi A^{\mathrm{T}} & \Lambda(0) - C\Pi C^{\mathrm{T}} \end{bmatrix} \geq 0 \tag{7.26}$$

Note that if there exists $\Pi > 0$ that satisfies $M(\Pi) \geq 0$, then from (7.21), we have (Q, R, S) satisfying (7.22). Thus $Z(z)$ of (7.11) becomes positive real.

Theorem 7.1. *Suppose that $(A,\ C,\ \bar{C}^{\mathrm{T}})$ is minimal, and A is stable. Let Π be a solution of the LMI (7.26), and let a factorization of $M(\Pi)$ be given by*

$$M(\Pi) = \begin{bmatrix} B \\ D \end{bmatrix} \begin{bmatrix} B \\ D \end{bmatrix}^{\mathrm{T}} \geq 0 \tag{7.27}$$

where $\begin{bmatrix} B \\ D \end{bmatrix}$ has full column rank. In terms of B and D, we further define

$$W(z) = D + C(zI - A)^{-1}B \tag{7.28}$$

Then, $W(z)$ is a minimal spectral factor of the spectral density matrix $\Phi(z)$ that satisfies

$$\Phi(z) = W(z)W^{\mathrm{T}}(z^{-1}) \tag{7.29}$$

Conversely, suppose that there exists a stable minimal spectral factor satisfying (7.29). Then, there exists a solution $\Pi > 0$ satisfying (7.26).

Proof. [107] Let $\Pi > 0$ be a solution of (7.26). It is clear that $B,\ D$ satisfying (7.27) are unique up to orthogonal transforms. From the (1,1)-block of the equality of (7.27), we have the Lyapunov equation

$$\Pi - A\Pi A^{\mathrm{T}} = BB^{\mathrm{T}} \tag{7.30}$$

where Π is positive definite and A is stable. Thus it follows from Lemma 3.5 that $(A,\ B)$ is reachable, implying that $W(z)$ of (7.28) is minimal and stable.

Now, from (7.28), we have

$$\begin{aligned} W(z)W^{\mathrm{T}}(z^{-1}) &= [D + C(zI - A)^{-1}B][D^{\mathrm{T}} + B^{\mathrm{T}}(z^{-1}I - A^{\mathrm{T}})^{-1}C^{\mathrm{T}}] \\ &= DD^{\mathrm{T}} + C(zI - A)^{-1}BB^{\mathrm{T}}(z^{-1}I - A^{\mathrm{T}})^{-1}C^{\mathrm{T}} \\ &\quad + C(zI - A)^{-1}BD^{\mathrm{T}} + DB^{\mathrm{T}}(z^{-1}I - A^{\mathrm{T}})^{-1}C^{\mathrm{T}} \end{aligned} \tag{7.31}$$

From (7.30), we have the identity

$$BB^{\mathrm{T}} = (zI - A)\Pi(z^{-1}I - A^{\mathrm{T}}) + (zI - A)\Pi A^{\mathrm{T}} + A\Pi(z^{-1}I - A^{\mathrm{T}})$$

Substituting this identity into (7.31) and rearranging the terms using $Q = BB^{\mathrm{T}}$, $S = BD^{\mathrm{T}}$, $R = DD^{\mathrm{T}}$ and (7.21) yield

$$\begin{aligned} W(z)W^{\mathrm{T}}(z^{-1}) &= DD^{\mathrm{T}} + C\Pi C^{\mathrm{T}} + C(zI - A)^{-1}(A\Pi C^{\mathrm{T}} + BD^{\mathrm{T}}) \\ &\quad + (C\Pi A^{\mathrm{T}} + DB^{\mathrm{T}})(z^{-1}I - A^{\mathrm{T}})^{-1}C^{\mathrm{T}} \\ &= \Lambda(0) + C(zI - A)^{-1}\bar{C}^{\mathrm{T}} + \bar{C}(z^{-1}I - A^{\mathrm{T}})^{-1}C^{\mathrm{T}} \\ &= Z(z) + Z^{\mathrm{T}}(z^{-1}) = \Phi(z) \end{aligned} \tag{7.32}$$

Thus we conclude that $W(z)$ is a stable minimal spectral factor of $\Phi(z)$.

Conversely, suppose that $W(z) = D + C_1(zI - A_1)^{-1}B$ is a minimal spectral factor. Since $(A, C, \bar{C}, \Lambda(0))$ are given data, we can set $A_1 = A$ and $C_1 = C$ as in (7.19). Now using $W(z) = D + C(zI - A)^{-1}B$, we get

$$W(z)W^{\mathrm{T}}(z^{-1}) = [C(zI - A)^{-1} \quad I] \begin{bmatrix} BB^{\mathrm{T}} & BD^{\mathrm{T}} \\ DB^{\mathrm{T}} & DD^{\mathrm{T}} \end{bmatrix} \begin{bmatrix} (zI - A^{\mathrm{T}})^{-1}C^{\mathrm{T}} \\ I \end{bmatrix} \tag{7.33}$$

Since (A, B) is reachable, there exists a unique solution $\breve{\Pi} > 0$ for the Lyapunov equation $BB^{\mathrm{T}} = \breve{\Pi} - A\breve{\Pi}A^{\mathrm{T}}$, from which

$$BB^{\mathrm{T}} = (zI - A)\breve{\Pi}(z^{-1}I - A^{\mathrm{T}}) + (zI - A)\breve{\Pi}A^{\mathrm{T}} + A\breve{\Pi}(z^{-1}I - A^{\mathrm{T}})$$

Substituting this identity into (7.33) yields

$$\begin{aligned} W(z)W^{\mathrm{T}}(z^{-1}) &= DD^{\mathrm{T}} + C\breve{\Pi}C^{\mathrm{T}} + C(zI - A)^{-1}(A\breve{\Pi}C^{\mathrm{T}} + BD^{\mathrm{T}}) \\ &\quad + (C\breve{\Pi}A^{\mathrm{T}} + DB^{\mathrm{T}})(z^{-1}I - A^{\mathrm{T}})^{-1}C^{\mathrm{T}} \end{aligned}$$

But from (7.11) and (7.12),

$$\begin{aligned} W(z)W^{\mathrm{T}}(z^{-1}) &= \Phi(z) = Z(z) + Z^{\mathrm{T}}(z^{-1}) \\ &= C(zI - A)^{-1}\bar{C}^{\mathrm{T}} + \Lambda(0) + \bar{C}(z^{-1}I - A^{\mathrm{T}})^{-1}C^{\mathrm{T}} \end{aligned}$$

holds. Since (C, A) is observable, all the columns of $C(zI - A)^{-1}$ are independent. Thus, comparing two expressions for $W(z)W^{\mathrm{T}}(z^{-1})$ above gives

$$DD^{\mathrm{T}} = \Lambda(0) - C\breve{\Pi}C^{\mathrm{T}}, \qquad BD^{\mathrm{T}} = \bar{C}^{\mathrm{T}} - A\breve{\Pi}C^{\mathrm{T}}$$

By using these relations in (7.33), we get

$$\begin{aligned} W(z)W^{\mathrm{T}}(z^{-1}) &= [C(zI - A)^{-1} \quad I] \begin{bmatrix} \breve{\Pi} - A\breve{\Pi}A^{\mathrm{T}} & \bar{C}^{\mathrm{T}} - A\breve{\Pi}C^{\mathrm{T}} \\ \bar{C} - C\breve{\Pi}A^{\mathrm{T}} & \Lambda(0) - C\breve{\Pi}C^{\mathrm{T}} \end{bmatrix} \\ &\qquad \times \begin{bmatrix} (zI - A^{\mathrm{T}})^{-1}C^{\mathrm{T}} \\ I \end{bmatrix} \\ &= [C(zI - A)^{-1} \quad I]M(\breve{\Pi}) \begin{bmatrix} (zI - A^{\mathrm{T}})^{-1}C^{\mathrm{T}} \\ I \end{bmatrix} \end{aligned}$$

Clearly, the right-hand side of the above equation is nonnegative definite for $z = e^{j\theta}$, $-\pi < \theta \le \pi$, so that $M(\breve{\Pi}) \ge 0$. Thus the triplet $(\breve{\Pi}, B, D)$ satisfies (7.27), implying that the LMI (7.26) has a solution $\breve{\Pi} > 0$. □

Now we examine the size of spectral factor $W(z)$ under the assumption that $\begin{bmatrix} B \\ D \end{bmatrix}$ has full column rank and that $R(\Pi) := \Lambda(0) - C\Pi C^{\mathrm{T}} > 0$[2]. Recall that the factorization formula for the block matrix of the form (see Problem 2.2)

$$\begin{bmatrix} X & Y \\ Z & V \end{bmatrix} = \begin{bmatrix} I & YV^{-1} \\ 0 & I \end{bmatrix} \begin{bmatrix} X - YV^{-1}Z & 0 \\ 0 & V \end{bmatrix} \begin{bmatrix} I & 0 \\ V^{-1}Z & I \end{bmatrix}$$

Comparing the above expression with $M(\Pi)$ of (7.26), we get $X = \Pi - A\Pi A^{\mathrm{T}}$, $Y = \bar{C}^{\mathrm{T}} - A\Pi C^{\mathrm{T}} = Z^{\mathrm{T}}$ and $V = R(\Pi)$. Thus it can be shown that

$$M(\Pi) = \begin{bmatrix} I & K \\ 0 & I \end{bmatrix} \begin{bmatrix} \Pi - A\Pi A^{\mathrm{T}} - KR(\Pi)K^{\mathrm{T}} & 0 \\ 0 & R(\Pi) \end{bmatrix} \begin{bmatrix} I & 0 \\ K^{\mathrm{T}} & I \end{bmatrix}$$

where K is exactly the Kalman gain given by [see (5.74)]

$$K = (\bar{C}^{\mathrm{T}} - A\Pi C^{\mathrm{T}})R^{-1}(\Pi)$$

We define

$$\mathrm{Ric}(\Pi) := A\Pi A^{\mathrm{T}} - \Pi + KR(\Pi)K^{\mathrm{T}} \tag{7.34}$$

Then, under the assumption that $R(\Pi) > 0$, we see that

$$M(\Pi) \ge 0 \quad \Leftrightarrow \quad \mathrm{Ric}(\Pi) \le 0$$

This implies that $M(\Pi) \ge 0$ if and only if $R(\Pi) > 0$ and the algebraic Riccati inequality (ARI)

$$A\Pi A^{\mathrm{T}} - \Pi + (\bar{C}^{\mathrm{T}} - A\Pi C^{\mathrm{T}})(\Lambda(0) - C\Pi C^{\mathrm{T}})^{-1}(\bar{C} - C\Pi A^{\mathrm{T}}) \le 0 \tag{7.35}$$

holds. Moreover, we have

$$m := \operatorname{rank} M(\Pi) = p + \operatorname{rank} \mathrm{Ric}(\Pi) \tag{7.36}$$

Hence, if Π is a solution of the ARE

$$\Pi = A\Pi A^{\mathrm{T}} + (\bar{C}^{\mathrm{T}} - A\Pi C^{\mathrm{T}})(\Lambda(0) - C\Pi C^{\mathrm{T}})^{-1}(\bar{C} - C\Pi A^{\mathrm{T}}) \tag{7.37}$$

then the corresponding spectral factor $W(z)$ is a $p \times p$ square matrix. Otherwise, we have $m > p$, so that the spectral factor $W(z)$ becomes a wide rectangular matrix.

It follows from Theorem 7.1 and the above argument that a Markov model of y is given by

$$x(t+1) = Ax(t) + B\nu(t) \tag{7.38a}$$

$$y(t) = Cx(t) + D\nu(t) \tag{7.38b}$$

where B and D are solutions of the LMI, and ν is a white noise with mean zero and covariance matrix I_m with $m \ge p$.

[2]Note that the latter condition cannot be avoided in the following development.

Lemma 7.2. *Suppose that* $(A, \bar{C}^{\mathrm{T}})$ *is reachable. Then all solutions of the ARE (7.37) are positive definite. (Note that* $\Pi = E\{x(t)x^{\mathrm{T}}(t)\} \geq 0$.*)*

Proof. Suppose that Π is not positive definite. Then, there exists $\eta \in \mathbb{R}^n$ with $\eta^{\mathrm{T}}\Pi = 0$. Thus, pre-multiplying (7.37) by η^{T} and post-multiplying by η yield

$$\eta^{\mathrm{T}} A\Pi A^{\mathrm{T}}\eta + \eta^{\mathrm{T}}(\bar{C}^{\mathrm{T}} - A\Pi C^{\mathrm{T}})(\Lambda(0) - C\Pi C^{\mathrm{T}})^{-1}(\bar{C} - C\Pi A^{\mathrm{T}})\eta = 0$$

Since both terms in the left-hand side are nonnegative, we have

$$\eta^{\mathrm{T}} A\Pi = 0, \qquad \eta^{\mathrm{T}}\bar{C}^{\mathrm{T}} = 0$$

Define $\eta_1^{\mathrm{T}} := \eta^{\mathrm{T}} A$. Then, from the above, $\eta_1^{\mathrm{T}}\Pi = 0$ holds, and hence we get

$$\eta_1^{\mathrm{T}} A\Pi = 0, \quad \eta_1^{\mathrm{T}}\bar{C}^{\mathrm{T}} = 0 \quad \Rightarrow \quad \eta^{\mathrm{T}} A^2\Pi = 0, \quad \eta^{\mathrm{T}} A\bar{C}^{\mathrm{T}} = 0$$

Repeating this procedure, we have eventually

$$\eta^{\mathrm{T}}[\bar{C}^{\mathrm{T}} \quad A\bar{C}^{\mathrm{T}} \quad A^2\bar{C}^{\mathrm{T}} \quad \cdots \quad] = 0$$

This is a contradiction that $(A, \bar{C}^{\mathrm{T}})$ is reachable, implying that $\Pi > 0$. □.

The ARE of (7.37) is the same as that of (5.75) satisfied by the state covariance equation of the stationary Kalman filter. Hence, we see that the square spectral factor is closely related to the stationary Kalman filter as shown in Example 7.1 below.

7.3.2 Simple Examples

Example 7.1. Suppose that $A = 1/3$, $C = 2$, $\bar{C} = 2/3$, $\Lambda(0) = 9/4$ are given. From (7.11) and (7.12), the spectral density is given by

$$\Phi(z) = Z(z) + Z^{\mathrm{T}}(z^{-1}) = \frac{4/3}{z - 1/3} + \frac{9}{8} + \frac{4/3}{z^{-1} - 1/3} + \frac{9}{8}$$

It is easy to see that

$$Z(z) = \frac{9}{8} + \frac{4/3}{z - 1/3}$$

is strictly positive real. Also, the ARI (7.35) becomes

$$\mathrm{Ric}(\Pi) = \frac{1}{9}\Pi - \Pi + \left(\frac{2}{3} - \frac{2}{3}\Pi\right)^2\left(\frac{9}{4} - 4\Pi\right)^{-1} \leq 0$$

where $9/4 - 4\Pi > 0$ by the assumption. It therefore follows that

$$4\Pi^2 - \frac{26}{9}\Pi + \frac{4}{9} = 4\left(\Pi - \frac{1}{2}\right)\left(\Pi - \frac{2}{9}\right) \leq 0$$

Hence, we have $\Pi_* = 2/9$ and $\Pi^* = 1/2$, and any solution Π of the Riccati inequality satisfies $\Pi_* \leq \Pi \leq \Pi^*$. It should be noted that these are boundary solutions of the LMI as well.

i) Consider the case where $\Pi = \Pi_* = 2/9$. From the LMI of (7.26),

$$M(2/9) = \begin{bmatrix} 16/81 & 14/27 \\ 14/27 & 49/36 \end{bmatrix} = \begin{bmatrix} 4/9 \\ 7/6 \end{bmatrix} \begin{bmatrix} 4/9 & 7/6 \end{bmatrix}$$

Thus we have $B = 4/9$, $D = 7/6$, so that, from (7.28), the spectral factor is given by

$$W_*(z) = \frac{7}{6} + \frac{8/9}{z - 1/3} = \left(\frac{7}{6}\right) \frac{z + 3/7}{z - 1/3}$$

The corresponding Markov model is given by the innovation model

$$x(t+1) = \frac{1}{3}x(t) + \frac{4}{9}\nu(t) \tag{7.39a}$$

$$y(t) = 2x(t) + \frac{7}{6}\nu(t) \tag{7.39b}$$

where ν is a white noise with mean 0 and variance 1.

ii) Consider the case where $\Pi = \Pi^* = 1/2$. In this case, we have

$$M(1/2) = \begin{bmatrix} 4/9 & 1/3 \\ 1/3 & 1/4 \end{bmatrix} = \begin{bmatrix} 2/3 \\ 1/2 \end{bmatrix} \begin{bmatrix} 2/3 & 1/2 \end{bmatrix}$$

Thus we get $B = 2/3$, $D = 1/2$, so that the spectral factor is given by

$$W^*(z) = \frac{1}{2} + \frac{4/3}{z - 1/3} = \left(\frac{1}{2}\right) \frac{z + 7/3}{z - 1/3}$$

so that the Markov model becomes

$$x(t+1) = \frac{1}{3}x(t) + \frac{2}{3}\nu(t) \tag{7.40a}$$

$$y(t) = 2x(t) + \frac{1}{2}\nu(t) \tag{7.40b}$$

We observe that $W_*(z)$ and its inverse $W_*^{-1}(z)$ are stable, implying that this is a minimal phase function. But, the inverse of $W^*(z)$ is unstable, so that this is a non-minimal phase function. □

The next example is concerned with a spectral factor for Π satisfying the Riccati inequality, *i.e.*, $2/9 < \Pi < 1/2$.

Example 7.2. For simplicity, let $\Pi = 1/4$. Then, we have

$$M(1/4) = \begin{bmatrix} 2/9 & 1/2 \\ 1/2 & 5/4 \end{bmatrix} = \begin{bmatrix} b_1 & b_2 \\ d & 0 \end{bmatrix} \begin{bmatrix} b_1 & d \\ b_2 & 0 \end{bmatrix} = \begin{bmatrix} b_1^2 + b_2^2 & b_1 d \\ d b_1 & d^2 \end{bmatrix}$$

Though there are other solutions to the above equation, we pick a particular solution $d = \sqrt{5}/2$, $b_1 = 1/\sqrt{5}$, $b_2 = 1/\sqrt{45}$. Thus the spectral factor is given by

$$W(z) = \begin{bmatrix} \dfrac{\sqrt{5}}{2} + \dfrac{2/\sqrt{5}}{z-1/3} & \dfrac{2/\sqrt{45}}{z-1/3} \end{bmatrix}$$

In this case, since $m = \operatorname{rank} M(1/4) = 2$, the spectral factor becomes rectangular, and the corresponding Markov model is given by

$$x(t+1) = \frac{1}{3}x(t) + \frac{1}{\sqrt{5}}\nu_1(t) + \frac{1}{\sqrt{45}}\nu_2(t) \tag{7.41a}$$

$$y(t) = 2x(t) + \frac{\sqrt{5}}{2}\nu_1(t) \tag{7.41b}$$

where ν_1 and ν_2 are mutually independent white noises with mean zero and unit variance. That $\Pi = 1/4$ implies that the variance of the state x of (7.41) is $1/4$.

Now we construct the stationary Kalman filter for the system (7.41). Since $A = 1/3$, $C = 2$, $Q = 2/9$, $S = 1/2$, $R = 5/4$, the ARE of (5.67) becomes

$$P = \frac{1}{9}P - \left(\frac{2P}{3} + \frac{1}{2}\right)^2 \left(4P + \frac{5}{4}\right)^{-1} + \frac{2}{9}$$

Rearranging the above equation yields

$$36P^2 + 8P - \frac{1}{4} = (4P+1)\left(9P - \frac{1}{4}\right) = 0$$

Thus we have $P = 1/36$ since $P \geq 0$, so that from (5.68), the Kalman gain is given by $K = 8/21$. This implies that the stationary Kalman filter has the form

$$\hat{x}(t+1 \mid t) = \frac{1}{3}\hat{x}(t \mid t-1) + \frac{8}{21}e(t) \tag{7.42a}$$

$$y(t) = 2\hat{x}(t \mid t-1) + e(t) \tag{7.42b}$$

where e is the innovation process with zero mean and covariance (see Lemma 5.7)

$$\operatorname{cov}\{e\} = C^2P + R = (7/6)^2$$

It can easily be shown that the innovation process e is related to ν of (7.39) via $e(t) = (7/6)\nu(t)$, so that (7.39) and (7.42) are equivalent under this relation. Also, the transfer function of the stationary Kalman filter from e to y becomes

$$T_{ye}(z) = 1 + \frac{16/21}{z-1/3} = \frac{z+3/7}{z-1/3} = \frac{6}{7}W_*(z)$$

We see that $T_{ye}(z)$ equals the minimal phase spectral factor obtained in Example 7.1 up to a constant factor $6/7$.

Also, it can be shown that the stationary Kalman filter for a state space model (7.40) is the same as the one derived above. This fact implies that the spectral factor corresponding to Π_* is of minimal phase, and its state space model is given by a stationary Kalman filter. □

Example 7.3. Consider the case where $A = 1/3$, $C = C$, $\bar{C} = 2/3$ and $\Lambda(0) = 2$. Note that $(A, \bar{C}, C)$ are the same as in Example 7.1, while $\Lambda(0)$ is reduced by $1/4$. Thus we have

$$Z(z) = 1 + \frac{4/3}{z - 1/3} \quad \Rightarrow \quad 0 \leq \mathfrak{Re} Z(e^{j\omega}) \leq 3$$

Thus $Z(z)$ is positive real, but not strictly positive real. Under the assumption that $R(\Pi) = 2 - 4\Pi > 0$, it follows from (7.35) that

$$\mathrm{Ric}(\Pi) = \frac{1}{9}\Pi - \Pi + \left(\frac{2}{3} - \frac{2}{3}\Pi\right)^2 \left(2 - 4\Pi\right)^{-1} \leq 0$$

Rearranging the above inequality yields

$$9\Pi^2 - 6\Pi + 1 \leq 0 \quad \Rightarrow \quad (3\Pi - 1)^2 \leq 0$$

Obviously, this inequality has only one degenerate solution $\Pi = \Pi_* = \Pi^* = 1/3$, so that

$$M(1/3) = \begin{bmatrix} 8/27 & 4/9 \\ 4/9 & 2/3 \end{bmatrix} \quad \Rightarrow \quad B = \frac{2\sqrt{2}}{3\sqrt{3}}, \quad D = \frac{\sqrt{2}}{\sqrt{3}}$$

Thus the spectral factor is given by

$$W(z) = \frac{\sqrt{2}}{\sqrt{3}} + \frac{2\sqrt{2}}{3\sqrt{3}} \frac{2}{z - 1/3} = \sqrt{\frac{2}{3}} \frac{z + 1}{z - 1/3}$$

We see that if the data $(A, C, \bar{C}, \Lambda(0))$ do not satisfy the strictly positive real condition, the Riccati inequality degenerates, and the spectral factor $W(z)$ has zeros on the unit circle. □

From above examples, we see that under the assumption that $R(\Pi) > 0$, there exist the maximum and minimum solutions (Π_*, Π^*) of the LMI (7.26), and that all other solutions of the LMI are bounded $(\Pi_* \leq \Pi \leq \Pi^*)$. If we can show that this observation holds for general matrix cases, then we can completely solve the stochastic realization problem. The rest of this chapter is devoted to the studies in this direction.

7.4 Positivity and Existence of Markov Models

7.4.1 Positive Real Lemma

Let $\mathcal{P}$ be the set of solutions of the LMI (7.26), *i.e.*,

$$\mathcal{P} = \{\Pi \mid M(\Pi) \geq 0, \ \ \Pi^{\mathrm{T}} = \Pi, \ \ \Pi \geq 0\}$$

where it may be noted that the condition that Π is positive definite is not imposed here. However, eventually, we can prove that all $\Pi \in \mathcal{P}$ are positive definite under the minimality assumption in Subsection 7.4.2.

Given a positive definite $\varPi \in \mathcal{P}$, we can find B and D by the factorization of (7.27) and hence we get $Q = BB^{\mathrm{T}}$, $S = BD^{\mathrm{T}}$, $R = DD^{\mathrm{T}}$. Thus, associated with $\varPi \in \mathcal{P}$, there exists a Markov model for y given by (7.14) [or (7.38)]. In this section, we prove some results that characterize the set $\mathcal{P}$.

Lemma 7.3. *The set $\mathcal{P}$ defined above is closed and convex.*

Proof. To prove the closedness, we consider a sequence $\varPi_1, \varPi_2, \cdots \in \mathcal{P}$ such that $\lim_{k\to\infty} \varPi_k = \varPi$, *i.e.*, $\lim_{k\to\infty} \|\varPi_k - \varPi\| = 0$. Since $M(\cdot)$ is continuous from Problem 7.4, we get

$$0 \le \lim_{k\to\infty} M(\varPi_k) = M(\lim_{k\to\infty} \varPi_k) = M(\varPi)$$

Clearly, $\varPi$ is symmetric and nonnegative definite, so that $\varPi \in \mathcal{P}$ holds.

Now suppose that $\varPi_1, \varPi_2 \in \mathcal{P}$. Then, for $\alpha + \beta = 1$ with $\alpha, \beta \ge 0$, it can easily be shown that

$$M(\alpha\varPi_1 + \beta\varPi_2) = \alpha M(\varPi_1) + \beta M(\varPi_2) \ge 0$$

Thus, $\alpha\varPi_1 + \beta\varPi_2 \in \mathcal{P}$. This completes the proof. □

Definition 7.2. *For $u(i) \in \mathbb{R}^p$, $i = -1, -2, \cdots$, we define the infinite dimensional vector*

$$\boldsymbol{u} := \begin{bmatrix} u(-1) \\ u(-2) \\ \vdots \end{bmatrix}$$

Also, associated with the Toeplitz matrix T_+ of (7.5), we define a quadratic form

$$\boldsymbol{u}^{\mathrm{T}} T_+ \boldsymbol{u} := \sum_{k,l=-\infty}^{-1} u^{\mathrm{T}}(k) \Lambda(l-k) u(l) \tag{7.43}$$

where it may be noted that k, l take negative integers. In this case, if $\boldsymbol{u}^{\mathrm{T}} T_+ \boldsymbol{u} \ge 0$ holds for any $\boldsymbol{u}$, then T_+ is referred to as positive real. Also, if $\boldsymbol{u} = 0$ follows from $\boldsymbol{u}^{\mathrm{T}} T_+ \boldsymbol{u} = 0$, then T_+ is called strictly positive real. Moreover, if the condition

$$\boldsymbol{u}^{\mathrm{T}} T_+ \boldsymbol{u} = \rho \|\boldsymbol{u}\|^2, \qquad \exists\, \rho > 0 \tag{7.44}$$

holds, T_+ is called coercive. □

It can be shown that the Toeplitz matrix T_+ is positive real if and only if all finite block Toeplitz matrices are positive definite, *i.e.*,

$$T_+(N) = \begin{bmatrix} \Lambda(0) & \Lambda^{\mathrm{T}}(1) & \cdots & \Lambda^{\mathrm{T}}(N-1) \\ \Lambda(1) & \Lambda(0) & \cdots & \Lambda^{\mathrm{T}}(N-2) \\ \vdots & \vdots & \ddots & \vdots \\ \Lambda(N-1) & \Lambda(N-2) & \cdots & \Lambda(0) \end{bmatrix} > 0, \quad \forall\, N \tag{7.45}$$

holds. Also, define the block anti-diagonal matrix

$$\tilde{J} = \begin{bmatrix} 0 & & & I_p \\ & & I_p & \\ & \cdot^{\cdot^{\cdot}} & & \\ & I_p & & 0 \\ I_p & & & \end{bmatrix} \in \mathbb{R}^{Np \times Np}$$

Then, for the finite block Toeplitz matrix

$$T_-(N) = \begin{bmatrix} \Lambda(0) & \Lambda(1) & \cdots & \Lambda(N-1) \\ \Lambda^{\mathrm{T}}(1) & \Lambda(0) & \cdots & \Lambda(N-2) \\ \vdots & \vdots & \ddots & \vdots \\ \Lambda^{\mathrm{T}}(N-1) & \Lambda^{\mathrm{T}}(N-2) & \cdots & \Lambda(0) \end{bmatrix} \tag{7.46}$$

we have $T_-(N) = \tilde{J} T_+(N) \tilde{J}$. Thus, we see that T_+ is positive real if and only if T_- is positive real.

The following theorem gives a necessary and sufficient condition such that the set $\mathcal{P}$ is non-empty.

Theorem 7.2. *(Positive real lemma) The set $\mathcal{P}$ is non-empty if and only if the Toeplitz operator T_+ of (7.5) is positive real.* □

For a proof of this theorem, we need the following lemma, which gives a useful identity satisfied by the matrices Π, Q, R, S.

Lemma 7.4. *Suppose that Π, Q, R, S are solutions of (7.21). Then, we have*

$$\boldsymbol{u}^{\mathrm{T}} T_+ \boldsymbol{u} = \xi^{\mathrm{T}}(0) \Pi \xi(0) + \sum_{t=-\infty}^{-1} [\xi^{\mathrm{T}}(t) \ \ u^{\mathrm{T}}(t)] \begin{bmatrix} Q & S \\ S^{\mathrm{T}} & R \end{bmatrix} \begin{bmatrix} \xi(t) \\ u(t) \end{bmatrix} \tag{7.47}$$

where ξ is given by

$$\xi(t+1) = A^{\mathrm{T}} \xi(t) + C^{\mathrm{T}} u(t), \qquad \xi(-\infty) = 0 \tag{7.48}$$

Proof. A proof is deferred in Appendix of Section 7.10. □

Proof of Theorem 7.2 If Π, Q, R, S satisfy (7.21) and (7.22), then the right-hand side of (7.47) is nonnegative. Thus we see that T_+ is positive real.

Conversely, suppose that T_+ is positive real, and we show that $\mathcal{P} \neq \phi$. To this end, define

$$\mathcal{S}(\xi) = \left\{ \boldsymbol{u} \;\middle|\; \xi = \sum_{t=-\infty}^{-1} (A^{\mathrm{T}})^{-t-1} C^{\mathrm{T}} u(t) \right\}, \qquad \xi \in \mathbb{R}^n$$

and a nonnegative matrix Π^* by[3]

$$\xi^{\mathrm{T}} \Pi^* \xi = \min_{\boldsymbol{u} \in \mathcal{S}(\xi)} \sum_{k,l=-\infty}^{-1} u^{\mathrm{T}}(k) \Lambda(l-k) u(l) = \min_{\boldsymbol{u} \in \mathcal{S}(\xi)} \boldsymbol{u}^{\mathrm{T}} T_+ \boldsymbol{u} \tag{7.49}$$

In terms of Π^*, we define [see (7.21)]

$$Q^* = \Pi^* - A\Pi^* A^{\mathrm{T}}, \quad S^* = \bar{C}^{\mathrm{T}} - A\Pi^* C^{\mathrm{T}}, \quad R^* = \Lambda(0) - C\Pi^* C^{\mathrm{T}}$$

It is easy to see that if we can prove

$$\begin{bmatrix} Q^* & S^* \\ (S^*)^{\mathrm{T}} & R^* \end{bmatrix} \geq 0 \tag{7.50}$$

then we have $\mathcal{P} \neq \phi$, and hence the proof is completed.

To prove (7.50), we consider the system of (7.48):

$$\xi(t+1) = A^{\mathrm{T}} \xi(t) + C^{\mathrm{T}} u(t), \qquad \xi(-\infty) = 0$$

Let $\boldsymbol{u} = (\,\cdots, u(-2), u(-1)\,)$ be a control vector that brings the state vector to $\xi(0) = \xi$ at $t = 0$. It then follows from Lemma 7.4 that

$$\boldsymbol{u}^{\mathrm{T}} T_+ \boldsymbol{u} = \xi^{\mathrm{T}} \Pi^* \xi + \sum_{t=-\infty}^{-1} [\xi^{\mathrm{T}}(t) \;\; u^{\mathrm{T}}(t)] \begin{bmatrix} Q^* & S^* \\ (S^*)^{\mathrm{T}} & R^* \end{bmatrix} \begin{bmatrix} \xi(t) \\ u(t) \end{bmatrix} \tag{7.51}$$

Also, let $\boldsymbol{v} = (\cdots, v(-2), v(-1)\,)$ be defined by

$$v(t) := u(t+1), \qquad t = -2, -3, \cdots$$

where $v(-1)$ is not specified. Let the corresponding states be given by ξ_v. From the definition of the control vector $\boldsymbol{v}$, it follows that

$$\xi_v(t) = \xi(t+1), \quad t = -2, -3, \cdots; \quad \xi_v(-1) = \xi(0) = \xi$$

with the boundary conditions $\xi_v(-\infty) = \xi(-\infty) = 0$. Define $\xi_v(0) = \zeta$. Then, we see from Lemma 7.4 that

$$\begin{aligned}
\boldsymbol{v}^{\mathrm{T}} T_+ \boldsymbol{v} &= \zeta^{\mathrm{T}} \Pi^* \zeta + \sum_{t=-\infty}^{-1} [\xi_v^{\mathrm{T}}(t) \;\; v^{\mathrm{T}}(t)] \begin{bmatrix} Q^* & S^* \\ (S^*)^{\mathrm{T}} & R^* \end{bmatrix} \begin{bmatrix} \xi_v(t) \\ v(t) \end{bmatrix} \\
&= \zeta^{\mathrm{T}} \Pi^* \zeta + \sum_{t=-\infty}^{-1} [\xi^{\mathrm{T}}(t+1) \;\; u^{\mathrm{T}}(t+1)] \begin{bmatrix} Q^* & S^* \\ (S^*)^{\mathrm{T}} & R^* \end{bmatrix} \begin{bmatrix} \xi(t+1) \\ u(t+1) \end{bmatrix} \\
&= \zeta^{\mathrm{T}} \Pi^* \zeta + \sum_{t=-\infty}^{-1} [\xi^{\mathrm{T}}(t) \;\; u^{\mathrm{T}}(t)] \begin{bmatrix} Q^* & S^* \\ (S^*)^{\mathrm{T}} & R^* \end{bmatrix} \begin{bmatrix} \xi(t) \\ u(t) \end{bmatrix} \\
&\quad + [\xi^{\mathrm{T}}(0) \;\; u^{\mathrm{T}}(0)] \begin{bmatrix} Q^* & S^* \\ (S^*)^{\mathrm{T}} & R^* \end{bmatrix} \begin{bmatrix} \xi(0) \\ u(0) \end{bmatrix}
\end{aligned} \tag{7.52}$$

[3]This is an optimal control problem that minimizes a generalized energy with the terminal condition $\xi(0) = \xi$. We show in Subsection 7.4.2 that the right-hand side of (7.49) is quadratic in ξ, and Π^* is positive definite.

Since $\xi(0) = \xi$ and $u(0) = v(-1)$, we see from (7.51) and (7.52) that

$$\boldsymbol{v}^{\mathrm{T}} T_+ \boldsymbol{v} - \boldsymbol{u}^{\mathrm{T}} T_+ \boldsymbol{u} = \zeta^{\mathrm{T}} \Pi^* \zeta - \xi^{\mathrm{T}} \Pi^* \xi + [\xi^{\mathrm{T}} \;\; v^{\mathrm{T}}(-1)] \begin{bmatrix} Q^* & S^* \\ (S^*)^{\mathrm{T}} & R^* \end{bmatrix} \begin{bmatrix} \xi \\ v(-1) \end{bmatrix}$$

Hence, we get

$$[\xi^{\mathrm{T}} \;\; v^{\mathrm{T}}(-1)] \begin{bmatrix} Q^* & S^* \\ (S^*)^{\mathrm{T}} & R^* \end{bmatrix} \begin{bmatrix} \xi \\ v(-1) \end{bmatrix} = (\boldsymbol{v}^{\mathrm{T}} T_+ \boldsymbol{v} - \zeta^{\mathrm{T}} \Pi^* \zeta) - (\boldsymbol{u}^{\mathrm{T}} T_+ \boldsymbol{u} - \xi^{\mathrm{T}} \Pi^* \xi) \tag{7.53}$$

From the definition of Π^*, we see that $\boldsymbol{v}^{\mathrm{T}} T_+ \boldsymbol{v} - \zeta^{\mathrm{T}} \Pi^* \zeta \geq 0$ and that $\boldsymbol{u}^{\mathrm{T}} T_+ \boldsymbol{u} - \xi^{\mathrm{T}} \Pi^* \xi$ can be made arbitrarily small by a proper choice of $\boldsymbol{u}$, and hence the right-hand side of (7.53) becomes nonnegative. Since ξ and $v(-1)$ are arbitrary, we have proved (7.50). □

Theorem 7.3. *The following statements are equivalent.*

(i) The Toeplitz matrix T_+ of (7.5) is positive real.

(ii) The transfer matrix $Z(z)$ of (7.11) is positive real.

Proof. From (7.11), a state space model corresponding to $Z^{\mathrm{T}}(z)$ is given by

$$\bar{x}(t+1) = A^{\mathrm{T}} \bar{x}(t) + C^{\mathrm{T}} u(t), \qquad \bar{x}(-\infty) = 0 \tag{7.54a}$$

$$y(t) = \bar{C} \bar{x}(t) + \frac{1}{2} \Lambda(0) u(t) \tag{7.54b}$$

From (7.54), we see that

$$y(t) = \sum_{k=-\infty}^{t-1} \bar{C} (A^{\mathrm{T}})^{t-k-1} C^{\mathrm{T}} u(k) + \frac{1}{2} \Lambda(0) u(t)$$

$$= \sum_{k=-\infty}^{t-1} \Lambda^{\mathrm{T}}(t-k) u(k) + \frac{1}{2} \Lambda(0) u(t)$$

and hence

$$y^{\mathrm{T}}(t) u(t) = \sum_{k=-\infty}^{t-1} u^{\mathrm{T}}(k) \Lambda(t-k) u(t) + \frac{1}{2} u^{\mathrm{T}}(t) \Lambda(0) u(t)$$

Taking the sum of both sides of the above equation yields [see (7.87), Section 7.10]

$$\sum_{t=-\infty}^{-1} y^{\mathrm{T}}(t) u(t) = \sum_{t=-\infty}^{-1} \sum_{k=-\infty}^{t-1} u^{\mathrm{T}}(k) \Lambda(t-k) u(t) + \frac{1}{2} \sum_{t=-\infty}^{-1} u^{\mathrm{T}}(t) \Lambda(0) u(t)$$

$$= \frac{1}{2} \sum_{t=-\infty}^{-1} \sum_{k=-\infty}^{-1} u^{\mathrm{T}}(k) \Lambda(t-k) u(t) = \frac{1}{2} \boldsymbol{u}^{\mathrm{T}} T_+ \boldsymbol{u}$$

Since $y = Z^{\mathrm{T}}(z)u$, it follows from Lemma 3.4 (ii) that

$$\sum_{t=-\infty}^{-1} y^{\mathrm{T}}(t)u(t) = \frac{1}{2\pi}\int_{-\pi}^{\pi} u^{\mathrm{T}}(e^{j\omega})Z(e^{j\omega})u(e^{-j\omega})d\omega$$

$$= \frac{1}{4\pi}\int_{-\pi}^{\pi} u^{\mathrm{T}}(e^{j\omega})\left[Z(e^{j\omega}) + Z^{\mathrm{H}}(e^{j\omega})\right]u(e^{-j\omega})d\omega$$

The right-hand side of this equation is nonnegative for any $u(e^{j\omega})$ if and only if

$$Z(e^{j\omega}) + Z^{\mathrm{H}}(e^{j\omega}) \geq 0, \quad -\pi < \omega \leq \pi$$

Hence, we have

$$\boldsymbol{u}^{\mathrm{T}}T_{+}\boldsymbol{u} = 2\sum_{t=-\infty}^{-1} y^{\mathrm{T}}(t)u(t) \geq 0 \quad \Leftrightarrow \quad Z(z): \text{ positive real} \tag{7.55}$$

This completes the proof of this theorem. □

Theorem 7.4. *Suppose that the Toeplitz matrix T_{+} of (7.5) is positive real. Then, $\mathcal{P}$ is bounded, closed and convex, and there exist the maximum Π^* and the minimum Π_* such that for any $\Pi \in \mathcal{P}$,*

$$\Pi_* \leq \Pi \leq \Pi^* \tag{7.56}$$

holds, where the inequality $A \geq B$ means that $A - B$ is nonnegative definite.

Proof. That $\mathcal{P}$ is a closed convex set is already shown in Lemma 7.3. Thus it suffices to show that (7.56) holds.

First we show $\Pi \leq \Pi^*$, $\forall\, \Pi \in \mathcal{P}$. From the definition of Π^* of (7.49) and Lemma 7.4 that

$$\xi^{\mathrm{T}}\Pi^*\xi = \min_{\boldsymbol{u}\in\mathcal{S}(\xi)}\left\{\xi^{\mathrm{T}}\Pi\xi + \sum_{t=-\infty}^{-1}\left[\xi^{\mathrm{T}}(t)\;\; u^{\mathrm{T}}(t)\right]\begin{bmatrix} Q & S \\ S^{\mathrm{T}} & R\end{bmatrix}\begin{bmatrix}\xi(t)\\ u(t)\end{bmatrix}\right\} \geq 0$$

where $\xi(0) = \xi$. It therefore follows that for any $\Pi \in \mathcal{P}$,

$$\xi^{\mathrm{T}}\Pi^*\xi - \xi^{\mathrm{T}}\Pi\xi = \min_{\boldsymbol{u}\in\mathcal{S}(\xi)}\sum_{t=-\infty}^{-1}\left[\xi^{\mathrm{T}}(t)\;\; u^{\mathrm{T}}(t)\right]\begin{bmatrix} Q & S \\ S^{\mathrm{T}} & R\end{bmatrix}\begin{bmatrix}\xi(t)\\ u(t)\end{bmatrix} \geq 0$$

Since ξ is arbitrary, $\Pi \leq \Pi^*$ holds.

To show $\Pi \geq \Pi_*$, we define the backward process $\bar{y}$ by

$$\bar{y}(t) = y(-t), \qquad t = 0, \pm 1, \cdots \tag{7.57}$$

Clearly, the process $\bar{y}$ is stationary, since

$$\bar{\Lambda}(l) = E\{\bar{y}(t+l)\bar{y}^{\mathrm{T}}(t)\} = E\{y(-t-l)y^{\mathrm{T}}(-t)\} = \Lambda(-l)$$

Let $\bar{\mathcal{P}}$ be the set of covariance matrices $\bar{\Pi}$ of the backward Markov models associated with the covariance matrices $\{\Lambda(k)\}$. As shown in Section 7.2, there exists a one-to-one correspondence between $\mathcal{P}$ and $\bar{\mathcal{P}}$ in the sense that

$$\Pi \in \mathcal{P} \quad \Leftrightarrow \quad \bar{\Pi} = \Pi^{-1} \in \bar{\mathcal{P}}$$

Now let $\bar{\Pi}^*$ be the maximum element in the set $\bar{\mathcal{P}}$. In fact, we can show that $\bar{\Pi}^*$ exists by using the same technique used in the proof of the first part. Then, by the above one-to-one correspondence, $\Pi_* := (\bar{\Pi}^*)^{-1}$ becomes the minimum element in $\mathcal{P}$. Hence, for any $\Pi \in \mathcal{P}$, we see that $\Pi_* \leq \Pi$ holds. Therefore, if $\mathcal{P}$ is not empty, it follows that (7.56) holds. □

7.4.2 Computation of Extremal Points

We have shown that Π^* and Π_* are respectively the maximum and the minimum in the set $\mathcal{P}$, and that for any $\Pi \in \mathcal{P}$, the inequality $\Pi_* \leq \Pi \leq \Pi^*$ is satisfied. In this subsection, we provide methods of computing the extreme points Π^* and Π_*, and show that these extreme points respectively coincide with the extreme solutions of the Riccati inequality defined in Section 7.3. First, we compute Π^* as a limit of solutions of finite dimensional optimization problems derived from (7.49).

We assume that $(A, C, \bar{C}^{\mathrm{T}})$ is minimal, and define the vector

$$\boldsymbol{u}_k := \begin{bmatrix} u(-1) \\ \vdots \\ u(-k) \end{bmatrix} \in \mathbb{R}^{kp}$$

and also for $\xi \in \mathbb{R}^n$, we define the set

$$\mathcal{S}_k(\xi) = \left\{ \boldsymbol{u}_k \;\middle|\; \xi = \sum_{t=-k}^{-1} (A^{\mathrm{T}})^{-t-1} C^{\mathrm{T}} u(t) \right\} = \left\{ \boldsymbol{u}_k \;\middle|\; \xi = \mathcal{O}_k^{\mathrm{T}} \boldsymbol{u}_k \right\}$$

Then we consider a finite dimensional optimization problem of the form

$$\xi^{\mathrm{T}} \Pi_k \xi := \min_{\boldsymbol{u}_k \in \mathcal{S}_k(\xi)} \boldsymbol{u}_k^{\mathrm{T}} T_+(k)\, \boldsymbol{u}_k \geq 0 \tag{7.58}$$

where $T_+(k)$ is the block Toeplitz matrix of (7.45), and $\mathcal{O}_k$ is the extended observability matrix with rank n. Since $\mathcal{S}_k(\xi) \subset \mathcal{S}_{k+1}(\xi) \subset \mathcal{S}(\xi)$, it follows that $\Pi_k \geq \Pi_{k+1} \geq \Pi^*$. Also, $\Pi^* \geq 0$ holds by definition, so that Π_k is decreasing and bounded below. Thus Π_k converges to Π^*, a solution to the original infinite dimensional problem.

The finite dimensional problem of (7.58) is a quadratic problem with a linear constraint, so that it can be solved via the Lagrange method. In fact, define

$$\mathcal{L} = \frac{1}{2} \boldsymbol{u}_k^{\mathrm{T}} T_+(k)\, \boldsymbol{u}_k + \lambda_k^{\mathrm{T}} (\xi - \mathcal{O}_k^{\mathrm{T}} \boldsymbol{u}_k)$$

Then, from the optimality condition, we have

$$\frac{\partial \mathcal{L}}{\partial \boldsymbol{u}_k} = 0 \quad \Rightarrow \quad T_+(k)\,\boldsymbol{u}_k - \mathcal{O}_k \lambda_k = 0$$

and $\xi = \mathcal{O}_k^{\mathrm{T}} \boldsymbol{u}_k$. Hence, we see that $\lambda_k = (\mathcal{O}_k^{\mathrm{T}} T_+^{-1}(k)\mathcal{O}_k)^{-1}\xi$, where the inverse exists since $T_+^{-1}(k)$ and $\mathcal{O}_k$ have full rank. Thus, we see that the optimal solution is given by

$$\boldsymbol{u}_k = T_+^{-1}(k)\mathcal{O}_k(\mathcal{O}_k^{\mathrm{T}} T_+^{-1}(k)\mathcal{O}_k)^{-1}\xi \quad \Rightarrow \quad \Pi_k = (\mathcal{O}_k^{\mathrm{T}} T_+^{-1}(k)\mathcal{O}_k)^{-1}$$

For simplicity, we define

$$\bar{\Omega}_k := \Pi_k^{-1} = \mathcal{O}_k^{\mathrm{T}} T_+^{-1}(k)\mathcal{O}_k$$

where recall that

$$\mathcal{O}_k = \begin{bmatrix} C \\ CA \\ \vdots \\ CA^{k-1} \end{bmatrix} \in \mathbb{R}^{pk \times n}$$

We now derive a recursive equation satisfied by $\bar{\Omega}_k$ and $\bar{\Omega}_{k+1}$. We see from (7.45) and (7.7) that

$$\begin{aligned} \bar{\Omega}_{k+1} &= \mathcal{O}_{k+1}^{\mathrm{T}} T_+^{-1}(k+1)\mathcal{O}_{k+1} \\ &= [C^{\mathrm{T}} \;\; A^{\mathrm{T}}\mathcal{O}_k^{\mathrm{T}}] \begin{bmatrix} \Lambda(0) & \bar{C}\mathcal{O}_k^{\mathrm{T}} \\ \mathcal{O}_k \bar{C}^{\mathrm{T}} & T_+(k) \end{bmatrix}^{-1} \begin{bmatrix} C \\ \mathcal{O}_k A \end{bmatrix} \end{aligned} \tag{7.59}$$

By the inversion results for the block matrix [see Problem 2.3 (c)],

$$\begin{aligned} &\begin{bmatrix} \Lambda(0) & \bar{C}\mathcal{O}_k^{\mathrm{T}} \\ \mathcal{O}_k \bar{C}^{\mathrm{T}} & T_+(k) \end{bmatrix}^{-1} \\ &\quad = \begin{bmatrix} \Delta_k & -\Delta_k \bar{C}\mathcal{O}_k^{\mathrm{T}} T_+^{-1}(k) \\ -T_+^{-1}(k)\mathcal{O}_k \bar{C}^{\mathrm{T}} \Delta_k & T_+^{-1}(k) + T_+^{-1}(k)\mathcal{O}_k \bar{C}^{\mathrm{T}} \Delta_k \bar{C}\mathcal{O}_k^{\mathrm{T}} T_+^{-1}(k) \end{bmatrix} \end{aligned}$$

where

$$\Delta_k := [\Lambda(0) - \bar{C}\mathcal{O}_k^{\mathrm{T}} T_+^{-1}(k)\mathcal{O}_k \bar{C}^{\mathrm{T}}]^{-1} = (\Lambda(0) - \bar{C}\bar{\Omega}_k \bar{C}^{\mathrm{T}})^{-1}$$

It therefore follows from (7.59) that

$$\begin{aligned} \bar{\Omega}_{k+1} &= C^{\mathrm{T}}\Delta_k C - A^{\mathrm{T}}\mathcal{O}_k^{\mathrm{T}} T_+^{-1}(k)\mathcal{O}_k \bar{C}^{\mathrm{T}}\Delta_k C - C^{\mathrm{T}}\Delta_k \bar{C}\mathcal{O}_k^{\mathrm{T}} T_+^{-1}(k)\mathcal{O}_k A \\ &\quad + A^{\mathrm{T}}\mathcal{O}_k^{\mathrm{T}} T_+^{-1}(k)\mathcal{O}_k A + A^{\mathrm{T}}\mathcal{O}_k^{\mathrm{T}} T_+^{-1}(k)\mathcal{O}_k \bar{C}^{\mathrm{T}}\Delta_k \bar{C}\mathcal{O}_k^{\mathrm{T}} T_+^{-1}(k)\mathcal{O}_k A \\ &= A^{\mathrm{T}}\bar{\Omega}_k A + (C^{\mathrm{T}} - A^{\mathrm{T}}\bar{\Omega}_k \bar{C}^{\mathrm{T}})(\Lambda(0) - \bar{C}\bar{\Omega}_k \bar{C}^{\mathrm{T}})^{-1}(C - \bar{C}\bar{\Omega}_k A) \end{aligned}$$

From the above result, we have a recursive algorithm for computing Π^*.

Algorithm 1 Compute the solution of the discrete-time Riccati equation

$$\bar{\Omega}_{k+1} = A^{\mathrm{T}} \bar{\Omega}_k A + (C^{\mathrm{T}} - A^{\mathrm{T}} \bar{\Omega}_k \bar{C}^{\mathrm{T}})(\Lambda(0) - \bar{C} \bar{\Omega}_k \bar{C}^{\mathrm{T}})^{-1}(C - \bar{C} \bar{\Omega}_k A) \quad (7.60)$$

with the initial condition $\bar{\Omega}_0 = 0$ to get $\bar{\Omega}_\infty = \lim_{k\to\infty} \bar{\Omega}_k$. Then, the maximum Π^* in $\mathcal{P}$ and the associated covariance matrices are given by

$$\Pi^* = \bar{\Omega}_\infty^{-1}, \qquad Q^* = \Pi^* - A\Pi^* A^{\mathrm{T}}$$

$$S^* = \bar{C}^{\mathrm{T}} - A\Pi^* C^{\mathrm{T}}, \qquad R^* = \Lambda(0) - C\Pi^* C^{\mathrm{T}}$$

We see from the dual of Lemma 7.2 that the limit point $\bar{\Omega}_\infty$ is positive definite. □

Remark 7.1. It follows from (7.25) that the dual LMI for (7.26) is given by

$$M(\bar{\Pi}) := \begin{bmatrix} \bar{\Pi} - A^{\mathrm{T}} \bar{\Pi} A & C^{\mathrm{T}} - A^{\mathrm{T}} \bar{\Pi} \bar{C}^{\mathrm{T}} \\ C - \bar{C} \bar{\Pi} A & \Lambda(0) - \bar{C} \bar{\Pi} \bar{C}^{\mathrm{T}} \end{bmatrix} \geq 0, \qquad \bar{\Pi} > 0 \quad (7.61)$$

Thus, Algorithm 1 recursively computes the minimum solution of the dual Riccati equation associated with (7.61), thereby giving the minimal covariance matrix of the backward Markov model. Thus, the inverse of the limit gives the maximum solution Π^* to the ARI associated with the LMI (7.26), and hence to the LMI (7.26) itself. □

By using the discrete-time Riccati equation associated with the LMI (7.26), we readily derive the following algorithm.

Algorithm 2 Compute the solution of the discrete-time Riccati equation

$$\Omega_{k+1} = A\Omega_k A^{\mathrm{T}} + (\bar{C}^{\mathrm{T}} - A\Omega_k C^{\mathrm{T}})(\Lambda(0) - C\Omega_k C^{\mathrm{T}})^{-1}(\bar{C} - C\Omega_k A^{\mathrm{T}}) \quad (7.62)$$

with the initial condition $\Omega_0 = 0$ to get $\Omega_\infty = \lim_{k\to\infty} \Omega_k$. Then, the minimum Π_* in $\mathcal{P}$ and the associated covariance matrices are given by

$$\Pi_* = \Omega_\infty, \qquad Q_* = \Pi_* - A\Pi_* A^{\mathrm{T}}$$

$$S_* = \bar{C}^{\mathrm{T}} - A\Pi_* C^{\mathrm{T}}, \qquad R_* = \Lambda(0) - C\Pi_* C^{\mathrm{T}}$$

It can be shown from Lemma 7.2 that the limit Ω_∞ is positive definite. □

Remark 7.2. The discrete-time Riccati equation in Algorithm 2 is the same as the discrete-time Riccati equation of (5.62). It is not difficult to show that

$$\Omega_k = \mathcal{C}_k T_-^{-1}(k) \mathcal{C}_k^{\mathrm{T}}$$

satisfies (7.62) [see Problem 7.6], where

$$\mathcal{C}_k = [\bar{C}^{\mathrm{T}} \ A\bar{C}^{\mathrm{T}} \ \cdots \ A^{k-1}\bar{C}^{\mathrm{T}}]$$

Also, (7.62) is a recursive algorithm that computes the minimum solution of the ARE (7.37). Hence, the solution Π_* of Algorithm 2 gives the minimum solution to the ARI associated with the LMI, so that Π_* is the minimum solution to the LMI (7.26). □

The existence of maximum and minimum solutions of the LMI (7.26) has been proved and their computational methods are established. This implies that the stochastic minimal realization problem stated by Faurre is now completely solved. The most difficult task in the above procedure is, however, to obtain a minimal realization $(A, C, \bar{C}^{\mathrm{T}})$ by the deterministic realization algorithm. In Chapter 8, we shall show that this difficulty is resolved by means of the approach based on the canonical correlation analysis (CCA).

7.5 Algebraic Riccati-like Equations

We derive algebraic Riccati-like equations satisfied by the difference $\Theta := \Pi^* - \Pi_*$ between the maximum and minimum solutions. Lemmas in this section are useful for proving Theorem 7.5 below.

Consider the ARE of (7.37):

$$\Pi = A\Pi A^{\mathrm{T}} + (\bar{C}^{\mathrm{T}} - A\Pi C^{\mathrm{T}})(\Lambda(0) - C\Pi C^{\mathrm{T}})^{-1}(\bar{C} - C\Pi A^{\mathrm{T}}) \tag{7.63}$$

It should be noted that this ARE has the same form as the ARE of (5.75), where the minimum solution Π_* equals Σ of (5.75).

Let the stationary Kalman gain be defined by

$$K = (\bar{C}^{\mathrm{T}} - A\Pi C^{\mathrm{T}})(\Lambda(0) - C\Pi C^{\mathrm{T}})^{-1} \tag{7.64}$$

and let $A_K := A - KC$ be the closed-loop matrix. Then, we have the following lemma.

Lemma 7.5. *The ARE of (7.63) is expressed as*

$$\Pi = A_K \Pi A_K^{\mathrm{T}} - K\Lambda(0)K^{\mathrm{T}} + K\bar{C} + \bar{C}^{\mathrm{T}}K^{\mathrm{T}} \tag{7.65}$$

Proof. Use (7.64) and the definition of A_K. See Problem 7.7. □

Let Π^* and Π_* be the maximum and minimum solutions of (7.63), respectively. Thus, in terms of these solutions, we define

$$K^* := (\bar{C}^{\mathrm{T}} - A\Pi^* C^{\mathrm{T}})(\Lambda(0) - C\Pi^* C^{\mathrm{T}})^{-1}$$
$$K_* := (\bar{C}^{\mathrm{T}} - A\Pi_* C^{\mathrm{T}})(\Lambda(0) - C\Pi_* C^{\mathrm{T}})^{-1}$$

and $A^* := A - K^*C$, $A_* := A - K_*C$. It then follows from Lemma 7.5 that

$$\Pi^* = A^*\Pi^*(A^*)^{\mathrm{T}} - K^*\Lambda(0)(K^*)^{\mathrm{T}} + K^*\bar{C} + \bar{C}^{\mathrm{T}}(K^*)^{\mathrm{T}} \tag{7.66}$$

$$\Pi_* = A_*\Pi_*A_*^{\mathrm{T}} - K_*\Lambda(0)(K_*)^{\mathrm{T}} + K_*\bar{C} + \bar{C}^{\mathrm{T}}(K_*)^{\mathrm{T}} \tag{7.67}$$

The following lemma derives the Riccati-like equations satisfied by the difference between the maximum and the minimum solutions of the ARE (7.63).

Lemma 7.6. *Let $\Theta := \Pi^* - \Pi_*$. Then, Θ satisfies the following algebraic Riccati-like equations*

$$\Theta = A_* \Theta A_*^{\mathrm{T}} + (K_* - K^*)(\Lambda(0) - C\Pi^* C^{\mathrm{T}})(K_* - K^*)^{\mathrm{T}} \tag{7.68}$$

and

$$\Theta = A_* \Theta A_*^{\mathrm{T}} + A_* \Theta C^{\mathrm{T}}(\Lambda(0) - C\Pi^* C^{\mathrm{T}})^{-1} C\Theta A_*^{\mathrm{T}} \tag{7.69}$$

Proof. Since $A^* = A_* + (K_* - K^*)C$, it follows from (7.66) that

$$\begin{aligned}\Pi^* &= (A_* + (K_* - K^*)C)\Pi^*(A_* + (K_* - K^*)C)^{\mathrm{T}} \\ &\quad - K^*\Lambda(0)(K^*)^{\mathrm{T}} + K^*\bar{C} + \bar{C}^{\mathrm{T}}(K^*)^{\mathrm{T}} \\ &= A_*\Pi^* A_*^{\mathrm{T}} + (K_* - K^*)C\Pi^* A_*^{\mathrm{T}} + A_*\Pi^* C^{\mathrm{T}}(K_* - K^*)^{\mathrm{T}} \\ &\quad + (K_* - K^*)C\Pi^* C^{\mathrm{T}}(K_* - K^*)^{\mathrm{T}} - K^*\Lambda(0)(K^*)^{\mathrm{T}} \\ &\quad + K^*\bar{C} + \bar{C}^{\mathrm{T}}(K^*)^{\mathrm{T}}\end{aligned} \tag{7.70}$$

Also, from $A_* = A - K_*C$,

$$\begin{aligned}A_*\Pi^* C^{\mathrm{T}} &= A\Pi^* C^{\mathrm{T}} - K_* C\Pi^* C^{\mathrm{T}} \\ &= \bar{C}^{\mathrm{T}} - K^*(\Lambda(0) - C\Pi^* C^{\mathrm{T}}) - K_* C\Pi^* C^{\mathrm{T}} \\ &= \bar{C}^{\mathrm{T}} - (K_* - K^*)C\Pi^* C^{\mathrm{T}} - K^*\Lambda(0)\end{aligned} \tag{7.71}$$

Thus, using (7.71), we see that (7.70) becomes

$$\begin{aligned}\Pi^* &= A_*\Pi^* A_*^{\mathrm{T}} - (K_* - K^*)C\Pi^* C^{\mathrm{T}}(K_* - K^*)^{\mathrm{T}} + K_*\bar{C} + \bar{C}^{\mathrm{T}}(K_*)^{\mathrm{T}} \\ &\quad + K^*\Lambda(0)(K^*)^{\mathrm{T}} - K_*\Lambda(0)(K^*)^{\mathrm{T}} - K^*\Lambda(0)(K_*)^{\mathrm{T}}\end{aligned}$$

Taking the difference between the above equation and (7.67) gives

$$\begin{aligned}\Theta &= A_*\Theta A_*^{\mathrm{T}} - (K_* - K^*)C\Pi^* C^{\mathrm{T}}(K_* - K^*)^{\mathrm{T}} \\ &\quad + K^*\Lambda(0)(K^*)^{\mathrm{T}} - K_*\Lambda(0)(K^*)^{\mathrm{T}} - K^*\Lambda(0)(K_*)^{\mathrm{T}} + K_*\Lambda(0)(K_*)^{\mathrm{T}}\end{aligned}$$

Rearranging the terms yields (7.68). Similarly to the derivation of (7.71), we have

$$A_*\Pi_* C^{\mathrm{T}} = (A - K_*C)\Pi_* C^{\mathrm{T}} = \bar{C}^{\mathrm{T}} - K_*\Lambda(0) \tag{7.72}$$

Taking the difference between (7.71) and (7.72) yields

$$A_*\Theta C^{\mathrm{T}} = A_*(\Pi^* - \Pi_*)C^{\mathrm{T}} = (K_* - K^*)(\Lambda(0) - C\Pi^* C^{\mathrm{T}})$$

Applying this relation to (7.68) leads to (7.69). □

Lemma 7.7. *If Θ is nonsingular, the inverse Θ^{-1} satisfies*

$$\Theta^{-1} = A_*^{\mathrm{T}}\Theta^{-1}A_* + C^{\mathrm{T}}(\Lambda(0) - C\Pi_* C^{\mathrm{T}})^{-1}C \tag{7.73}$$

Proof. From (7.69), we see that if Θ is nonsingular, so is A_*. Thus it follows that

$$A_*^{-1}\Theta(A_*^{\mathrm{T}})^{-1} = \Theta + \Theta C^{\mathrm{T}}(\Lambda(0) - C\Pi_* C^{\mathrm{T}} - C\Theta C^{\mathrm{T}})^{-1}C\Theta$$

Application of the matrix inversion lemma of (5.10) gives

$$A_*^{\mathrm{T}}\Theta^{-1}A_* = \Theta^{-1} - C^{\mathrm{T}}(\Lambda(0) - C\Pi_* C^{\mathrm{T}})^{-1}C$$

This completes the proof of (7.73). □

We shall consider the strictly positive real conditions and a related theorem due to Faurre [47] in the next section.

7.6 Strictly Positive Real Conditions

In this section, we show equivalent conditions for strict positive realness, which will be used for proving some results related to reduced stochastic realization. In the following, we assume that there exist the maximum and minimum solutions Π^* and Π_* of the ARE (7.63).

Definition 7.3. *(Faurre [47]) For the minimum solution Π_*, we define*

$$Q_* = \Pi_* - A\Pi_* A^{\mathrm{T}}, \quad S_* = \bar{C}^{\mathrm{T}} - A\Pi_* C^{\mathrm{T}}, \quad R_* = \Lambda(0) - C\Pi_* C^{\mathrm{T}}$$

Suppose that the following inequality

$$R_* := \Lambda(0) - C\Pi_* C^{\mathrm{T}} > 0 \tag{7.74}$$

holds. Then, the stochastic realization problem is called regular. □

Theorem 7.5. *Let $(A, C, \bar{C}^{\mathrm{T}})$ be a minimal realization. Then the following (i) ∼ (iv) are equivalent conditions for strictly positive realness.*

(i) $Z(z)$ is strictly positive real, or T_+ is coercive.
(ii) $\Theta = \Pi^* - \Pi_*$ is positive definite.
(iii) $R_* > 0$, and $A_* := A - S_* R_*^{-1} C$ is stable.
(iv) The interior of $\mathcal{P}$ is non-void. In other words, there exists $\Pi \in \mathcal{P}$ such that the corresponding covariance matrices are positive definite, i.e.,

$$\begin{bmatrix} Q & S \\ S^{\mathrm{T}} & R \end{bmatrix} > 0$$

Proof. 1° (i) → (ii). From (7.44), there exists $\rho > 0$ such that the operator $T_+ - \rho I$ corresponding to $(A,\ C,\ \bar{C}, \Lambda(0) - \rho I_p)$ is positive real. Let $\mathcal{P}_\rho$ be the set of solutions of the LMI with $(A,\ C,\ \bar{C}, \Lambda(0) - \rho I_p)$. Then, we see that $\mathcal{P}_\rho \subset \mathcal{P}$. Let $\Pi_0 \in \mathcal{P}_\rho$, and define

$$Q_0 = \Pi_0 - A\Pi_0 A^{\mathrm{T}}, \quad S_0 = \bar{C}^{\mathrm{T}} - A\Pi_0 C^{\mathrm{T}}, \quad R_0 = \Lambda(0) - C\Pi_0 C^{\mathrm{T}}$$

It follows from Lemma 7.4 and (7.49) that

$$\xi^{\mathrm{T}}\Pi^*\xi = \xi\Pi_0\xi + \min_{u\in\mathcal{S}(\xi)} \sum_{t=-\infty}^{-1} \left\{ [\xi^{\mathrm{T}}(t) \ \ u^{\mathrm{T}}(t)] \begin{bmatrix} Q_0 & S_0 \\ S_0^{\mathrm{T}} & R_0 - \rho I_p \end{bmatrix} \begin{bmatrix} \xi(t) \\ u(t) \end{bmatrix} + \rho u^{\mathrm{T}}(t)u(t) \right\}$$

The constraint equation for the above optimization problem is given by

$$\xi(t+1) = A^{\mathrm{T}}\xi(t) + C^{\mathrm{T}}u(t), \qquad \xi(0) = \xi, \qquad \xi(-\infty) = 0$$

Then, from $\begin{bmatrix} Q_0 & S_0 \\ S_0^{\mathrm{T}} & R_0 - \rho I \end{bmatrix} \geq 0$, we see that

$$\xi^{\mathrm{T}}(\Pi^* - \Pi_0)\xi \geq \rho \min_{u\in\mathcal{S}(\xi)} \sum_{t=-\infty}^{-1} u^{\mathrm{T}}(t)u(t) \tag{7.75}$$

Referring to the derivation of Algorithm 1 in Subsection 7.4.2, we observe that the optimal solution to the minimization problem in (7.75) becomes

$$\min_{u\in\mathcal{S}(\xi)} \sum_{t=-\infty}^{-1} u^{\mathrm{T}}(t)u(t) = \xi^{\mathrm{T}}(\mathcal{O}^{\mathrm{T}}\mathcal{O})^{-1}\xi > 0, \qquad \xi \neq 0 \tag{7.76}$$

where, since A is stable and $(C,\ A)$ is observable,

$$\mathcal{O}^{\mathrm{T}}\mathcal{O} = \sum_{i=0}^{\infty}(A^{\mathrm{T}})^i C^{\mathrm{T}} C A^i > 0$$

Therefore we have

$$\xi^{\mathrm{T}}\Theta\xi \geq \xi^{\mathrm{T}}(\Pi^* - \Pi_0)\xi \geq \rho\,\xi^{\mathrm{T}}(\mathcal{O}^{\mathrm{T}}\mathcal{O})^{-1}\xi > 0, \qquad \xi \neq 0$$

This completes the proof of (ii).

2° Next we show (ii) → (iii). For Π^* and Π_*, we define $R^* = \Lambda(0) - C\Pi^* C^{\mathrm{T}}$ and $R_* = \Lambda(0) - C\Pi_* C^{\mathrm{T}}$. Suppose that $\Pi^* - \Pi_* > 0$ holds, but R_* is not positive definite. Then there exists $\eta \in \mathbb{R}^p$ such that $R_*\eta = 0$, $\eta \neq 0$. Hence,

$$(\Lambda(0) - C\Pi_* C^{\mathrm{T}})\eta = 0$$

holds. Noting that $\Lambda(0) > 0$, we have $C^{\mathrm{T}}\eta \neq 0$. Since $R_* \geq R^* \geq 0$, it also follows that $R^*\eta = 0$. Thus we have

$$(R_* - R^*)\eta = 0 \quad \Rightarrow \quad (\Pi^* - \Pi_*)C^{\mathrm{T}}\eta = 0$$

However, since $C^{\mathrm{T}}\eta \neq 0$, we see that $\Pi^* - \Pi_*$ is not positive definite, a contradiction. Thus we have $R_* > 0$. It follows from Lemma 7.7 that if $\Theta = \Pi^* - \Pi_* > 0$, the inverse Θ^{-1} satisfies

$$\Theta^{-1} = A_*^{\mathrm{T}}\Theta^{-1}A_* + C^{\mathrm{T}}(\Lambda(0) - C\Pi_* C^{\mathrm{T}})^{-1}C \tag{7.77}$$

Since $(C,\ A_*)$ is observable, the Lyapunov theorem implies that if $\Theta^{-1} > 0$, A_* is stable.

3° We prove (iii) → (iv); this part is somewhat involved. By the hypothesis, A_* is stable and $R_* > 0$. Let $V > 0$, $V \in \mathbb{R}^{n \times n}$, and consider the Lyapunov equation

$$X = A_*^{\mathrm{T}} X A_* + C^{\mathrm{T}}(\Lambda(0) - C\Pi_* C^{\mathrm{T}})^{-1}C + V \tag{7.78}$$

Obviously, we have $X > 0$, and hence X^{-1} exists. In fact, even if $V = 0$, we have $X > 0$ due to the observability of $(C,\ A_*)$.

We derive the equation satisfied by the inverse X^{-1}. From (7.78), we get

$$X - C^{\mathrm{T}}(\Lambda(0) - C\Pi_* C^{\mathrm{T}})^{-1}C = A_*^{\mathrm{T}} X A_* + V$$

Applying the matrix inversion lemma of (5.10) to both sides of the above equation yields

$$\begin{aligned} &X^{-1} + X^{-1}C^{\mathrm{T}}(\Lambda(0) - C[\Pi_* + X^{-1}]C^{\mathrm{T}})^{-1}CX^{-1} \\ &\quad = A_*^{-1}X^{-1}A_*^{-\mathrm{T}} - A_*^{-1}X^{-1}A_*^{-\mathrm{T}}(A_*^{-1}X^{-1}A_*^{-\mathrm{T}} + V^{-1})^{-1}A_*^{-1}X^{-1}A_*^{-\mathrm{T}} \end{aligned}$$

where A_* is assumed to be nonsingular. Pre-multiplying the above equation by A_*, and post-multiplying by A_*^{T}, we have

$$\begin{aligned} A_*X^{-1}A_*^{\mathrm{T}} + {} & A_*X^{-1}C^{\mathrm{T}}(\Lambda(0) - C[\Pi_* + X^{-1}]C^{\mathrm{T}})^{-1}CX^{-1}A_*^{\mathrm{T}} \\ &= X^{-1} - (X + XA_*V^{-1}A_*^{\mathrm{T}}X)^{-1} \end{aligned}$$

Thus the inverse X^{-1} satisfies

$$\begin{aligned} X^{-1} = A_*X^{-1}A_*^{\mathrm{T}} &+ A_*X^{-1}C^{\mathrm{T}}(\Lambda(0) - C[\Pi_* + X^{-1}]C^{\mathrm{T}})^{-1}CX^{-1}A_*^{\mathrm{T}} \\ &+ (X + XA_*V^{-1}A_*^{\mathrm{T}}X)^{-1} \end{aligned} \tag{7.79}$$

Note that if $V \to 0$, then it follows that $X^{-1} \to \Theta = \Pi^* - \Pi_*$, and hence the above equation reduces to (7.69).

Now define $\Pi := \Pi_* + X^{-1}$. Then the covariance matrices associated with Π are given by

$$\begin{aligned} Q &= \Pi_* + X^{-1} - A(\Pi_* + X^{-1})A^{\mathrm{T}} \\ S &= \bar{C}^{\mathrm{T}} - A(\Pi_* + X^{-1})C^{\mathrm{T}} \\ R &= \Lambda(0) - C(\Pi_* + X^{-1})C^{\mathrm{T}} \end{aligned}$$

We show that (Q, S, R) satisfy the condition (iv). If $V \to I \cdot \infty$, then $X^{-1} \to 0$, so that for a sufficiently large V, we get $R > 0$. It thus suffices to show that

$$-\mathrm{Ric}(\Pi) := Q - SR^{-1}S^{\mathrm{T}} > 0$$

Since Π_* satisfies the ARE of (7.63), and since $A = A_* + K_*C$, we have

$$\begin{aligned} Q &= (\Pi_* - A\Pi_*A^{\mathrm{T}}) + (X^{-1} - AX^{-1}A^{\mathrm{T}}) \\ &= K_*(\Lambda(0) - C\Pi_*C^{\mathrm{T}})K_*^{\mathrm{T}} + (X^{-1} - (A_* + K_*C)X^{-1}(A_* + K_*C)^{\mathrm{T}}) \end{aligned}$$

Moreover, from (7.79), it can be shown that

$$\begin{aligned} Q &= K_*(\Lambda(0) - C\Pi_*C^{\mathrm{T}})K_*^{\mathrm{T}} + A_*X^{-1}C^{\mathrm{T}}R^{-1}CX^{-1}A_*^{\mathrm{T}} \\ &\quad + (X + XA_*V^{-1}A_*^{\mathrm{T}}X)^{-1} - A_*X^{-1}C^{\mathrm{T}}K_*^{\mathrm{T}} - K_*CX^{-1}A_*^{\mathrm{T}} \\ &\quad - K_*CX^{-1}C^{\mathrm{T}}K_*^{\mathrm{T}} \\ &= K_*RK_*^{\mathrm{T}} + A_*X^{-1}C^{\mathrm{T}}R^{-1}CX^{-1}A_*^{\mathrm{T}} \\ &\quad + (X + XA_*V^{-1}A_*^{\mathrm{T}}X)^{-1} - A_*X^{-1}C^{\mathrm{T}}K_*^{\mathrm{T}} - K_*CX^{-1}A_*^{\mathrm{T}} \\ &= (K_*R - A_*X^{-1}C^{\mathrm{T}})R^{-1}(K_*R - A_*X^{-1}C^{\mathrm{T}})^{\mathrm{T}} \\ &\quad + (X + XA_*V^{-1}A_*^{\mathrm{T}}X)^{-1} \end{aligned} \tag{7.80}$$

Also, by utilizing (7.72), we have

$$\begin{aligned} S &= \bar{C}^{\mathrm{T}} - (A_* + K_*C)(\Pi_* + X^{-1})C^{\mathrm{T}} \\ &= K_*\Lambda(0) - A_*X^{-1}C^{\mathrm{T}} - K_*C\Pi_*C^{\mathrm{T}} - K_*CX^{-1}C^{\mathrm{T}} \\ &= K_*R - A_*X^{-1}C^{\mathrm{T}} \end{aligned}$$

and hence

$$SR^{-1}S^{\mathrm{T}} = (K_*R - A_*X^{-1}C^{\mathrm{T}})R^{-1}(K_*R - A_*X^{-1}C^{\mathrm{T}})^{\mathrm{T}}$$

Subtracting the above equation from (7.80) yields

$$Q - SR^{-1}S^{\mathrm{T}} = (X + XA_*V^{-1}A_*^{\mathrm{T}}X)^{-1} > 0$$

This completes a proof of (iii) → (iv).

4° We finally prove (iv) → (i). From the assumption, there exists an interior point $\Pi_0 \in \mathcal{P}$ and $\rho > 0$ such that

$$\begin{bmatrix} Q_0 & S_0 \\ S_0^{\mathrm{T}} & R_0 \end{bmatrix} \geq \rho I_{n+p} \quad \Rightarrow \quad \begin{bmatrix} Q_0 & S_0 \\ S_0^{\mathrm{T}} & R_0 - \rho I_p \end{bmatrix} \geq 0$$

Since Π_0 is a solution of the LMI with $(A,\, C,\, \bar{C},\, \Lambda(0) - \rho I_p)$, we see that T_+ is coercive. □

We present without proof a lemma related to Theorem 7.5 (ii), which will be used in Section 8.6.

Lemma 7.8. *Suppose that $\Lambda(0) > 0$ holds in $Z(z)$ in (7.11), but we do not assume that $(A, C, \bar{C}^{\mathrm{T}})$ is minimal. Then, if the LMI of (7.26) has two positive definite solutions Π_1 and Π_2, and if $\Pi_2 - \Pi_1 > 0$, then $Z(z)$ is strictly positive real.*

Proof. If $(A, C, \bar{C}^{\mathrm{T}})$ is minimal, the result is obvious from Theorem 7.5. A proof of the non-minimal case is reduced to the minimal case; see [106]. □

7.7 Stochastic Realization Algorithm

By using the deterministic realization algorithm of Lemma 6.1, we have the following stochastic realization algorithm.

Lemma 7.9. *(Stochastic realization algorithm [15])*

Step 1: For given covariance matrices $\{\Lambda(l),\ l = 0, 1, \cdots, L\}$, we form the block Hankel matrix

$$H_{k,k} = \begin{bmatrix} \Lambda(1) & \Lambda(2) & \cdots & \Lambda(k) \\ \Lambda(2) & \Lambda(3) & \cdots & \Lambda(k+1) \\ \vdots & \vdots & \ddots & \vdots \\ \Lambda(k) & \Lambda(k+1) & \cdots & \Lambda(2k-1) \end{bmatrix} \in \mathbb{R}^{kp \times kp}$$

where $2k - 1 \leq L$ and $k > n$.

Step 2: Compute the SVD of $H_{k,k}$ such that

$$H_{k,k} = [U_s \ \ U_n] \begin{bmatrix} \Sigma_s & 0 \\ 0 & \Sigma_n \end{bmatrix} \begin{bmatrix} V_s^{\mathrm{T}} \\ V_n^{\mathrm{T}} \end{bmatrix} \simeq U_s \Sigma_s V_s^{\mathrm{T}} \tag{7.81}$$

where Σ_s contains the largest n singular values of $H_{k,k}$, and the other singular values are small, i.e., $\sigma_1 \geq \sigma_2 \geq \cdots \geq \sigma_n \gg \sigma_{n+1} \geq \cdots$.

Step 3: Based on the SVD of (7.81), the extended observability and reachability matrices are defined by

$$\mathcal{O}_k = U_s \Sigma_s^{1/2}, \qquad \mathcal{C}_k = \Sigma_s^{1/2} V_s^{\mathrm{T}} \tag{7.82}$$

Step 4: Compute the matrices A, C, $\bar{C}^{\mathrm{T}}$ by

$$A = \mathcal{O}_{k-1}^{\dagger} \overline{\mathcal{O}}_k, \quad C = \mathcal{O}_k(1:p, 1:n), \quad \bar{C}^{\mathrm{T}} = \mathcal{C}_k(1:n, 1:p) \tag{7.83}$$

where $\overline{\mathcal{O}}_k = \mathcal{O}_k(p+1:kp, 1:n)$.

Step 5: By using $(A, C, \bar{C}, \Lambda(0))$ so obtained, we define the ARE

$$\Pi = A \Pi A^{\mathrm{T}} + (\bar{C}^{\mathrm{T}} - A \Pi C^{\mathrm{T}})(\Lambda(0) - C \Pi C^{\mathrm{T}})^{-1}(\bar{C} - C \Pi A^{\mathrm{T}}) \tag{7.84}$$

Compute the minimum (or stabilizing) solution $\Pi_* \geq 0$ by the method described in Subsection 7.4.2, or by the method of Lemma 5.14, to obtain the Kalman gain

$$K = (\bar{C}^{\mathrm{T}} - A\Pi_* C^{\mathrm{T}})(\Lambda(0) - C\Pi_* C^{\mathrm{T}})^{-1} \tag{7.85}$$

Then we have an innovation model

$$x(t+1) = Ax(t) + Ke(t) \tag{7.86a}$$

$$y(t) = Cx(t) + e(t) \tag{7.86b}$$

where $\mathrm{cov}\{e(t)\} = \Lambda(0) - C\Pi_* C^{\mathrm{T}}$. □

Since the covariance matrix of the innovation process e of (7.86) is not a unit matrix, note that the Markov model of (7.86) is different from the Markov model of (7.38). In fact, K in (7.86) is expressed as $K = BD^{-1}$ using B and D in (7.38).

A crucial problem in this algorithm is how we can compute accurate estimates of covariance matrices based on given finite measured data. To get good estimates, we need a large amount of data. If the accuracy of estimates of covariance matrices is lost, then data $(A, C, \bar{C}, \Lambda(0))$ may not be positive real, and hence there may be a possibility that there exist no stabilizing solutions for the ARE of (7.84); see [58, 106, 154].

7.8 Notes and References

- By using the deterministic realization theory together with the LMI and AREs, Faurre [45–47] has developed a complete theory of stochastic realization. Other relevant references in this chapter are Aoki [15], Van Overschee and De Moor [163, 165] and Lindquist and Picci [106, 107].
- In Section 7.1, as preliminaries, we have introduced the covariance matrices and spectral density matrices of a stationary process, and positive real matrices. In Section 7.2, we have defined the problem of stochastic realization for a stationary process based on [46].
- In Section 7.3, by using the results of [46, 107], we have shown that the stochastic realization problem can be solved by means of the LMI and associated ARI and ARE. Also, some simple numerical examples are included to illustrate the procedure of stochastic realization, including solutions of the associated ARIs, spectral factors and innovation models.
- Section 7.4 deals with the positivity of covariance data and the existence of Markov models. By using the fact that there exists a one-to-one correspondence between a solution of LMI and a Markov model, we have shown that the set of all solutions of the LMI is a closed bounded convex set, and there exist the minimum and maximum elements in it under the assumption that the covariance data is positive real; the proofs are based on the solutions of related optimal control problem due to Faurre [45, 47]. Also, two recursive methods to compute the maximum and minimum solutions of the ARE are provided.

- In Section 7.5, we have introduced algebraic Riccati-like equations satisfied by the difference of the maximum and minimum solutions of the LMI, together with proofs. Section 7.6 provides some equivalent conditions such that the given data are strictly positive real. A different proof of the positive real lemma based on convexity theory is found in [135].
- In Section 7.7, a stochastic subspace identification algorithm is presented by use of the deterministic realization algorithm of Lemma 6.1. A program of this algorithm is provided in Table D.3. However, there is a possibility that this algorithm does not work since the estimated finite covariance sequence may not be positive real; see [38, 106] for details. In Section 7.10, a proof of Lemma 7.4 is included.

7.9 Problems

7.1 Let $Z(z) = B(z)/A(z)$, and let

$$A(e^{j\omega}) := a(\omega) + jb(\omega), \qquad B(e^{j\omega}) := c(\omega) + jd(\omega)$$

Derive a condition such that $Z(z)$ is positive real in terms of $a(\omega)$, $b(\omega)$, $c(\omega)$, $d(\omega)$. Also, derive a positive real condition for the function

$$Z(z) = \frac{bz + c}{z + a}$$

7.2 Find the condition such that the second-order transfer function

$$A(z) = 1 + a_1 z^{-1} + a_2 z^{-2}$$

is positive real.

7.3 Find the condition such that

$$Z(z) = \frac{1}{A(z)} - \frac{1}{2} = \frac{1}{1 + a_1 z^{-1} + a_2 z^{-2}} - \frac{1}{2}$$

is positive real. Note that this condition appears in the convergence analysis of the recursive extended least-squares algorithm for ARMA models [109, 145].

7.4 Prove the following estimate for the matrix norm.

$$\left\| \begin{matrix} A & B \\ C & D \end{matrix} \right\| \le \|A\| + \|B\| + \|C\| + \|D\|$$

From this estimate, we can prove the continuity of $M(\cdot)$ of Lemma 7.3.

7.5 Let $A = 1/3$, $C = \bar{C} = \sqrt{2}/3$, $\Lambda(0) = 2/3$. Show that the LMI of (7.26) is given by

$$M(\Pi) = \begin{bmatrix} \frac{8}{9}\Pi & \frac{\sqrt{2}}{3}\left(1 - \frac{1}{3}\Pi\right) \\ \frac{\sqrt{2}}{3}\left(1 - \frac{1}{3}\Pi\right) & \frac{2}{3}\left(1 - \frac{1}{3}\Pi\right) \end{bmatrix} \ge 0$$

Compute the spectral factors corresponding to Π_* and Π^*.

7.6 Show that $\Omega_k = \mathcal{C}_k T_-^{-1}(k)\mathcal{C}_k^{\mathrm{T}}$ satisfies (7.62).

7.7 Derive (7.65) from (7.63).

7.8 Solve the optimal control problem (7.76) in the proof of Theorem 7.5.

7.10 Appendix: Proof of Lemma 7.4

Define the positive definite function $V(t) = \xi^{\mathrm{T}}(t)\Pi\xi(t)$ and compute the difference. It follows from (7.48) that

$$\begin{aligned} V(t+1) - V(t) &= [\xi^{\mathrm{T}}(t)A + u^{\mathrm{T}}(t)C]\Pi[A^{\mathrm{T}}\xi(t) + C^{\mathrm{T}}u(t)] - \xi^{\mathrm{T}}(t)\Pi\xi(t) \\ &= \xi^{\mathrm{T}}(t)(A\Pi A^{\mathrm{T}} - \Pi)\xi(t) + u^{\mathrm{T}}(t)C\Pi C^{\mathrm{T}}u(t) \\ &\quad + \xi^{\mathrm{T}}(t)A\Pi C^{\mathrm{T}}u(t) + u^{\mathrm{T}}(t)C\Pi A^{\mathrm{T}}\xi(t) \end{aligned}$$

Moreover, from (7.21),

$$\begin{aligned} V(t+1) - V(t) &= -\xi^{\mathrm{T}}(t)Q\xi(t) + u^{\mathrm{T}}(t)[\Lambda(0) - R]u(t) \\ &\quad + \xi^{\mathrm{T}}(t)(\bar{C}^{\mathrm{T}} - S)u(t) + u^{\mathrm{T}}(t)(\bar{C} - S^{\mathrm{T}})\xi(t) \\ &= -[\xi^{\mathrm{T}}(t)\ \ u^{\mathrm{T}}(t)]\begin{bmatrix} Q & S \\ S^{\mathrm{T}} & R \end{bmatrix}\begin{bmatrix} \xi(t) \\ u(t) \end{bmatrix} \\ &\quad + u^{\mathrm{T}}(t)\Lambda(0)u(t) + \xi^{\mathrm{T}}(t)\bar{C}^{\mathrm{T}}u(t) + u^{\mathrm{T}}(t)\bar{C}\xi(t) \end{aligned}$$

Taking the sum of the both sides of the above equation over $(-\infty, -1]$ yields

$$\begin{aligned} &\xi^{\mathrm{T}}(0)\Pi\xi(0) + \sum_{t=-\infty}^{-1} [\xi^{\mathrm{T}}(t)\ \ u^{\mathrm{T}}(t)]\begin{bmatrix} Q & S \\ S^{\mathrm{T}} & R \end{bmatrix}\begin{bmatrix} \xi(t) \\ u(t) \end{bmatrix} \\ &\qquad = \sum_{t=-\infty}^{-1} u^{\mathrm{T}}(t)\Lambda(0)u(t) + \sum_{t=-\infty}^{-1} \xi^{\mathrm{T}}(t)\bar{C}^{\mathrm{T}}u(t) + \sum_{t=-\infty}^{-1} u^{\mathrm{T}}(t)\bar{C}\xi(t) \\ &\qquad =: I_1 + I_2 + I_3 \end{aligned}$$

where we have used the boundary conditions $V(0) = \xi^{\mathrm{T}}(0)\Pi\xi(0)$ and $V(-\infty) = 0$.

Since, from (7.48), $\xi(t) = \sum_{k=-\infty}^{t-1} (A^{\mathrm{T}})^{t-1-k}C^{\mathrm{T}}u(k)$, we see that

$$\begin{aligned} I_2 &= \sum_{t=-\infty}^{-1} \xi^{\mathrm{T}}(t)\bar{C}^{\mathrm{T}}u(t) = \sum_{t=-\infty}^{-1}\sum_{k=-\infty}^{t-1} u^{\mathrm{T}}(k)CA^{t-k-1}\bar{C}^{\mathrm{T}}u(t) \\ &= \sum_{t=-\infty}^{-1}\sum_{k=-\infty}^{t-1} u^{\mathrm{T}}(k)\Lambda(t-k)u(t) \end{aligned}$$

Also, noting that $I_3 = I_2^{\mathrm{T}}$ holds, we change the order of sums in the above equation, and then interchange t and k to get

$$
\begin{aligned}
I_3 &= \sum_{t=-\infty}^{-1} \sum_{k=-\infty}^{t-1} u^{\mathrm{T}}(t) \Lambda^{\mathrm{T}}(t-k) u(k) \\
&= \sum_{k=-\infty}^{-2} \sum_{t=k+1}^{-1} u^{\mathrm{T}}(t) \Lambda(k-t) u(k) \\
&= \sum_{t=-\infty}^{-2} \sum_{k=t+1}^{-1} u^{\mathrm{T}}(k) \Lambda(t-k) u(t) \\
&= \sum_{t=-\infty}^{-1} \sum_{k=t+1}^{-1} u^{\mathrm{T}}(k) \Lambda(t-k) u(t) \qquad (7.87)
\end{aligned}
$$

The last equality is due to the fact that for $t = -1$, the second sum $\sum_{k=t+1}^{-1}$ becomes void. Hence, it follows that

$$
\begin{aligned}
I_1 + I_2 + I_3 &= \sum_{t=-\infty}^{-1} \left(u^{\mathrm{T}}(t) \Lambda(0) + \sum_{k=-\infty}^{t-1} u^{\mathrm{T}}(k) \Lambda(t-k) \right. \\
&\qquad \left. + \sum_{k=t+1}^{-1} u^{\mathrm{T}}(k) \Lambda(t-k) \right) u(t) \\
&= \sum_{t=-\infty}^{-1} \sum_{k=-\infty}^{-1} u^{\mathrm{T}}(k) \Lambda(t-k) u(t) = \boldsymbol{u}^{\mathrm{T}} T_+ \boldsymbol{u}
\end{aligned}
$$

This completes the proof. □

8

Stochastic Realization Theory (2)

This chapter presents the stochastic realization theory due to Akaike [2, 3]. First, we briefly review the method of canonical correlation analysis (CCA). We define the future and past spaces of a stationary stochastic process, and introduce two predictor spaces in terms of the orthogonal projection of the future onto the past, and *vice versa*. Based on these predictor spaces, we derive forward and backward innovation representations of a stationary process. We also discuss a stochastic balanced realization problem based on the CCA, including a model reduction of stochastic systems. Finally, presented are subspace algorithms to obtain stochastic state space models based on finite observed data. Some numerical results are included.

8.1 Canonical Correlation Analysis

The canonical correlation analysis (CCA) is a technique of multivariate statistical analysis that clarifies the mutual dependence between two sets of variables by finding a new coordinate system in the space of each set of variables.

Let x and y be two vectors of zero mean random variables defined by

$$x = \begin{bmatrix} x_1 \\ x_2 \\ \vdots \\ x_k \end{bmatrix} \in \mathbb{R}^k, \qquad y = \begin{bmatrix} y_1 \\ y_2 \\ \vdots \\ y_l \end{bmatrix} \in \mathbb{R}^l$$

Let the linear spaces spanned by x and y respectively be given by

$$\mathbb{X} = \operatorname{span}\{x_1, \cdots, x_k\}, \qquad \mathbb{Y} = \operatorname{span}\{y_1, \cdots, x_l\}$$

First, we find vectors $w_1 \in \mathbb{X}$ and $z_1 \in \mathbb{Y}$ with the maximum mutual correlation, and define (w_1, z_1) as the first coordinates in the new system. Then we find $w_2 \in \mathbb{X}$ and $z_2 \in \mathbb{Y}$ such that their correlation is maximum under the assumption that they are uncorrelated with the first coordinates (w_1, z_1). This procedure is continued until two new coordinate systems are determined.

Let the covariance matrices of two vectors x and y be given by

$$\Sigma = E\left\{\begin{bmatrix} x \\ y \end{bmatrix}\begin{bmatrix} x^{\mathrm{T}} & y^{\mathrm{T}} \end{bmatrix}\right\} = \begin{bmatrix} \Sigma_{xx} & \Sigma_{xy} \\ \Sigma_{yx} & \Sigma_{yy} \end{bmatrix} \in \mathbb{R}^{(k+l)\times(k+l)} \tag{8.1}$$

where it is assumed for simplicity that $\Sigma_{xx} > 0$ and $\Sigma_{yy} > 0$. Also, without loss of generality, we assume that $k \leq l$. We define two scalar variables

$$w_1 = a^{\mathrm{T}} x = \sum_{i=1}^{k} \alpha_i x_i, \qquad z_1 = b^{\mathrm{T}} y = \sum_{j=1}^{l} \beta_j y_j$$

by using two vectors $a \in \mathbb{R}^k$ and $b \in \mathbb{R}^l$, respectively. We wish to find the vectors a and b that maximize the correlation between w_1 and z_1, which is expressed as

$$\rho = \frac{\mathrm{cov}\{a^{\mathrm{T}} x,\, b^{\mathrm{T}} y\}}{\sqrt{\mathrm{cov}\{a^{\mathrm{T}} x\}}\sqrt{\mathrm{cov}\{b^{\mathrm{T}} y\}}} = \frac{a^{\mathrm{T}} \Sigma_{xy} b}{\sqrt{(a^{\mathrm{T}} \Sigma_{xx} a)\,(b^{\mathrm{T}} \Sigma_{yy} b)}}$$

Note that if a pair $(a,\, b)$ maximizes ρ, then the pair $(c_1 a,\, c_2 b)$ also maximizes ρ for all non-zero scalars $c_1,\, c_2$. Thus, we impose the following conditions

$$a^{\mathrm{T}} \Sigma_{xx} a = 1, \qquad b^{\mathrm{T}} \Sigma_{yy} b = 1 \tag{8.2}$$

The problem of maximizing ρ under the constraint of (8.2) is solved by means of the Lagrange method. Let the Lagrangian be given by

$$\mathfrak{L} = a^{\mathrm{T}} \Sigma_{xy} b + \frac{1}{2}\lambda_1 (1 - a^{\mathrm{T}} \Sigma_{xx} a) + \frac{1}{2}\lambda_2 (1 - b^{\mathrm{T}} \Sigma_{yy} b)$$

Then, the optimality conditions satisfied by the vectors a and b are

$$\frac{\partial \mathfrak{L}}{\partial a} = \Sigma_{xy} b - \lambda_1 \Sigma_{xx} a = 0, \qquad \frac{\partial \mathfrak{L}}{\partial b} = \Sigma_{yx} a - \lambda_2 \Sigma_{yy} b = 0 \tag{8.3}$$

Pre-multiplying the first equation of (8.3) by a^{T} and the second by b^{T} and using (8.2), we have

$$a^{\mathrm{T}} \Sigma_{xy} b = b^{\mathrm{T}} \Sigma_{yx} a = \lambda_1 = \lambda_2$$

Letting $\lambda_1 = \lambda_2 = \rho$, it follows from (8.3) that

$$\Sigma_{xy} b - \rho \Sigma_{xx} a = 0, \qquad \Sigma_{yx} a - \rho \Sigma_{yy} b = 0 \tag{8.4}$$

Since $\Sigma_{yy} > 0$, we can eliminate b from the above equations to get

$$(\Sigma_{xy} \Sigma_{yy}^{-1} \Sigma_{yx} - \rho^2 \Sigma_{xx}) a = 0, \qquad a \neq 0 \tag{8.5}$$

This is a GEP since $\Sigma_{xx} \neq I$.

We see that a necessary and sufficient condition that a has a non-trivial solution is given by

$$\det(\Sigma_{xy} \Sigma_{yy}^{-1} \Sigma_{yx} - \rho^2 \Sigma_{xx}) = 0 \tag{8.6}$$

where this is a kth-order polynomial in ρ^2 since $\det(\Sigma_{xx}) \neq 0$. Let square root matrices of Σ_{xx} and Σ_{yy} respectively be $\Sigma_{xx}^{1/2}$ and $\Sigma_{yy}^{1/2}$, satisfying

$$\Sigma_{xx} = \Sigma_{xx}^{1/2}\Sigma_{xx}^{\mathrm{T}/2}, \qquad \Sigma_{yy} = \Sigma_{yy}^{1/2}\Sigma_{yy}^{\mathrm{T}/2}$$

It therefore follows from (8.6) that

$$\det\left(\Sigma_{xx}^{-1/2}\Sigma_{xy}\Sigma_{yy}^{-\mathrm{T}/2}\Sigma_{yy}^{-1/2}\Sigma_{yx}\Sigma_{xx}^{-\mathrm{T}/2} - \rho^2 I_k\right) = 0$$

Define $\Xi := \Sigma_{xx}^{-1/2}\Sigma_{xy}\Sigma_{yy}^{-\mathrm{T}/2} \in \mathbb{R}^{k\times l}$. Then, we have

$$\det(\Xi\Xi^{\mathrm{T}} - \rho^2 I_k) = 0 \tag{8.7}$$

This implies that ρ^2 is an eigenvalue of $\Xi\Xi^{\mathrm{T}} \in \mathbb{R}^{k\times k}$, and that k eigenvalues of $\Xi\Xi^{\mathrm{T}}$ are nonnegative. Let $\rho_1 \geq \rho_2 \geq \cdots \geq \rho_k \geq 0$ be the positive square roots of the eigenvalues of $\Xi\Xi^{\mathrm{T}}$, and let $a_1, a_2, \cdots, a_k \in \mathbb{R}^n$ be the corresponding eigenvectors obtained from (8.5). Then, we define the matrix

$$L = [a_1 \;\; a_2 \;\; \cdots \;\; a_k] \in \mathbb{R}^{k\times k}$$

Similarly, eliminating a from (8.4) yields

$$(\Sigma_{yx}\Sigma_{xx}^{-1}\Sigma_{xy} - \rho^2\Sigma_{yy})b = 0, \qquad b \neq 0$$

and hence

$$\det(\Sigma_{yx}\Sigma_{xx}^{-1}\Sigma_{xy} - \rho^2\Sigma_{yy}) = 0 \tag{8.8}$$

Since $\Sigma_{yy} > 0$, (8.8) is equivalent to $\det(\Xi^{\mathrm{T}}\Xi - \rho^2 I_l) = 0$. Since $\Xi^{\mathrm{T}}\Xi \in \mathbb{R}^{l\times l}$ is nonnegative definite, it has l nonnegative eigenvalues. Let $\rho_1 \geq \rho_2 \geq \cdots \geq \rho_l \geq 0$ be the positive square roots of the eigenvalues of $\Xi^{\mathrm{T}}\Xi$[1]. Let the corresponding eigenvectors be given by $b_1, b_2, \cdots, b_l \in \mathbb{R}^l$, and define the matrix

$$M = [b_1 \;\; b_2 \;\; \cdots \;\; b_l] \in \mathbb{R}^{l\times l}$$

Definition 8.1. *The maximum correlation ρ_1 is called the first canonical correlation. In terms of corresponding two vectors a_1 and b_1, we have two scalars*

$$w_1 := a_1^{\mathrm{T}} x, \qquad z_1 := b_1^{\mathrm{T}} y$$

These variables are called the first canonical variables. Similarly, ρ_i is called the ith canonical correlation, and $w_i = a_i^{\mathrm{T}} x$ and $z_i = b_i^{\mathrm{T}} y$ are the ith canonical variables. Also, two vectors $w = L^{\mathrm{T}} x$ and $z = M^{\mathrm{T}} y$ are called canonical vectors. □

The following lemma shows that L and M are the square root inverses of the covariance matrices $\Sigma_{xx} \in \mathbb{R}^{k\times k}$ and $\Sigma_{yy} \in \mathbb{R}^{l\times l}$, respectively.

[1] Since nonzero eigenvalues of $\Xi\Xi^{\mathrm{T}}$ and $\Xi^{\mathrm{T}}\Xi$ are equal, we use the same symbol for them.

Lemma 8.1. *Let L and M be defined as above. Then, we have*

$$L^{\mathrm{T}} \Sigma_{xx} L = I_k, \qquad M^{\mathrm{T}} \Sigma_{yy} M = I_l \tag{8.9}$$

and

$$L^{\mathrm{T}} \Sigma_{xy} M = D = \begin{bmatrix} \rho_1 & & & \\ & \rho_2 & & 0 \\ & & \ddots & \\ & & & \rho_k \end{bmatrix} \in \mathbb{R}^{k \times l} \tag{8.10}$$

where $1 \geq \rho_1 \geq \cdots \geq \rho_k \geq 0$.

Proof. We prove (8.9) under the assumption that $\rho_i \neq \rho_j$ for $i \neq j$. Let (a_i, b_i) and (a_j, b_j) be pairs of eigenvectors corresponding to ρ_i and ρ_j, respectively. From (8.5), we have

$$\Sigma_{xy}\Sigma_{yy}^{-1}\Sigma_{yx}a_i = \rho_i^2 \Sigma_{xx} a_i, \qquad \Sigma_{xy}\Sigma_{yy}^{-1}\Sigma_{yx}a_j = \rho_j^2 \Sigma_{xx} a_j$$

Pre-multiplying the first and the second equations by a_j^{T} and a_i^{T}, respectively, and subtracting both sides of resulting equations yield

$$(\rho_i^2 - \rho_j^2) a_j^{\mathrm{T}} \Sigma_{xx} a_i = 0, \qquad i \neq j$$

Thus we see that $a_j^{\mathrm{T}} \Sigma_{xx} a_i = 0,\ i \neq j$. In view of (8.2), this fact implies that $L^{\mathrm{T}} \Sigma_{xx} L = I_k$. We can also prove $M^{\mathrm{T}} \Sigma_{yy} M = I_l$ by using b_i and b_j.

It follows from (8.4) that $\Sigma_{xy} b_i = \rho_i \Sigma_{xx} a_i$. Pre-multiplying this equation by a_i^{T} yields

$$a_i^{\mathrm{T}} \Sigma_{xy} b_i = \rho_i a_i^{\mathrm{T}} \Sigma_{xx} a_i = \rho_i$$

Similarly, pre-multiplying $\Sigma_{xy} b_i = \rho_i \Sigma_{xx} a_i$ by a_j^{T} $(j \neq i)$ gives

$$a_j^{\mathrm{T}} \Sigma_{xy} b_i = \rho_i a_j^{\mathrm{T}} \Sigma_{xx} a_i = 0, \qquad j \neq i$$

These equations prove (8.10).

Finally we show that $\rho_i^2 \leq 1$. Let θ be a scalar, and consider

$$\det\Big(\theta \Sigma_{xx} - (\Sigma_{xx} - \Sigma_{xy}\Sigma_{yy}^{-1}\Sigma_{yx})\Big) = 0$$

Since $\Sigma_{xx} > 0$ and $\Sigma_{xx} - \Sigma_{xy}\Sigma_{yy}^{-1}\Sigma_{yx} \geq 0$, we get $\theta \geq 0$. This can be proved by using the technique of simultaneous diagonalization of two nonnegative definite matrices. Thus, we have

$$\det\Big((1-\theta)\Sigma_{xx} - \Sigma_{xy}\Sigma_{yy}^{-1}\Sigma_{yx}\Big) = 0$$

Comparing this with (8.6) gives $\theta = 1 - \rho^2 \geq 0$, and hence $\rho^2 \leq 1$. ◻

Let $w = L^{\mathrm{T}}x$ and $z = M^{\mathrm{T}}y$. Then, we see from Lemma 8.1 that

$$E\{ww^{\mathrm{T}}\} = I_k, \qquad E\{zz^{\mathrm{T}}\} = I_l$$

and

$$E\{wz^{\mathrm{T}}\} = D = \begin{bmatrix} \rho_1 & & & \\ & \rho_2 & & 0 \\ & & \ddots & \\ & & & \rho_k \end{bmatrix} \tag{8.11}$$

The elements of canonical vectors w and z, which are respectively obtained by linear transforms of x and y, are white noises with mean zero and unit variance, and they are arranged in descending order of mutual correlations as shown in Table 8.1. Thus, both whitening and correlating two vectors can be performed by the CCA.

Table 8.1. Canonical correlation analysis

$$x \xrightarrow{L^{\mathrm{T}}} \begin{bmatrix} w_1 \\ \vdots \\ w_k \\ \\ \\ \end{bmatrix} \begin{matrix} \xleftrightarrow{\rho_1} \\ \vdots \\ \xleftrightarrow{\rho_k} \\ \\ \\ \end{matrix} \begin{bmatrix} z_1 \\ \vdots \\ z_k \\ \vdots \\ z_l \end{bmatrix} \xleftarrow{M^{\mathrm{T}}} y$$

We see from (8.7) that the canonical correlations $\rho_1 \geq \rho_2 \geq \cdots \geq \rho_k$ are the singular values of Ξ, so that they are computed as follows.

Lemma 8.2. *Suppose that the covariance matrices of x and y are given by (8.1). Then, the canonical correlations are computed by the SVD*

$$\Xi = \Sigma_{xx}^{-1/2}\Sigma_{xy}\Sigma_{yy}^{-\mathrm{T}/2} = UDV^{\mathrm{T}} \tag{8.12}$$

where D is defined by (8.11). Also, the canonical vectors are given by

$$w = L^{\mathrm{T}}x = U^{\mathrm{T}}\Sigma_{xx}^{-1/2}x, \qquad z = M^{\mathrm{T}}y = V^{\mathrm{T}}\Sigma_{yy}^{-1/2}y$$

Proof. It follows from (8.12) that $(U^{\mathrm{T}}\Sigma_{xx}^{-1/2})\Sigma_{xy}(\Sigma_{yy}^{-\mathrm{T}/2}V) = D$. Comparing this with (8.10) gives the desired results. □

8.2 Stochastic Realization Problem

We consider the same stochastic realization problem treated in Chapter 7. Suppose that $\{y(t),\ t = 0, \pm 1, \cdots\}$ is a regular full rank p-dimensional stationary process. We assume that the mean of y is zero and the covariance matrix is given by

$$\Lambda(l) = E\{y(t+l)y^{\mathrm{T}}(t)\}, \qquad l = 0, \pm 1, \cdots \tag{8.13}$$

Suppose that the covariance matrices satisfy the summability condition

$$\sum_{l=-\infty}^{\infty} \|\Lambda(l)\| < \infty \tag{8.14}$$

Then, the spectral density matrix of y is defined by

$$\Phi(z) = \sum_{l=-\infty}^{\infty} \Lambda(l) z^{-l} \tag{8.15}$$

Given the covariance matrices (or equivalently the spectral density matrix) of a stationary process y, the stochastic realization problem is to find a Markov model of the form

$$x(t+1) = Ax(t) + w(t) \tag{8.16a}$$

$$y(t) = Cx(t) + v(t) \tag{8.16b}$$

where $x \in \mathbb{R}^n$ is a state vector, and $w \in \mathbb{R}^n$ and $v \in \mathbb{R}^p$ are white noises with mean zero and covariance matrices

$$E\left\{ \begin{bmatrix} w(t) \\ v(t) \end{bmatrix} \begin{bmatrix} w^{\mathrm{T}}(s) & v^{\mathrm{T}}(s) \end{bmatrix} \right\} = \begin{bmatrix} Q & S \\ S^{\mathrm{T}} & R \end{bmatrix} \delta_{ts} \tag{8.17}$$

In the following, it is assumed that we have an infinite sequence of data, from which we can compute the true covariance matrices.

Let t be the present time. Define infinite dimensional future and past vectors

$$f(t) := \begin{bmatrix} y(t) \\ y(t+1) \\ \vdots \\ \\ \end{bmatrix}, \qquad p(t) := \begin{bmatrix} y(t-1) \\ y(t-2) \\ \vdots \\ \\ \end{bmatrix}$$

Then, the cross-covariance matrix of the future and past is given by

$$H = E\{f(t)p^{\mathrm{T}}(t)\} = \begin{bmatrix} \Lambda(1) & \Lambda(2) & \Lambda(3) & \cdots \\ \Lambda(2) & \Lambda(3) & \Lambda(4) & \cdots \\ \Lambda(3) & \Lambda(4) & \Lambda(5) & \cdots \\ \vdots & \vdots & \vdots & \ddots \end{bmatrix} \tag{8.18}$$

and the covariance matrices of the future and the past are respectively given by

$$T_{+} = E\{f(t)f^{\mathrm{T}}(t)\} = \begin{bmatrix} \Lambda(0) & \Lambda^{\mathrm{T}}(1) & \Lambda^{\mathrm{T}}(2) & \cdots \\ \Lambda(1) & \Lambda(0) & \Lambda^{\mathrm{T}}(1) & \cdots \\ \Lambda(2) & \Lambda(1) & \Lambda(0) & \cdots \\ \vdots & \vdots & \vdots & \ddots \end{bmatrix} \tag{8.19}$$

and

$$T_- = E\{p(t)p^{\mathrm{T}}(t)\} = \begin{bmatrix} \Lambda(0) & \Lambda(1) & \Lambda(2) & \cdots \\ \Lambda^{\mathrm{T}}(1) & \Lambda(0) & \Lambda(1) & \cdots \\ \Lambda^{\mathrm{T}}(2) & \Lambda^{\mathrm{T}}(1) & \Lambda(0) & \cdots \\ \vdots & \vdots & \vdots & \ddots \end{bmatrix} \tag{8.20}$$

It should be noted that H is an infinite block Hankel matrix, and $T_\pm$ are infinite block Toeplitz matrices.

Let $\mathcal{Y} = \overline{\mathrm{span}}\{y(t) \mid t = 0, \pm 1, \cdots\}$ be a Hilbert space generated by all the linear functionals of the second-order stationary stochastic process y. Let $\mathcal{Y}_t^+$ and $\mathcal{Y}_t^-$ respectively be linear spaces generated by the future $f(t)$ and the past $p(t)$, *i.e.*,

$$\mathcal{Y}_t^+ = \overline{\mathrm{span}}\{y(t),\, y(t+1),\, \cdots\}, \qquad \mathcal{Y}_t^- = \overline{\mathrm{span}}\{y(t-1),\, y(t-2),\, \cdots\}$$

We assume that these spaces are closed with respect to the mean-square norm, so that $\mathcal{Y}_t^+$ and $\mathcal{Y}_t^-$ are subspaces of the Hilbert space $\mathcal{Y}$.

8.3 Akaike's Method

In this section, we shall study the CCA-based stochastic realization method due to Akaike [2, 3].

8.3.1 Predictor Spaces

A necessary and sufficient condition that y has a finite dimensional stochastic realization is that the Hankel matrix of (8.18) has a finite rank, *i.e.*, rank $(H) < \infty$. In order to show this fact by means of the CCA technique, we begin with the definition of forward and backward predictor spaces.

Definition 8.2. *Let the orthogonal projection of the future $\mathcal{Y}_t^+$ onto the past $\mathcal{Y}_t^-$ be defined by*

$$\begin{aligned} \mathcal{X}_t := \hat{E}\{\mathcal{Y}_t^+ \mid \mathcal{Y}_t^-\} &= \overline{\mathrm{span}}\Big\{\hat{E}\{y(t+h) \mid \mathcal{Y}_t^-\} \,\Big|\, h = 0,\, 1,\, \cdots\Big\} \\ &= \overline{\mathrm{span}}\Big\{\hat{y}(t+h \mid t-) \,\Big|\, h = 0,\, 1,\, \cdots\Big\} \end{aligned}$$

And, let the orthogonal projection of the past $\mathcal{Y}_t^-$ onto the future $\mathcal{Y}_t^+$ be given by

$$\begin{aligned} \check{\mathcal{X}}_t := \hat{E}\{\mathcal{Y}_t^- \mid \mathcal{Y}_t^+\} &= \overline{\mathrm{span}}\Big\{\hat{E}\{y(t-l) \mid \mathcal{Y}_t^+\} \,\Big|\, l = 1,\, 2,\, \cdots\Big\} \\ &= \overline{\mathrm{span}}\Big\{\check{y}(t-l \mid t+) \,\Big|\, l = 1,\, 2,\, \cdots\Big\} \end{aligned}$$

Then the spaces $\mathcal{X}_t$ and $\check{\mathcal{X}}_t$ are called the forward and the backward predictor spaces, respectively. □

The generators $\hat{y}(t+h \mid t-)$ of the forward predictor space are the minimum variance estimates of the future $y(t+h),\ h=0,\,1,\,\cdots$ based on the past $\mathcal{Y}_t^-$, and the generators $\check{y}(t-l \mid t+)$ of the backward predictor space are the minimum variance estimates of the past $y(t-l),\ l=1,\,2,\,\cdots$ based on the future $\mathcal{Y}_t^+$. The notations $\hat{y}(t+h \mid t-)$ and $\check{y}(t-l \mid t+)$ are used in this section only; in fact, the optimal forward estimates should be written as $\hat{y}(t+h \mid t-1)$ by using the notation defined in Chapter 5.

The optimality conditions for the forward and backward estimates are that the respective estimation errors are orthogonal to the data spaces $\mathcal{Y}_t^-$ and $\mathcal{Y}_t^+$, respectively. Thus the optimality conditions are expressed as

$$E\Big\{\Big[y(t+h)-\hat{y}(t+h \mid t-)\Big]y^{\mathrm{T}}(t-l)\Big\}=0$$

and

$$E\Big\{\Big[y(t-l)-\check{y}(t-l \mid t+)\Big]y^{\mathrm{T}}(t+h)\Big\}=0$$

where $h=0,\,1,\,\cdots$ and $l=1,\,2,\,\cdots$.

Lemma 8.3. *Suppose that* $\operatorname{rank}(H)<\infty$. *Then, the two predictor spaces* $\mathcal{X}_t$ *and* $\check{\mathcal{X}}_t$ *are finite dimensional, and are respectively written as*

$$\mathcal{X}_t=\operatorname{span}\Big\{\hat{y}(t+h \mid t-)\,\Big|\,h=0,\,1,\,\cdots,\,r-1\Big\}$$

and

$$\check{\mathcal{X}}_t=\operatorname{span}\Big\{\check{y}(t-l \mid t+)\,\Big|\,l=1,\,2,\,\cdots,\,r\Big\}$$

where r *is a positive integer determined by the factorization of given covariance matrices* $\{\Lambda(k),\ k=1,\,2,\,\cdots\}$.

Proof. Since $\operatorname{rank}(H)<\infty$, the covariance matrix $\Lambda(k)$ has a factorization given by (7.7). Thus it follows from Theorem 3.13 that there exist an integer $r>0$ and scalars $\alpha_1,\,\cdots,\,\alpha_r\in\mathbb{R}$ such that

$$\Lambda(r+k)+\sum_{i=1}^{r}\alpha_i\Lambda(r+k-i)=0,\qquad k=1,\,2,\,\cdots \tag{8.21}$$

This is a set of linear equations satisfied by the covariance matrices. From the definition of covariance matrices, it can be shown that (8.21) is rewritten as

$$E\Big\{\Big[y(t+r+h)+\sum_{i=1}^{r}\alpha_i y(t+r+h-i)\Big]y^{\mathrm{T}}(t-l)\Big\}=0 \tag{8.22}$$

and

$$E\Big\{\Big[y(t-r-l)+\sum_{i=1}^{r}\alpha_i y(t-r-l+i)\Big]y^{\mathrm{T}}(t+h)\Big\}=0 \tag{8.23}$$

where $h=0,\,1,\,\cdots$ and $l=1,\,2,\,\cdots$. We see that (8.22) is equivalent to

$$\hat{E}\Big\{y(t+r+h)+\sum_{i=1}^{r}\alpha_i y(t+r+h-i)\,\Big|\,\mathcal{Y}_t^-\Big\}=0, \qquad h=0,1,\cdots$$

so that we have

$$\hat{y}(t+r+h\mid t-)=-\sum_{i=1}^{r}\alpha_i\hat{y}(t+r+h-i\mid t-), \qquad h=0,1,\cdots \tag{8.24}$$

Similarly, from (8.23),

$$\hat{E}\Big\{y(t-r-l)+\sum_{i=1}^{r}\alpha_i y(t-r-l-i)\,\Big|\,\mathcal{Y}_t^+\Big\}=0, \qquad l=1,2,\cdots$$

This implies that

$$\check{y}(t-r-l\mid t+)=-\sum_{i=1}^{r}\alpha_i\check{y}(t-r-l+i\mid t+), \qquad l=1,2,\cdots \tag{8.25}$$

From (8.24) and (8.25), we see that the predictor spaces $\mathcal{X}_t=\hat{E}\{\mathcal{Y}_t^+\mid\mathcal{Y}_t^-\}$ and $\check{\mathcal{X}}_t=\hat{E}\{\mathcal{Y}_t^-\mid\mathcal{Y}_t^+\}$ are finite dimensional, and the former is generated by the forward predictors $\hat{y}(t+h\mid t-)=\hat{E}\{y(t+h)\mid\mathcal{Y}_t^-\}$, $h=0,1,\cdots,r-1$, and the latter the backward predictors $\check{y}(t-l\mid t+)=\hat{E}\{y(t-l)\mid\mathcal{Y}_t^+\}$, $l=1,2,\cdots,r$. □

In the above lemma, the positive integer r may not be minimum. But, applying the CCA described in Section 8.1 to the following two vectors

$$\phi:=\begin{bmatrix}\hat{y}(t\mid t-)\\ \hat{y}(t+1\mid t-)\\ \vdots\\ \hat{y}(t+r-1\mid t-)\end{bmatrix}, \qquad \check{\phi}:=\begin{bmatrix}\check{y}(t-1\mid t+)\\ \check{y}(t-2\mid t+)\\ \vdots\\ \check{y}(t-r\mid t+)\end{bmatrix}$$

we obtain the minimal dimensional orthonormal basis vectors $x(t)$ and $\check{x}(t)$ for the predictor spaces $\mathcal{X}_t$ and $\check{\mathcal{X}}_t$, respectively. Being orthonormal, we have $\operatorname{cov}\{x(t)\}=I_n=\operatorname{cov}\{\check{x}(t)\}$. It should be noted that since $y(t)$ is a stationary process, $x(t)$ and $\check{x}(t)$, the orthonormal bases of $\mathcal{X}_t$ and $\check{\mathcal{X}}_t$, are jointly stationary.

For the transition from time t to $t+1$, we see that the predictor space evolves from $\mathcal{X}_t$ to $\mathcal{X}_{t+1}=\hat{E}\{\mathcal{Y}_{t+1}^+\mid\mathcal{Y}_{t+1}^-\}$. Since $\mathcal{Y}_{t+1}^-=\mathcal{Y}_t^-\vee\operatorname{span}\{y(t)\}$, the space $\mathcal{Y}_{t+1}^-$ has the orthogonal decomposition

$$\mathcal{Y}_{t+1}^-=\mathcal{Y}_t^-\oplus\operatorname{span}\{\tilde{y}(t)\} \tag{8.26}$$

where $\tilde{y}(t):=y(t)-\hat{y}(t\mid t-)$ is the forward innovation for $y(t)$. Thus, it follows that

$$\begin{aligned}\mathcal{X}_{t+1}&=\hat{E}\{\mathcal{Y}_{t+1}^+\mid\mathcal{Y}_{t+1}^-\}=\hat{E}\{\mathcal{Y}_{t+1}^+\mid\mathcal{Y}_t^-\oplus\operatorname{span}\{\tilde{y}(t)\}\}\\&=\hat{E}\{\mathcal{Y}_{t+1}^+\mid\mathcal{Y}_t^-\}+\hat{E}\{\mathcal{Y}_{t+1}^+\mid\operatorname{span}\{\tilde{y}(t)\}\}\end{aligned}$$

Since $\mathcal{Y}_{t+1}^+ \subset \mathcal{Y}_t^+$, we see that $\hat{E}\{\mathcal{Y}_{t+1}^+ \mid \mathcal{Y}_t^-\} \subset \mathcal{X}_t$. Hence, the first term in the right-hand side of the above equation gets smaller than $\mathcal{X}_t$, but by the addition of new information $\tilde{y}(t)$, a transition from $\mathcal{X}_t$ to $\mathcal{X}_{t+1}$ is made.

Definition 8.3. *[105] Suppose that a subspace $\mathcal{S}_t$ ($\subset \mathcal{Y}_t^-$) satisfies*

$$\hat{E}\{\mathcal{Y}_t^+ \mid \mathcal{S}_t\} = \hat{E}\{\mathcal{Y}_t^+ \mid \mathcal{Y}_t^-\}$$

Then, $\mathcal{S}_t$ is called a splitting subspace for $(\mathcal{Y}_t^+, \mathcal{Y}_t^-)$. □

Lemma 8.4. *The predictor space $\mathcal{X}_t = \hat{E}\{\mathcal{Y}_t^+ \mid \mathcal{Y}_t^-\}$ is a minimal splitting subspace for $(\mathcal{Y}_t^+, \mathcal{Y}_t^-)$.*

Proof. We note that $\mathcal{X}_t = \hat{E}\{\mathcal{X}_t \mid \mathcal{X}_t\}$ and $\mathcal{X}_t \subset \mathcal{Y}_t^-$ hold. It thus follows from the property of orthogonal projection that

$$\begin{aligned}\hat{E}\{\mathcal{Y}_t^+ \mid \mathcal{Y}_t^-\} = \mathcal{X}_t = \hat{E}\{\mathcal{X}_t \mid \mathcal{X}_t\} &= \hat{E}\{\hat{E}\{\mathcal{Y}_t^+ \mid \mathcal{Y}_t^-\} \mid \mathcal{X}_t\} \\ &= \hat{E}\{\mathcal{Y}_t^+ \mid \mathcal{X}_t\}\end{aligned}$$

The last equality implies that $\mathcal{X}_t$ is a splitting subspace for $(\mathcal{Y}_t^+, \mathcal{Y}_t^-)$. Also, it can be shown that if a subspace $\mathcal{S}_t$ of $\mathcal{Y}_t^-$ satisfies

$$\hat{E}\{\mathcal{Y}_t^+ \mid \mathcal{S}_t\} = \hat{E}\{\mathcal{Y}_t^+ \mid \mathcal{Y}_t^-\}$$

then we have $\mathcal{S}_t \supset \mathcal{X}_t$. Hence, $\mathcal{X}_t$ is the minimal splitting subspace for $(\mathcal{Y}_t^+, \mathcal{Y}_t^-)$. □

This lemma shows that $\mathcal{X}_t$ contains the minimal necessary information to predict the future of the output y based on the past $\mathcal{Y}_t^-$. We can also show that the backward predictor space $\check{\mathcal{X}}_t = \hat{E}\{\mathcal{Y}_t^- \mid \mathcal{Y}_t^+\}$ is the minimal splitting subspace for $(\mathcal{Y}_t^-, \mathcal{Y}_t^+)$, contained in the future $\mathcal{Y}_t^+$. Thus, two predictor spaces defined above can be viewed as basic interfaces between the past and the future in stochastic systems. It will be shown that we can derive a Markov model for the stationary process $y(t)$ by using either $\check{x}(t)$ or $x(t)$.

8.3.2 Markovian Representations

The stochastic realization technique due to Faurre considered in Section 7.3 is based on the deterministic realization method that computes $(A, C, \bar{C}, \Lambda(0))$ from the given covariance matrices and then finds solutions $(\Pi > 0, Q, R, S)$ of LMI (7.26). On the other hand, the method to be developed here is based on the CCA technique, so that it is completely different from that of Section 7.3. By deriving a basis vector of the predictor space by the CCA, we obtain a Markov model with a state vector given by the basis vector.

We first derive a stochastic realization based on the basis vector $\check{x}(t) \in \check{\mathcal{X}}_t$. Recall that $\check{x}(t)$ has zero mean and covariance matrix I_n.

Theorem 8.1. *In terms of the basis vector* $\check{x}(t) \in \mathcal{Y}_t^+$, *a Markov model for the stationary process* y *is given by*

$$\check{x}(t+1) = A\check{x}(t) + \check{w}(t) \tag{8.27a}$$

$$y(t) = C\check{x}(t) + \check{v}(t) \tag{8.27b}$$

where $A \in \mathbb{R}^{n\times n}$, $C \in \mathbb{R}^{p\times n}$, $\bar{C} \in \mathbb{R}^{p\times n}$ *satisfy*

$$\begin{aligned} A &= E\{\check{x}(t+1)\check{x}^{\mathrm{T}}(t)\} \\ C &= E\{y(t)\check{x}^{\mathrm{T}}(t)\} \\ \bar{C} &= E\{y(t)\check{x}^{\mathrm{T}}(t+1)\} \end{aligned} \tag{8.28}$$

Also, $\check{w}$ *and* $\check{v}$ *are white noise vectors with mean zero and covariance matrices*

$$E\left\{\begin{bmatrix}\check{w}(t)\\ \check{v}(t)\end{bmatrix}\begin{bmatrix}\check{w}^{\mathrm{T}}(s) & \check{v}^{\mathrm{T}}(s)\end{bmatrix}\right\} = \begin{bmatrix} Q & S \\ S^{\mathrm{T}} & R\end{bmatrix}\delta_{ts}$$

where $\operatorname{cov}\{\check{x}(t)\} = \Pi = I_n$, *and*

$$Q = I_n - AA^{\mathrm{T}}, \qquad S = \bar{C}^{\mathrm{T}} - AC^{\mathrm{T}}, \qquad R = \Lambda(0) - CC^{\mathrm{T}} \tag{8.29}$$

Proof. 1° Let the basis vector of $\check{\mathcal{X}}_t$ be given by $\check{x}(t)$. Since $\check{\mathcal{X}}_t = \hat{E}\{\mathcal{Y}_t^- \mid \mathcal{Y}_t^+\} \subset \mathcal{Y}_t^+$, we see that $\check{x}(t)$ is included in $\mathcal{Y}_t^+$. Also, we get $\check{x}(t+1) \in \mathcal{Y}_{t+1}^+ \subset \mathcal{Y}_t^+$. Hence, we can decompose $\check{x}(t+1)$ as the sum of orthogonal projections onto $\operatorname{span}\{\check{x}(t)\}$ and its complementary space $(\operatorname{span}\{\check{x}(t)\})^\perp \cap \mathcal{Y}_t^+$, *i.e.*,

$$\check{x}(t+1) = \hat{E}\{\check{x}(t+1) \mid \check{x}(t)\} + \hat{E}\{\check{x}(t+1) \mid (\operatorname{span}\{\check{x}(t)\})^\perp \cap \mathcal{Y}_t^+\}$$

The first term in the right-hand side of the above equation is expressed as

$$\hat{E}\{\check{x}(t+1) \mid \check{x}(t)\} = E\{\check{x}(t+1)\check{x}^{\mathrm{T}}(t)\}(E\{\check{x}(t)\check{x}^{\mathrm{T}}(t)\})^{-1}\check{x}(t) = A\check{x}(t)$$

Define $\check{w}(t) := \check{x}(t+1) - A\check{x}(t)$. Then, $\check{w}(t) \in \mathcal{Y}_t^+$, but $\check{w}(t) \perp \check{\mathcal{X}}_t$. We show that $\check{w}(t)$ is orthogonal to $\mathcal{Y}_t^-$. Let $\xi \in \mathcal{Y}_t^-$. Then, we have $\hat{E}\{\xi \mid \mathcal{Y}_t^+\} \in \check{\mathcal{X}}_t$. Also, by definition, $\xi - \hat{E}\{\xi \mid \mathcal{Y}_t^+\} \perp \mathcal{Y}_t^+$, and hence $\xi - \hat{E}\{\xi \mid \mathcal{Y}_t^+\} \perp \check{w}(t)$. Since $\hat{E}\{\xi \mid \mathcal{Y}_t^+\} \perp \check{w}(t)$, we obtain $\xi \perp \check{w}(t)$ for any $\xi \in \mathcal{Y}_t^-$. This proves the desired result.

Thus, $\check{w}(t+l)$ is orthogonal to $\mathcal{Y}_{t+l}^-$. Since $\mathcal{Y}_t^- \subset \mathcal{Y}_{t+l}^-$, $l = 0, 1, \cdots$, we see that $\check{w}(t+l) \perp \mathcal{Y}_t^-$ holds. However, since $\check{w}(t+l) \in \mathcal{Y}_t^+$, it follows that $\check{w}(t+l)$ is orthogonal to $\check{x}(t)$, implying that

$$\check{w}(t+l) \perp \check{x}(t),\ y(t-1), \qquad l = 0, 1, \cdots \tag{8.30}$$

2° Since $y(t) \in \mathcal{Y}_t^+$ and $\check{\mathcal{X}}_t \subset \mathcal{Y}_t^+$, the output $y(t)$ has a unique decomposition

$$\begin{aligned} y(t) &= \hat{E}\{y(t) \mid \operatorname{span}\{\check{x}(t)\}\} + \hat{E}\{y(t) \mid (\operatorname{span}\{\check{x}(t)\})^\perp \cap \mathcal{Y}_t^+\} \\ &= C\check{x}(t) + \check{v}(t) \end{aligned}$$

This shows that (8.27b) holds. By the definition of $\check{v}(t)$, we have $\check{v}(t) \in \mathcal{Y}_t^+$ and $\check{v}(t) \perp \check{\mathcal{X}}_t$. As in the proof of $\check{w}(t) \perp \mathcal{Y}_t^-$ in 1°, we can show that $\check{v}(t) \perp \mathcal{Y}_t^-$, and hence $\check{v}(t+h) \perp \mathcal{Y}_t^-$, $h = 0, 1, \cdots$. Moreover, since $\check{v}(t+h) \in \mathcal{Y}_{t+h}^+ \subset \mathcal{Y}_t^+$ holds, it follows that $\check{v}(t+h)$ is orthogonal to $\check{\mathcal{X}}_t = \hat{E}\{\mathcal{Y}_t^- \mid \mathcal{Y}_t^+\}$, *i.e.*, $\check{v}(t+h) \perp \check{x}(t)$, $h = 0, 1, \cdots$. This implies that

$$\check{v}(t+l) \perp \check{x}(t),\, y(t-1), \qquad l = 0, 1, \cdots \tag{8.31}$$

3° We see from (8.30) that

$$E\{\check{w}(t+l)\check{w}^{\mathrm{T}}(t)\} = E\{\check{w}(t+l)[\check{x}(t+1) - A\check{x}(t)]^{\mathrm{T}}\} = 0, \qquad l = 1, 2, \cdots$$

This implies that $\check{w}$ is a white noise. Also, it follows from (8.31) that $\check{v}$ is a white noise. Again, from (8.30) and (8.31), we have

$$E\{\check{v}(t+l)\check{w}^{\mathrm{T}}(t)\} = E\{\check{v}(t+l)[\check{x}(t+1) - A\check{x}(t)]^{\mathrm{T}}\} = 0, \qquad l = 1, 2, \cdots$$

and

$$E\{\check{w}(t+l)\check{v}^{\mathrm{T}}(t)\} = E\{\check{w}(t+l)[y(t) - C\check{x}(t)]^{\mathrm{T}}\} = 0, \qquad l = 1, 2, \cdots$$

This shows that $(\check{w}, \check{v})$ are jointly white noises.

Finally, it can be shown from (8.28) that

$$\begin{aligned} Q &= E\{[\check{x}(t+1) - A\check{x}(t)][\check{x}(t+1) - A\check{x}(t)]^{\mathrm{T}}\} = I_n - AA^{\mathrm{T}} \\ S &= E\{[\check{x}(t+1) - A\check{x}(t)][y(t) - C\check{x}(t)]^{\mathrm{T}}\} = \bar{C}^{\mathrm{T}} - AC^{\mathrm{T}} \\ R &= E\{[y(t) - C\check{x}(t)][y(t) - C\check{x}(t)]^{\mathrm{T}}\} = \Lambda(0) - CC^{\mathrm{T}} \end{aligned}$$

This completes the proof of (8.29). □

We now show that the orthogonal projection of (8.27a) onto $\mathcal{Y}_{t+1}^-$ yields another Markov model for y. It should be noted that projecting the state vector of (8.27a) onto the past $\mathcal{Y}_{t+1}^-$ is equivalent to constructing the stationary Kalman filter for the system described by (8.27).

Since $\hat{E}\{\mathcal{Y}_t^+ \mid \mathcal{Y}_t^-\} = \mathcal{X}_t = \mathrm{span}\{x(t)\}$, we see from the proof of Lemma 8.4 that

$$\hat{E}\{\mathcal{Y}_t^+ \mid \mathcal{Y}_t^-\} = \hat{E}\{\mathcal{Y}_t^+ \mid \mathcal{X}_t\} = \hat{E}\{\mathcal{Y}_t^+ \mid \mathrm{span}\{x(t)\,\}$$

Thus, noting that $\check{x}(t) \in \mathcal{Y}_t^+$, it follows that

$$\begin{aligned} \hat{E}\{\check{x}(t) \mid \mathcal{Y}_t^-\} &= \hat{E}\{\check{x}(t) \mid x(t)\} \\ &= E\{\check{x}(t)x^{\mathrm{T}}(t)\}(E\{x(t)x^{\mathrm{T}}(t)\})^{-1}x(t) = \Upsilon x(t) \end{aligned} \tag{8.32}$$

where $\Upsilon = E\{\check{x}(t)x^{\mathrm{T}}(t)\} = \mathrm{diag}(\rho_1, \rho_2, \cdots, \rho_n)$.

The next theorem gives the second Markovian model due to Akaike [2].

Theorem 8.2. *In terms of* $z(t) = \Upsilon x(t) \in \mathcal{Y}_t^-$, *a Markov model for* y *is given by*

$$z(t+1) = Az(t) + w(t) \tag{8.33a}$$

$$y(t) = Cz(t) + v(t) \tag{8.33b}$$

where the covariance matrices satisfy $\mathrm{cov}\{z(t)\} = \Pi = \Upsilon^2$ *and*

$$Q = \Upsilon^2 - A\Upsilon^2 A^{\mathrm{T}}, \quad S = \bar{C}^{\mathrm{T}} - A\Upsilon^2 C^{\mathrm{T}}, \quad R = \Lambda(0) - C\Upsilon^2 C^{\mathrm{T}} \tag{8.34}$$

Proof. Similarly to (8.32), we have

$$\hat{E}\{\breve{x}(t+1) \mid \mathcal{Y}_{t+1}^-\} = \Upsilon x(t+1) = z(t+1)$$

By using the orthogonal decomposition $\mathcal{Y}_{t+1}^- = \mathcal{Y}_t^- \oplus \mathrm{span}\{\tilde{y}(t)\}$ defined in (8.26), we project the right-hand side of (8.27a) onto $\mathcal{Y}_{t+1}^-$ to get

$$\begin{aligned} z(t+1) &= \hat{E}\{A\breve{x}(t) + \breve{w}(t) \mid \mathcal{Y}_{t+1}^-\} \\ &= \hat{E}\{A\breve{x}(t) + \breve{w}(t) \mid \mathcal{Y}_t^- \oplus \mathrm{span}\{\tilde{y}(t)\}\} \\ &= \hat{E}\{A\breve{x}(t) + \breve{w}(t) \mid \mathcal{Y}_t^-\} + \hat{E}\{\breve{x}(t+1) \mid \tilde{y}(t)\} \end{aligned}$$

From (8.32) and the fact that $\breve{w}(t) \perp \mathcal{Y}_t^-$, the first term in the right-hand side of the above equation becomes $\hat{E}\{A\breve{x}(t) + \breve{w}(t) \mid \mathcal{Y}_t^-\} = Az(t)$. Defining the second term as $w(t) := \hat{E}\{\breve{x}(t+1) \mid \tilde{y}(t)\}$, we have (8.33a). Since $y(t) \in \mathcal{Y}_{t+1}^-$, it has a unique decomposition

$$\begin{aligned} y(t) &= \hat{E}\{C\breve{x}(t) + \breve{v}(t) \mid \mathcal{Y}_{t+1}^-\} \\ &= \hat{E}\{C\breve{x}(t) + \breve{v}(t) \mid \mathcal{Y}_t^- \oplus \mathrm{span}\{\tilde{y}(t)\}\} \\ &= \hat{E}\{C\breve{x}(t) + \breve{v}(t) \mid \mathcal{Y}_t^-\} + \hat{E}\{y(t) \mid \tilde{y}(t)\} \end{aligned}$$

where we see that $\hat{E}\{\breve{v}(t) \mid \mathcal{Y}_t^-\} = 0$ and $\hat{E}\{y(t) \mid \tilde{y}(t)\} = \tilde{y}(t)$. Hence, defining $\tilde{y}(t) = v(t)$, we have (8.33b). We can prove (8.34) similarly to (8.29). □

Since $w(t) := \hat{E}\{\breve{x}(t+1) \mid \tilde{y}(t)\}$ belongs to the space spanned by $v(t) = \tilde{y}(t)$, we have

$$\hat{E}\{w(t) \mid v(t)\} = E\{w(t)v^{\mathrm{T}}(t)\}(E\{v(t)v^{\mathrm{T}}(t)\})^{-1}v(t) = SR^{-1}v(t)$$

Thus, by putting $K = SR^{-1}$, the Markov model of (8.33) is reduced to

$$z(t+1) = Az(t) + Kv(t) \tag{8.35a}$$

$$y(t) = Cz(t) + v(t) \tag{8.35b}$$

This is the stationary Kalman filter for the system described by (8.27), since

$$z(t) = \hat{E}\{\breve{x}(t) \mid \mathcal{Y}_t^-\} = \Upsilon x(t)$$

is the one-step predicted estimate of the state vector $\check{x}(t)$. Also, the state covariance matrix of (8.35a) is given by $E\{z(t)z^{\mathrm{T}}(t)\} = \Pi = \Upsilon^2$, and the error covariance matrix is given by $P := E\{[\check{x}(t) - z(t)][\check{x}(t) - z(t)]^{\mathrm{T}}\} = I - \Upsilon^2$.

We see that the Markov model of (8.27) is a forward model with the maximum state covariance matrix ($\Pi = I_n$) for given data $(A,\ C,\ \bar{C},\ \Lambda(0))$, while the forward Markov model of (8.33) has the minimum state covariance matrix ($\Pi = \Upsilon^2$).

8.4 Canonical Correlations Between Future and Past

In this section, we consider the canonical correlations between the future and the past of the stationary stochastic process y of (8.16). To this end, we recall two AREs associated with the stationary Kalman filter (5.75) and the stationary backward Kalman filter (5.88), *i.e.*,

$$\Sigma = A\Sigma A^{\mathrm{T}} + (\bar{C}^{\mathrm{T}} - A\Sigma C^{\mathrm{T}})(\Lambda(0) - C\Sigma C^{\mathrm{T}})^{-1}(\bar{C} - C\Sigma A^{\mathrm{T}}) \tag{8.36}$$

and

$$\bar{\Sigma} = A^{\mathrm{T}}\bar{\Sigma}A + (C^{\mathrm{T}} - A^{\mathrm{T}}\bar{\Sigma}\bar{C}^{\mathrm{T}})(\Lambda(0) - \bar{C}\bar{\Sigma}\bar{C}^{\mathrm{T}})^{-1}(C - \bar{C}\bar{\Sigma}A) \tag{8.37}$$

It is easy to see that the stabilizing solution Σ of (8.36) is equal to the minimum solution $\Omega_\infty = \Pi_*$, which is computable by Algorithm 2 of Subsection 7.4.2, and hence we have $\Sigma = \Pi_*$, the minimum solution of (7.37). But, the stabilizing solution $\bar{\Sigma}$ of (8.37) is equal to the minimum solution $\bar{\Omega}_\infty = (\Pi^*)^{-1}$, which is computable by Algorithm 1, so that we have $\bar{\Sigma} = (\Pi^*)^{-1}$. Therefore, in terms of stabilizing solutions Σ and $\bar{\Sigma}$, the inequality of Theorem 7.4 is expressed as

$$\Sigma \le \Pi \le \bar{\Sigma}^{-1}$$

In the following, we show that the square roots of eigenvalues of the product $\Sigma\bar{\Sigma}$ are the canonical correlations of the future and the past of the stationary process y. This is quite analogous to the fact that the Hankel singular values of a deterministic system (A, B, C) are given by the square roots of the eigenvalues of the product of the reachability and observability Gramians (see Section 3.8).

Theorem 8.3. *The canonical correlations of the future and the past of the stationary process y are given by the square roots of eigenvalues of the product $\Sigma\bar{\Sigma}$. If* $\mathrm{rank}(H) = n$, *then the canonical correlations between the future and the past are given by* $(\sigma_1, \cdots, \sigma_n, 0, \cdots, 0)$.

Proof. Define the finite future and past by

$$f_k(t) = \begin{bmatrix} y(t) \\ y(t+1) \\ \vdots \\ y(t+k-1) \end{bmatrix}, \qquad p_k(t) = \begin{bmatrix} y(t-1) \\ y(t-2) \\ \vdots \\ y(t-k) \end{bmatrix}$$

and also define $H_{k,k} := E\{f_k(t)p_k^{\mathrm{T}}(t)\}$, $T_+(k) := E\{f_k(t)f_k^{\mathrm{T}}(t)\}$ and $T_-(k) := E\{p_k(t)p_k^{\mathrm{T}}(t)\}$. Then, we see that

$$\lim_{k\to\infty} H_{k,k} = H, \qquad \lim_{k\to\infty} T_+(k) = T_+, \qquad \lim_{k\to\infty} T_-(k) = T_-$$

From Algorithm 1 of Subsection 7.4.2, it follows that

$$\bar{\Sigma} = (\Pi^*)^{-1} = \bar{\Omega}_\infty = \lim_{k\to\infty} \bar{\Omega}_k = \lim_{k\to\infty} \mathcal{O}_k^{\mathrm{T}} T_+^{-1}(k)\mathcal{O}_k$$

Also, from Algorithm 2,

$$\Sigma = \Pi_* = \Omega_\infty = \lim_{k\to\infty} \Omega_k = \lim_{k\to\infty} \mathcal{C}_k T_-^{-1}(k)\mathcal{C}_k^{\mathrm{T}}$$

Let the Cholesky factorization of block Toeplitz matrices be $T_+(k) = L_k L_k^{\mathrm{T}}$ and $T_-(k) = M_k M_k^{\mathrm{T}}$. Then it follows that

$$\begin{aligned}
\lambda(\Omega_k \bar{\Omega}_k) &= \lambda\big(\mathcal{C}_k T_-^{-1}(k)\mathcal{C}_k^{\mathrm{T}} \mathcal{O}_k^{\mathrm{T}} T_+^{-1}(k)\mathcal{O}_k\big) \\
&= \lambda\big((M_k M_k^{\mathrm{T}})^{-1} H_{k,k}^{\mathrm{T}} (L_k L_k^{\mathrm{T}})^{-1} H_{k,k}\big) \\
&= \lambda\Big((L_k^{-1} H_{k,k} M_k^{-\mathrm{T}})^{\mathrm{T}} (L_k^{-1} H_{k,k} M_k^{-\mathrm{T}})\Big) = \sigma^2(L_k^{-1} H_{k,k} M_k^{-\mathrm{T}})
\end{aligned}$$

where $H_{k,k} = \mathcal{O}_k \mathcal{C}_k$ and $\lambda(AB) = \lambda(BA)$ except for zero eigenvalues are used. It follows from Lemma 8.2 that the singular values of $L_k^{-1} H_{k,k} M_k^{-\mathrm{T}}$ are the canonical correlations between $f_k(t)$ and $p_k(t)$. Thus taking the limit,

$$\lambda(\Sigma\bar{\Sigma}) = \lim_{k\to\infty} \lambda(\Omega_k \bar{\Omega}_k) = \lim_{k\to\infty} \sigma^2(L_k^{-1} H_{k,k} M_k^{-\mathrm{T}}) = \sigma^2(L^{-1} H M^{-\mathrm{T}})$$

where L and M are respectively the Cholesky factors of the matrices T_+ and T_-. Thus we see that the square root of the ith eigenvalue of $\Sigma\bar{\Sigma}$ equals the ith canonical correlation between the future $f(t)$ and past $p(t)$, as was to be proved. □

8.5 Balanced Stochastic Realization

In this section, we consider a balanced stochastic realization based on the CCA. From the previous section, we see that in the balanced stochastic realization, the state covariance matrices of both forward and backward realizations are equal to the diagonal matrix $\Sigma = \mathrm{diag}(\sigma_1, \cdots, \sigma_n)$; see also Definition 3.9.

8.5.1 Forward and Backward State Vectors

We assume that $\mathrm{rank}(H) = n$. Let the Cholesky factorization of block Toeplitz matrices T_+ and T_-, defined by (8.19) and (8.20), be given by $T_+ = LL^{\mathrm{T}}$ and $T_- =$

MM^{T}, respectively[2]. Then, as shown above, the canonical correlations between the future and past are given by the SVD of the normalized H, *i.e.*,

$$L^{-1}HM^{-\mathrm{T}} = U\Sigma V^{\mathrm{T}}$$

so that we have

$$H = LU\Sigma V^{\mathrm{T}} M^{\mathrm{T}} \tag{8.38}$$

From the assumption that $\operatorname{rank}(H) = n$, it follows that $\Sigma = \operatorname{diag}(\sigma_1, \cdots, \sigma_n)$, $1 \geq \sigma_1 \geq \cdots \geq \sigma_n > 0$, and $U^{\mathrm{T}}U = I_n$, $V^{\mathrm{T}}V = I_n$.

According to Lemma 8.2, we define two n-dimensional canonical vectors

$$\alpha(t) := V^{\mathrm{T}} M^{-1} p(t), \qquad \beta(t) := U^{\mathrm{T}} L^{-1} f(t)$$

Then, it can be shown that $E\{\alpha(t)\alpha^{\mathrm{T}}(t)\} = E\{\beta(t)\beta^{\mathrm{T}}(t)\} = I_n$, and

$$E\{\beta(t)\alpha^{\mathrm{T}}(t)\} = \operatorname{diag}(\sigma_1, \cdots, \sigma_n)$$

Thus we see that $(\sigma_1, \cdots, \sigma_n)$ are canonical correlations between $f(t)$ and $p(t)$.

It therefore follows that the orthogonal projection of the future $f(t)$ onto the past $\mathcal{Y}_t^-$ is expressed as

$$\begin{aligned}\hat{E}\{f(t) \mid \mathcal{Y}_t^-\} &= E\{f(t)p^{\mathrm{T}}(t)\}(E\{p(t)p^{\mathrm{T}}(t)\})^{-1}p(t) = HT_-^{-1}p(t) \\ &= LU\Sigma V^{\mathrm{T}} M^{\mathrm{T}}(MM^{\mathrm{T}})^{-1}p(t) = LU\Sigma\alpha(t)\end{aligned} \tag{8.39}$$

Hence, we see that the canonical vector $\alpha(t)$ is the orthonormal basis of the forward predictor space $\mathcal{X}_t = \hat{E}\{\mathcal{Y}_t^+ \mid \mathcal{Y}_t^-\}$. Similarly, the orthogonal projection of the past $p(t)$ onto the future space $\mathcal{Y}_t^+$ is given by

$$\hat{E}\{p(t) \mid \mathcal{Y}_t^+\} = H^{\mathrm{T}} T_+^{-1} f(t) = MV\Sigma\beta(t) \tag{8.40}$$

This implies that the canonical vector $\beta(t)$ is the orthonormal basis of the backward predictor space $\check{\mathcal{X}}_t = \hat{E}\{\mathcal{Y}_t^- \mid \mathcal{Y}_t^+\}$.

Let the extended observability and reachability matrices be defined by

$$\mathcal{O} := LU\Sigma^{1/2}, \qquad \mathcal{C} := \Sigma^{1/2}V^{\mathrm{T}}M^{\mathrm{T}} \tag{8.41}$$

Then, from (8.38), the block Hankel matrix H has a decomposition

$$H = (LU\Sigma^{1/2})(\Sigma^{1/2}V^{\mathrm{T}}M^{\mathrm{T}}) = \mathcal{O}\mathcal{C} \tag{8.42}$$

where $\operatorname{rank}(\mathcal{O}) = \operatorname{rank}(\mathcal{C}) = n$.

Let $x(t)$ and $x_b(t-1)$ respectively be given by

$$x(t) := \Sigma^{1/2}\alpha(t) = \mathcal{C}T_-^{-1}p(t) \tag{8.43}$$

---

[2]Since H, T_+, T_- are infinite dimensional, it should be noted that the manipulation of these matrices are rather formal. An operator theoretic treatment of infinite dimensional matrices is beyond the scope of this book; see Chapter 12 of [183].

and

$$x_b(t-1) := \Sigma^{1/2}\beta(t) = \mathcal{O}^{\mathrm{T}} T_{+}^{-1} f(t) \tag{8.44}$$

The former is called a forward state vector, and the latter a backward state vector. By definition, we see that

$$E\{x(t)x^{\mathrm{T}}(t)\} = \Sigma = E\{x_b(t-1)x_b^{\mathrm{T}}(t-1)\} \tag{8.45}$$

It follows from (8.41) and (8.43) that (8.39) is rewritten as

$$\hat{E}\{f(t) \mid \mathcal{Y}_t^-\} = \mathcal{O}x(t) \tag{8.46}$$

This implies that the past data necessary for predicting the future $f(t)$ is compressed as the forward state vector $x(t)$. Similarly, from (8.44), we see that (8.40) is expressed as

$$\hat{E}\{p(t) \mid \mathcal{Y}_t^+\} = \mathcal{C}^{\mathrm{T}} x_b(t-1) \tag{8.47}$$

so that $x_b(t-1)$ is the backward state vector that is needed to predict the past $p(t)$ by means of the future data.

In the next subsection, we show that a forward (backward) Markov model for the output vector y is derived by using the state vector $x(t)$ ($x_b(t-1)$). From (8.45), it can be shown that the state covariance matrices of both forward and backward Markov models are equal to the canonical correlation matrix, so that these Markov models are called balanced stochastic realizations.

8.5.2 Innovation Representations

We derive innovation representations for a stationary process by means of the vectors $x(t)$ and $x_b(t-1)$ obtained by using the CCA, and show that these representations are balanced.

Theorem 8.4. *In terms of the state vector defined by (8.43), a forward innovation model for y is given by*

$$x(t+1) = Ax(t) + Ke(t) \tag{8.48a}$$

$$y(t) = Cx(t) + e(t) \tag{8.48b}$$

where e is the innovation process defined by

$$e(t) = y(t) - \hat{E}\{y(t) \mid \mathcal{Y}_t^-\} \tag{8.49}$$

which is a white noise with mean zero. Moreover, it can be shown that the matrices A, C, $\bar{C}^{\mathrm{T}}$, K, R are given by

$$
\begin{aligned}
A &= \mathcal{C}^{\leftarrow}\mathcal{C}^{\dagger} = \mathcal{O}^{\dagger}\mathcal{O}^{\uparrow} \\
&= \Sigma^{-1/2}U^{\mathrm{T}}L^{-1}H^{\leftarrow}M^{-\mathrm{T}}V\Sigma^{-1/2} \in \mathbb{R}^{n\times n} \quad (8.50)\\
C &= \mathcal{O}(1:p, 1:n) \in \mathbb{R}^{p\times n} \quad (8.51)\\
\bar{C}^{\mathrm{T}} &= \mathcal{C}(1:n, 1:p) \in \mathbb{R}^{n\times p} \quad (8.52)\\
R &= \Lambda(0) - C\Sigma C^{\mathrm{T}} \in \mathbb{R}^{p\times p} \quad (8.53)\\
K &= (\bar{C}^{\mathrm{T}} - A\Sigma C^{\mathrm{T}})R^{-1} \in \mathbb{R}^{n\times p} \quad (8.54)
\end{aligned}
$$

where $(\cdot)^{\leftarrow}$ *and* $(\cdot)^{\uparrow}$ *denote the operations that remove the first block column and the first block row, respectively. Also,* Σ *is a stabilizing solution of the ARE*

$$
\Sigma = A\Sigma A^{\mathrm{T}} + (\bar{C}^{\mathrm{T}} - A\Sigma C^{\mathrm{T}})(\Lambda(0) - C\Sigma C^{\mathrm{T}})^{-1}(\bar{C}^{\mathrm{T}} - A\Sigma C^{\mathrm{T}})^{\mathrm{T}} \quad (8.55)
$$

and $A_K = A - KC$ *is stable.*

Proof. 1° Define $w(t)$ as

$$
w(t) := x(t+1) - \hat{E}\{x(t+1) \mid x(t)\} \quad (8.56)
$$

By the definition of orthogonal projection,

$$
\hat{E}\{x(t+1) \mid x(t)\} = E\{x(t+1)x^{\mathrm{T}}(t)\}(E\{x(t)x^{\mathrm{T}}(t)\})^{-1}x(t) = Ax(t)
$$

where, from (8.43),

$$
A = \mathcal{C}T_{-}^{-1}E\{p(t+1)p^{\mathrm{T}}(t)\}T_{-}^{-1}\mathcal{C}^{\mathrm{T}}\Sigma^{-1}
$$

Also, by the definition of $p(t)$, it can be shown that

$$
E\{p(t+1)p^{\mathrm{T}}(t)\} = \begin{bmatrix} \Lambda(1) & \Lambda(2) & \Lambda(3) & \cdots \\ \Lambda(0) & \Lambda(1) & \Lambda(2) & \cdots \\ \Lambda^{\mathrm{T}}(1) & \Lambda(0) & \Lambda(1) & \cdots \\ \vdots & \vdots & \vdots & \ddots \end{bmatrix} = T_{-}^{\leftarrow}
$$

From the decomposition $T_{-} = MM^{\mathrm{T}}$, we have $T_{-}^{\leftarrow} = M(M^{\mathrm{T}})^{\leftarrow}$, so that

$$
\begin{aligned}
A &= \mathcal{C}(MM^{\mathrm{T}})^{-1}M(M^{\mathrm{T}})^{\leftarrow}(MM^{\mathrm{T}})^{-1}\mathcal{C}^{\mathrm{T}}\Sigma^{-1} \\
&= \Sigma^{1/2}V^{\mathrm{T}}M^{\mathrm{T}}(MM^{\mathrm{T}})^{-1}M(M^{\mathrm{T}})^{\leftarrow}(MM^{\mathrm{T}})^{-1}MV\Sigma^{1/2}\Sigma^{-1} \\
&= \Sigma^{1/2}V^{\mathrm{T}}(M^{\mathrm{T}})^{\leftarrow}(M^{\mathrm{T}})^{-1}V\Sigma^{-1/2} = \mathcal{C}^{\leftarrow}\mathcal{C}^{\dagger} = \mathcal{O}^{\dagger}\mathcal{O}^{\uparrow} \quad (8.57)
\end{aligned}
$$

The last equality in the above equation is obtained from $\mathcal{O}\mathcal{C}^{\leftarrow} = \mathcal{O}^{\uparrow}\mathcal{C}$ of Theorem 6.1 (iv). Moreover, we see from (8.38) that $\Sigma^{-1/2}U^{\mathrm{T}}L^{-1}H = \Sigma^{1/2}V^{\mathrm{T}}M^{\mathrm{T}}$, so that

$$
\Sigma^{-1/2}U^{\mathrm{T}}L^{-1}H^{\leftarrow} = \Sigma^{1/2}V^{\mathrm{T}}(M^{\mathrm{T}})^{\leftarrow}
$$

Thus, we have (8.50) from (8.57).

2° From (8.42) and (8.43),

$$
\begin{aligned}
\hat{E}\{y(t) \mid \mathcal{Y}_t^-\} &= E\{y(t)p^{\mathrm{T}}(t)\}(E\{p(t)p^{\mathrm{T}}(t)\})^{-1}p(t) \\
&= E\{y(t)[y^{\mathrm{T}}(t-1) \ \ y^{\mathrm{T}}(t-2) \ \cdots]\}(T_-)^{-1}p(t) \\
&= [\Lambda(1) \ \ \Lambda(2) \ \ \cdots](T_-)^{-1}p(t) = H(1:p,:)(T_-)^{-1}p(t) \\
&= \mathcal{O}(1:p, 1:n)\mathcal{C}(1:n,:)(T_-)^{-1}p(t) \\
&= \mathcal{O}(1:p, 1:n)x(t) = Cx(t)
\end{aligned}
$$

Thus, we see that (8.48b) and (8.51) hold. Suppose that $l \geq t$. Since

$$
e(l) = y(l) - Cx(l) \perp \mathcal{Y}_l^-, \qquad x(t) \in \mathcal{Y}_t^- \subset \mathcal{Y}_l^-
$$

we have $e(l) \perp x(t)$, $l = t, t+1, \cdots$, implying that $e(t)$ is a white noise. Also, computing the covariance matrices of both sides of (8.48b) yields (8.53).

3° From (8.43) and (8.56), we have

$$
w(t) = x(t+1) - Ax(t) \in \mathcal{Y}_{t+1}^-, \qquad w(t) \perp \mathcal{Y}_t^-
$$

and $\mathcal{Y}_{t+1}^- = \mathcal{Y}_t^- \oplus \operatorname{span}\{e(t)\}$. Thus, $w(t)$ belongs to $\operatorname{span}\{e(t)\}$. This implies that $w(t)$ can be expressed in terms of the innovation process $e(t)$, so that

$$
w(t) = \hat{E}\{w(t) \mid e(t)\} = E\{w(t)e^{\mathrm{T}}(t)\}R^{-1}e(t) = Ke(t)
$$

However, since $w(t) \perp x(t)$, we get

$$
E\{w(t)e^{\mathrm{T}}(t)\} = E\{w(t)[y(t) - Cx(t)]^{\mathrm{T}}\} = E\{w(t)y^{\mathrm{T}}(t)\}
$$

Hence, from (8.43) and (8.56),

$$
\begin{aligned}
E\{w(t)e^{\mathrm{T}}(t)\} &= E\{[\mathcal{C}T_-^{-1}p(t+1) - Ax(t)]y^{\mathrm{T}}(t)\} \\
&= \mathcal{C}T_-^{-1}E\{p(t+1)y^{\mathrm{T}}(t)\} - AE\{x(t)[Cx(t) + e(t)]^{\mathrm{T}}\} \qquad (8.58)
\end{aligned}
$$

From the definition of T_-, the first term in the right-hand side of the above equation becomes

$$
\begin{aligned}
\mathcal{C}T_-^{-1}E\{p(t+1)y^{\mathrm{T}}(t)\} &= \mathcal{C}T_-^{-1}\begin{bmatrix} \Lambda(0) \\ \Lambda^{\mathrm{T}}(1) \\ \vdots \end{bmatrix} = \mathcal{C}\begin{bmatrix} I_p \\ 0 \\ \vdots \end{bmatrix} \\
&= \mathcal{C}(:, 1:p) =: \bar{C}^{\mathrm{T}}
\end{aligned}
$$

Also, the second term of (8.58) is equal to $A\Sigma C^{\mathrm{T}}$, so that we have (8.54) and (8.52). Thus the state equation (8.48a) is derived. Moreover, computing the covariance matrices of both sides of (8.48a), we get the ARE (8.55). Finally, the stability of $A_K = A - KC$ follows from Lemma 5.14; see also Theorem 5.4. □

We see that the stochastic realization results of Theorem 8.4 provide a forward Markov model based on the canonical vector $\alpha(t)$. We can also derive a backward Markov model for the stationary process y in terms of the canonical vector $\beta(t)$.

Theorem 8.5. *By means of the state vector defined by (8.44), we have the following backward Markov model*

$$x_b(t-1) = A^{\mathrm{T}} x_b(t) + \bar{K}^{\mathrm{T}} e_b(t) \tag{8.59a}$$

$$y(t) = \bar{C} x_b(t) + e_b(t) \tag{8.59b}$$

where the innovation process e_b defined by

$$e_b(t) = y(t) - \hat{E}\{y(t) \mid \mathcal{Y}_{t+1}^{+}\}$$

is a white noise with mean zero and covariance matrix $\bar{R}$, where $\bar{R}$ and $\bar{K}^{\mathrm{T}}$ are respectively given by

$$\bar{R} = \Lambda(0) - \bar{C}\Sigma\bar{C}^{\mathrm{T}} \in \mathbb{R}^{p\times p} \tag{8.60}$$

and

$$\bar{K}^{\mathrm{T}} = (C^{\mathrm{T}} - A^{\mathrm{T}}\Sigma\bar{C}^{\mathrm{T}})\bar{R}^{-1} \in \mathbb{R}^{n\times p} \tag{8.61}$$

Moreover, the covariance matrix Σ for the backward model satisfies the ARE

$$\Sigma = A^{\mathrm{T}}\Sigma A + (C^{\mathrm{T}} - A^{\mathrm{T}}\Sigma\bar{C}^{\mathrm{T}})(\Lambda(0) - \bar{C}\Sigma\bar{C}^{\mathrm{T}})^{-1}(C - \bar{C}\Sigma A) \tag{8.62}$$

and $A^{\mathrm{T}} - \bar{K}^{\mathrm{T}}\bar{C}$ is stable.

Proof. We can prove the theorem by using the same technique used in the proof of Theorem 8.4, but here we derive the result from (8.44) by a direct calculation. From the definition of T_+ and (8.44),

$$\begin{aligned} x_b(t-1) &= \mathcal{O}^{\mathrm{T}} T_+^{-1} f(t) \\ &= [C^{\mathrm{T}} \;\; A^{\mathrm{T}}\mathcal{O}^{\mathrm{T}}] \begin{bmatrix} \Lambda(0) & \bar{C}\mathcal{O}^{\mathrm{T}} \\ \mathcal{O}\bar{C}^{\mathrm{T}} & T_+ \end{bmatrix}^{-1} \begin{bmatrix} y(t) \\ f(t+1) \end{bmatrix} \end{aligned} \tag{8.63}$$

The inverse of the block matrix in (8.63) is given by

$$\begin{bmatrix} \Lambda(0) & \bar{C}\mathcal{O}^{\mathrm{T}} \\ \mathcal{O}\bar{C}^{\mathrm{T}} & T_+ \end{bmatrix}^{-1} = \begin{bmatrix} V & -V\bar{C}\mathcal{O}^{\mathrm{T}}T_+^{-1} \\ -T_+^{-1}\mathcal{O}\bar{C}^{\mathrm{T}}V & T_+^{-1} + T_+^{-1}\mathcal{O}\bar{C}^{\mathrm{T}}V\bar{C}\mathcal{O}^{\mathrm{T}}T_+^{-1} \end{bmatrix}$$

where, from (8.45),

$$V := (\Lambda(0) - \bar{C}\mathcal{O}^{\mathrm{T}}T_+^{-1}\mathcal{O}\bar{C}^{\mathrm{T}})^{-1} = (\Lambda(0) - \bar{C}\Sigma\bar{C}^{\mathrm{T}})^{-1}$$

Thus, computing the right-hand side of (8.63) yields

$$\begin{aligned} x_b(t-1) &= (C^{\mathrm{T}} - A^{\mathrm{T}}\mathcal{O}^{\mathrm{T}}T_+^{-1}\mathcal{O}\bar{C}^{\mathrm{T}})Vy(t) + A^{\mathrm{T}}\mathcal{O}^{\mathrm{T}}T_+^{-1}f(t+1) \\ &\quad + A^{\mathrm{T}}\mathcal{O}^{\mathrm{T}}T_+^{-1}\mathcal{O}\bar{C}^{\mathrm{T}}V\bar{C}\mathcal{O}^{\mathrm{T}}T_+^{-1}f(t+1) - C^{\mathrm{T}}V\bar{C}\mathcal{O}^{\mathrm{T}}T_+^{-1}f(t+1) \end{aligned}$$

By using $\mathcal{O}^{\mathrm{T}} T_+^{-1} \mathcal{O} = \Sigma$ and $\mathcal{O}^{\mathrm{T}} T_+^{-1} f(t+1) = x_b(t)$,

$$x_b(t-1) = A^{\mathrm{T}} x_b(t) + (C^{\mathrm{T}} - A^{\mathrm{T}} \Sigma \bar{C}^{\mathrm{T}}) V[y(t) - \bar{C}^{\mathrm{T}} x_b(t)]$$

Define $\bar{R}$ and $\bar{K}^{\mathrm{T}}$ as in (8.60) and (8.61), respectively. Then, we immediately obtain (8.59a). Also, we see from (8.47) that

$$\hat{E}\{p(t+1) \mid \mathcal{Y}_{t+1}^{+}\} = \mathcal{C}^{\mathrm{T}} x_b(t)$$

From the first p rows of the above expression, we get $\hat{E}\{y(t) \mid \mathcal{Y}_{t+1}^{+}\} = \bar{C} x_b(t)$, so that we have (8.59b). By definition, $e_b(t) \perp \mathcal{Y}_{t+1}^{+}$, and hence $e_b(t) \perp \mathcal{Y}_{t+l}^{+}$ for $l = 1, 2, \cdots$. Also, since $x_b(t+l) \in \mathcal{Y}_{t+l}^{+}$ for $l = 1, 2, \cdots$, it follows that

$$E\{e_b(t) e_b^{\mathrm{T}}(t+l)\} = E\{e_b(t)[y(t+l) - \bar{C} x_b(t+l)]^{\mathrm{T}}\} = 0$$

holds for $l = 1, 2, \cdots$, implying that e_b is a backward white noise. Thus, its covariance matrix is given by (8.60). Finally, computing the covariance matrices of both sides of (8.59a) yields (8.62). □

The state covariance matrices of two stochastic realizations given in Theorems 8.4 and 8.5 are equal and are given by $\mathrm{cov}\{x(t)\} = \Sigma = \mathrm{cov}\{x_b(t-1)\}$. Also, two AREs (8.55) and (8.62) have the common solution Σ, a diagonal matrix with canonical correlations as its diagonals. It follows from Algorithms 1 and 2 shown in Subsection 7.4.2 that Σ is the minimum solution of both AREs. Thus, two systems defined by (8.48) and (8.59) are respectively the forward and backward Markov models for the stationary process y with the same state covariance matrix. In this sense, a pair of realizations (8.48) and (8.59) are called stochastically balanced.

8.6 Reduced Stochastic Realization

In Sections 4.8 and 7.2, we have introduced a backward Markov model as a dual of the forward Markov model.

In this section, we first derive a forward Markov model corresponding to the backward model of (8.59). This gives a forward model for the stationary process y with the maximum state covariance matrix $\Sigma^* = \Sigma^{-1}$.

Lemma 8.5. *Let* $x^*(t) := \Sigma^{-1} x_b(t-1)$. *Let the state space model with* $x^*(t)$ *as the state vector be given by*

$$x^*(t+1) = A x^*(t) + K^* e^*(t) \tag{8.64a}$$

$$y(t) = C x^*(t) + e^*(t) \tag{8.64b}$$

Then, the above realization is a forward Markov model with $\mathrm{cov}\{x^*(t)\} = \Sigma^{-1}$, *which satisfies the ARE*

$$\Sigma^{-1} = A\Sigma^{-1}A^{\mathrm{T}} + (\bar{C}^{\mathrm{T}} - A\Sigma^{-1}C^{\mathrm{T}}) \\ \times (\Lambda(0) - C\Sigma^{-1}C^{\mathrm{T}})^{-1}(\bar{C} - C\Sigma^{-1}A^{\mathrm{T}}) \tag{8.65}$$

Also, the covariance matrix of e^* *and the gain matrix* K^* *are respectively given by*

$$R^* = \Lambda(0) - C\Sigma^{-1}C^{\mathrm{T}} \tag{8.66}$$

and

$$K^* = (\bar{C}^{\mathrm{T}} - A\Sigma^{-1}C^{\mathrm{T}})(R^*)^{-1} \tag{8.67}$$

Proof. A proof is deferred in Appendix of Section 8.11. □

We see that (8.65) is the same as (8.55), and Σ and Σ^{-1} are respectively the minimum and maximum solution of the ARE. Since the elements of the diagonal matrix Σ are the canonical correlations between the future and the past, they lie in the interval $[0, 1]$. By assumption, we have $\sigma_n > 0$, so that if $\sigma_1 < 1$, then $\Sigma^{-1} - \Sigma > 0$ holds. It therefore follows from Theorem 7.5 (ii), (iii) that

$$A_K := A - KC$$

is stable, implying that the inverse system of (8.48) is stable. It is also shown in [69, 107] that, under the assumption of $\Lambda(0) > 0$, the condition that $\sigma_1 < 1$ implies that $\Phi(\omega) > 0, \ -\pi < \omega < \pi$.

We now consider a reduction of a Markov model constructed in Theorem 8.4. We partition the covariance matrix of the state vector as $\Sigma_1 = \mathrm{diag}(\sigma_1, \cdots, \sigma_r)$ and $\Sigma_2 = \mathrm{diag}(\sigma_{r+1}, \cdots, \sigma_n)$. Accordingly, we define

$$A = \begin{bmatrix} A_{11} & A_{12} \\ A_{21} & A_{22} \end{bmatrix}, \qquad C = [C_1 \ \ C_2], \qquad \bar{C} = [\bar{C}_1 \ \ \bar{C}_2] \tag{8.68}$$

Also, we consider the transfer matrix defined by (7.11)

$$Z(z) = C(zI - A)^{-1}\bar{C}^{\mathrm{T}} + \frac{1}{2}\Lambda(0) \tag{8.69}$$

and its reduced model

$$Z_r(z) = C_1(zI_r - A_{11})^{-1}\bar{C}_1^{\mathrm{T}} + \frac{1}{2}\Lambda(0) \tag{8.70}$$

The following lemma gives conditions such that the reduced model $Z_r(z)$ becomes (strictly) positive real.

Lemma 8.6. *Suppose that* $(A, C, \bar{C}^{\mathrm{T}})$ *is minimal and balanced. Then, if* $Z(z)$ *is (strictly) positive real, so is* $Z_r(z)$*. Moreover, if* $\sigma_r > \sigma_{r+1}$*, the reduced model* $Z_r(z)$ *is minimal*[3].

Proof. (i) Suppose first that $Z(z)$ is positive real. From (8.55),

---

[3]In general, this is not a balanced model.

$$M(\Sigma) = \begin{bmatrix} \Sigma - A\Sigma A^{\mathrm{T}} & \bar{C}^{\mathrm{T}} - A\Sigma C^{\mathrm{T}} \\ \bar{C} - C\Sigma A^{\mathrm{T}} & \Lambda(0) - C\Sigma C^{\mathrm{T}} \end{bmatrix} = \begin{bmatrix} K \\ I_p \end{bmatrix} R[K^{\mathrm{T}} \ \ I_p] \geq 0 \qquad (8.71)$$

By using the partitions in (8.68), $M(\Sigma)$ is expressed as

$$\begin{bmatrix} \Sigma_1 - A_{11}\Sigma_1 A_{11}^{\mathrm{T}} - A_{12}\Sigma_2 A_{12}^{\mathrm{T}} & -A_{11}\Sigma_1 A_{21}^{\mathrm{T}} - A_{12}\Sigma_2 A_{22}^{\mathrm{T}} & \bar{C}_1^{\mathrm{T}} - A_{11}\Sigma_1 C_1^{\mathrm{T}} - A_{12}\Sigma_2 C_2^{\mathrm{T}} \\ -A_{21}\Sigma_1 A_{11}^{\mathrm{T}} - A_{22}\Sigma_2 A_{12}^{\mathrm{T}} & \Sigma_2 - A_{21}\Sigma_1 A_{21}^{\mathrm{T}} - A_{22}\Sigma_2 A_{22}^{\mathrm{T}} & \bar{C}_2^{\mathrm{T}} - A_{21}\Sigma_1 C_1^{\mathrm{T}} - A_{22}\Sigma_2 C_2^{\mathrm{T}} \\ \bar{C}_1 - C_1\Sigma_1 A_{11}^{\mathrm{T}} - C_2\Sigma_2 A_{12}^{\mathrm{T}} & \bar{C}_2 - C_1\Sigma_1 A_{21}^{\mathrm{T}} - C_2\Sigma_2 A_{22}^{\mathrm{T}} & \Lambda(0) - C_1\Sigma_1 C_1^{\mathrm{T}} - C_2\Sigma_2 C_2^{\mathrm{T}} \end{bmatrix}$$

Deleting the second block row and column from the above matrix gives

$$\begin{bmatrix} \Sigma_1 - A_{11}\Sigma_1 A_{11}^{\mathrm{T}} - A_{12}\Sigma_2 A_{12}^{\mathrm{T}} & \bar{C}_1^{\mathrm{T}} - A_{11}\Sigma_1 C_1^{\mathrm{T}} - A_{12}\Sigma_2 C_2^{\mathrm{T}} \\ \bar{C}_1 - C_1\Sigma_1 A_{11}^{\mathrm{T}} - C_2\Sigma_2 A_{12}^{\mathrm{T}} & \Lambda(0) - C_1\Sigma_1 C_1^{\mathrm{T}} - C_2\Sigma_2 C_2^{\mathrm{T}} \end{bmatrix} \geq 0 \quad (*)$$

In terms of $(A_{11}, C_1, \bar{C}_1, \Lambda(0))$, we define

$$M_1(\Pi) := \begin{bmatrix} \Pi - A_{11}\Pi A_{11}^{\mathrm{T}} & \bar{C}_1^{\mathrm{T}} - A_{11}\Pi C_1^{\mathrm{T}} \\ \bar{C}_1 - C_1\Pi A_{11}^{\mathrm{T}} & \Lambda(0) - C_1\Pi C_1^{\mathrm{T}} \end{bmatrix}, \qquad \Pi \in \mathbb{R}^{r\times r}$$

Then, it follows from $(*)$ that

$$M_1(\Sigma_1) \geq \begin{bmatrix} A_{12} \\ C_2 \end{bmatrix} \Sigma_2 [A_{12}^{\mathrm{T}} \ \ C_2^{\mathrm{T}}]$$

Since $\Sigma_2 > 0$, we have $M_1(\Sigma_1) \geq 0$ with $\Sigma_1 > 0$.

We show that A_{11} is stable. Since (8.71) gives a full rank decomposition of $M(\Sigma)$, we see from Theorem 7.1 that (A, K) is reachable. By replacing B by $KR^{1/2}$ in the proof of Lemma 3.7 (i), it can be shown that A_{11} is stable. Since, as shown above, $M_1(\Sigma_1) \geq 0$ with $\Sigma_1 > 0$, this implies that $Z_r(z)$ is a positive real matrix; see the comment following (7.26) in Subsection 7.3.1.

(ii) Suppose that $Z(z)$ is strictly positive real. Since $(A, C, \bar{C})$ is minimal, it follows from Theorem 7.5 (ii), (iii) that $\Sigma^{-1} - \Sigma > 0$ and $R = \Lambda(0) - C\Sigma C^{\mathrm{T}} > 0$. Since Σ_1 is a submatrix of Σ, we see that $\Sigma_1^{-1} - \Sigma_1 > 0$ and $\Lambda(0) - C_1\Sigma_1 C_1^{\mathrm{T}} > 0$. As already shown in (i), we have $M_1(\Sigma_1) \geq 0$, and $M_1(\Sigma_1^{-1}) \geq 0$ since both Σ and Σ^{-1} satisfy the same ARE. In other words, there exist two positive definite solutions Σ_1^{-1} and Σ_1 satisfying the LMI and

$$\Sigma_1^{-1} - \Sigma_1 > 0$$

It therefore follows from Lemma 7.8 that $Z_r(z)$ becomes strictly positive real.

(iii) Suppose that $\sigma_r > \sigma_{r+1}$ holds. Then, it follows from Lemma 3.7 (ii) that $Z_r(z)$ is a minimal realization. □

Thus we have shown that the reduced model $Z_r(z)$ is (strictly) positive real and minimal, but not balanced. It should be noted that the minimal solution Π_* of $M_1(\Pi) \geq 0$ satisfies the ARE

$$\begin{aligned} \Pi_* = A_{11}\Pi_* A_{11}^{\mathrm{T}} + (\bar{C}_1^{\mathrm{T}} - A_{11}\Pi_* C_1^{\mathrm{T}}) \\ \times (\Lambda(0) - C_1\Pi_* C_1^{\mathrm{T}})^{-1}(\bar{C}_1^{\mathrm{T}} - A_{11}\Pi_* C_1^{\mathrm{T}})^{\mathrm{T}} \end{aligned} \tag{8.72}$$

so that the gain matrix is expressed as

$$K_{1*} = (\bar{C}_1^{\mathrm{T}} - A_{11}\Pi_* C_1^{\mathrm{T}})(\Lambda(0) - C_1\Pi_* C_1^{\mathrm{T}})^{-1} \tag{8.73}$$

Then the reduced order Markov model of (8.48) is given by

$$x_1(t+1) = A_{11}x_1(t) + K_{1*}\hat{e}(t) \tag{8.74a}$$

$$y(t) = C_1 x_1(t) + \hat{e}(t) \tag{8.74b}$$

where $\mathrm{cov}\{\hat{e}(t)\} = \Lambda(0) - C_1\Pi_* C_1^{\mathrm{T}}$, and $A_{11} - K_{1*}C_1$ is stable.

Corollary 8.1. *The reduced order Markov model of (8.74) is stable, and inversely stable.* □

It should be noted that K_{1*} of (8.73) is different from the gain matrix

$$K_1 = (\bar{C}_1^{\mathrm{T}} - A_{11}\Sigma_1 C_1^{\mathrm{T}} - A_{12}\Sigma_2 C_2^{\mathrm{T}})(\Lambda(0) - C_1\Sigma_1 C_1^{\mathrm{T}} - C_2\Sigma_2 C_2^{\mathrm{T}})^{-1}$$

which is the first block element of K obtained from (8.71). Moreover, the reduced Markov model with $(A_{11}, C_1, \bar{C}_1, K_1)$ is not necessarily of minimal phase, since $A_{11} - K_1 C_1$ may not be stable.

A remaining issue is, therefore, that if there exists a model reduction procedure that keeps positivity and balancedness simultaneously. The answer to this question is affirmative. In fact, according to Lemma 3.8, we can define

$$A_r = A_{11} + A_{12}(\alpha I - A_{22})^{-1}A_{21} \tag{8.75a}$$

$$C_r = C_1 + C_2(\alpha I - A_{22})^{-1}A_{21} \tag{8.75b}$$

$$\bar{C}_r = \bar{C}_1 + \bar{C}_2(\alpha I - A_{22}^{\mathrm{T}})^{-1}A_{12}^{\mathrm{T}} \tag{8.75c}$$

$$\Lambda_r(0) = \Lambda(0) + C_2(\alpha I - A_{22})^{-1}\bar{C}_2^{\mathrm{T}} + \bar{C}_2(\alpha I - A_{22}^{\mathrm{T}})^{-1}C_2^{\mathrm{T}} \tag{8.75d}$$

where $|\alpha| = 1$. Then, we have the following lemma.

Lemma 8.7. *Suppose that $Z(z) = (A, C, \bar{C}, \frac{1}{2}\Lambda(0))$ is strictly positive real, minimal and balanced. If $\sigma_r > \sigma_{r+1}$ holds, then $Z_r(z) = (A_r, C_r, \bar{C}_r, \frac{1}{2}\Lambda_r(0))$ of (8.75) is strictly positive real, minimal and balanced.*

Proof. See [106, 108]. It should be noted that the expression of $\Lambda_r(0)$ is different from that of Lemma 3.8 in order to keep it symmetric. □

So far we have considered the stochastic realization problem under the assumption that an infinite time series data is available. In the next section, we shall derive algorithms of identifying state space models based on given finite observed data.

8.7 Stochastic Realization Algorithms

Let a finite collection of data be given by $\{y(t),\, t = 0,\, 1,\, \cdots, N + 2k - 2\}$, where $k > 0$ and N is sufficiently large. We assume that the given data is a finite sample from a stationary process. We define the block Toeplitz matrix[4]

$$\check{Y}_{0|k-1} := \begin{bmatrix} y(k-1) & y(k) & \cdots & y(N+k-2) \\ y(k-2) & y(k-1) & \cdots & y(N+k-3) \\ \vdots & \vdots & & \vdots \\ y(0) & y(1) & \cdots & y(N-1) \end{bmatrix} \in \mathbb{R}^{kp\times N}$$

and the block Hankel matrix

$$Y_{k|2k-1} := \begin{bmatrix} y(k) & y(k+1) & \cdots & y(k+N-1) \\ y(k+1) & y(k+2) & \cdots & y(k+N) \\ \vdots & \vdots & & \vdots \\ y(2k-1) & y(2k) & \cdots & y(N+2k-2) \end{bmatrix} \in \mathbb{R}^{kp\times N}$$

where $k > n$, and the number of columns of block matrices is N.

Let k be the present time. As before, we write $Y_p = \check{Y}_{0|k-1}$ and $Y_f = Y_{k|2k-1}$, respectively. The sample covariance matrices of the given data are defined by

$$\frac{1}{N}\begin{bmatrix} Y_p \\ Y_f \end{bmatrix}\begin{bmatrix} Y_p^{\mathrm{T}} & Y_f^{\mathrm{T}} \end{bmatrix} = \begin{bmatrix} \Sigma_{pp} & \Sigma_{pf} \\ \Sigma_{fp} & \Sigma_{ff} \end{bmatrix}$$

Also, consider the LQ decomposition of the form

$$\frac{1}{\sqrt{N}}\begin{bmatrix} Y_p \\ Y_f \end{bmatrix} = \begin{bmatrix} L_{11} & 0 \\ L_{21} & L_{22} \end{bmatrix}\begin{bmatrix} Q_1^{\mathrm{T}} \\ Q_2^{\mathrm{T}} \end{bmatrix} \tag{8.76}$$

Then, it follows that

$$\Sigma_{fp} = L_{21}L_{11}^{\mathrm{T}}, \qquad \Sigma_{ff} = L_{21}L_{21}^{\mathrm{T}} + L_{22}L_{22}^{\mathrm{T}}, \qquad \Sigma_{pp} = L_{11}L_{11}^{\mathrm{T}}$$

We see that the above sample covariance matrices Σ_{fp}, Σ_{ff}, Σ_{pp} are finite dimensional approximations to the infinite matrices H, T_+ and T_- of (8.18), (8.19) and (8.20), respectively.

The following numerical algorithm is based on the theory of balanced stochastic realization of Theorem 8.4.

Stochastic Balanced Realization – Algorithm A

Step 1: Compute square root matrices L and M of the covariance matrices Σ_{ff} and Σ_{pp} such that

$$\Sigma_{ff} = LL^{\mathrm{T}}, \qquad \Sigma_{pp} = MM^{\mathrm{T}} \tag{8.77}$$

---

[4]It should be noted that although $Y_{0|k-1}$ is a Hankel matrix, $\check{Y}_{0|k-1}$ is a Toeplitz matrix.

Step 2: Compute the SVD of the normalized covariance matrix Σ_{fp} such that

$$L^{-1}\Sigma_{fp}M^{-\mathrm{T}} = U\Sigma V^{\mathrm{T}} \simeq \hat{U}\hat{\Sigma}\hat{V}^{\mathrm{T}} \tag{8.78}$$

where $\hat{\Sigma}$ is given by neglecting sufficiently small singular values of Σ, and hence the dimension of the state vector is given by $n = \dim \hat{\Sigma}$.

Step 3: Define the extended observability and reachability matrices as

$$\mathcal{O}_k = L\hat{U}\hat{\Sigma}^{1/2}, \qquad \mathcal{C}_k = \hat{\Sigma}^{1/2}\hat{V}^{\mathrm{T}}M^{\mathrm{T}} \tag{8.79}$$

Step A4: Compute the estimates of A, C, $\bar{C}^{\mathrm{T}}$ as

$$A = \underline{\mathcal{O}}_k^{\dagger}\overline{\mathcal{O}}_k, \qquad C = \mathcal{O}_k(1:p,:), \qquad \bar{C}^{\mathrm{T}} = \mathcal{C}_k(:,1:p) \tag{8.80}$$

where $\underline{\mathcal{O}}_k = \mathcal{O}_k(1:(k-1)p,:)$ and $\overline{\mathcal{O}}_k = \mathcal{O}_k(p+1:kp,:)$.

Step A5: Let $\Lambda(0) = \Sigma_{ff}(1:p,1:p)$. Then, the Kalman gain is given by

$$K = (\bar{C}^{\mathrm{T}} - A\hat{\Sigma}C^{\mathrm{T}})(\Lambda(0) - C\hat{\Sigma}C^{\mathrm{T}})^{-1} \tag{8.81}$$

Step A6: By the formula in Theorem 8.4, we have an innovation representation of the form

$$x(t+1) = Ax(t) + Ke(t)$$
$$y(t) = Cx(t) + e(t)$$

where, from (8.53), the covariance matrix of the innovation process is given by $R = \Lambda(0) - C\hat{\Sigma}C^{\mathrm{T}}$.

Remark 8.1. Since Algorithm A is based on the stochastic balanced realization of Theorem 8.4, we observe that this algorithm is quite different from Algorithm 2 of Van Overschee and De Moor ([165], p. 87). In fact, in the latter algorithm, based on the obtained $(A, C, \bar{C}, \Lambda(0))$, it is necessary to solve the ARE of (7.84) to derive the Kalman gain K from (7.85). However, as stated in Chapter 7, there may be a possibility that the estimate $(A, C, \bar{C}, \Lambda(0))$ obtained above is not positive real, and hence the ARE of (7.84) does not have a stabilizing solution. In Algorithm A, however, we exploit the fact that $\hat{\Sigma}$ obtained by the CCA is an approximate solution of the ARE of (8.55), so that we always get an innovation model. □

We present an alternative algorithm that utilizes the estimate of state vector. The algorithm is the same as Algorithm A until *Step* 3.

Stochastic Balanced Realization – Algorithm B

Step B4: Compute the estimate of the state vector

$$\bar{X}_k = \hat{\Sigma}^{1/2}\hat{V}^{\mathrm{T}}M^{-1}\check{Y}_{0|k-1} \in \mathbb{R}^{n\times N}$$

and define the matrices with $N-1$ columns as

$$\hat{X}_{k+1} = \bar{X}_k(:, 2 : N), \quad \hat{X}_k = \bar{X}_k(:, 1 : N-1), \quad \hat{Y}_{k|k} = Y_{k|k}(:, 1 : N-1)$$

Step B5: Compute the estimate of (A, C) by applying the least-squares method to

$$\begin{bmatrix} \hat{X}_{k+1} \\ \hat{Y}_{k|k} \end{bmatrix} = \begin{bmatrix} A \\ C \end{bmatrix} \hat{X}_k + \begin{bmatrix} \rho_w \\ \rho_v \end{bmatrix}$$

where $\rho_w \in \mathbb{R}^{n\times(N-1)}$ and $\rho_v \in \mathbb{R}^{p\times(N-1)}$ are residuals.

Step B6: Compute the sample covariance matrices of residuals

$$\begin{bmatrix} \hat{Q} & \hat{S} \\ \hat{S}^{\mathrm{T}} & \hat{R} \end{bmatrix} = \frac{1}{N-1} \begin{bmatrix} \rho_w\rho_w^{\mathrm{T}} & \rho_w\rho_v^{\mathrm{T}} \\ \rho_v\rho_w^{\mathrm{T}} & \rho_v\rho_v^{\mathrm{T}} \end{bmatrix} \tag{8.82}$$

Then, we solve the ARE associated with the Kalman filter [see (5.67)]

$$P = APA^{\mathrm{T}} - (APC^{\mathrm{T}} + \hat{S})(CPC^{\mathrm{T}} + \hat{R})^{-1}(APC^{\mathrm{T}} + \hat{S})^{\mathrm{T}} + \hat{Q} \tag{8.83}$$

to get a stabilizing solution $P \geq 0$. Thus the Kalman gain is given by

$$K = (APC^{\mathrm{T}} + \hat{S})(CPC^{\mathrm{T}} + \hat{R})^{-1}$$

Step B7: The identified innovation model is given by

$$\begin{aligned} \hat{x}(t+1) &= A\hat{x}(t) + K\hat{e}(t) \\ y(t) &= C\hat{x}(t) + \hat{e}(t) \end{aligned}$$

where $\mathrm{var}\{\hat{e}(t)\} = CPC^{\mathrm{T}} + \hat{R}$.

Remark 8.2. In Algorithm B, the covariance matrix obtained by (8.82) is always nonnegative definite, so that the stabilizing solution $P \geq 0$ of the ARE of (8.83) exists. Thus we can always derive an approximate balanced innovation model because $\mathrm{cov}\{\bar{X}_k\} = \hat{\Sigma}$.

In Algorithm 3 of Van Overschee and De Moor ([165], p. 90), however, one must solve the Lyapunov equation

$$\Sigma^s = A\Sigma^s A^{\mathrm{T}} + \hat{Q}$$

under the assumption that A obtained in *Step* B5 is stable. By using the solution $\Sigma^s > 0$, the matrices $\bar{C}$ and $\Lambda(0)$ are then computed by

$$\bar{C} = C\Sigma^s A^{\mathrm{T}} + \hat{S}^{\mathrm{T}}, \qquad \Lambda(0) = \hat{R} + C\Sigma^s C^{\mathrm{T}}$$

to get the data $(A,\ C,\ \bar{C},\ \Lambda(0))$. The rest of Algorithm 3 is to solve the ARE of (7.84) to obtain the Kalman gain as stated in Remark 8.1. It should be noted that Algorithm 3 works under the assumption that the estimated A is stable; otherwise it does not provide any estimates. Hence, Algorithm B derived here is somewhat different from Algorithm 3 [165]. □

8.8 Numerical Results

We show some simulation results using simple models; the first one is a 2nd-order ARMA model, and the second one a 3rd-order state space model.

Example 8.1. Consider the ARMA model described by

$$y(t) - 1.5y(t-1) + 0.7y(t-2) = e(t) - 0.5e(t-1) + 0.3e(t-2)$$

where e is a zero mean white Gaussian noise with unit variance. We have generated time series data under zero initial conditions. By using Algorithm A with $k = 10$, we have identified innovation models, from which transfer functions are computed. Table 8.2 shows canonical correlations $\sigma_i,\ i = 1, \cdots, 6$ between the future and past *vs*. the number of data N, where $N = \infty$ means that the exact canonical correlations are computed by using the relation $\sigma_i = \sqrt{\lambda_i(\Sigma\bar{\Sigma})}$ derived in Theorem 8.3. We observe from Table 8.2 that, though the values of the first two canonical correlations σ_1 and σ_2 do not change very much, other canonical correlations $\sigma_3, \sigma_4, \cdots$ get smaller as the number of data N increases. For smaller $N \leq 1000$, we find that σ_3 and σ_4 are rather large, so that it is not easy to estimate the order of the ARMA model. However, as the number of data increases, the difference between σ_2 and σ_3 becomes larger, so that we can correctly estimate the order $n = 2$.

Table 8.2. Canonical correlations between the future and past

N	σ_1	σ_2	σ_3	σ_4	σ_5	σ_6
500	0.9047	0.4916	0.2328	0.2302	0.1687	0.0865
1000	0.9095	0.5028	0.1685	0.1638	0.1322	0.0776
2000	0.9189	0.5121	0.0997	0.0781	0.0488	0.0415
5000	0.9170	0.5108	0.0760	0.0404	0.0392	0.0311
10000	0.9165	0.5087	0.0468	0.0297	0.0253	0.0224
20000	0.9137	0.5137	0.0342	0.0288	0.0217	0.0129
50000	0.9130	0.5070	0.0142	0.0122	0.0116	0.0064
∞	0.9133	0.5036	0.0000	0.0000	0.0000	0.0000

Now we consider the case where the number of data is fixed as $N = 10000$. If we take $k = 80$, then the first six canonical correlations are given by

$$\Sigma = \text{diag}(0.9171,\ 0.5144,\ 0.1403,\ 0.1380,\ 0.1313,\ 0.1304)$$

Compared with $\Sigma = \text{diag}(0.9165,\ 0.5087,\ 0.0468,\ 0.0297,\ 0.0253,\ 0.0224)$ in Table 8.2 ($N = 10000$), we see that though the first two canonical correlations σ_1 and σ_2 are not significantly affected, the values of $\sigma_3, \sigma_4, \cdots$ are quite changed by taking a large value of k. This may be caused by the following reason; for a fixed N, the sample cross-covariance matrix Σ_{fp} (or covariance matrices Σ_{ff} and

Table 8.3. Parameter estimation by Algorithm A

N	a_1 (-1.5)	a_2 (0.7)	c_1 (-0.5)	c_2 (0.3)
500	-1.3621	0.6147	-0.3842	0.4376
1000	-1.4146	0.6487	-0.4301	0.3795
2000	-1.4723	0.6742	-0.4418	0.3182
5000	-1.5109	0.7077	-0.4791	0.2865
10000	-1.5107	0.7035	-0.4900	0.2899
20000	-1.5152	0.7096	-0.5074	0.2859
50000	-1.4991	0.6989	-0.4940	0.3026

Σ_{pp}) computed from (8.76) will loose the block Hankel (Toeplitz) property of true covariance matrices as the number of block rows k increases.

Table 8.3 displays the estimated parameters of the 2nd-order ARMA model. In the identification problem of Example 6.7 where both the input and output data are available, we have obtained very good estimation results based on small number of data, say, $N = 100$. However, as we can see from Table 8.3, we need a large number of data for the identification of time series model where only the output data is available. This is also true when we use the stochastic realization algorithm given in Lemma 7.9, because we need accurate covariance data to get good estimates for Markov models. Although not included here, quite similar results are obtained by using Algorithm B, which is based on the estimate of state vectors. □

Example 8.2. We show some simulation results for the 3rd-order state space model used in [165], which is given by

$$x(t+1) = \begin{bmatrix} 0.6 & 0.6 & 0 \\ -0.6 & 0.6 & 0 \\ 0 & 0 & 0.4 \end{bmatrix} x(t) + \begin{bmatrix} 0.17 \\ -0.15 \\ 0.28 \end{bmatrix} e(t)$$

$$y(t) = [0.78 \quad 0.53 \quad 1.0] x(t) + e(t)$$

where e is a Gaussian white noise with mean zero and unit variance. As in Example 8.1, we have used Algorithm A to compute canonical correlations and estimates of transfer functions, where $k = 10$. The simulation results are shown in Tables 8.4 and 8.5. We observe that, except that the estimate of the parameter c_3 is rather poor, the simulation results for the 3rd-order system are similar to those of the 2nd-order system treated in Example 8.1. Also, we see that as the increase of number of data N, the canonical correlations are getting closer to the true values. □

It should be noted that the above results depend heavily on the simulation conditions, so that they are to be understood as "examples." Also, there are possibilities that the stochastic subspace methods developed in the literature may fail; detailed analyses of stochastic subspace methods are found in [38, 58, 154].

Table 8.4. Canonical correlations between the future and past

N	σ_1	σ_2	σ_3	σ_4	σ_5	σ_6
2000	0.4042	0.2063	0.1299	0.0873	0.0805	0.0542
5000	0.4038	0.2010	0.0970	0.0739	0.0432	0.0339
10000	0.3899	0.2063	0.1095	0.0483	0.0299	0.0293
20000	0.3801	0.2181	0.1043	0.0340	0.0267	0.0214
50000	0.3840	0.2216	0.1060	0.0139	0.0114	0.0112
∞	0.3820	0.2244	0.1030	0.0000	0.0000	0.0000

Table 8.5. Parameter estimation by Algorithm A

	a_1	a_2	a_3	c_1	c_2	c_3
N	(-1.6)	(1.2)	(-0.288)	(-1.2669)	(0.6866)	(-0.024)
2000	-1.5809	1.1360	-0.2214	-1.2068	0.5735	0.0748
5000	-1.4937	1.0825	-0.2228	-1.1214	0.5476	0.0461
10000	-1.5641	1.1693	-0.2611	-1.2081	0.6414	0.0190
20000	-1.6151	1.2161	-0.2901	-1.2812	0.7050	-0.0269
50000	-1.6046	1.2082	-0.2825	-1.2655	0.6890	-0.0120

8.9 Notes and References

- This chapter has re-considered the stochastic realization problem based on the canonical correlation analysis (CCA) due to Akaike [2,3]. Also, we have derived forward and backward innovation representations of a stationary process, and discussed a stochastic balanced realization problem, including a model reduction of stochastic systems.
- In Section 8.1 we have reviewed the basic idea of the CCA based on [14, 136]. The stochastic realization problem is restated in Section 8.2. The development of Section 8.3 is based on the pioneering works of Akaike [2–4]. In Section 8.4, we have discussed canonical correlations between the future and the past of a stationary process. We have shown that they are determined by the square roots of eigenvalues of the product of two state covariance matrices of the forward and the backward innovation models; see Table 4.1 and [39].
- Section 8.5 is devoted to balanced stochastic realizations based on Desai *et al.* [42,43], Aoki [15], and Lindquist and Picci [106,107]. By extending the results of [106, 107], we have also developed stochastic subspace identification algorithms based on the LQ decomposition in Hilbert space of time series [151, 152], and stochastic balanced realizations on a finite interval [153, 154].
- Section 8.6 has considered reduced stochastic realizations based on [43, 106]. An earlier result on model reduction is due to [50], and a survey on model reduction is given in [18]. The relation between the CCA and phase matching problems has

been discueed in [64,77]. Some applications to economic time series analysis are found in [16].

- In Section 8.7, the stochastic subspace identification algorithms are derived based on the balanced realization theorem (Theorem 8.4); see also the algorithm in [112]. Section 8.8 includes some numerical results, showing that a fairly large number of data is needed to obtain good estimates of time series models. Moreover, Appendix includes a proof of Lemma 8.5.

8.10 Problems

8.1 Show that the result of Lemma 8.1 is compactly written as

$$\begin{bmatrix} L^{\mathrm{T}} & 0 \\ 0 & M^{\mathrm{T}} \end{bmatrix} \begin{bmatrix} \Sigma_{xx} & \Sigma_{xy} \\ \Sigma_{yx} & \Sigma_{yy} \end{bmatrix} \begin{bmatrix} L & 0 \\ 0 & M \end{bmatrix} = \begin{bmatrix} I & D \\ D^{\mathrm{T}} & I \end{bmatrix}$$

and we have

$$\det \begin{bmatrix} I & D \\ D^{\mathrm{T}} & I \end{bmatrix} = (1-\rho_1^2)\cdots(1-\rho_k^2)$$

8.2 Let $\mathcal{Y}$ be a Hilbert space. Let $\mathcal{B} = \mathrm{span}\{b\}$ be a subspace of $\mathcal{Y}$. Show that the orthogonal projection of $a \in \mathcal{Y}$ onto $\mathcal{B}$ is equivalent to finding K such that $\|a - Kb\|_2$ is minimized with respect to K, and that the optimal K is given by

$$K = E\{ab^{\mathrm{T}}\}(E\{bb^{\mathrm{T}}\})^{-1}$$

8.3 Compute the covariance matrices of the output process y for three realizations given in Theorems 8.4, 8.5 and Lemma 8.5, and show that these are all given by (7.7).

8.4 [43] Suppose that y is scalar in Theorem 8.4, and that canonical correlations $\{\sigma_i,\ i = 1, \cdots, n\}$ are different. Prove that there exists a matrix $S = \mathrm{diag}(\pm 1, \cdots, \pm 1)$ such that

$$A = SA^{\mathrm{T}}S, \qquad \bar{C} = CS$$

8.5 In Subsection 8.5.1, consider the following two factorizations $T_+ = L_1 L_1^{\mathrm{T}} = L_2 L_2^{\mathrm{T}}$ and $T_- = M_1 M_1^{\mathrm{T}} = M_2 M_2^{\mathrm{T}}$. Then, the SVD of the normalized block Hankel matrix gives

$$H = L_1 U_1 \Sigma V_1^{\mathrm{T}} M_1^{\mathrm{T}} = L_2 U_2 \Sigma V_2^{\mathrm{T}} M_2^{\mathrm{T}}$$

Suppose that the canonical correlations $\{\sigma_i,\ i = 1, \cdots, n\}$ are different. Let the two realizations of y be given by $(A_j, C_j, \bar{C}_j, K_j, R_j)$, $j = 1, 2$. Show that there exists a matrix $S = \mathrm{diag}(\pm 1, \cdots, \pm 1)$ such that

$$A_2 = SA_1S, \quad C_2 = C_1S, \quad \bar{C}_2 = \bar{C}_1S, \quad K_2 = SK, \quad R_2 = R_1$$

8.11 Appendix: Proof of Lemma 8.5

1° From (8.59a), the covariance matrix of x_b is given by

$$E\{x_b(t+l)x_b^{\mathrm{T}}(t)\} = \begin{cases} \Sigma A^l, & l = 0, 1, \cdots \\ (A^{\mathrm{T}})^{-l}\Sigma, & l = -1, -2, \cdots \end{cases} \tag{8.84}$$

Define $w^*(t) := x^*(t+1) - Ax^*(t)$. Note that $x_b(t-1) \in \hat{E}\{\mathcal{Y}_t^- \mid \mathcal{Y}_t^+\} \subset \mathcal{Y}_t^+$ and $x_b(t) \in \hat{E}\{\mathcal{Y}_{t+1}^- \mid \mathcal{Y}_{t+1}^+\} \subset \mathcal{Y}_{t+1}^+ \subset \mathcal{Y}_t^+$. Then, since $x^*(t) = \Sigma^{-1}x_b(t-1)$, we get

$$w^*(t) = \Sigma^{-1}x_b(t) - A\Sigma^{-1}x_b(t-1) \in \mathcal{Y}_t^+$$

and hence for $l = 0, 1, \cdots$,

$$w^*(t+l) = \Sigma^{-1}x_b(t+l) - A\Sigma^{-1}x_b(t+l-1) \in \mathcal{Y}_{t+l}^+ \subset \mathcal{Y}_t^+ \tag{8.85}$$

Also, from (8.84),

$$\begin{aligned} &E\{w^*(t+l)(x^*(t))^{\mathrm{T}}\} \\ &\quad = E\{x^*(t+l+1)(x^*(t))^{\mathrm{T}}\} - AE\{x^*(t+l)(x^*(t))^{\mathrm{T}}\} \\ &\quad = \Sigma^{-1}E\{x_b(t+l)x_b^{\mathrm{T}}(t-1)\}\Sigma^{-1} \\ &\qquad - A\Sigma^{-1}E\{x_b(t+l-1)x_b^{\mathrm{T}}(t-1)\}\Sigma^{-1} \\ &\quad = \Sigma^{-1}\Sigma A^{l+1}\Sigma^{-1} - A\Sigma^{-1}\Sigma A^l \Sigma^{-1} = 0, \qquad l = 0, 1, \cdots \end{aligned}$$

Since $x^*(t) = \Sigma^{-1}x_b(t-1)$, it follows that

$$E\{w^*(t+l)x_b^{\mathrm{T}}(t-1)\} = 0, \qquad l = 0, 1, \cdots$$

implying that $w^*(t+l) \perp \mathrm{span}\{x_b(t-1)\} = \hat{E}\{\mathcal{Y}_t^- \mid \mathcal{Y}_t^+\}$. This together with (8.85) show that $w^*(t+l) \perp \mathcal{Y}_t^-$. Hence, the following relation holds.

$$w^*(t+l) \perp x^*(t),\, y(t-1), \qquad l = 0, 1, \cdots \tag{8.86}$$

2° Define $e^*(t) := y(t) - Cx^*(t)$. Since $x^*(t) = \Sigma^{-1}x_b(t-1) \in \mathcal{Y}_t^+$, we have $e^*(t) \in \mathcal{Y}_t^+$. Also, from (8.59),

$$\begin{aligned} E\{y(t)(x^*(t))^{\mathrm{T}}\} &= E\{[\bar{C}x_b(t) + e_b(t)]x_b^{\mathrm{T}}(t-1)\}\Sigma^{-1} \\ &= \bar{C}E\{x_b(t)x_b^{\mathrm{T}}(t-1)\}\Sigma^{-1} + E\{e_b(t)x_b^{\mathrm{T}}(t-1)\}\Sigma^{-1} \\ &= \bar{C}\Sigma A\Sigma^{-1} + E\{e_b(t)[x_b^{\mathrm{T}}(t)A + e_b^{\mathrm{T}}(t)\bar{K}]\}\Sigma^{-1} \\ &= \bar{C}\Sigma A\Sigma^{-1} + E\{e_b(t)e_b^{\mathrm{T}}(t)\}\bar{K}\Sigma^{-1} \\ &= \bar{C}\Sigma A\Sigma^{-1} + (C - \bar{C}\Sigma A)\Sigma^{-1} = C\Sigma^{-1} \end{aligned}$$

Hence, we have

$$E\{e^*(t)(x^*(t))^{\mathrm{T}}\} = E\{y(t)(x^*(t))^{\mathrm{T}}\} - CE\{x^*(t)(x^*(t))^{\mathrm{T}}\}$$
$$= C\Sigma^{-1} - C\Sigma^{-1} = 0$$

This implies that $e^*(t) \perp x^*(t)$. Thus, for any t, we have $e^*(t) \perp \hat{E}\{\mathcal{Y}_t^- \mid \mathcal{Y}_t^+\}$ and $e^*(t) \in \mathcal{Y}_t^+$. Hence, similarly to the proof in 1°, we get

$$e^*(t+l) \perp \mathcal{Y}_{t+l}^- \quad \Rightarrow \quad e^*(t+l) \perp \mathcal{Y}_t^-, \qquad l = 0, 1, \cdots$$

By definition, $e^*(t+l) \in \mathcal{Y}_{t+l}^+ \subset \mathcal{Y}_t^+$, it follows that $e^*(t+l) \perp x^*(t)$, $l = 0, 1, \cdots$. Thus, summarizing above, the following relation holds.

$$e^*(t+l) \perp x^*(t),\ y(t-1), \qquad l = 0, 1, \cdots \tag{8.87}$$

3° It follows from (8.86) and (8.87) that for $h = 1, 2, \cdots$,

$$E\{w^*(t+h)(w^*(t))^{\mathrm{T}}\} = E\{w^*(t+h)[x^*(t+1) - Ax^*(t)]^{\mathrm{T}}\} = 0$$
$$E\{w^*(t+h)(e^*(t))^{\mathrm{T}}\} = E\{w^*(t+h)[y(t) - Cx^*(t)]^{\mathrm{T}}\} = 0$$
$$E\{e^*(t+h)(e^*(t))^{\mathrm{T}}\} = E\{e^*(t+h)[y(t) - Cx^*(t)]^{\mathrm{T}}\} = 0$$
$$E\{e^*(t+h)(w^*(t))^{\mathrm{T}}\} = E\{e^*(t+h)[x^*(t+1) - Ax^*(t)]^{\mathrm{T}}\} = 0$$

This implies that the joint process $(w^*(t),\ e^*(t))$ is white noise.

Now we compute the covariance matrices of $w^*(t)$ and $e^*(t)$. We see that the covariance matrix of $w^*(t)$ is given by

$$Q^* = E\{w^*(t)(w^*(t))^{\mathrm{T}}\}$$
$$= E\{[x^*(t+1) - Ax^*(t)][x^*(t+1) - Ax^*(t)]^{\mathrm{T}}\}$$
$$= \Sigma^{-1} - A\Sigma^{-1}A^{\mathrm{T}}$$

Noting that $x^*(t) \perp e^*(t)$, we have

$$S^* = E\{w^*(t)(e^*(t))^{\mathrm{T}}\} = E\{[x^*(t+1) - Ax^*(t)](e^*(t))^{\mathrm{T}}\}$$
$$= E\{x^*(t+1)[y(t) - Cx^*(t)]^{\mathrm{T}}\}$$
$$= \Sigma^{-1}E\{x_b(t)[\bar{C}x_b(t) + e_b(t) - C\Sigma^{-1}x_b(t-1)]^{\mathrm{T}}\}$$
$$= \Sigma^{-1}E\{x_b(t)x_b^{\mathrm{T}}(t)\}\bar{C}^{\mathrm{T}} - \Sigma^{-1}E\{x_b(t)x_b^{\mathrm{T}}(t-1)\}\Sigma^{-1}C^{\mathrm{T}}$$
$$= \Sigma^{-1}\Sigma\bar{C}^{\mathrm{T}} - \Sigma^{-1}\Sigma A\Sigma^{-1}C^{\mathrm{T}} = \bar{C}^{\mathrm{T}} - A\Sigma^{-1}C^{\mathrm{T}}$$

Also, the covariance matrix of $e^*(t)$ is given by

$$R^* = E\{e^*(t)(e^*(t))^{\mathrm{T}}\} = E\{[y(t) - Cx^*(t)][y(t) - Cx^*(t)]^{\mathrm{T}}\}$$
$$= \Lambda(0) - C\Sigma^{-1}C^{\mathrm{T}}$$

It therefore follows that

$$\begin{bmatrix} Q^* & S^* \\ (S^*)^{\mathrm{T}} & R^* \end{bmatrix} = \begin{bmatrix} \Sigma^{-1} - A\Sigma^{-1}A^{\mathrm{T}} & \bar{C}^{\mathrm{T}} - A\Sigma^{-1}C^{\mathrm{T}} \\ (\bar{C}^{\mathrm{T}} - A\Sigma^{-1}C^{\mathrm{T}})^{\mathrm{T}} & \Lambda(0) - C\Sigma^{-1}C^{\mathrm{T}} \end{bmatrix} \tag{8.88}$$

4° Finally, by using the ARE of (8.62), *i.e.*,

$$\Sigma = A^{\mathrm{T}}\Sigma A + (C^{\mathrm{T}} - A^{\mathrm{T}}\Sigma\bar{C}^{\mathrm{T}})(\Lambda(0) - \bar{C}\Sigma\bar{C}^{\mathrm{T}})^{-1}(C^{\mathrm{T}} - A^{\mathrm{T}}\Sigma\bar{C}^{\mathrm{T}})^{\mathrm{T}} \tag{8.89}$$

we can derive, as shown below, the ARE of (8.65):

$$\begin{aligned} \Sigma^{-1} = A\Sigma^{-1}A^{\mathrm{T}} + (\bar{C}^{\mathrm{T}} - A\Sigma^{-1}C^{\mathrm{T}})(\Lambda(0) - C\Sigma^{-1}C^{\mathrm{T}})^{-1} \\ \times (\bar{C} - C\Sigma^{-1}A^{\mathrm{T}}) \end{aligned} \tag{8.90}$$

This equation implies that the block matrix of (8.88) is degenerate, so that there exists a linear relation between $w^*(t)$ and $e^*(t)$. Hence, we have

$$\begin{aligned} w^*(t) &= \hat{E}\{w^*(t) \mid e^*(t)\} \\ &= E\{w^*(t)(e^*(t))^{\mathrm{T}}\}(E\{e^*(t)(e^*(t))^{\mathrm{T}}\})^{-1}e^*(t) \\ &= S^*(R^*)^{-1}e^*(t) = K^*e^*(t) \end{aligned}$$

This completes a proof of Lemma 7.5.

Derivation of Equation (8.90) As shown in Section 5.8 (see Problem 5.7), the ARE of (8.89) is expressed as

$$\Sigma = F^{\mathrm{T}}\Sigma F + F^{\mathrm{T}}\Sigma\bar{C}^{\mathrm{T}}(\Lambda(0) - \bar{C}\Sigma\bar{C}^{\mathrm{T}})^{-1}\bar{C}\Sigma F + C^{\mathrm{T}}\Lambda^{-1}(0)C$$

where $F := A - \bar{C}^{\mathrm{T}}\Lambda^{-1}(0)C$. Using the matrix inversion lemma of (5.10) yields

$$\begin{aligned} \Sigma - C^{\mathrm{T}}\Lambda^{-1}(0)C &= F^{\mathrm{T}}[\Sigma + \Sigma\bar{C}^{\mathrm{T}}(\Lambda(0) - \bar{C}\Sigma\bar{C}^{\mathrm{T}})^{-1}\bar{C}\Sigma]F \\ &= F^{\mathrm{T}}[\Sigma^{-1} - \bar{C}^{\mathrm{T}}\Lambda^{-1}(0)\bar{C}]^{-1}F \end{aligned} \tag{8.91}$$

Again, using the matrix inversion lemma, the inverse of the left-hand side of (8.91) becomes

$$[\Sigma - C^{\mathrm{T}}\Lambda^{-1}(0)C]^{-1} = \Sigma^{-1} + \Sigma^{-1}C^{\mathrm{T}}(\Lambda(0) - C\Sigma^{-1}C^{\mathrm{T}})^{-1}C\Sigma^{-1}$$

Suppose that F is invertible. Then, by computing the inverse of the right-hand side of (8.91), and rearranging the terms,

$$\Sigma^{-1} = F\Sigma^{-1}F^{\mathrm{T}} + F\Sigma^{-1}C^{\mathrm{T}}(\Lambda(0) - C\Sigma^{-1}C^{\mathrm{T}})^{-1}C\Sigma^{-1}F^{\mathrm{T}} + \bar{C}^{\mathrm{T}}\Lambda^{-1}(0)\bar{C}$$

This is equivalent to (8.90). □

Part III

Subspace Identification

9

Subspace Identification (1) – ORT

This chapter deals with the stochastic realization problem for a stationary stochastic process with exogenous inputs. As preliminaries, we review projections in a Hilbert space, and explain the feedback-free conditions and the PE condition to be satisfied by input signals. The output process is then decomposed into the deterministic and stochastic components; the former is obtained by the orthogonal projection of the output process onto the Hilbert space spanned by the exogenous inputs, while the latter is obtained by the complementary projection. By a geometric procedure, we develop a minimal state space model of the output process with a very natural block structure, in which the plant and the noise model are independently parametrized. Subspace algorithms are then derived based on this convenient model structure. Some numerical results are included to show the applicability of the present algorithm.

9.1 Projections

We briefly review projections in a Hilbert space and present some related facts that will be used in the following. Let $x \in \mathbb{R}^n$ be a random vector. Let the second-order moment of x be defined by

$$E\{\|x\|^2\} = \sum_{i=1}^{n} E\{x_i^2\}$$

where $E\{\cdot\}$ denotes the mathematical expectation. Let a set of random vectors with finite second-order moments be defined by

$$\mathcal{H} = E\Big\{x \;\Big|\; E\{\|x\|^2\} < \infty\Big\}$$

Then the mean-square norm of $x \in \mathcal{H}$ is given by $\|x\|_{\mathcal{H}} = \sqrt{E\{\|x\|^2\}}$. It is well known that $\mathcal{H}$ is a linear space, and by completing the linear space with this norm, we have a Hilbert space generated by random vectors with finite second moments.

Let a, b be elements of $\mathcal{H}$, and let $\mathcal{A}$, $\mathcal{B}$ be subspaces of $\mathcal{H}$. If $E\{ab^{\mathrm{T}}\} = 0$, we say that a and b are orthogonal. Also, if $E\{ab^{\mathrm{T}}\} = 0$ holds for all $a \in \mathcal{A}$ and $b \in \mathcal{B}$, then we say that $\mathcal{A}$ and $\mathcal{B}$ are orthogonal, and we write $\mathcal{A} \perp \mathcal{B}$. Moreover, $\mathcal{A} \vee \mathcal{B}$ denotes the vector sum such that $\{a + b \mid a \in \mathcal{A},\ b \in \mathcal{B}\}$, $\mathcal{A} + \mathcal{B}$ denotes the direct sum ($\mathcal{A} \cap \mathcal{B} = \{0\}$), and $\mathcal{A} \oplus \mathcal{B}$ the orthogonal sum ($\mathcal{A} \perp \mathcal{B}$). The symbol $\mathcal{A}^{\perp}$ denotes the orthogonal complement of the subspace $\mathcal{A}$ in $\mathcal{H}$, and $\mathrm{span}\{a, b\}$ denotes the Hilbert space generated by all the linear combinations of random vectors a and b. If infinite random vectors are involved, we write $\overline{\mathrm{span}}\{a_1, a_2, \cdots\}$.

Let $\mathcal{A}$ and $\mathcal{B}$ be subspaces of $\mathcal{H}$. Then, the orthogonal projection of $a \in \mathcal{A}$ onto $\mathcal{B}$ is denoted by $\hat{E}\{a \mid \mathcal{B}\}$. If $\mathcal{B} = \mathrm{span}\{b\}$, the orthogonal projection is written as

$$\hat{E}\{a \mid \mathcal{B}\} = E\{ab^{\mathrm{T}}\}E\{bb^{\mathrm{T}}\}^{\dagger}b$$

where $(\cdot)^{\dagger}$ denotes the pseudo-inverse. The orthogonal projection onto the orthogonal complement $\mathcal{B}^{\perp}$ is denoted by $\hat{E}\{a \mid \mathcal{B}^{\perp}\} = a - \hat{E}\{a \mid \mathcal{B}\}$. Also, the orthogonal projection of the space $\mathcal{A}$ onto $\mathcal{B}$ is denoted by $\hat{E}\{\mathcal{A} \mid \mathcal{B}\}$.

Lemma 9.1. *Let $\mathcal{B}$, $\mathcal{C} \subset \mathcal{H}$, and suppose that $a \in \mathcal{B} \vee \mathcal{C}$ and $\mathcal{B} \cap \mathcal{C} = \{0\}$ hold. Then, we have the decomposition formula*

$$\hat{E}\{a \mid \mathcal{B} \vee \mathcal{C}\} = \hat{E}_{\|\mathcal{C}}\{a \mid \mathcal{B}\} + \hat{E}_{\|\mathcal{B}}\{a \mid \mathcal{C}\} \tag{9.1}$$

where $\hat{E}_{\|\mathcal{C}}\{a \mid \mathcal{B}\}$ is the oblique projection of a onto $\mathcal{B}$ along $\mathcal{C}$, and $\hat{E}_{\|\mathcal{B}}\{a \mid \mathcal{C}\}$ the oblique projection of a onto $\mathcal{C}$ along $\mathcal{B}$ as in Figure 9.1. □

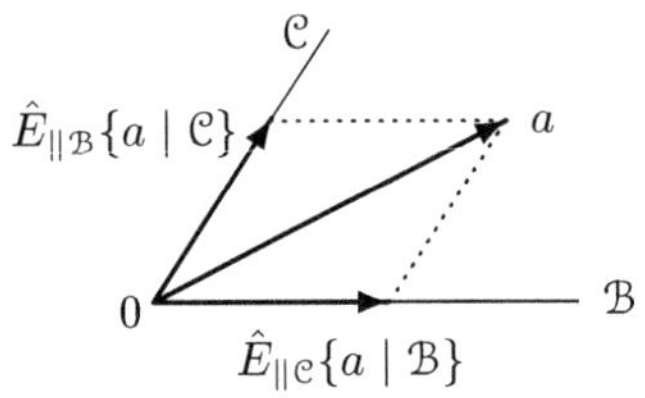

Figure 9.1. Oblique projections

We write the oblique projection of $\mathcal{A}$ onto $\mathcal{B}$ along $\mathcal{C}$ as $\hat{E}_{\|\mathcal{C}}\{\mathcal{A} \mid \mathcal{B}\}$. If $\mathcal{B} \perp \mathcal{C}$, then the oblique projection reduces to the orthogonal projection onto $\mathcal{B}$.

Definition 9.1. *Suppose that $a \in \mathcal{A}$ and $b \in \mathcal{B}$ satisfy the orthogonality condition*

$$E\{(a - \hat{E}\{a \mid \mathcal{C}\})(b - \hat{E}\{b \mid \mathcal{C}\})^{\mathrm{T}}\} = 0, \qquad \mathcal{C} \subset \mathcal{H} \tag{9.2}$$

Then, we say that a and b are conditionally orthogonal with respect to $\mathcal{C}$. If (9.2) holds for all $a \in \mathcal{A}$ and $b \in \mathcal{B}$, then we say that $\mathcal{A}$ and $\mathcal{B}$ are conditionally orthogonal with respect to $\mathcal{C}$. This is simply denoted by $\mathcal{A} \perp \mathcal{B} \mid \mathcal{C}$. □

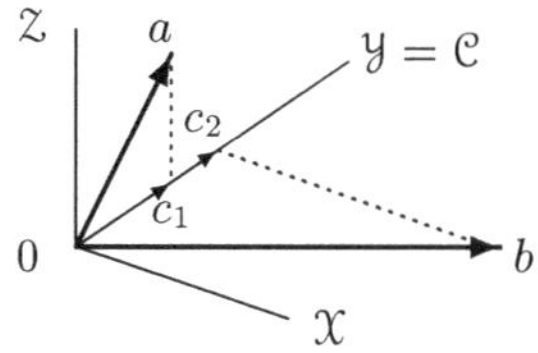

Figure 9.2. Conditional orthogonality ($a \in \mathcal{Z} \vee \mathcal{C}$, $b \in \mathcal{X} \vee \mathcal{C}$)

In Figure 9.2, let the orthogonal projections of a and b onto $\mathcal{Y} = \mathcal{C}$ be denoted by c_1 and c_2, respectively. Then, the condition (9.2) implies that $a - c_1$ and $b - c_2$ are orthogonal.

Lemma 9.2. *The conditional orthogonality* $\mathcal{A} \perp \mathcal{B} \mid \mathcal{C}$ *is equivalent to the following condition*

$$\hat{E}\{\mathcal{B} \mid \mathcal{A} \vee \mathcal{C}\} = \hat{E}\{\mathcal{B} \mid \mathcal{C}\} \tag{9.3}$$

Proof. The conditional orthogonality implies that (9.2) holds for all $a \in \mathcal{A}$, $b \in \mathcal{B}$. Since $(b - \hat{E}\{b \mid \mathcal{C}\}) \perp \mathcal{C}$ and $\hat{E}\{a \mid \mathcal{C}\} \in \mathcal{C}$, it follows from (9.2) that

$$E\{a(b - \hat{E}\{b \mid \mathcal{C}\})^{\mathrm{T}}\} = E\{(a + c)(b - \hat{E}\{b \mid \mathcal{C}\})^{\mathrm{T}}\} = 0 \tag{9.4}$$

holds for all $a \in \mathcal{A}$, $b \in \mathcal{B}$ and $c \in \mathcal{C}$. Since $\mathcal{A} \vee \mathcal{C} = \{a + c \mid a \in \mathcal{A},\ c \in \mathcal{C}\}$, we see from (9.4) that $\mathcal{B} - \hat{E}\{\mathcal{B} \mid \mathcal{C}\} \perp \mathcal{A} \vee \mathcal{C}$. Hence,

$$\hat{E}\{\mathcal{B} \mid \mathcal{A} \vee \mathcal{C}\} = \hat{E}\{\hat{E}\{\mathcal{B} \mid \mathcal{C}\} \mid \mathcal{A} \vee \mathcal{C}\} \tag{9.5}$$

However, since $\mathcal{C} \subset \mathcal{A} \vee \mathcal{C}$, the right-hand side equals $\hat{E}\{\mathcal{B} \mid \mathcal{C}\}$. Conversely, if (9.3) holds, then we have (9.5), so that $\mathcal{B} - \hat{E}\{\mathcal{B} \mid \mathcal{C}\} \perp \mathcal{A} \vee \mathcal{C}$. This implies that (9.4) holds for all a, b, c, and hence (9.2) holds. □

9.2 Stochastic Realization with Exogenous Inputs

Consider a discrete-time stochastic system shown in Figure 9.3, where $u \in \mathbb{R}^m$ is the input vector, $y \in \mathbb{R}^p$ the output vector, $\xi \in \mathbb{R}^q$ the noise vector. We assume that u and y are second-order stationary random processes with mean zero and that they

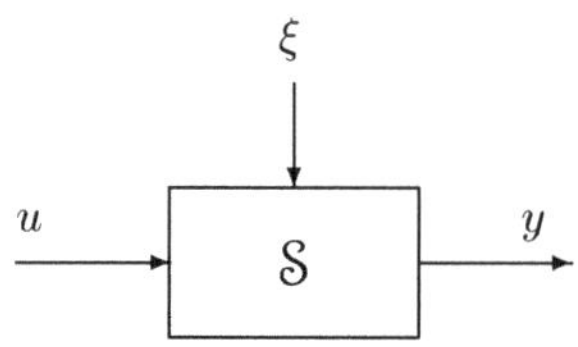

Figure 9.3. Stochastic system with exogenous input

are available for identification. The output y is also excited by the external noise ξ, which is not directly accessible. We need the following assumption, whose meaning is explained in the next section.

Assumption 9.1. *There is no feedback from the output y to the input u. This is called a feedback-free condition.* □

The stochastic realization problem with exogenous inputs is stated as follows.

Stochastic Realization Problem

Suppose that infinite data $\{u(t),\ y(t),\ t = 0,\ \pm 1,\ \cdots\}$ are given. The problem is to define a suitable state vector x with minimal dimension and to derive a state space model with the input vector u and the output vector y of the form

$$x(t+1) = Ax(t) + Bu(t) + Ke(t) \tag{9.6a}$$

$$y(t) = Cx(t) + Du(t) + e(t) \tag{9.6b}$$

where e is the innovation process defined below (see Lemma 9.6). It should be noted that the stochastic realization problems considered in Chapters 7 and 8 are realizations of stationary processes without exogenous inputs.

Consider the joint input-output process $w = \begin{bmatrix} y \\ u \end{bmatrix} \in \mathbb{R}^d$, where $d := p + m$. Since we are given infinite input-output data, the exact covariance matrices of w are given by

$$\Lambda_{ww}(l) = E\{w(t+l)w^{\mathrm{T}}(t)\} = \begin{bmatrix} \Lambda_{yy}(l) & \Lambda_{yu}(l) \\ \Lambda_{uy}(l) & \Lambda_{uu}(l) \end{bmatrix}, \quad l = 0,\ \pm 1,\ \cdots$$

and the spectral density matrix is

$$\Phi_{ww}(z) = \sum_{l=-\infty}^{\infty} \Lambda_{ww}(l) z^{-l} = \begin{bmatrix} \Phi_{yy}(z) & \Phi_{yu}(z) \\ \Phi_{uy}(z) & \Phi_{uu}(z) \end{bmatrix}$$

We consider the prediction problem of $w(t+k)$, $k = 1, 2, \cdots$ based on the present and past observations $w(t), w(t-1), \cdots$ such that

$$\Sigma_k = \min_{\{f_i\}} \operatorname{cov}\left\{ w(t+k) - \sum_{i=0}^{\infty} f_i w(t-i) \right\}$$

where $f_i \in \mathbb{R}^{d\times d}$ are coefficients. In general, for the minimum prediction error covariance matrices, we have $0 \le \Sigma_1 \le \Sigma_2 \le \cdots$. If $\Sigma_1 > 0$, then the process w is regular. If $\det \Sigma_1 = 0$, w is called singular [138]; see also Section 4.5. If the spectral density matrix $\Phi_{ww}(z)$ has full rank, we simply say that w has full rank. The regularity and full rank conditions imply that the joint process w does not degenerate.

Let the Hilbert space generated by w be defined by

$$\mathcal{W} = \overline{\text{span}}\{w(\tau) \mid \tau = 0, \pm 1, \cdots\}$$

This space contains all linear combinations of the history of the process w. Also, we define Hilbert subspaces generated by u and y as

$$\mathcal{U} = \overline{\text{span}}\{u(\tau) \mid \tau = 0, \pm 1, \cdots\}$$
$$\mathcal{Y} = \overline{\text{span}}\{y(\tau) \mid \tau = 0, \pm 1, \cdots\}$$

Let t be the present time, and define subspaces generated by the past and future of u and y as

$$\mathcal{U}_t^- = \overline{\text{span}}\{u(\tau) \mid \tau < t\}, \qquad \mathcal{Y}_t^- = \overline{\text{span}}\{y(\tau) \mid \tau < t\}$$
$$\mathcal{U}_t^+ = \overline{\text{span}}\{u(\tau) \mid \tau \geq t\}, \qquad \mathcal{Y}_t^+ = \overline{\text{span}}\{y(\tau) \mid \tau \geq t\}$$

It may be noted that the present time t is included in the future and not in the past by convention.

9.3 Feedback-Free Processes

There exists a quite long history of studies on the feedback between two observed processes [24–26, 53, 63]. In this section, we provide the definition of feedback-free and consider some equivalent conditions for it.

Suppose that the joint process is regular and of full rank. It therefore follows from the Wold decomposition theorem (Theorem 4.3) that the joint process w is expressed as a moving average representation

$$\begin{bmatrix} y(t) \\ u(t) \end{bmatrix} = \sum_{i=0}^{\infty} \begin{bmatrix} A_i & B_i \\ C_i & D_i \end{bmatrix} \begin{bmatrix} \nu(t-i) \\ \eta(t-i) \end{bmatrix} \tag{9.7}$$

where $A_i \in \mathbb{R}^{p \times p}$, $B_i \in \mathbb{R}^{p \times m}$, $C_i \in \mathbb{R}^{m \times p}$, $D_i \in \mathbb{R}^{m \times m}$ are constant matrices, and $\nu \in \mathbb{R}^p$ and $\eta \in \mathbb{R}^m$ are zero mean white noise vectors with covariance matrices

$$E\left\{ \begin{bmatrix} \nu(t) \\ \eta(t) \end{bmatrix} \begin{bmatrix} \nu^{\mathrm{T}}(s) & \eta^{\mathrm{T}}(s) \end{bmatrix} \right\} = \bar{Q}\delta_{ts}, \qquad \bar{Q} > 0$$

Define $d \times d$ matrices $\Gamma_i := \begin{bmatrix} A_i & B_i \\ C_i & D_i \end{bmatrix}$, $i = 0, 1, \cdots$ and the $d \times d$ transfer matrix

$$\Gamma(z) = \sum_{i=0}^{\infty} \Gamma_i z^{-i} = \begin{bmatrix} A(z) & B(z) \\ C(z) & D(z) \end{bmatrix}$$

Theorems 4.3 and 4.4 assert that $\{\Gamma_i, i = 0, 1, \cdots\}$ are square summable and $\Gamma^{-1}(z)$ is analytic in $|z| > 1$. In the following, we further assume that both $\Gamma(z)$ and $\Gamma^{-1}(z)$ are stable, *i.e.* $\Gamma(z)$ is of minimal phase. Also, we assume that $\Gamma(z)$ is a rational matrix [24, 25], so that we consider only a class of finitely generated stationary processes, which is a subclass of regular full rank stationary processes. The condition of rationality of $\Gamma(z)$ is, however, relaxed in [26].

Now we provide the definition that there is no feedback from the output y to the input u, and introduce some equivalent feedback-free conditions. The following definition is called the strong feedback-free property [26]; however for simplicity, we call it the feedback-free property.

Definition 9.2. *There is no feedback from y to u for the joint process of (9.7), if the following conditions are satisfied.*

(i) The covariance matrix $\bar{Q}$ is block diagonal with

$$\bar{Q} = \begin{bmatrix} \bar{Q}_1 & 0 \\ 0 & \bar{Q}_2 \end{bmatrix}, \qquad \bar{Q}_1 \in \mathbb{R}^{p \times p}, \qquad \bar{Q}_2 \in \mathbb{R}^{m \times m}$$

(ii) The moving average representation (9.7) is expressed as

$$\begin{bmatrix} y(t) \\ u(t) \end{bmatrix} = \sum_{i=0}^{\infty} \begin{bmatrix} A_i & B_i \\ 0 & D_i \end{bmatrix} \begin{bmatrix} \nu(t-i) \\ \eta(t-i) \end{bmatrix} \tag{9.8}$$

so that the transfer matrix $\Gamma(z) = \begin{bmatrix} A(z) & B(z) \\ 0 & D(z) \end{bmatrix}$ is block upper triangular. □

Theorem 9.1. *Suppose that the joint process w is regular and of full rank. Then, the following conditions (i) ∼ (v) are equivalent.*

(i) There is no feedback from the output vector y to the input vector u.

(ii) The smoothed estimate of $y(t)$ based on the whole input data is causal, i.e.

$$\hat{E}\{y(t) \mid \mathcal{U}\} = \hat{E}\{y(t) \mid \mathcal{U}_{t+1}^{-}\}, \qquad t = 0, \pm 1, \cdots \tag{9.9}$$

(iii) In terms of the input u, the output process y is expressed as

$$y(t) = \sum_{i=0}^{\infty} K_i u(t-i) + \sum_{i=0}^{\infty} L_i \nu(t-i) \tag{9.10}$$

where

$$K(z) = \sum_{i=0}^{\infty} K_i z^{-i}, \qquad L(z) = \sum_{i=0}^{\infty} L_i z^{-i}$$

are rational matrices such that $L(z)$ has full rank, and $K(z)$, $L(z)\ L^{-1}(z)$ are stable, and the processes u and ν are uncorrelated, where $\nu \in \mathbb{R}^p$ is a zero mean white noise vector.

(iv) *Given the past of u, the future of u is uncorrelated with the past of y, so that the conditional orthogonality condition*

$$\mathcal{U}_t^+ \perp \mathcal{Y}_t^- \mid \mathcal{U}_t^-, \qquad t = 0, \pm 1, \cdots \tag{9.11}$$

holds. This condition is equivalent to

$$\hat{E}\{\mathcal{U}_t^+ \mid \mathcal{Y}_t^- \vee \mathcal{U}_t^-\} = \hat{E}\{\mathcal{U}_t^+ \mid \mathcal{U}_t^-\}, \quad t = 0, \pm 1, \cdots \tag{9.12}$$

which implies that the past of y is irrelevant for the prediction of the future of u given the past of u. This condition is due to Granger [63].

Proof. A proof is deferred in Appendix of Subsection 9.10.1. □

9.4 Orthogonal Decomposition of Output Process

9.4.1 Orthogonal Decomposition

Suppose that Assumption 9.1 holds. Putting $\mathcal{A} = \mathcal{U}_{t+1}^+$, $\mathcal{B} = \mathcal{Y}_{t+1}^-$ and $\mathcal{C} = \mathcal{U}_{t+1}^-$, we get $\mathcal{A} \vee \mathcal{C} = \mathcal{U}$. It then follows from Lemma 9.2 and (9.11) that

$$\hat{E}\{\mathcal{Y}_{t+1}^- \mid \mathcal{U}\} = \hat{E}\{\mathcal{Y}_{t+1}^- \mid \mathcal{U}_{t+1}^-\}, \qquad t = 0, \pm 1, \cdots \tag{9.13}$$

Since $y(t) \in \mathcal{Y}_{t+1}^-$, we have

$$\hat{E}\{y(t) \mid \mathcal{U}\} = \hat{E}\{y(t) \mid \mathcal{U}_{t+1}^-\}, \qquad t = 0, \pm 1, \cdots \tag{9.14}$$

This is the same as the condition (ii) of Theorem 9.1.

We now define the orthogonal decomposition of the output process y.

Definition 9.3. *Consider the orthogonal projection of $y(t)$ onto $\mathcal{U}$ such that*

$$y_d(t) = \hat{E}\{y(t) \mid \mathcal{U}\} = \hat{E}\{y(t) \mid \mathcal{U}_{t+1}^-\} \tag{9.15}$$

Then, y_d is called the deterministic component of y. Also, the complementary projection

$$\begin{aligned} y_s(t) &= y(t) - \hat{E}\{y(t) \mid \mathcal{U}_{t+1}^-\} \\ &= y(t) - \hat{E}\{y(t) \mid \mathcal{U}\} = \hat{E}\{y(t) \mid \mathcal{U}^\perp\} \end{aligned} \tag{9.16}$$

is called the stochastic component of y. □

The deterministic component y_d is the orthogonal projection of the output y onto the Hilbert space $\mathcal{U}$ spanned by the input process u, so that y_d is the part of y that is linearly related to the input process u. On the other hand, the stochastic component y_s is the part of y that is orthogonal to the data space $\mathcal{U}$; see Figure 9.4. Thus, though y_s is causal, it is orthogonal to the whole input space.

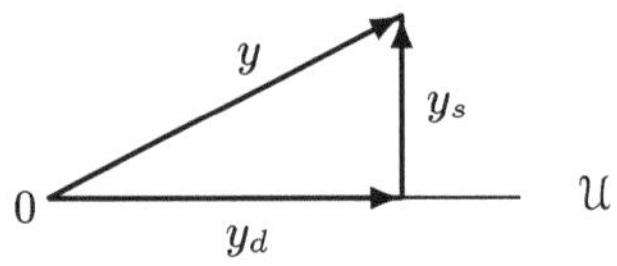

Figure 9.4. Orthogonal decomposition

Lemma 9.3. *Under Assumption 9.1, the output process y is decomposed into the deterministic component y_d and the stochastic component y_s. In fact,*

$$y(t) = y_d(t) + y_s(t), \qquad t = 0, \pm 1, \cdots \tag{9.17}$$

where $y_s(t) \perp y_d(\tau)$ holds for all t, $\tau = 0, \pm 1, \cdots$.

Proof. Immediate from (9.15) and (9.16). □

From this lemma, we see that if there is no feedback from the output to the input, a state space model for y is expressed as the orthogonal sum of state space models for the deterministic component y_d and the stochastic component y_s. It follows from Theorem 9.1 (iii) that y_d and y_s correspond to the first- and the second-term of the right-hand side of (9.10), respectively.

9.4.2 PE Condition

In this subsection, we consider the PE condition to be satisfied by the input process, which is one of the important conditions in system identification [109, 145]; see also Appendix B.

Assumption 9.2. *For each t, the input space $\mathcal{U}$ has the direct sum decomposition*

$$\mathcal{U} = \mathcal{U}_t^- \dotplus \mathcal{U}_t^+ \tag{9.18}$$

where $\mathcal{U}_t^- \cap \mathcal{U}_t^+ = \{0\}$. □

The condition $\mathcal{U}_t^- \cap \mathcal{U}_t^+ = \{0\}$ is equivalent to the fact that the spectral density function of u is positive definite on the unit circle [106], *i.e.*,

$$\Phi_{uu}(\omega) \geq cI_m, \qquad \exists\ c > 0 \tag{9.19}$$

In this case, the input u has PE condition with order infinity.

The condition (9.19) is equivalent to the fact that all the canonical angles between the past and future spaces of the input are positive. It also follows from [64, 69] that a necessary and sufficient condition that the canonical angles between $\mathcal{U}_t^+$ and $\mathcal{U}_t^-$ is zero is that $\Phi_{uu}(z)$ has some zeros on the unit circle.

Remark 9.1. The above assumption is too restrictive for many practical cases, and we could instead assume the PE condition of sufficiently high order and the finite dimensionality of the underlying "true" system. The reason for choosing the above condition is that it allows very simple proofs in the mathematical statements below, and it does not require the finite dimensionality assumption on the "true" systems. □

Lemma 9.4. *Under Assumptions 9.1 and 9.2, the deterministic component y_d of (9.15) is expressed as*

$$y_d(t) = \sum_{i=0}^{\infty} G_i u(t-i) = \sum_{i=-\infty}^{t} G_{t-i} u(i) \tag{9.20}$$

where $G_i \in \mathbb{R}^{p \times m}$ are constant matrices with $G(z) = \sum_{i=0}^{\infty} G_i z^{-i}$ stable.

Proof. This fact is essentially shown in the proof of Theorem 9.1 (iii). From (9.15), we have $y_d(t) \in \mathcal{U}_{t+1}^-$, so that it is expressed as a linear combination of the present and past inputs as in (9.20).

The stability of $G(z)$ can also be proved as follows. The optimality condition for $\{G_i, \ i = 0, 1, \cdots\}$ is given by

$$y(t) - \sum_{i=0}^{\infty} G_i u(t-i) \perp u(t-j), \qquad j = 0, 1, \cdots$$

Thus it follows that

$$\Lambda_{yu}(j) = \sum_{i=0}^{\infty} G_i \Lambda_{uu}(j-i), \qquad j = 0, 1, \cdots$$

Since the above equation is a discrete-time Wiener-Hopf equation, we can solve it by using the spectral factorization technique. Suppose that the spectral density matrix $\Phi_{uu}(z)$ is factored as

$$\Phi_{uu}(z) = \Theta(z)\Theta^{\mathrm{T}}(z^{-1})$$

where $\Theta(z)$ is of minimal phase. It then follows from [11] that the optimal transfer matrix $G(z)$ is given by

$$G(z) = [\Phi_{yu}(z)\Theta^{-\mathrm{T}}(z^{-1})]_+ \Theta^{-1}(z) \tag{9.21}$$

where $[\,\cdot\,]_+$ denotes the operation that extracts the causal part of the transfer matrices. Thus the stability of $G(z)$ follows from the definition of (9.21). □

Lemma 9.5. *The deterministic component y_d and the stochastic component y_s of (9.15) and (9.16) are mutually uncorrelated second-order stationary processes, and are regular and of full rank.*

Proof. Lemma 9.3 shows that two components are uncorrelated. We see from (9.20) that $y_d = G(z)u$. However, since u is second-order stationary and since $G(z)$ is stable, y_d is a second-order stationary process. Thus $y_s := y - y_d$ is also second-order stationary. Moreover it may be noted that y and u are regular and of full rank, so are y_d and y_s. □

Remark 9.2. A finite sum approximation of y_d of (9.20) is given by

$$y_d(t) = \sum_{i=0}^{k-1} G_i u(t-i)$$

for a sufficiently large $k > 0$. It can be shown that this is easily computed by means of LQ decomposition (see Section A.2). □

9.5 State Space Realizations

In order to obtain a realization of the stochastic component y_s, we can employ the results of Chapter 8. For the deterministic component y_d, however, the mathematical development of a state space realization is slightly involved, since we must employ oblique projections due to the existence of the exogenous input u.

We begin with a realization of the stochastic component.

9.5.1 Realization of Stochastic Component

Let the Hilbert space generated by y_s be defined by

$$\tilde{\mathcal{Y}} = \overline{\text{span}}\{y_s(\tau) \mid \tau = 0, \pm 1, \cdots\} \subset \mathcal{U}^\perp$$

and let Hilbert subspaces generated by the past and future of y_s be defined by

$$\tilde{\mathcal{Y}}_t^- = \overline{\text{span}}\{y_s(\tau) \mid \tau < t\}, \qquad \tilde{\mathcal{Y}}_t^+ = \overline{\text{span}}\{y_s(\tau) \mid \tau \geq t\}$$

It follows from the stochastic realization results in Chapter 8 that a necessary and sufficient condition that the stochastic component has a finite dimensional realization is that the predictor space

$$\tilde{\mathcal{X}}_t^{+/-} = \hat{E}\{\tilde{\mathcal{Y}}_t^+ \mid \tilde{\mathcal{Y}}_t^-\} \tag{9.22}$$

is finite dimensional.

Theorem 9.2. *Define* $\dim(\tilde{\mathcal{X}}_t^{+/-}) = \tilde{n}$. *Then, the minimal dimension of realization is* $\tilde{n}$. *In terms of a basis vector* x_s *of the predictor space, a state space realization of the stochastic component* y_s *is given by*

$$x_s(t+1) = A_s x_s(t) + K_s e_s(t) \tag{9.23a}$$

$$y_s(t) = C_s x_s(t) + e_s(t) \tag{9.23b}$$

where e_s *is the innovation process for* y_s*, or the one-step prediction error defined by*

$$e_s(t) := y_s(t) - \hat{E}\{y_s(t) \mid \tilde{\mathcal{Y}}_t^-\}$$

Proof. Immediate from Theorem 8.4 in Subsection 8.5.2. □

The innovation representation of (9.23) is called the stochastic subsystem. The following lemma shows that the innovation process e_s is the same as the conditional innovation process e of y.

Lemma 9.6. *The innovation process e_s is expressed as*

$$e_s(t) = e(t) := y(t) - \hat{E}\{y(t) \mid \mathcal{U}_{t+1}^- \vee \mathcal{Y}_t^-\} \tag{9.24}$$

Proof. By using the property of orthogonal projection, we can show that

$$\begin{aligned} e(t) &= y(t) - \hat{E}\{y(t) \mid \mathcal{U}_{t+1}^- \vee \mathcal{Y}_t^-\} = y(t) - \hat{E}\{y(t) \mid \mathcal{U}_{t+1}^- \oplus \tilde{\mathcal{Y}}_t^-\} \\ &= (y(t) - \hat{E}\{y(t) \mid \mathcal{U}_{t+1}^-\}) - \hat{E}\{y(t) \mid \tilde{\mathcal{Y}}_t^-\} \\ &= y_s(t) - \hat{E}\{y_s(t) + y_d(t) \mid \tilde{\mathcal{Y}}_t^-\} \\ &= y_s(t) - \hat{E}\{y_s(t) \mid \tilde{\mathcal{Y}}_t^-\} = e_s(t) \end{aligned}$$

where we used the fact that $y_d(t) \perp \tilde{\mathcal{Y}}_t^-$. □

9.5.2 Realization of Deterministic Component

A state space realization of the deterministic component should be a state space model with the input process u and the output process y_d. In the following, we use the idea of N4SID described in Section 6.6 to construct a state space realization for the deterministic component of the output process.

As usual, let the Hilbert space generated by y_d be defined by

$$\hat{\mathcal{Y}} = \overline{\text{span}}\{y_d(\tau) \mid \tau = 0, \pm 1, \cdots\} \subset \mathcal{U}$$

and let Hilbert spaces spanned by the future and past of the deterministic component y_d be defined by

$$\hat{\mathcal{Y}}_t^+ = \overline{\text{span}}\{y_d(\tau) \mid \tau \geq t\}, \qquad \hat{\mathcal{Y}}_t^- = \overline{\text{span}}\{y_d(\tau) \mid \tau < t\}$$

Definition 9.4. For any t, if a subspace $\mathcal{S}_t$ ($\subset \mathcal{U}_t^-$) satisfies the relation

$$\hat{E}_{\|\mathcal{U}_t^+}\{\hat{\mathcal{Y}}_t^+ \mid \mathcal{U}_t^-\} = \hat{E}_{\|\mathcal{U}_t^+}\{\hat{\mathcal{Y}}_t^+ \mid \mathcal{S}_t\} \tag{9.25}$$

then $\mathcal{S}_t$ is called an oblique splitting subspace for the pair $(\hat{\mathcal{Y}}_t^+, \mathcal{U}_t^-)$. □

The space $\mathcal{S}_t$ satisfying (9.25) carries the information contained in $\mathcal{U}_t^-$ that is needed to predict the future outputs $y_d(t + l)$, $l = 0, 1, \cdots$, so that it is a candidate of the state space. Also, the oblique predictor space for the deterministic component

$$\mathcal{X}_t^{+/-} = \hat{E}_{\|\mathcal{U}_t^+}\{\hat{\mathcal{Y}}_t^+ \mid \mathcal{U}_t^-\} \tag{9.26}$$

is obviously contained in $\mathcal{U}_t^-$ and is oblique splitting. This can be proved similarly to the proof of Lemma 8.4. In fact, since $\mathcal{X}_t^{+/-} \subset \mathcal{U}_t^-$, it follows from the property of projection that

$$\begin{aligned}\hat{E}_{\|\mathcal{U}_t^+}\{\hat{\mathcal{Y}}_t^+ \mid \mathcal{U}_t^-\} = \mathcal{X}_t^{+/-} &= \hat{E}_{\|\mathcal{U}_t^+}\{\mathcal{X}_t^{+/-} \mid \mathcal{X}_t^{+/-}\} \\ &= \hat{E}_{\|\mathcal{U}_t^+}\{\hat{E}_{\|\mathcal{U}_t^+}\{\hat{\mathcal{Y}}_t^+ \mid \mathcal{U}_t^-\} \mid \mathcal{X}_t^{+/-}\} \\ &= \hat{E}_{\|\mathcal{U}_t^+}\{\hat{\mathcal{Y}}_t^+ \mid \mathcal{X}_t^{+/-}\}\end{aligned}$$

This shows that $\mathcal{X}_t^{+/-}$ is an oblique splitting from (9.25). Thus all the information in $\hat{\mathcal{U}}_t^-$ that is related to the future is contained in the predictor space $\mathcal{X}_t^{+/-}$.

We further define the extended space by

$$\bar{\mathcal{Y}}_t^+ = \hat{\mathcal{Y}}_t^+ \vee \mathcal{X}_t^{+/-} \tag{9.27}$$

Then, we have the following basic result.

Lemma 9.7. *The predictor space $\mathcal{X}_t^{+/-}$ defined by (9.26) is an oblique splitting subspace for $(\bar{\mathcal{Y}}_t^+, \mathcal{U}_t^-)$, and*

$$\mathcal{X}_t^{+/-} = \bar{\mathcal{Y}}_t^+ \cap \mathcal{U}_t^- \tag{9.28}$$

holds. Moreover, under Assumption 9.2, we have the following direct sum decomposition

$$\bar{\mathcal{Y}}_t^+ = (\bar{\mathcal{Y}}_t^+ \cap \mathcal{U}_t^-) + (\bar{\mathcal{Y}}_t^+ \cap \mathcal{U}_t^+) \tag{9.29}$$

Proof. A proof is deferred to Appendix of Subsection 9.10.2. □

Now we assume that $\dim(\mathcal{X}_t^{+/-}) = n$ holds. Let the subspace generated by $u(t)$ be defined by

$$\mathcal{U}_t := \operatorname{span}\{u(t)\} \subset \mathcal{U}_t^+$$

It follows from Assumption 9.2 that $\mathcal{U}_{t+1}^- = \mathcal{U}_t^- + \mathcal{U}_t$ ($\mathcal{U}_t^- \cap \mathcal{U}_t = \{0\}$). Hence, we have a direct sum decomposition

$$\mathcal{X}_{t+1}^{+/-} = \bar{\mathcal{Y}}_{t+1}^+ \cap \mathcal{U}_{t+1}^- = (\bar{\mathcal{Y}}_{t+1}^+ \cap \mathcal{U}_t^-) + (\bar{\mathcal{Y}}_{t+1}^+ \cap \mathcal{U}_t) \tag{9.30}$$

where $\bar{\mathcal{Y}}_{t+1}^+ \subset \bar{\mathcal{Y}}_t^+$ holds, so that

$$(\bar{\mathcal{Y}}_{t+1}^+ \cap \mathcal{U}_t^-) \subset (\bar{\mathcal{Y}}_t^+ \cap \mathcal{U}_t^-) = \mathcal{X}_t^{+/-}$$

Since $(\bar{\mathcal{Y}}_{t+1}^+ \cap \mathcal{U}_t) \subset \mathcal{U}_t$ holds, we see from (9.30) that

$$\mathcal{X}_{t+1}^{+/-} \subset \mathcal{X}_t^{+/-} + \mathcal{U}_t \tag{9.31}$$

It should be noted here that the right-hand side of the above equation is a direct sum, since $\mathcal{X}_t^{+/-} \cap \mathcal{U}_t = \{0\}$ holds from $\mathcal{X}_t^{+/-} \subset \mathcal{U}_t^-$.

Let $x_d(t) \in \mathbb{R}^n$ be a basis vector for $\mathcal{X}_t^{+/-}$, and $x_d(t+1)$ be the shifted version of it. We see from (9.28) that $x_d(t+1)$ is a basis for the space $\mathcal{X}_{t+1}^{+/-} = \bar{\mathcal{Y}}_{t+1}^+ \cap \mathcal{U}_{t+1}^-$.

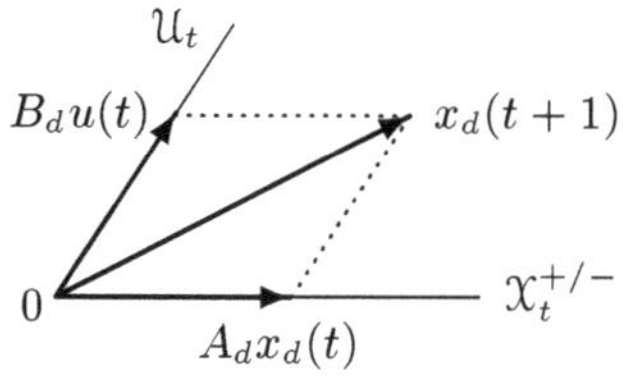

Figure 9.5. Direct sum decomposition of $x_d(t+1)$

As in Figure 9.5, the projection of $x_d(t+1) \in \mathcal{X}_{t+1}^{+/-}$ onto the subspaces in the right-hand side of (9.31) gives the state space equation

$$x_d(t+1) = A_d x_d(t) + B_d u(t) \tag{9.32}$$

where $A_d \in \mathbb{R}^{n\times n}$ and $B_d \in \mathbb{R}^{n\times m}$. Note that, since the right-hand side of (9.31) is a direct sum, (9.32) is a unique direct sum decomposition.

Since $y_d(t) \in \hat{\mathcal{Y}}_t^+ \cap \mathcal{U}_{t+1}^-$, it follows from (9.27) that

$$y_d(t) \in \bar{\mathcal{Y}}_t^+ \cap \mathcal{U}_{t+1}^- = (\bar{\mathcal{Y}}_t^+ \cap \mathcal{U}_t^-) + (\bar{\mathcal{Y}}_t^+ \cap \mathcal{U}_t) \subset \mathcal{X}_t^{+/-} + \mathcal{U}_t$$

Hence, the projection of $y_d(t)$ onto the two subspaces in the right-hand side of the above equation yields a unique output equation[1]

$$y_d(t) = C_d x_d(t) + D_d u(t) \tag{9.33}$$

where $C_d \in \mathbb{R}^{p\times n}$ and $D_d \in \mathbb{R}^{p\times m}$ are constant matrices.

Since the predictor space $\mathcal{X}_t^{+/-}$ is included in $\mathcal{U}_t^-$, we see that $x_d(t) \in \mathcal{U}_t^-$. As in Lemma 9.4, it follows that $x_d(t)$ is expressed as

$$x_d(t) = \sum_{i=1}^{\infty} F_i u(t-i) \quad \Rightarrow \quad x_d = F(z)u$$

where $F(z)$ is stable and $F_i \in \mathbb{R}^{n\times m}$. Since u is regular and stationary, so is x_d. Thus it follows from (9.32) that

$$x_d = (zI - A_d)^{-1} B_d u = F(z)u \quad \Rightarrow \quad F(z) = (zI - A_d)^{-1} B_d$$

Since $F(z)$ is stable, if (A_d, B_d) is reachable, all the eigenvalues of A_d must be inside the unit disk, so that A_d is stable.

Summarizing above results, we have the following theorem.

Theorem 9.3. *Suppose that the joint process w has a rational spectral density matrix and that the input process u satisfies Assumptions 9.1 and 9.2. Then, the predictor*

---

[1]The decomposition of $y_d(t)$ is obtained by replacing $x_d(t+1) \to y_d(t)$, $A_d x_d(t) \to C_d x_d(t)$, $B_d x_d(t) \to D_d u(t)$ in Figure 9.5.

space $\mathcal{X}_t^{+/-}$ *has dimension* n*, and a state space model with the state vector* $x_d(t) \in \mathcal{X}_t^{+/-}$ *is given by*

$$x_d(t+1) = A_d x_d(t) + B_d u(t) \tag{9.34a}$$

$$y_d(t) = C_d x_d(t) + D_d u(t) \tag{9.34b}$$

where A_d *is stable. This is called the deterministic subsystem. Moreover, let* $\bar{\mathcal{X}}_t$ *be the state space of another realization of* y_d*, and let* $\dim(\bar{\mathcal{X}}_t) = \bar{n}$*. Then we have* $\bar{n} \geq n$*.*

Proof. The first half of the theorem is obvious from above. We shall prove the result for the dimension of the state spaces. Let $\bar{x} \in \mathbb{R}^{\bar{n}}$ be a state vector, and let a realization for y_d be given by

$$\bar{x}(t+1) = \bar{A}\bar{x}(t) + \bar{B}u(t)$$

$$y_d(t) = \bar{C}\bar{x}(t) + \bar{D}u(t)$$

The impulse response matrices of the system are defined by

$$W_t = \begin{cases} \bar{D}, & t = 0 \\ \bar{C}\bar{A}^{t-1}\bar{B}, & t = 1, 2, \cdots \end{cases}$$

The following proof is related to that of the second half of Lemma 9.7; see Subsection 9.10.2. In terms of impulse responses, we have

$$\begin{aligned} y_d(t+k) &= \sum_{i=-\infty}^{t-1} W_{t+k-i}u(i) + \sum_{i=t}^{t+k} W_{t+k-i}u(i) \\ &=: y_d^-(t+k) + y_d^+(t+k), \qquad k = 0, 1, \cdots \end{aligned} \tag{9.35}$$

where $y_d^-(t+k) \in \mathcal{U}_t^-$ and $y_d^+(t+k) \in \mathcal{U}_t^+$. Since $\mathcal{U}_t^- \cap \mathcal{U}_t^+ = \{0\}$, we see that $y_d^-(t+k)$ is the oblique projection of $y_d(t+k)$ onto $\mathcal{U}_t^-$ along $\mathcal{U}_t^+$, so that

$$y_d^-(t+k) = \hat{E}_{\|\mathcal{U}_t^+}\{y_d(t+k) \mid \mathcal{U}_t^-\} \tag{9.36}$$

and hence $\{y_d^-(t+k) \mid k = 0, 1, \cdots\}$ generates $\mathcal{X}_t^{+/-}$ [see (9.26)]. Similarly, $y_d^+(t+k)$ is the oblique projection of $y_d(t+k)$ onto $\mathcal{U}_t^+$ along $\mathcal{U}_t^-$, and also by definition, $\{y_d^+(t+k) \mid k = 0, 1, \cdots\}$ generates $\hat{\mathcal{Y}}_t^+$. It therefore follows that

$$y_d^+(t+k) \in (\hat{\mathcal{Y}}_t^+ \cap \mathcal{U}_t^+) \subset (\bar{\mathcal{Y}}_t^+ \cap \mathcal{U}_t^+)$$

By combining this with (9.36), we get

$$y_d(t+k) = y_d^-(t+k) + y_d^+(t+k) \in (\bar{\mathcal{Y}}_t^+ \cap \mathcal{U}_t^-) + (\bar{\mathcal{Y}}_t^+ \cap \mathcal{U}_t^+) = \bar{\mathcal{Y}}_t^+$$

This implies that the two terms in the right-hand side of (9.35) are respectively the oblique projections of $y_d(t+k)$ onto the two subspaces of (9.29). Moreover, from

(9.35), we can write $y_d^-(t+k) = \bar{C}\bar{A}^k \bar{x}(t)$, $k = 0, 1, \cdots$, so that the space $\mathcal{X}_t^{+/-}$ is generated by $\{\bar{C}\bar{A}^k \bar{x}(t) \mid k = 0, 1, \cdots\}$. However, the latter is contained in $\bar{\mathcal{X}}_t := \text{span}\{\bar{x}(t)\}$. Thus it follows that $\mathcal{X}_t^{+/-} \subset \bar{\mathcal{X}}_t$, implying that $\dim(x_d) \leq \dim(\bar{x})$. □

This theorem shows that the state space $\bar{\mathcal{X}}_t$ of a realization of y_d includes the state space $\mathcal{X}_t^{+/-}$, so that $\mathcal{X}_t^{+/-}$ is a state space with minimal dimension.

9.5.3 The Joint Model

As mentioned above, to obtain a state space realization for y, it suffices to combine the two realizations for the deterministic and stochastic components. From Theorems 9.2 and 9.3, we have the basic result of this chapter.

Theorem 9.4. *A state space realization of y is given by*

$$\begin{bmatrix} x_d(t+1) \\ x_s(t+1) \end{bmatrix} = \begin{bmatrix} A_d & 0 \\ 0 & A_s \end{bmatrix} \begin{bmatrix} x_d(t) \\ x_s(t) \end{bmatrix} + \begin{bmatrix} B_d \\ 0 \end{bmatrix} u(t) + \begin{bmatrix} 0 \\ K_s \end{bmatrix} e(t) \tag{9.37a}$$

$$y(t) = [C_d \ \ C_s] \begin{bmatrix} x_d(t) \\ x_s(t) \end{bmatrix} + D_d u(t) + e(t) \tag{9.37b}$$

Proof. A proof is immediate from Theorems 9.2 and 9.3 by using the fact that $e_s = e$ (Lemma 9.6). □

We see from (9.37) that the state-space model for the output process y has a very natural block structure. The realization (9.37) is a particular form of the state space model of (9.6) in that A- and B-matrix have block structure, so that the state vector x_d of the deterministic subsystem is not reachable from the innovation process e, while the state vector x_s of the stochastic subsystem is not reachable from the input vector u. Also, if the deterministic and stochastic subsystems have some common dynamics, *i.e.*, if A_d and A_s have some common eigenvalues, then the system of (9.37) is not observable, so that it is not minimal.

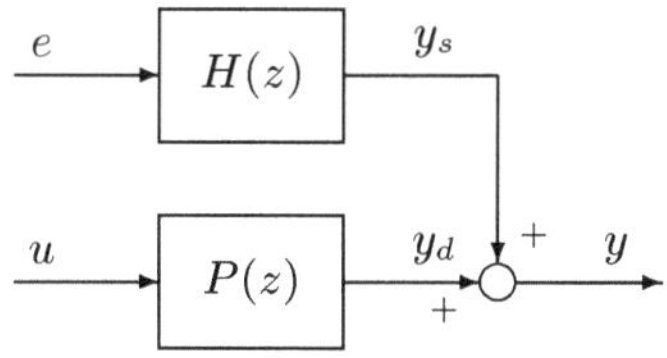

Figure 9.6. Multivariable transfer function model

Thus, as shown in Figure 9.6, the output y is described by

$$y = P(z)u + H(z)e \tag{9.38}$$

where $P(z)$ and $H(z)$ are respectively defined by

$$P(z) = D_d + C_d(zI_n - A_d)^{-1}B_d$$

and

$$H(z) = I_p + C_s(zI_{\tilde{n}} - A_s)^{-1}K_s$$

It should be noted that by the orthogonal decomposition method, we have a multivariable input-output model where the plant transfer matrix $P(z)$ and the noise transfer matrix $H(z)$ have independent parametrizations.

Up to now, we have considered the ideal case where an infinite input-output data is available. In the following sections, we derive subspace identification methods by adapting the present realization results to given finite input-output data.

9.6 Realization Based on Finite Data

In practice, we observe a finite collection of input-output data. In this section, we consider the realization based on finite data. Suppose that we have a finite input-output data $\{u(t),\, y(t),\, t = 0,\, 1,\, \cdots,\, T\}$. Let the linear spaces generated by u and y be denoted by

$$\mathcal{U}_{[0,T]} = \text{span}\{u(t) \mid t = 0,\, 1,\, \cdots,\, T\}$$
$$\mathcal{Y}_{[0,T]} = \text{span}\{y(t) \mid t = 0,\, 1,\, \cdots,\, T\}$$

Also define the orthogonal complement of $\mathcal{U}_{[0,T]}$ on the joint space $\mathcal{U}_{[0,T]} \vee \mathcal{Y}_{[0,T]}$, which is denoted by $\mathcal{Z}_{[0,T]}$. Therefore, we have

$$\mathcal{U}_{[0,T]} \oplus \mathcal{Z}_{[0,T]} = \mathcal{U}_{[0,T]} \vee \mathcal{Y}_{[0,T]}$$

Lemma 9.8. *For the deterministic subsystem of (9.34), we define*

$$\hat{y}_d(t) := \hat{E}\{y_d(t) \mid \mathcal{U}_{[0,T]}\} = \hat{E}\{y(t) \mid \mathcal{U}_{[0,T]}\}$$

Then the projected output $\hat{y}_d(t)$ is described by the state space model

$$\hat{x}_d(t+1) = A_d\hat{x}_d(t) + B_d u(t) \tag{9.39a}$$
$$\hat{y}_d(t) = C_d\hat{x}_d(t) + D_d u(t) \tag{9.39b}$$
$$\hat{x}_d(0) = \hat{E}\{x_d(0) \mid \mathcal{U}_{[0,T]}\} \tag{9.39c}$$

where $\hat{x}_d(t) := \hat{E}\{x_d(t) \mid \mathcal{U}_{[0,T]}\}$. It should be noted that the above equation is the same as (9.34), but the initial condition $\hat{x}_d(0)$ is different from $x_d(0)$ as shown in (9.39c).

Proof. Since $\mathcal{U} \supset \mathcal{U}_{[0,T]}$, we have (9.39) by projecting (9.34) onto $\mathcal{U}_{[0,T]}$. □

Lemma 9.8 shows that we can identify the system matrices (A_d, B_d, C_d, D_d) based on the data $\{u(t),\ \hat{y}_d(t),\ t = 0, 1, \cdots, T\}$ by using the subspace method described in the next section. Moreover, from the identified matrices and (9.39), we have

$$\hat{y}_d(t) = C_d A_d^t \hat{x}_d(0) + \sum_{i=0}^{t-1} C_d A_d^{t-i-1} B_d u(i) + D_d u(t) \tag{9.40}$$

for $t = 0, 1, \cdots, T$. Since the estimates of (A_d, B_d, C_d, D_d) are given, we can obtain the estimate of the initial state vector $\hat{x}_d(0)$ of (9.39c) by applying the least-squares method to (9.40).

We turn our attention to realization of the stochastic subsystem. In this case, we need to compute the orthogonal projection of the output data onto the orthogonal complement $\mathcal{Z}_{[0,T]}$, which is written as

$$\hat{y}_s(t) := y(t) - \hat{E}\{y(t) \mid \mathcal{U}_{[0,T]}\}, \qquad t = 0, 1, \cdots, T \tag{9.41}$$

The next lemma clarifies the relation between the stochastic components y_s and $\hat{y}_s$ defined above.

Lemma 9.9. *Define the estimation error*

$$\tilde{y}_d(t) = y_d(t) - \hat{y}_d(t), \qquad t = 0, 1, \cdots, T$$

Then we have

$$\hat{y}_s(t) = y_s(t) + \tilde{y}_d(t), \qquad t = 0, 1, \cdots, T \tag{9.42}$$

Proof. We first note that

$$y(t) = y_s(t) + y_d(t), \qquad y_s(t) \perp \mathcal{U}$$

Since $\mathcal{U} \supset \mathcal{U}_{[0,T]}$, it follows from (9.41) that

$$\begin{aligned}\hat{y}_s(t) &= y_s(t) + y_d(t) - \hat{E}\{y_s(t) + y_d(t) \mid \mathcal{U}_{[0,T]}\} \\ &= y_s(t) + \Big(y_d(t) - \hat{E}\{y_d(t) \mid \mathcal{U}_{[0,T]}\}\Big) = y_s(t) + \tilde{y}_d(t)\end{aligned}$$

as was to be proved, where we used the fact that $\hat{E}\{y_s(t) \mid \mathcal{U}_{[0,T]}\} = 0$. □

Lemma 9.9 means that for a finite data case, the output $\hat{y}_s(t)$ projected onto the complement $\mathcal{Z}_{[0,T]}$ is different from the true stochastic component defined by $y_s(t) = \hat{E}\{y(t) \mid \mathcal{U}^{\perp}\}$, because the former is perturbed by the smoothed error $\tilde{y}_d(t)$ of the deterministic component.

Define the vector $\tilde{x}_d(t) := x_d(t) - \hat{x}_d(t)$. It then follows from (9.34) and (9.39) that

$$\tilde{x}_d(t+1) = A_d \tilde{x}_d(t), \qquad \tilde{y}_d(t) = C_d \tilde{x}_d(t)$$

so that the term acting on the stochastic component $y_s(t)$ is expressed as

$$\tilde{y}_d(t) = C_d A_d^t \tilde{x}_d(0), \qquad t = 0, 1, \cdots, T \tag{9.43}$$

Thus the estimation error $\tilde{x}_d(0)$ does influence on the stochastic component as well as the deterministic component. If we ignore the additive term $\tilde{y}_d(t)$ in (9.42), there are possibilities that we may identify stochastic subsystems with higher dimensions than the true order $\tilde{n}$. Hence it is desirable to filtering out this additive terms by some means. For more detail, see [30, 131].

9.7 Subspace Identification Method – ORT Method

In this section, we develop subspace methods for identifying state space models based on finite input-output data. In the sequel, the subspace method is called the ORT method, since the identification methods developed in this section are based on the orthogonal decomposition technique introduced in Section 9.4.

Suppose that the input-output data $\{u(t), y(t), t = 0, 1, \cdots, N + 2k - 2\}$ are given with N sufficiently large and $k > n$. Based on the input-output data, we define as usual block Hankel matrices as

$$U_{0|k-1} = \begin{bmatrix} u(0) & u(1) & \cdots & u(N-1) \\ u(1) & u(2) & \cdots & u(N) \\ \vdots & \vdots & \ddots & \vdots \\ u(k-1) & u(k) & \cdots & u(N+k-2) \end{bmatrix} \in \mathbb{R}^{km \times N}$$

and

$$U_{k|2k-1} = \begin{bmatrix} u(k) & u(k+1) & \cdots & u(k+N-1) \\ u(k+1) & u(k+2) & \cdots & u(k+N) \\ \vdots & \vdots & \ddots & \vdots \\ u(2k-1) & u(2k) & \cdots & u(N+2k-2) \end{bmatrix} \in \mathbb{R}^{km \times N}$$

Similarly, we define $Y_{0|k-1}, Y_{k|2k-1} \in \mathbb{R}^{kp \times N}$, and also

$$U_{0|2k-1} := \begin{bmatrix} U_{0|k-1} \\ U_{k|2k-1} \end{bmatrix}, \qquad Y_{0|2k-1} := \begin{bmatrix} Y_{0|k-1} \\ Y_{k|2k-1} \end{bmatrix}$$

9.7.1 Subspace Identification of Deterministic Subsystem

Consider the subspace identification of the deterministic subsystem of (9.39). We define the extended observability matrix as

$$\mathcal{O}_k = \begin{bmatrix} C_d \\ C_d A_d \\ \vdots \\ C_d A_d^{k-1} \end{bmatrix} \in \mathbb{R}^{kp \times n}, \qquad k > n$$

and the block lower triangular Toeplitz matrix as

$$\Psi_k(D_d, B_d) = \begin{bmatrix} D_d & & & & \\ C_d B_d & D_d & & & \\ C_d A_d B_d & C_d B_d & D_d & & \\ \vdots & \vdots & \ddots & \ddots & \\ C_d A_d^{k-2} B_d & \cdots & \cdots & C_d B_d & D_d \end{bmatrix} \in \mathbb{R}^{kp \times km}$$

By iteratively using (9.39a) and (9.39b), we obtain a matrix input-output equation of the form [see (6.23)]

$$\hat{Y}^d_{k|2k-1} = \mathcal{O}_k \hat{X}^d_k + \Psi_k U_{k|2k-1} \tag{9.44}$$

where $\hat{Y}^d_{k|2k-1}$ is the block Hankel matrix generated by $\hat{y}_d$ and $\hat{X}^d_k$ is defined by

$$\hat{X}^d_k = [\hat{x}_d(k) \quad \hat{x}_d(k+1) \quad \cdots \quad \hat{x}_d(k+N-1)]$$

We need the following assumption for applying the subspace method to the system of (9.44); see Section 6.3.

Assumption 9.3. *A1)* $\operatorname{rank}(\hat{X}^d_k) = n$.
A2) $\operatorname{rank}(U_{0|2k-1}) = 2km$.
A3) $\operatorname{span}(\hat{X}^d_k) \cap \operatorname{span}(U_{k|2k-1}) = \{0\}$. □

As mentioned in Section 6.3, Assumption A1) implies that the state vectors $\hat{x}_d(k), \hat{x}_d(k+1), \cdots$ are sufficiently excited. The condition A2) implies that the input process u has the PE condition of order $2k$, and A3) is guaranteed by the feedback-free condition.

First we present a method of computing the deterministic component of the output process. According to the idea of the MOESP method of Section 6.5, we consider the LQ decomposition:

$$\begin{bmatrix} U_{0|2k-1} \\ Y_{0|2k-1} \end{bmatrix} = \begin{bmatrix} R_{11} & 0 \\ R_{21} & R_{22} \end{bmatrix} \begin{bmatrix} \bar{Q}_1^{\mathrm{T}} \\ \bar{Q}_2^{\mathrm{T}} \end{bmatrix} \tag{9.45}$$

where $R_{11} \in \mathbb{R}^{2km \times 2km}$, $R_{22} \in \mathbb{R}^{2kp \times 2kp}$ are block lower triangular matrices, and $\bar{Q}_1 \in \mathbb{R}^{N \times 2km}$, $\bar{Q}_2 \in \mathbb{R}^{N \times 2kp}$ are orthogonal matrices. It follows from A2) of Assumption 9.3 that R_{11} is nonsingular, so that we have

$$Y_{0|2k-1} = R_{21}\bar{Q}_1^{\mathrm{T}} + R_{22}\bar{Q}_2^{\mathrm{T}} = R_{21}R_{11}^{-1}U_{0|2k-1} + R_{22}\bar{Q}_2^{\mathrm{T}}$$

We see that $R_{21}\bar{Q}_1^{\mathrm{T}}$ belongs to the row space of $U_{0|2k-1}$, and $R_{22}\bar{Q}_2^{\mathrm{T}}$ is orthogonal to it since $\bar{Q}_1^{\mathrm{T}}\bar{Q}_2 = 0$. Hence, $R_{21}\bar{Q}_1^{\mathrm{T}}$ is the orthogonal projection of $Y_{0|2k-1}$ onto the rowspace $U_{0|2k-1}$. Thus, the deterministic component is given by

$$\hat{Y}^d_{0|2k-1} = R_{21}\bar{Q}_1^{\mathrm{T}} = R_{21}R_{11}^{-1}U_{0|2k-1} \tag{9.46}$$

and hence the stochastic component is

$$\hat{Y}^s_{0|2k-1} := Y_{0|2k-1} - \hat{Y}^d_{0|2k-1} = R_{22}\bar{Q}_2^{\mathrm{T}} \tag{9.47}$$

Bearing the above facts in mind, we consider a related LQ decomposition

$$\begin{bmatrix} U_{k|2k-1} \\ U_{0|k-1} \\ Y_{0|k-1} \\ Y_{k|2k-1} \end{bmatrix} = \begin{bmatrix} L_{11} & 0 & 0 & 0 \\ L_{21} & L_{22} & 0 & 0 \\ L_{31} & L_{32} & L_{33} & 0 \\ L_{41} & L_{42} & L_{43} & L_{44} \end{bmatrix} \begin{bmatrix} Q_1^{\mathrm{T}} \\ Q_2^{\mathrm{T}} \\ Q_3^{\mathrm{T}} \\ Q_4^{\mathrm{T}} \end{bmatrix} \tag{9.48}$$

where $L_{11}, L_{22} \in \mathbb{R}^{km \times km}$, $L_{33}, L_{44} \in \mathbb{R}^{kp \times kp}$ are block lower triangular matrices, and $Q_1, Q_2 \in \mathbb{R}^{N \times km}$, $Q_3, Q_4 \in \mathbb{R}^{N \times kp}$ are orthogonal matrices. In view of (9.46) and (9.48), the deterministic component $\hat{Y}^d_{k|2k-1}$ is given by

$$\hat{Y}^d_{k|2k-1} = L_{41}Q_1^{\mathrm{T}} + L_{42}Q_2^{\mathrm{T}}$$

On the other hand, from (9.44) and (9.48), we have

$$\Psi_k L_{11} Q_1^{\mathrm{T}} + \mathcal{O}_k \hat{X}^d_k = L_{41}Q_1^{\mathrm{T}} + L_{42}Q_2^{\mathrm{T}} \tag{9.49}$$

Post-multiplying (9.49) by Q_2 yields $\mathcal{O}_k \hat{X}^d_k Q_2 = L_{42}$. Since $\hat{X}^d_k Q_2$ has full row rank from A1) of Assumption 9.3, we have

$$\mathrm{Im}(\mathcal{O}_k) = \mathrm{Im}(L_{42}) \tag{9.50}$$

Similarly, pre-multiplying (9.49) by a matrix $(\mathcal{O}_k^{\perp})^{\mathrm{T}}$ satisfying $(\mathcal{O}_k^{\perp})^{\mathrm{T}}\mathcal{O}_k = 0$, and post-multiplying by Q_1 yield

$$(\mathcal{O}_k^{\perp})^{\mathrm{T}} L_{41} = (\mathcal{O}_k^{\perp})^{\mathrm{T}} \Psi_k(B_d, D_d) L_{11} \tag{9.51}$$

Making use of (9.50) and (9.51), we can derive a subspace method of identifying the deterministic subsystem. In the following, we assume that LQ decomposition of (9.48) is given.

Subspace Identification of Deterministic Subsystem – ORT Method

Step 1: Compute the SVD of L_{42}:

$$L_{42} = [\hat{U} \ \ \bar{U}] \begin{bmatrix} \hat{S} & 0 \\ 0 & \bar{S} \end{bmatrix} \begin{bmatrix} \hat{V}^{\mathrm{T}} \\ \bar{V}^{\mathrm{T}} \end{bmatrix} \simeq \hat{U}\hat{S}\hat{V}^{\mathrm{T}} \tag{9.52}$$

where $\hat{S}$ is obtained by neglecting smaller singular values, so the dimension of the state vector equals $\dim(\hat{S})$. Thus, the decomposition

$$\hat{U}\hat{S}\hat{V}^{\mathrm{T}} = (\hat{U}\hat{S}^{1/2})(\hat{S}^{1/2}\hat{V}^{\mathrm{T}})$$

gives the extended observability matrix $\mathcal{O}_k = \hat{U}\hat{S}^{1/2}$.

Step 2: Compute the estimates of A_d and C_d by

$$A_d = \mathcal{O}_{k-1}^{\dagger}\overline{\mathcal{O}}_k, \qquad C_d = \mathcal{O}_k(1:p,:) \tag{9.53}$$

where $\overline{\mathcal{O}}_k$ denotes the matrix obtained by deleting the first p rows from $\mathcal{O}_k$.

Step 3: Given the estimates of A_d and C_d, the Toeplitz matrix Ψ_k becomes linear with respect to B_d and D_d. By using $\bar{U}^{\mathrm{T}}$ of (9.52) for $(\mathcal{O}_k^{\perp})^{\mathrm{T}}$, it follows from (9.51) that

$$\bar{U}^{\mathrm{T}} L_{41} L_{11}^{-1} = \bar{U}^{\mathrm{T}} \Psi_k(B_d, D_d) \tag{9.54}$$

Then we can obtain the least-squares estimates of B_d and D_d by rewriting the above equation as (6.44) in the MOESP algorithm.

Remark 9.3. In [171] (Theorems 2 and 4), the LQ decomposition of (9.48) is used to develop the PO-MOESP algorithm. More precisely, the following relations

$$\mathrm{Im}(\mathcal{O}_k) = \mathrm{Im}\,[L_{42} \;\; L_{43}]$$

and

$$\bar{U}^{\mathrm{T}}[L_{31} \;\; L_{32} \;\; L_{41}] = \bar{U}^{\mathrm{T}} \Psi_k(B_d, D_d)[L_{21} \;\; L_{22} \;\; L_{11}]$$

are employed. We see that these relations are different from (9.50) and (9.51); this is a point where the ORT method is different from the PO-MOESP method. □

9.7.2 Subspace Identification of Stochastic Subsystem

We derive a subspace identification algorithm for the stochastic subsystem. For data matrices, we use the same notation as in Subsection 9.7.1. As shown in (9.47), the stochastic component is given by

$$\hat{Y}^s_{0|2k-1} = Y_{0|2k-1} - \hat{Y}^d_{0|2k-1}$$

It follows from (9.48) that

$$\hat{Y}^s_{0|2k-1} = \begin{bmatrix} L_{33} & 0 \\ L_{43} & L_{44} \end{bmatrix} \begin{bmatrix} Q_3^{\mathrm{T}} \\ Q_4^{\mathrm{T}} \end{bmatrix} \tag{9.55}$$

so that we define

$$\tilde{Y}_{0|k-1} = L_{33} Q_3^{\mathrm{T}}, \qquad \tilde{Y}_{k|2k-1} = L_{43} Q_3^{\mathrm{T}} + L_{44} Q_4^{\mathrm{T}}$$

The sample covariance matrices of stochastic components are then given by

$$\begin{bmatrix} \Sigma_{pp} & \Sigma_{pf} \\ \Sigma_{fp} & \Sigma_{ff} \end{bmatrix} = \frac{1}{N} \begin{bmatrix} \tilde{Y}_{0|k-1} \\ \tilde{Y}_{k|2k-1} \end{bmatrix} \left[(\tilde{Y}_{0|k-1})^{\mathrm{T}} \;\; (\tilde{Y}_{k|2k-1})^{\mathrm{T}} \right]$$

Thus we have

$$\Sigma_{fp} = \frac{1}{N} L_{43} L_{33}^{\mathrm{T}}, \quad \Sigma_{ff} = \frac{1}{N}\left(L_{43} L_{43}^{\mathrm{T}} + L_{44} L_{44}^{\mathrm{T}} \right), \quad \Sigma_{pp} = \frac{1}{N} L_{33} L_{33}^{\mathrm{T}}$$

The following algorithms are based on the stochastic subspace identification algorithms derived in Section 8.7.

Subspace Identification of Stochastic Subsystem – ORT Method
Algorithm A

Step 1: Compute square roots of covariance matrices Σ_{ff} and Σ_{pp} satisfying

$$\Sigma_{ff} = LL^{\mathrm{T}}, \qquad \Sigma_{pp} = MM^{\mathrm{T}}$$

Step 2: Compute the SVD of the normalized covariance matrix

$$L^{-1}\Sigma_{fp}M^{-\mathrm{T}} = U\Sigma V^{\mathrm{T}} \simeq \tilde{U}\tilde{\Sigma}\tilde{V}^{\mathrm{T}}$$

where $\tilde{\Sigma}$ is obtained by deleting sufficiently small singular values of Σ, so that the dimension of the state vector becomes $\tilde{n} = \dim \tilde{\Sigma}$.

Step 3: Compute the extended observability and reachability matrices by

$$\tilde{\mathcal{O}}_k = L\tilde{U}\tilde{\Sigma}^{1/2}, \qquad \tilde{\mathcal{C}}_k = \tilde{\Sigma}^{1/2}\tilde{V}^{\mathrm{T}}M^{\mathrm{T}}$$

Step A4: Compute the estimates of A_s, C_s and $\bar{C}_s$ by

$$A_s = \tilde{\mathcal{O}}_{k-1}^{\dagger}\overline{\mathcal{O}}_k, \qquad C_s = \tilde{\mathcal{O}}_k(1:p,:), \qquad \bar{C}_s^{\mathrm{T}} = \tilde{\mathcal{C}}_k(:,1:p)$$

where $\overline{\mathcal{O}}_k = \tilde{\mathcal{O}}_k(p+1:kp,:)$.

Step A5: The covariance matrix of y_s is given by $\Lambda_s(0) := \Sigma_{ff}(1:p,1:p)$. By using $(A_s,\, C_s,\, \bar{C}_s,\, \Lambda_s(0))$, the Kalman gain is given by

$$K_s = (\bar{C}_s^{\mathrm{T}} - A_s\tilde{\Sigma}C_s^{\mathrm{T}})(\Lambda_s(0) - C_s\tilde{\Sigma}C_s^{\mathrm{T}})^{-1}$$

so that the innovation model becomes

$$x_s(t+1) = A_s x_s(t) + K_s e(t)$$
$$y_s(t) = C_s x_s(t) + e(t)$$

where $\mathrm{var}\{e(t)\} = \Lambda_s(0) - C_s\tilde{\Sigma}C_s^{\mathrm{T}}$.

Now we present an algorithm that gives a state space model satisfying the positivity condition, where *Steps* 1–3 are the same as those of Algorithm A.

Algorithm B

Step B4: Compute the estimates of state vectors by

$$\tilde{X}_k = \tilde{\Sigma}^{1/2}\tilde{V}^{\mathrm{T}}M^{-1}\tilde{Y}_{0|k-1} \in \mathbb{R}^{n\times N}$$

and define matrices with $N-1$ columns

$$\hat{X}_{k+1} = \tilde{X}_k(:,2:N), \quad \hat{X}_k = \tilde{X}_k(:,1:N-1), \quad \hat{Y}_{k|k}^s = Y_{k|k}^s(:,1:N-1)$$

Step B5: Compute the estimates of matrices A_s and C_s applying the least-squares method to the overdetermined equations

$$\begin{bmatrix} \hat{X}_{k+1} \\ \hat{Y}^s_{k|k} \end{bmatrix} = \begin{bmatrix} A_s \\ C_s \end{bmatrix} \hat{X}_k + \begin{bmatrix} \rho_w \\ \rho_v \end{bmatrix}$$

where ρ_w and ρ_v are residuals.

Step B6: Compute the covariance matrices of residuals

$$\begin{bmatrix} \hat{Q} & \hat{S} \\ \hat{S}^{\mathrm{T}} & \hat{R} \end{bmatrix} = \frac{1}{N-1} \begin{bmatrix} \rho_w \rho_w^{\mathrm{T}} & \rho_w \rho_v^{\mathrm{T}} \\ \rho_v \rho_w^{\mathrm{T}} & \rho_v \rho_v^{\mathrm{T}} \end{bmatrix}$$

and solve the ARE [see (5.67)]

$$P = A_s P A_s^{\mathrm{T}} - (A_s P C_s^{\mathrm{T}} + \hat{S})(C_s P C_s^{\mathrm{T}} + \hat{R})^{-1}(A_s P C_s^{\mathrm{T}} + \hat{S})^{\mathrm{T}} + \hat{Q}$$

In terms of the stabilizing solution $P \geq 0$, we have the Kalman gain

$$K_s = (A_s P C_s^{\mathrm{T}} + \hat{S})(C_s P C_s^{\mathrm{T}} + \hat{R})^{-1}$$

Step B7: The innovation model is then given by

$$\hat{x}(t+1) = A_s \hat{x}(t) + K_s \hat{e}(t)$$
$$y(t) = C_s \hat{x}(t) + \hat{e}(t)$$

where $\mathrm{var}\{\hat{e}(t)\} = C_s P C_s^{\mathrm{T}} + \hat{R}$.

The above algorithm is basically the same as Algorithm B of stochastic balanced realization developed in Section 8.7. We see that since the covariance matrices of residuals obtained in *Step B6* is always nonnegative definite and since (C_s, A_s) is observable, the ARE has a unique stabilizing solution, from which we can compute the Kalman gain. Thus the present algorithm ensures the positivity condition of the stochastic subsystem; see also Remark 8.2.

9.8 Numerical Example

Some numerical results obtained by using the ORT and PO-MOESP methods are presented. A simulation model used is depicted in Figure 9.7, where the plant is a 5th-order model [175]

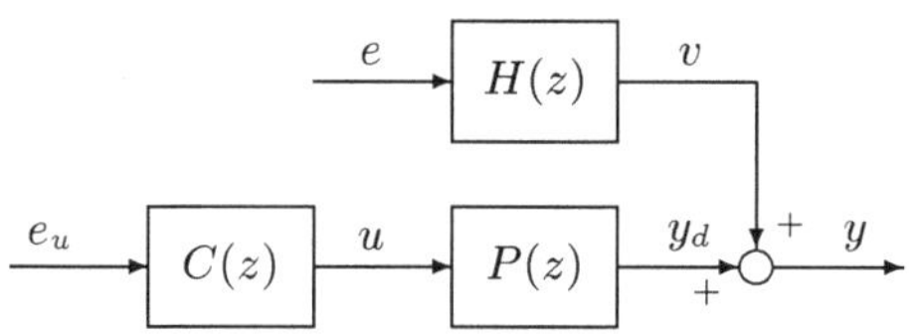

Figure 9.7. Simulation model

$$P(z) = \frac{0.0275z^{-4} + 0.0551z^{-5}}{1 - 2.3443z^{-1} + 3.081z^{-2} - 2.5274z^{-3} + 1.2415z^{-4} - 0.3686z^{-5}}$$

and where the input signal generation model $C(z)$ and the noise model $H(z)$ are respectively given by

$$C(z) = \frac{\sqrt{1-a^2}}{1-az^{-1}}, \qquad H(z) = \frac{1 - 0.2z^{-1} - 0.48z^{-2}}{1 + 0.4z^{-1} + 0.4z^{-2}}$$

The noises e and e_u are mutually uncorrelated white noises with $\mathcal{N}(0, \sigma^2)$ and $\mathcal{N}(0, 1)$, respectively. Thus the spectral density functions of u and v are given by

$$\Phi_u(\omega) = |C(e^{j\omega})|^2, \qquad \Phi_v(\omega) = \sigma^2 |H(e^{j\omega})|^2$$

so that their powers are proportional to the gain plots of $C(z)$ with $a = 0.9$ and $H(z)$ shown in Figure 9.8, respectively.

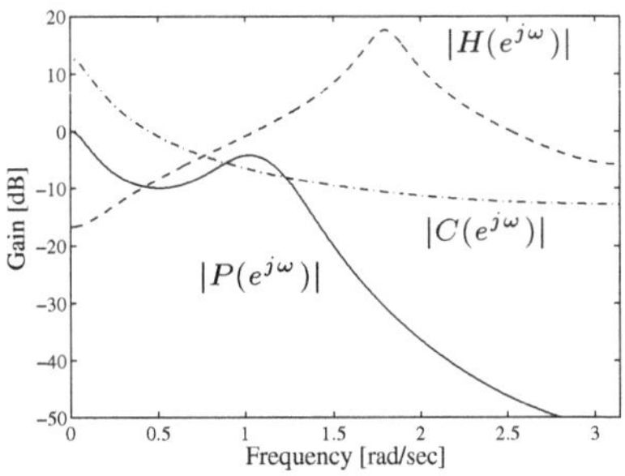

Figure 9.8. The Bode gain plots of $P(z)$, $C(z)$ and $H(z)$

In the present simulations, we consider the four cases according as u and/or v are white noises or colored noises, where the variance σ^2 of noise e is adjusted so that the variance of v becomes approximately $\sigma_v^2 = 0.01$. Also, the variance of the output $y_d = P(z)u$ changes according to the spectrum of u, so that the S/N ratio in the output becomes as

$$\sigma_d^2 / \sigma_v^2 = \begin{cases} 10.73 & \text{white noise } (a = 0) \\ 51.43 & \text{colored noise } (a = 0.9) \end{cases}$$

In each simulation, we take the number of data points $N = 1000$ and the number of block rows $k = 15$. We generated 30 data sets by using pseudo-random numbers. In order to compare simulation results, we have used the same pseudo-random numbers in each case.

Case 1: We show the simulation results for the case where the input signal u is a white noise. Figure 9.9 displays the Bode gain plots of the plant obtained by applying the ORT method, where v is a white noise in Figure 9.9(a), but is a colored noise in

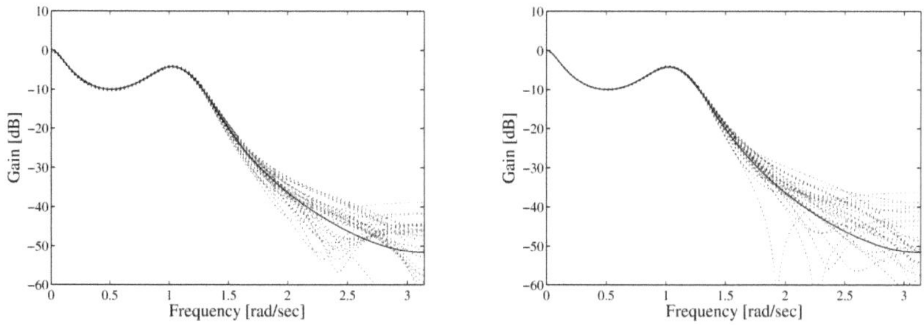

(a) (u = white, v = white) (b) (u = white, v = colored)

Figure 9.9. Identification results by ORT method

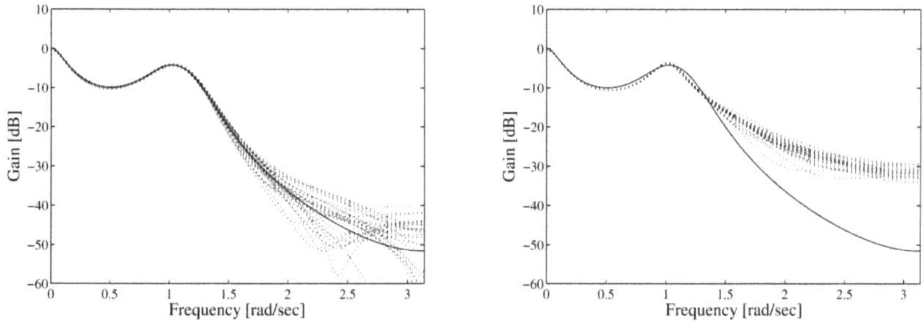

(a) (u = white, v = white) (b) (u = white, v = colored)

Figure 9.10. Identification results by PO-MOESP method

Figure 9.9(b), and where the real line denotes the Bode gain of the true plant. Also, Figure 9.10 displays the corresponding results by the PO-MOESP method. We see that there is no clear difference in the results of Figures 9.9(a) and 9.10(a). But, there are some differences in the results of Figures 9.9(b) and 9.10(b), in that the ORT method gives a slightly better result in the high frequency range and we observe a small bias in the low frequency range between 0.5 to 1 (rad) in the estimates by the PO-MOESP method.

Case 2: We consider the case where u is a colored noise. Figure 9.11 displays the results obtained by applying the ORT, and Figure 9.12 those obtained by using the PO-MOESP. Since the power of the input u decreases in the high frequency range as shown in Figure 9.8, the accuracy of the estimate degrades in the high frequency range.

Though the output S/N ratio for colored noise input is higher than for white noise input, the accuracy of identification is inferior to the white noise case. There are no clear difference in Figure 9.11(a) and 9.12(a) for white observation noise, but if v is a colored noise, there exist some appreciable differences in the results of the ORT and PO-MOESP methods in the low frequency range as well as in high frequency range as shown in Figures 9.11(b) and 9.12(b).

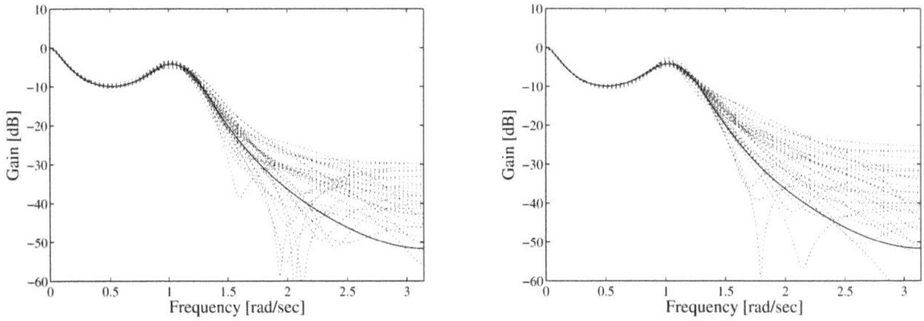

(a) (u = colored, v = white) (b) (u = colored, v = colored)

Figure 9.11. Identification results by ORT method

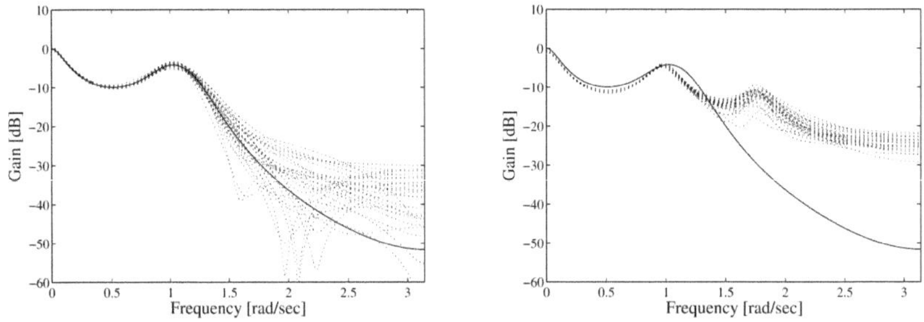

(a) (u = colored, v = white) (b) (u = colored, v = colored)

Figure 9.12. Identification results by PO-MOESP method

A reason why there is a considerable difference in the results of the ORT and PO-MOESP in the case of colored observation noise is that the noise v has a large power in the frequency range greater than 1.5 (rad) where the input u has a low power. In fact, the peak of the gain characteristic shown in Figure 9.12(b) is located around 1.8 (rad), which is the same as the peak of noise spectrum of Figure 9.8. In the PO-MOESP method, the frequency components beyond 1.5 (rad) are erroneously identified as those due to the input signal u rather than the noise v. On the other hand, in the ORT method, the data is first decomposed into the deterministic and stochastic components based on the preliminary orthogonal decomposition, and it seems that this decomposition is performed very well even if the noise is colored as long as the exogenous input satisfies a PE condition with sufficiently high order. For differences between the two algorithms; see Remark 9.3 and programs listed in Tables D.5 and D.7.

Further simulation studies showing a superiority of the ORT method based on the preliminary orthogonal decomposition are provided in Section 10.7, in which results of three subspace identification methods will be compared.

9.9 Notes and References

- Based on Picci and Katayama [130, 131], we have presented realization results for the stochastic system with exogenous inputs under the assumption that there is no feedback from the output to the input and the input satisfies a sufficiently high PE condition. The main idea is to decompose the output process into the deterministic and stochastic components, from which we have derived a state space model with a natural block structure.
- In Section 9.1, projections in a Hilbert space are reviewed and the property of conditional orthogonality is introduced. Then Section 9.2 has formulated the stochastic realization problem in the presence of exogenous inputs, and Section 9.3 discussed the feedback-free conditions in detail based on [24, 25, 53].
- In Section 9.4, we have considered the PE condition of the exogenous inputs, and developed a method of decomposing the output process into the deterministic component and the stochastic component. In Section 9.5, we have shown that a desired state space model can be obtained by combining realizations for the deterministic and stochastic components, resulting in a state space model with a block structure in which the plant and the noise model have independent parametrizations.
- In Section 9.6, a theoretical analysis of deterministic and stochastic realization methods based on finite input-output data is made, and in Section 9.7, the ORT method of identifying the deterministic and stochastic subsystems are developed by using the LQ decomposition and the SVD. Since the present algorithms are derived from the basic stochastic realization methods in the presence of exogenous inputs, they are different from those of MOESP [171–173] and N4SID [164, 165]. Some numerical results are included in Section 9.8; for further simulation studies, see [29, 30], [93], and for theoretical analyses of ill-conditioning of subspace estimates, see [31, 32].
- In Section 9.10, proofs of Theorem 9.1 and Lemma 9.7 are included.

9.10 Appendix: Proofs of Theorem and Lemma

9.10.1 Proof of Theorem 9.1

1° Proof of (i) → (ii). Rewriting (9.8), we have

$$\begin{bmatrix} y \\ u \end{bmatrix} = \begin{bmatrix} A(z) & B(z) \\ 0 & D(z) \end{bmatrix} \begin{bmatrix} \nu \\ \eta \end{bmatrix}$$

where ν and η are zero mean white noise vectors with covariance matrices $\bar{Q}_1$ and $\bar{Q}_2$, respectively. By assumption, $\Gamma(z)$ is stable, so are $A(z)$, $B(z)$ and $D(z)$. Also, we have

$$\Gamma^{-1}(z) = \begin{bmatrix} A^{-1}(z) & -A^{-1}(z)B(z)D^{-1}(z) \\ 0 & D^{-1}(z) \end{bmatrix}$$

Also, $\Gamma^{-1}(z)$ is stable, so are $A^{-1}(z)$ and $D^{-1}(z)$. Thus, in particular, $D(z)$ is of minimal phase. Hence, we see that $u = D(z)\eta$ is the innovation representation for u, and η is the innovation process for u.

Define

$$\mathcal{H} = \overline{\text{span}}\{\eta(\tau) \mid \tau = 0, \pm 1, \cdots\}, \qquad \mathcal{H}_t^- = \overline{\text{span}}\{\eta(\tau) \mid \tau < t\}$$

Since $D(z)$ is of minimal phase, we have $\mathcal{U} = \mathcal{H}$ and $\mathcal{U}_t^- = \mathcal{H}_t^-$. Moreover, we get

$$\begin{aligned}\hat{E}\left\{\sum_{i=0}^{\infty} B_i\eta(t-i) \,\middle|\, \mathcal{H}\right\} &= \sum_{i=0}^{\infty} B_i\eta(t-i) \\ &= \hat{E}\left\{\sum_{i=0}^{\infty} B_i\eta(t-i) \,\middle|\, \mathcal{H}_{t+1}^-\right\}\end{aligned}$$

Noting that $\nu \perp \mathcal{H}$, we have

$$\begin{aligned}\hat{E}\{y(t) \mid \mathcal{U}\} &= \hat{E}\left\{\sum_{i=0}^{\infty} A_i\nu(t-i) + \sum_{i=0}^{\infty} B_i\eta(t-i) \,\middle|\, \mathcal{H}\right\} \\ &= \sum_{i=0}^{\infty} B_i\eta(t-i) \\ &= \hat{E}\left\{\sum_{i=0}^{\infty} A_i\nu(t-i) + \sum_{i=0}^{\infty} B_i\eta(t-i) \,\middle|\, \mathcal{H}_{t+1}^-\right\} \\ &= \hat{E}\{y(t) \mid \mathcal{U}_{t+1}^-\}\end{aligned}$$

as was to be proved.

2° Proof of (ii) → (iii). From (ii), it follows that

$$y_d(t) := \hat{E}\{y(t) \mid \mathcal{U}_{t+1}^-\} = \sum_{i=0}^{\infty} B_i\eta(t-i) \quad \Rightarrow \quad y_d = B(z)\eta$$

so that from $u = D(z)\eta$, we have

$$y_d = B(z)D^{-1}(z)u =: K(z)u$$

Also, from the stability of $B(z)$ and the minimal phase property of $D(z)$, we see that $K(z)$ is stable.

We define $\zeta(t) := y(t) - \hat{E}\{y(t) \mid \mathcal{U}_{t+1}^-\}$. Then,

$$y(t) = \sum_{i=0}^{\infty} K_i u(t-i) + \zeta(t)$$

We show that $\zeta(t)$ is orthogonal to $\mathcal{U}$. Since $\hat{E}\{y(t) \mid \mathcal{U}_{t+1}^-\} = \hat{E}\{y(t) \mid \mathcal{U}\}$ and $\mathcal{U}_{t+1}^- \subset \mathcal{U}_{t+h}^- \subset \mathcal{U}$, $h = 1, 2, \cdots$, we get

$$\hat{E}\{y(t) \mid \mathcal{U}_{t+h}^-\} = \hat{E}\{y(t) \mid \mathcal{U}\}, \qquad h = 1, 2, \cdots$$

It therefore follows that

$$\begin{aligned}
&\hat{E}\left\{\sum_{i=0}^{\infty} K_i u(t-i) + \zeta(t) \,\middle|\, \mathcal{U}_{t+1}^-\right\} \\
&\qquad = \hat{E}\{y(t) \mid \mathcal{U}_{t+1}^-\} \\
&\qquad = \hat{E}\{y(t) \mid \mathcal{U}_{t+h}^-\} \\
&\qquad = \hat{E}\left\{\sum_{i=0}^{\infty} K_i u(t-i) + \zeta(t) \,\middle|\, \mathcal{U}_{t+h}^-\right\}
\end{aligned} \tag{9.56}$$

where $h = 1, 2, \cdots$. Note that

$$\begin{aligned}
\hat{E}\left\{\sum_{i=0}^{\infty} K_i u(t-i) \,\middle|\, \mathcal{U}_{t+1}^-\right\} &= \sum_{i=0}^{\infty} K_i u(t-i) \\
&= \hat{E}\left\{\sum_{i=0}^{\infty} K_i u(t-i) \,\middle|\, \mathcal{U}_{t+h}^-\right\}
\end{aligned}$$

Thus it follows from (9.56) that

$$\hat{E}\{\zeta(t) \mid \mathcal{U}_{t+h}^-\} = \hat{E}\{\zeta(t) \mid \mathcal{U}_{t+1}^-\} = 0, \qquad h = 1, 2, \cdots$$

where $\zeta(t) \perp \mathcal{U}_{t+1}^-$ is used. However, since $\mathcal{U}_{t+h}^- \subset \mathcal{U}_{t+1}^-$, $h = 0, -1, \cdots$, we have $\zeta(t) \perp \mathcal{U}_{t+h}^-$, $h = 0, -1, \cdots$. Thus it follows that

$$\hat{E}\{\zeta(t) \mid \mathcal{U}_{t+h}^-\} = 0, \qquad h = 0, \pm 1, \cdots$$

This implies that ζ is orthogonal to $\mathcal{U}$.

Recall that $\zeta := y - y_d$ is defined by the difference of two regular full rank stationary processes, so is ζ. Let the innovation representation of ζ be

$$\zeta(t) = \sum_{i=0}^{\infty} L_i \nu(t-i), \quad L_0 = I_p$$

where $\nu \in \mathbb{R}^p$ is a zero mean white noise vector with covariance matrix $\bar{Q}_1$, and where $L(z) := \sum_{i=0}^{\infty} L_i z^{-i}$ has full rank and minimal phase. Hence, y can be expressed as (9.10), where ν is uncorrelated with u due to the fact that $\zeta \perp u$. Since $K(z)$ is stable, and since $L(z)$ is of minimal phase, and $L(z)$ has full rank, the proof of (iii) is completed.

3° Proof of (iii) $\to$ (iv). Let the Hilbert space generated by the past of ζ be given by $\mathcal{Z}_t^- = \overline{\text{span}}\{\zeta(\tau) \mid \tau < t\}$. Since $\nu \perp u$,

$$\mathcal{U}_t^- \vee \mathcal{Y}_t^- = \mathcal{U}_t^- \oplus \mathcal{Z}_t^-$$

From (iii), it follows that $u(t+h) \perp \mathcal{Z}_t^-$, $h = 0, 1, \cdots$, so that

$$\begin{aligned}\hat{E}\{u(t+h) \mid \mathcal{U}_t^- \vee \mathcal{Y}_t^-\} &= \hat{E}\{u(t+h) \mid \mathcal{U}_t^- \oplus \mathcal{Z}_t^-\} \\ &= \hat{E}\{u(t+h) \mid \mathcal{U}_t^-\}, \qquad h = 0, 1, \cdots\end{aligned}$$

This implies that (9.12) holds.

4° Proof of (iv) → (i). We define $\zeta(t) = y(t) - \hat{E}\{y(t) \mid \mathcal{U}_{t+1}^-\}$. From (iv), we have

$$\begin{aligned}\hat{E}\{u(t+h) \mid \mathcal{U}_t^-\} &= \hat{E}\{u(t+h) \mid \mathcal{U}_t^- \vee \mathcal{Y}_t^-\} \\ &= \hat{E}\{u(t+h) \mid \mathcal{U}_t^- \oplus \mathcal{Z}_t^-\}, \qquad h = 0, 1, \cdots\end{aligned}$$

Thus it follows that $\hat{E}\{u(t+h) \mid \mathcal{Z}_t^-\} = 0$, $h = 0, 1, \cdots$. By the definition of $\zeta(t)$, $\hat{E}\{u(t+h) \mid \mathcal{Z}_t^-\} = 0$, $h = -1, -2, \cdots$ holds, implying that u is orthogonal to ζ.

Let the innovation representation of ζ be given by

$$\zeta(t) = \sum_{i=0}^{\infty} A_i \nu(t-i), \qquad A_0 = I_p$$

where ν is a zero mean white noise vector with covariance matrix $\bar{Q}_1$. Hence we have

$$\begin{aligned}y(t) &= \zeta(t) + \hat{E}\{y(t) \mid \mathcal{U}_{t+1}^-\} \\ &= \sum_{i=0}^{\infty} A_i \nu(t-i) + \sum_{i=0}^{\infty} K_i u(t-i)\end{aligned}$$

Let the innovation representation of the stationary process u be given by

$$u(t) = \sum_{i=0}^{\infty} D_i \eta(t-i), \qquad t = 0, \pm 1, \cdots \tag{9.57}$$

where $\eta \in \mathbb{R}^m$ is a zero mean white noise vector with covariance matrix $\bar{Q}_2$, where $D(z) = \sum_{i=0}^{\infty} D_i z^{-i}$ is of minimal phase. It follows from (9.57) that $u = D(z)\eta$, so that

$$y(t) = \sum_{i=0}^{\infty} A_i \nu(t-i) + \sum_{i=0}^{\infty} B_i \eta(t-i) \tag{9.58}$$

where $B(z) = K(z)D(z)$. From (9.57) and (9.58), $\Gamma(z)$ has a block upper triangular structure, implying that there is no feedback from y to u.

9.10.2 Proof of Lemma 9.7

First we show that $\mathcal{X}_t^{+/-}$ is an oblique splitting subspace for $(\bar{\mathcal{Y}}_t^+, \mathcal{U}_t^-)$. Clearly, from (9.27), any $\bar{y} \in \bar{\mathcal{Y}}_t^+$ is expressed as

$$\bar{y} = \hat{y} + \xi, \qquad \hat{y} \in \hat{\mathcal{Y}}_t^+, \qquad \xi \in \mathcal{X}_t^{+/-}$$

Since $\mathcal{X}_t^{+/-}$ is an oblique splitting subspace for $(\hat{\mathcal{Y}}_t^+, \mathcal{U}_t^-)$, and since $\xi \in \mathcal{X}_t^{+/-}$ is included in $\mathcal{U}_t^-$, we have

$$\begin{aligned}
\hat{E}_{\|\mathcal{U}_t^+}\{\bar{y} \mid \mathcal{U}_t^-\} &= \hat{E}_{\|\mathcal{U}_t^+}\{\hat{y} \mid \mathcal{U}_t^-\} + \hat{E}_{\|\mathcal{U}_t^+}\{\xi \mid \mathcal{U}_t^-\} \\
&= \hat{E}_{\|\mathcal{U}_t^+}\{\hat{y} \mid \mathcal{X}_t^{+/-}\} + \hat{E}_{\|\mathcal{U}_t^+}\{\xi \mid \mathcal{X}_t^{+/-}\} \\
&= \hat{E}_{\|\mathcal{U}_t^+}\{\bar{y} \mid \mathcal{X}_t^{+/-}\}
\end{aligned}$$

Hence

$$\begin{aligned}
\hat{E}_{\|\mathcal{U}_t^+}\{\bar{\mathcal{Y}}_t^+ \mid \mathcal{U}_t^-\} &= \overline{\text{span}}\left(\hat{E}_{\|\mathcal{U}_t^+}\{\bar{y} \mid \mathcal{U}_t^-\} \,\middle|\, \bar{y} \in \bar{\mathcal{Y}}_t^+\right) \\
&= \overline{\text{span}}\left(\hat{E}_{\|\mathcal{U}_t^+}\{\bar{y} \mid \mathcal{X}_t^{+/-}\} \,\middle|\, \bar{y} \in \bar{\mathcal{Y}}_t^+\right) \\
&= \hat{E}_{\|\mathcal{U}_t^+}\{\bar{\mathcal{Y}}_t^+ \mid \mathcal{X}_t^{+/-}\}
\end{aligned}$$

This proves the first statement.

From (9.26) and (9.27), we have $\mathcal{X}_t^{+/-} \subset \mathcal{U}_t^-$ and $\mathcal{X}_t^{+/-} \subset \bar{\mathcal{Y}}_t^+$, so that $\mathcal{X}_t^{+/-} \subset (\bar{\mathcal{Y}}_t^+ \cap \mathcal{U}_t^-)$ holds. Conversely, suppose that $\eta \in \bar{\mathcal{Y}}_t^+ \cap \mathcal{U}_t^-$ holds. Then, we have

$$\eta \in \bar{\mathcal{Y}}_t^+ = \hat{\mathcal{Y}}_t^+ \vee \mathcal{X}_t^{+/-}, \qquad \eta \in \mathcal{U}_t^-$$

Let $\eta = \eta_1 + \eta_2$ where $\eta_1 \in \hat{\mathcal{Y}}_t^+$ and $\eta_2 \in \mathcal{X}_t^{+/-} \subset \mathcal{U}_t^-$. Since $\eta \in \mathcal{U}_t^-$, we have $\eta_1 = \eta - \eta_2 \in \mathcal{U}_t^-$. However, since $\eta_1 \in \hat{\mathcal{Y}}_t^+$, it also follows from (9.26) that

$$\eta_1 = \hat{E}_{\|\mathcal{U}_t^+}\{\eta_1 \mid \mathcal{U}_t^-\} \in \mathcal{X}_t^{+/-}$$

Thus, we have $\eta = \eta_1 + \eta_2 \in \mathcal{X}_t^{+/-}$. This completes the proof of (9.28).

We now prove (9.29). The following inclusion relation clearly holds:

$$\bar{\mathcal{Y}}_t^+ \supset (\bar{\mathcal{Y}}_t^+ \cap \mathcal{U}_t^-) + (\bar{\mathcal{Y}}_t^+ \cap \mathcal{U}_t^+) \tag{9.59}$$

because the two sets in the right-hand side of the above equation are included in the set in the left-hand side. We show the converse. Since $\bar{\mathcal{Y}}_t^+ = \hat{\mathcal{Y}}_t^+ \vee \mathcal{X}_t^{+/-}$, it suffices to show that

$$\hat{\mathcal{Y}}_t^+ \subset (\bar{\mathcal{Y}}_t^+ \cap \mathcal{U}_t^-) + (\bar{\mathcal{Y}}_t^+ \cap \mathcal{U}_t^+) \tag{9.60}$$

$$\mathcal{X}_t^{+/-} \subset (\bar{\mathcal{Y}}_t^+ \cap \mathcal{U}_t^-) + (\bar{\mathcal{Y}}_t^+ \cap \mathcal{U}_t^+) \tag{9.61}$$

It follows from Lemma 9.4 that

$$y_d(t+h) = \sum_{i=-\infty}^{t+h} G_{t+h-i} u(i) = y_d^-(t+h) + y_d^+(t+h)$$

where

$$y_d^-(t+h) = \sum_{i=-\infty}^{t-1} G_{t+h-i}u(i) \in \mathcal{U}_t^- \tag{9.62a}$$

$$y_d^+(t+h) = \sum_{i=t}^{t+h} G_{t+h-i}u(i) \in \mathcal{U}_t^+ \tag{9.62b}$$

and where $\mathcal{U}_t^+ \cap \mathcal{U}_t^- = \{0\}$. Thus $y_d^-(t+h)$ is the oblique projection of $y_d(t+h)$ onto the past $\mathcal{U}_t^-$ along the future $\mathcal{U}_t^+$, so that it belongs to $\mathcal{X}_t^{+/-}$. Thus we get $y_d^-(t+h) \in \mathcal{X}_t^{+/-} = \bar{\mathcal{Y}}_t^+ \cap \mathcal{U}_t^-$. Also, we have

$$y_d^+(t+h) = y_d(t+h) - y_d^-(t+h) \in \bar{\mathcal{Y}}_t^+$$

Thus, from (9.62), the relation $y_d^+(t+h) \subset \bar{\mathcal{Y}}_t^+ \cap \mathcal{U}_t^+$ holds. Therefore, it follows that

$$y_d(t+h) = y_d^-(t+h) + y_d^+(t+h) \in (\bar{\mathcal{Y}}_t^+ \cap \mathcal{U}_t^-) + (\bar{\mathcal{Y}}_t^+ \cap \mathcal{U}_t^+)$$

Since $\overline{\text{span}}\{y_d(t+h) \mid h = 0, 1, \cdots\} = \hat{\mathcal{Y}}_t^+$, we have (9.60). Also, it follows that (9.61) trivially holds, since $\mathcal{X}_t^{+/-} = \bar{\mathcal{Y}}_t^+ \cap \mathcal{U}_t^-$ from (9.28).

This complete the proof of Lemma 9.7.

10

Subspace Identification (2) – CCA

In this chapter, we consider the stochastic realization problem in the presence of exogenous inputs by extending the CCA-based approach. The oblique projection of the future outputs on the space of the past observations along the space of the future inputs is factorized as a product of the extended observability matrix and the state vector. In terms of the state vector and the future inputs, we then derive an optimal predictor of the future outputs, which leads to a forward innovation model for the output process in the presence of exogenous inputs. The basic step of the realization procedure is a factorization of the conditional covariance matrix of the future outputs and the past input-output given future inputs; this factorization can easily be adapted to finite input-output data by using the LQ decomposition. We derive two stochastic subspace identification algorithms, of which relation to the N4SID method is explained. Some comparative simulation results with the ORT and PO-MOESP methods are also included.

10.1 Stochastic Realization with Exogenous Inputs

Consider a stochastic system shown in Figure 10.1, where $u \in \mathbb{R}^m$ is the exogenous input, $y \in \mathbb{R}^p$ the output vector, and ξ the stochastic disturbance, which is not observable. We assume that $\{u(t), y(t), t = 0, \pm 1, \cdots\}$ are zero mean second-order stationary stochastic processes, and that the joint input-output process (u, y) is of full rank and regular.

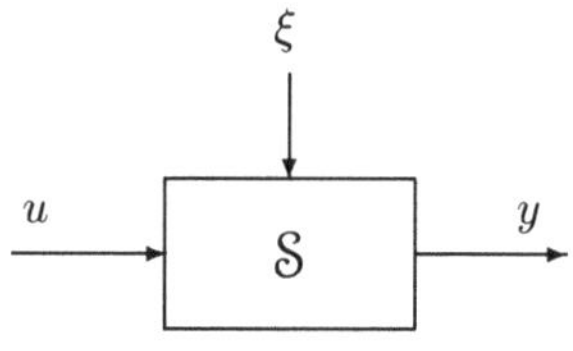

Figure 10.1. Stochastic system with exogenous input

The stochastic realization problem considered in this chapter is the same as the one studied in Chapter 9, which is restated below.

Stochastic Realization Problem

Under the assumption that the infinite data $\{u(t),\, y(t),\, t = 0,\, \pm 1,\, \cdots\}$ are given, we define a suitable state vector x with minimal dimension and derive a state space model with the input vector u and the output vector y of the form

$$x(t+1) = Ax(t) + Bu(t) + Ke(t) \tag{10.1a}$$

$$y(t) = Cx(t) + Du(t) + e(t) \tag{10.1b}$$

where e is an innovation process. □

In this chapter, we shall present a stochastic realization method in the presence of an exogenous input by means of the CCA-based approach. We extend the CCA method developed in Chapter 8 to the present case. Under the absence of feedback from y to u, we derive a predictor space for the output process y, leading to a minimal causal stationary realization of the process y with the exogenous process u as an input. A basic idea is to derive a multi-stage Wiener predictor of the future output in terms of the past input-output and the future inputs, where an important point is to define an appropriate causal state vector.

Let t be the present time, and k a positive integer. Then, we define the future vectors

$$f(t) := \begin{bmatrix} y(t) \\ y(t+1) \\ \vdots \\ y(t+k-1) \end{bmatrix}, \qquad u_+(t) := \begin{bmatrix} u(t) \\ u(t+1) \\ \vdots \\ u(t+k-1) \end{bmatrix}$$

and the past vectors

$$y_-(t) := \begin{bmatrix} y(t-1) \\ y(t-2) \\ \vdots \\ \end{bmatrix}, \qquad u_-(t) := \begin{bmatrix} u(t-1) \\ u(t-2) \\ \vdots \\ \end{bmatrix}$$

We further define

$$p(t) := \begin{bmatrix} w(t-1) \\ w(t-2) \\ \vdots \\ \end{bmatrix}, \qquad w(t) := \begin{bmatrix} y(t) \\ u(t) \end{bmatrix}$$

where $w \in \mathbb{R}^d$, $d := p + m$ is the joint input-output process. It should be noted that the future vectors $f(t) \in \mathbb{R}^{kp}$ and $u_+(t) \in \mathbb{R}^{km}$ are finite dimensional, but the past input-output vector $p(t)$ is infinite-dimensional.

The notation in this chapter is the same as that of Chapter 9. The linear spaces generated by the past of w, y and the future of u are respectively denoted by

$$\begin{aligned}\mathcal{P}_t^- &= \overline{\operatorname{span}}\{w(\tau) \mid \tau < t\} \\ \mathcal{Y}_t^- &= \overline{\operatorname{span}}\{y(\tau) \mid \tau < t\} \\ \mathcal{U}_t^+ &= \overline{\operatorname{span}}\{u(\tau) \mid \tau \geq t\}\end{aligned}$$

Also, we assume that these spaces are complete with respect to the mean-square norm $\|x\|_{\mathcal{H}} = \sqrt{E\{\|x\|^2\}}$, so that $\mathcal{P}_t^-$, $\mathcal{Y}_t^-$ and $\mathcal{U}_t^+$ are thought of as subspaces of an ambient Hilbert space $\mathcal{H} = \mathcal{U} \vee \mathcal{Y}$ that includes all linear functionals of the joint input-output process (u, y).

Let $\mathcal{B}$ be a subspace of the Hilbert space $\mathcal{H}$. Then the orthogonal projection of a vector $a \in \mathcal{H}$ onto $\mathcal{B}$ is denoted by $\hat{E}\{a \mid \mathcal{B}\}$. If $\mathcal{B}$ is generated by a vector b, then the orthogonal projection is expressed as

$$\begin{aligned}\hat{E}\{a \mid \mathcal{B}\} &= E\{ab^{\mathrm{T}}\}E\{bb^{\mathrm{T}}\}^{\dagger}b \\ &= \Sigma_{ab}\Sigma_{bb}^{\dagger}b =: \hat{E}(a \mid b)\end{aligned}$$

where $\Sigma_{ab} := E\{ab^{\mathrm{T}}\}$ is the covariance matrix of two random vectors a and b, and $(\cdot)^{\dagger}$ is the pseudo-inverse. Let $\mathcal{B}^{\perp}$ be the orthogonal complement of $\mathcal{B} \subset \mathcal{H}$. Then, the orthogonal projection of a onto $\mathcal{B}^{\perp}$ is given by

$$\hat{E}\{a \mid \mathcal{B}^{\perp}\} := a - \hat{E}\{a \mid \mathcal{B}\}$$

If $\mathcal{B}$ is generated by a random vector b, then we write $\hat{E}\{a \mid \mathcal{B}^{\perp}\} = \hat{E}(a \mid b^{\perp})$. For the oblique projection; see also Section 9.1.

We begin with a simple result on the conditional covariance matrices.

Lemma 10.1. *For three random vectors y, a, $b \in \mathcal{H}$, we define the conditional covariance matrix*

$$\Sigma_{ya|b} := E\{\hat{E}(y \mid b^{\perp})\hat{E}(a \mid b^{\perp})^{\mathrm{T}}\} \tag{10.2}$$

Then, it follows that

$$\Sigma_{ya|b} = \Sigma_{ya} - \Sigma_{yb}(\Sigma_{bb})^{-1}\Sigma_{ba} \tag{10.3}$$

where Σ_{bb} is assumed to be nonsingular.

Proof. By definition,

$$\hat{E}(y \mid b^{\perp}) = y - \Sigma_{yb}(\Sigma_{bb})^{-1}b, \qquad \hat{E}(a \mid b^{\perp}) = a - \Sigma_{ab}(\Sigma_{bb})^{-1}b$$

Substituting these relations into (10.2) yields

$$\Sigma_{ya|b} = E\{[y - \Sigma_{yb}(\Sigma_{bb})^{-1}b][a - \Sigma_{ab}(\Sigma_{bb})^{-1}b]^{\mathrm{T}}\}$$

Rearranging the terms, we get (10.3) [see also Lemma 5.1]. □

Lemma 10.2. *Consider three random vectors* $y,\ a,\ b \in \mathcal{H}$ *and two subspaces* $\mathcal{A} := \operatorname{span}\{a\}$ *and* $\mathcal{B} := \operatorname{span}\{b\}$ *with* $\mathcal{A} \cap \mathcal{B} = \{0\}$. *Then, we have*

$$\begin{aligned}\hat{E}\{y \mid \mathcal{A} \vee \mathcal{B}\} &= \hat{E}_{\|\mathcal{B}}\{y \mid \mathcal{A}\} + \hat{E}_{\|\mathcal{A}}\{y \mid \mathcal{B}\} \\ &=: \Pi(y)a + \Psi(y)b \qquad (10.4)\end{aligned}$$

Also, define the conditional covariance matrices by

$$\begin{aligned}\Sigma_{aa|b} &= E\{\hat{E}(a \mid b^{\perp})\hat{E}(a \mid b^{\perp})^{\mathrm{T}}\} \\ \Sigma_{ya|b} &= E\{\hat{E}(y \mid b^{\perp})\hat{E}(a \mid b^{\perp})^{\mathrm{T}}\} \\ \Sigma_{yb|a} &= E\{\hat{E}(y \mid a^{\perp})\hat{E}(b \mid a^{\perp})^{\mathrm{T}}\} \\ \Sigma_{bb|a} &= E\{\hat{E}(b \mid a^{\perp})\hat{E}(b \mid a^{\perp})^{\mathrm{T}}\}\end{aligned}$$

Then, $\Pi(y)$ *and* $\Psi(y)$ *satisfy the discrete Wiener-Hopf equations*

$$\Pi(y)\Sigma_{aa|b} = \Sigma_{ya|b}, \qquad \Psi(y)\Sigma_{bb|a} = \Sigma_{yb|a} \qquad (10.5)$$

where we note that if Σ_{aa} *and* Σ_{bb} *are positive definite, so are* $\Sigma_{aa|b}$ *and* $\Sigma_{bb|a}$.

Proof. We see that the orthogonal projection of (10.4) is given by

$$\hat{E}\{y \mid \mathcal{A} \vee \mathcal{B}\} = E\{y[a^{\mathrm{T}}\ \ b^{\mathrm{T}}]\}E\left\{\begin{bmatrix} a \\ b \end{bmatrix}[a^{\mathrm{T}}\ \ b^{\mathrm{T}}]\right\}^{-1}\begin{bmatrix} a \\ b \end{bmatrix}$$

Thus it follows that

$$\begin{aligned}\hat{E}\{y \mid \mathcal{A} \vee \mathcal{B}\} &= [\Sigma_{ya}\ \ \Sigma_{yb}]\begin{bmatrix} \Sigma_{aa} & \Sigma_{ab} \\ \Sigma_{ba} & \Sigma_{bb} \end{bmatrix}^{-1}\begin{bmatrix} a \\ 0 \end{bmatrix} \\ &\quad + [\Sigma_{ya}\ \ \Sigma_{yb}]\begin{bmatrix} \Sigma_{aa} & \Sigma_{ab} \\ \Sigma_{ba} & \Sigma_{bb} \end{bmatrix}^{-1}\begin{bmatrix} 0 \\ b \end{bmatrix} \qquad (10.6)\end{aligned}$$

We show that the first term in the right-hand side of the above equation is the oblique projection of y onto $\mathcal{A}$ along $\mathcal{B}$. Recall the inversion formula for a block matrix:

$$\begin{bmatrix} A & B \\ C & D \end{bmatrix}^{-1} = \begin{bmatrix} \Delta^{-1} & -\Delta^{-1}BD^{-1} \\ -D^{-1}C\Delta^{-1} & D^{-1} + D^{-1}C\Delta^{-1}BD^{-1} \end{bmatrix}$$

Putting $A = \Sigma_{aa}$, $B = C^{\mathrm{T}} = \Sigma_{ab}$ and $D = \Sigma_{bb}$, we have $\Delta := A - BD^{-1}C = \Sigma_{aa|b}$. Thus, from (10.4) and (10.6),

$$\begin{aligned}\Pi(y) &= [\Sigma_{ya}\ \ \Sigma_{yb}]\begin{bmatrix} \Sigma_{aa} & \Sigma_{ab} \\ \Sigma_{ba} & \Sigma_{bb} \end{bmatrix}^{-1} \\ &= [\Sigma_{ya}\ \ \Sigma_{yb}]\begin{bmatrix} \Sigma_{aa|b}^{-1} \\ -\Sigma_{bb}^{-1}\Sigma_{ba}\Sigma_{aa|b}^{-1} \end{bmatrix} \\ &= (\Sigma_{ya} - \Sigma_{yb}\Sigma_{bb}^{-1}\Sigma_{ba})\Sigma_{aa|b}^{-1} = \Sigma_{ya|b}\Sigma_{aa|b}^{-1}\end{aligned}$$

This proves the first relation in (10.5). Similarly, we can show the second relation holds.

Now let $v = Va$ for $V \in \mathbb{R}^{\bullet \times n_a}$. Note that $v \in \mathcal{A} = \text{span}\{a\}$. Then, we have

$$\Pi(v)a = \Sigma_{va|b}\Sigma_{aa|b}^{-1}a = V\Sigma_{aa|b}\Sigma_{aa|b}^{-1}a = Va = v$$

Hence, $\Pi(\cdot)$ is idempotent on $\mathcal{A}$. Also, putting $z = Zb$ for $Z \in \mathbb{R}^{\bullet \times n_b}$ and noting that $\Sigma_{ba|b} = 0$, we get

$$\Pi(z)a = \Sigma_{za|b}\Sigma_{aa|b}^{-1}a = Z\Sigma_{ba|b}\Sigma_{aa|b}^{-1}a = 0$$

so that $\Pi(\cdot)$ annihilates any element in $\mathcal{B} = \text{span}\{b\}$. It therefore follows that $\Pi(y)a$ is an oblique projection of y onto $\mathcal{A}$ along $\mathcal{B}$. In the same way, we can show that $\Psi(y)b$ is an oblique projection of y onto $\mathcal{B}$ along $\mathcal{A}$.

Suppose that Σ_{aa} and Σ_{bb} are positive definite. Then, the positivity of the conditional covariance matrices $\Sigma_{aa|b}$ and $\Sigma_{bb|a}$ is derived from $\mathcal{A} \cap \mathcal{B} = \{0\}$. In fact, if $\eta^{\mathrm{T}}\Sigma_{aa|b} = 0$, $\eta \neq 0$ holds, then

$$\begin{aligned} 0 &= \eta^{\mathrm{T}}\Sigma_{aa|b}\eta = \eta^{\mathrm{T}}E\{[a - \hat{E}\{a \mid b\}][a - \hat{E}\{a \mid b\}]^{\mathrm{T}}\}\eta \\ &= E\{[\eta^{\mathrm{T}}(a - \hat{E}\{a \mid b\})]^2\} \end{aligned}$$

This implies $\eta^{\mathrm{T}}a = \eta^{\mathrm{T}}\hat{E}\{a \mid b\} = \eta^{\mathrm{T}}\Sigma_{ab}\Sigma_{bb}^{-1}b \in \mathcal{B}$, a contradiction. This proves the positivity of $\Sigma_{aa|b}$. Similarly, we can prove the positivity of $\Sigma_{bb|a}$. □

10.2 Optimal Predictor

In this section, we shall consider the prediction problem of the future $f(t)$ by means of the past input-output $p(t)$ and the future input $u_+(t)$. In the following, we need two assumptions introduced in Chapter 9.

Assumption 10.1. *There is no feedback from the output y to the input u.* □

Assumption 10.2. *For each t, the input space $\mathcal{U}$ has the direct sum decomposition*

$$\mathcal{U} = \mathcal{U}_t^- + \mathcal{U}_t^+ \qquad (\mathcal{U}_t^- \cap \mathcal{U}_t^+ = \{0\}) \tag{10.7}$$

This is equivalent to the fact that the spectral density of the input u is positive definite on the unit circle, i.e.,

$$\Phi_{uu}(\omega) \geq cI_m, \qquad \exists\ c > 0 \tag{10.8}$$

holds. This is also equivalent to the fact that the canonical angles of the past and future are positive. □

As mentioned in Chapter 9, the condition of Assumption 10.2 may be too strong to be satisfied in practice; but it suffices to assume that the input has a PE condition of sufficiently high order and that the real system is finite dimensional.

The following theorem gives a solution to a multi-stage Wiener problem for predicting the future outputs based on the joint past input-output and the future inputs.

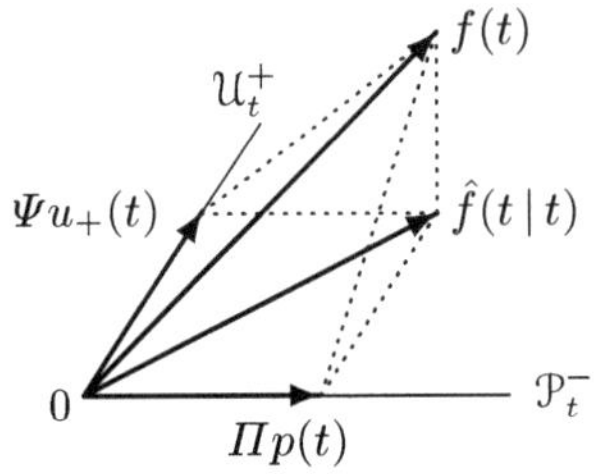

Figure 10.2. Oblique projection

Theorem 10.1. *Suppose that Assumptions 10.1 and 10.2 are satisfied. Then, the optimal predictor of the future $f(t)$ based on the past $p(t)$ and future $u_+(t)$ is given by*

$$\hat{f}(t \mid t) = \hat{E}\{f(t) \mid \mathcal{P}_t^- \vee \mathcal{U}_t^+\} = \Pi p(t) + \Psi u_+(t) \tag{10.9}$$

where $\Pi p(t)$ denotes the oblique projection of $f(t)$ onto $\mathcal{P}_t^-$ along $\mathcal{U}_t^+$, and $\Psi u_+(t)$ is the oblique projection of $f(t)$ onto $\mathcal{U}_t^+$ along $\mathcal{P}_t^-$ as shown in Figure 10.2. Moreover, Π and Ψ respectively satisfy the discrete Wiener-Hopf equations

$$\Pi \Sigma_{pp|u} = \Sigma_{fp|u}, \qquad \Psi \Sigma_{uu|p} = \Sigma_{fu|p} \tag{10.10}$$

Proof. If we can prove that $\mathcal{P}_t^- \cap \mathcal{U}_t^+ = \{0\}$, then we see from Lemma 10.1 that the orthogonal projection $\hat{E}\{f(t) \mid \mathcal{P}_t^- \vee \mathcal{U}_t^+\}$ is given by the direct sum of two oblique projections. Thus it suffices to show that $\mathcal{U}_t^+ \cap \mathcal{P}_t^- = \{0\}$.

Let $\zeta \in \mathcal{U}_t^+ \cap \mathcal{P}_t^-$. Then, we have $\zeta \in \mathcal{U}_t^+$ and $\zeta \in \mathcal{P}_t^- = \mathcal{Y}_t^- \vee \mathcal{U}_t^-$. From the latter condition, there exist $\eta \in \mathcal{Y}_t^-$ and $\nu \in \mathcal{U}_t^-$ such that $\zeta = \eta + \nu$. Since there is no feedback from y to u, it follows from (9.13) that

$$\hat{E}\{\mathcal{Y}_t^- \mid \mathcal{U}\} = \hat{E}\{\mathcal{Y}_t^- \mid \mathcal{U}_t^-\}, \qquad t = 0, \pm 1, \cdots$$

Hence, the orthogonal decomposition of $\eta \in \mathcal{Y}_t^-$ into the sum of deterministic and stochastic components gives $\eta = \eta_d + \eta_s$, where

$$\eta_d = \hat{E}\{\eta \mid \mathcal{U}\} = \hat{E}\{\eta \mid \mathcal{U}_t^-\}, \qquad \eta_s = \hat{E}\{\eta \mid \mathcal{U}^\perp\}$$

Thus, it follows that $\zeta = (\nu + \eta_d) + \eta_s$, $\nu + \eta_d \in \mathcal{U}_t^-$, $\eta_s \perp \mathcal{U}$. Since $\zeta \in \mathcal{U}_t^+$, we get $\hat{E}_{\|\mathcal{U}_t^+}\{\zeta \mid \mathcal{U}_t^-\} = 0$, implying that

$$\hat{E}_{\|\mathcal{U}_t^+}\{(\nu + \eta_d) + \eta_s \mid \mathcal{U}_t^-\} = 0$$

However, since $\nu + \eta_d \in \mathcal{U}_t^-$, and since $\eta_s \perp \mathcal{U}_t^-$,

$$\hat{E}_{\|\mathcal{U}_t^+}\{(\nu + \eta_d) + \eta_s \mid \mathcal{U}_t^-\} = \hat{E}_{\|\mathcal{U}_t^+}\{\nu + \eta_d \mid \mathcal{U}_t^-\} = \nu + \eta_d = 0$$

Thus, $\zeta \in \mathcal{U}_t^+$ satisfies $\zeta = \eta_s \perp \mathcal{U}$, so that $\zeta = 0$, as was to be proved. □

For convenience, we put an index k to denote that the matrices $\Pi \in \mathbb{R}^{kp\times\infty}$ and $\Psi \in \mathbb{R}^{kp\times km}$ have k block rows, so that we write them Π_k and Ψ_k, respectively. Thus, the terms in the right-hand side of (10.9) are expressed as

$$\Pi_k p(t) = \hat{E}_{\|\mathcal{U}_t^+}\{f(t) \mid \mathcal{P}_t^-\} \tag{10.11}$$

and

$$\Psi_k u_+(t) = \hat{E}_{\|\mathcal{P}_t^-}\{f(t) \mid \mathcal{U}_t^+\} \tag{10.12}$$

Next we show that by using the oblique projections of (10.11) and (10.12), we can construct a minimal dimensional state space model for the output y which is causal with respect to the input u. By causal here we mean that the oblique projection of (10.12) is causal, so that the operator Ψ_k has a block lower triangular form.

The following lemma will give a representation of the orthogonal projection in Theorem 10.1.

Lemma 10.3. *Suppose that Assumption 10.1 is satisfied. Then, from Theorem 10.1, we have*

$$\hat{E}\{y(t+h) \mid \mathcal{P}_t^- \vee \mathcal{U}_t^+\} = \hat{E}\{y(t+h) \mid \mathcal{P}_t^- \vee \mathcal{U}_{[t,\,t+h]}\} \tag{10.13}$$

where $h = 0, 1, \cdots$, *and where* $\mathcal{U}_{[t,\,t+h]} = \operatorname{span}\{u(t), \cdots, u(t+h)\}$. *Also,* Ψ_k *is given by*

$$\Psi_k = \begin{bmatrix} G_0 & & & & 0 \\ G_1 & G_0 & & & \\ G_2 & G_1 & G_0 & & \\ \vdots & \vdots & & \ddots & \\ G_{k-1} & G_{k-2} & \cdots & \cdots & G_0 \end{bmatrix} \in \mathbb{R}^{kp\times km} \tag{10.14}$$

where $(G_0, G_1, \cdots)$ *are impulse response matrices. Hence,* Ψ_k *becomes causal.*

Proof. First, we show the following relation for the conditional orthogonality:

$$\mathcal{A} \perp \mathcal{B} \mid \mathcal{C} \quad \Rightarrow \quad \mathcal{A} \perp \mathcal{B} \mid (\mathcal{A}_0 \vee \mathcal{C}), \quad \mathcal{A}_0 \subset \mathcal{A} \tag{10.15}$$

Indeed, since $\mathcal{A}_0 \subset \mathcal{A}$, we have $\mathcal{A} \perp \mathcal{B} \mid \mathcal{C} \Rightarrow \mathcal{A}_0 \perp \mathcal{B} \mid \mathcal{C}$. Thus two relations $\hat{E}\{\mathcal{B} \mid \mathcal{A} \vee \mathcal{C}\} = \hat{E}\{\mathcal{B} \mid \mathcal{C}\}$ and $\hat{E}\{\mathcal{B} \mid \mathcal{A}_0 \vee \mathcal{C}\} = \hat{E}\{\mathcal{B} \mid \mathcal{C}\}$ hold from Lemma 9.2. This implies that

$$\hat{E}\{\mathcal{B} \mid \mathcal{A} \vee \mathcal{C}\} = \hat{E}\{\mathcal{B} \mid \mathcal{C}\} = \hat{E}\{\mathcal{B} \mid \mathcal{A}_0 \vee \mathcal{C}\}$$

However, since the left-hand side can be written as $\hat{E}\{\mathcal{B} \mid \mathcal{A} \vee (\mathcal{A}_0 \vee \mathcal{C})\}$, it follows that $\hat{E}\{\mathcal{B} \mid \mathcal{A} \vee (\mathcal{A}_0 \vee \mathcal{C})\} = \hat{E}\{\mathcal{B} \mid \mathcal{A}_0 \vee \mathcal{C}\}$, as was to be proved.

Since there is no feedback from y to u, we see from Theorem 9.1 (iv) that

$$\mathcal{Y}_{t+h+1}^- \perp \mathcal{U}_{t+h+1}^+ \mid \mathcal{U}_{t+h+1}^-, \qquad h = 0, 1, \cdots \tag{10.16}$$

Putting $\mathcal{A}_0 = \mathcal{Y}_t^-$, $\mathcal{A} = \mathcal{Y}_{t+h+1}^-$, $\mathcal{B} = \mathcal{U}_{t+h+1}^+$ and $\mathcal{C} = \mathcal{U}_{t+h+1}^-$, it follows from (10.15) that

$$\mathcal{Y}_{t+h+1}^{-} \perp \mathcal{U}_{t+h+1}^{+} \mid (\mathcal{Y}_{t}^{-} \vee \mathcal{U}_{t+h+1}^{-}) \tag{10.17}$$

Also, putting $\mathcal{B} = \mathcal{Y}_{t+h+1}^{-}$, $\mathcal{A} = \mathcal{U}_{t+h+1}^{+}$, $\mathcal{C} = \mathcal{Y}_{t}^{-} \vee \mathcal{U}_{t+h+1}^{-}$ in (10.17), it follows from Lemma 9.2 that

$$\hat{E}\{\mathcal{Y}_{t+h+1}^{-} \mid \mathcal{Y}_{t}^{-} \vee \mathcal{U}\} = \hat{E}\{\mathcal{Y}_{t+h+1}^{-} \mid \mathcal{Y}_{t}^{-} \vee \mathcal{U}_{t+h+1}^{-}\}$$

Moreover, noting that $y(t+h) \in \mathcal{Y}_{t+h+1}^{-}$, we have

$$\begin{aligned}\hat{E}\{y(t+h) \mid \mathcal{Y}_{t}^{-} \vee \mathcal{U}\} &= \hat{E}\{y(t+h) \mid \mathcal{Y}_{t}^{-} \vee \mathcal{U}_{t+h+1}^{-}\} \\ &= \hat{E}\{y(t+h) \mid \mathcal{P}_{t}^{-} \vee \mathcal{U}_{[t,\,t+h]}\}\end{aligned}$$

This proves (10.13).

We show that Ψ_k is a causal operator. In fact, from (10.13),

$$\Psi_k u_+(t) = \begin{bmatrix} \hat{E}_{\|\mathcal{P}_t^-}\{y(t) \mid \mathcal{U}_{[t,\,t]}\} \\ \hat{E}_{\|\mathcal{P}_t^-}\{y(t+1) \mid \mathcal{U}_{[t,\,t+1]}\} \\ \vdots \\ \hat{E}_{\|\mathcal{P}_t^-}\{y(t+k-1) \mid \mathcal{U}_{[t,\,t+k-1]}\} \end{bmatrix}$$

Since $\mathcal{U}_{[t,\,t+h]} = \operatorname{span}\{u(t), \cdots, u(t+h)\}$, we have

$$\hat{E}_{\|\mathcal{P}_t^-}\{y(t+h) \mid \mathcal{U}_{[t,\,t+h]}\} = G_h u(t) + G_{h-1}u(t+1) + \cdots + G_0 u(t+h)$$

so that Ψ_k becomes a block Toeplitz matrix. The stationarity of Ψ_k follows from the stationarity of the joint process (u, y). □

In the next section, we shall define the state vector of the system with exogenous inputs in terms of the conditional CCA technique, where the conditional canonical correlations are defined as the canonical correlations between the future and the past after deleting the effects of future inputs from them.

10.3 Conditional Canonical Correlation Analysis

As shown in Chapters 7 and 8, the stochastic system without exogenous inputs is finite dimensional if and only if the covariance matrix of a block Hankel type has finite rank. It may, however, be noted that $\Sigma_{fp|u}$ is not a block Hankel matrix as shown in (10.18) below.

We introduce the conditional CCA in order to factorize the conditional covariance matrix $\Sigma_{fp|u}$ of the future and past given the future inputs. By stationarity, $\Sigma_{fp|u}$ is a $kp \times \infty$ semi-infinite dimensional block matrix whose rank is nondecreasing with respect to the future prediction horizon k. We then define a state vector to derive an innovation model for the system with exogenous inputs.

Suppose that the conditional covariance matrix $\Sigma_{fp|u}$ has finite rank, so that we assume $\operatorname{rank}(\Sigma_{fp|u}) = n$. It follows from Lemma 10.1 that the conditional covariance matrix is expressed as

$$\begin{aligned}\Sigma_{fp|u} &= E\{[f(t) - \hat{E}\{f(t) \mid \mathcal{U}_t^+\}][p(t) - \hat{E}\{p(t) \mid \mathcal{U}_t^+\}]^{\mathrm{T}}\} \\ &= E\{\hat{E}(f \mid u_+^{\perp})(t)(\hat{E}(p \mid u_+^{\perp})(t))^{\mathrm{T}}\} \end{aligned} \tag{10.18}$$

Also, the conditional covariance matrices $\Sigma_{ff|u}$ and $\Sigma_{pp|u}$ are defined similarly. Let the Cholesky factorizations be given by

$$\Sigma_{ff|u} = LL^{\mathrm{T}}, \qquad \Sigma_{pp|u} = MM^{\mathrm{T}}$$

and define

$$\varepsilon_+(t) := L^{-1}\hat{E}(f \mid u_+^{\perp})(t), \qquad \varepsilon_-(t) := M^{-1}\hat{E}(p \mid u_+^{\perp})(t)$$

Then, we have

$$E\{\varepsilon_+(t)\varepsilon_-^{\mathrm{T}}(t)\} = L^{-1}\Sigma_{fp|u}M^{-\mathrm{T}}$$

where the right-hand side of the above equation is the normalized conditional covariance matrix.

Consider the SVD of the normalized conditional covariance matrix

$$L^{-1}\Sigma_{fp|u}M^{-\mathrm{T}} = U\Sigma V^{\mathrm{T}} \tag{10.19}$$

where $U^{\mathrm{T}}U = I_n$, $V^{\mathrm{T}}V = I_n$, and where Σ is a diagonal matrix of the form

$$\Sigma = \operatorname{diag}(\sigma_1, \cdots, \sigma_n), \qquad 1 \geq \sigma_1 \geq \cdots \geq \sigma_n > 0$$

We define two n-dimensional vectors as

$$\alpha(t) := V^{\mathrm{T}}M^{-1}\hat{E}(p \mid u_+^{\perp})(t), \qquad \beta(t) := U^{\mathrm{T}}L^{-1}\hat{E}(f \mid u_+^{\perp})(t) \tag{10.20}$$

Then it can be shown that

$$E\{\alpha(t)\alpha^{\mathrm{T}}(t)\} = E\{\beta(t)\beta^{\mathrm{T}}(t)\} = I_n, \qquad E\{\beta(t)\alpha^{\mathrm{T}}(t)\} = \Sigma$$

Thus, comparing with the definition of canonical correlations in Subsection 8.5.1, we see that $\sigma_1, \cdots, \sigma_n$ are conditional canonical correlations between the future $f(t)$ and the past $p(t)$ given the future input $u_+(t)$. Also, $\alpha(t)$ and $\beta(t)$ are the corresponding conditional canonical vectors.

According to the method of Subsection 8.5.1, the extended observability and reachability matrices are defined by

$$\mathcal{O}_k := LU\Sigma^{1/2}, \qquad \mathcal{C}_\infty := \Sigma^{1/2}V^{\mathrm{T}}M^{\mathrm{T}} \tag{10.21}$$

where $\operatorname{rank}(\mathcal{O}_k) = \operatorname{rank}(\mathcal{C}_\infty) = n$. Thus from (10.19), the conditional covariance matrix $\Sigma_{fp|u}$ has a factorization

$$\Sigma_{fp|u} = (LU\Sigma^{1/2})(\Sigma^{1/2}V^{\mathrm{T}}M^{\mathrm{T}}) = \mathcal{O}_k\mathcal{C}_\infty$$

Define the n-dimensional state vector as

$$x(t) = \mathcal{C}_\infty \Sigma_{pp|u}^{-1} p(t) = \Sigma^{1/2}V^{\mathrm{T}}M^{-1}p(t) \tag{10.22}$$

Then, it can be shown that $x(t)$ is a basis vector of the predictor space

$$\mathcal{X}_t^{+/-} := \hat{E}_{\|\mathcal{U}_t^+}\{\mathcal{Y}_t^+ \mid \mathcal{P}_t^-\} \tag{10.23}$$

In fact, it follows from (10.11) and (10.22) that the oblique projection of the future $f(t)$ onto the past $\mathcal{P}_t^-$ along the future inputs $\mathcal{U}_t^+$ is given by

$$\hat{E}_{\|\mathcal{U}_t^+}\{f(t) \mid \mathcal{P}_t^-\} = \Pi_k p(t) = \Sigma_{fp|u}\Sigma_{pp|u}^{-1}p(t) = \mathcal{O}_k x(t) \tag{10.24}$$

where the covariance matrix of x is positive definite. Indeed, from (10.22),

$$\begin{aligned} E\{x(t)x^{\mathrm{T}}(t)\} &= \Sigma^{1/2}V^{\mathrm{T}}M^{-1}\Sigma_{pp}M^{-\mathrm{T}}V\Sigma^{1/2} \\ &\geq \Sigma^{1/2}V^{\mathrm{T}}M^{-1}\Sigma_{pp|u}M^{-\mathrm{T}}V\Sigma^{1/2} = \Sigma > 0 \end{aligned}$$

Note that if there are no exogenous inputs, the state covariance matrix is exactly the canonical correlation matrix as discussed in Section 8.5.

Using the state vector x defined above, the optimal predictor of (10.9) is then expressed as

$$\hat{f}(t \mid t) = \mathcal{O}_k x(t) + \Psi_k u_+(t) \tag{10.25}$$

This equation shows that given the future input $u_+(t)$, the state vector $x(t)$ carries information necessary to predict the future output $f(t)$ based on the past $\mathcal{P}_t^-$.

The property of x defined by (10.22) is summarized below.

Lemma 10.4. *Given the future inputs, the process $\{x(t),\ t = 0, \pm 1, \cdots\}$ defined above is a Markov process satisfying*

$$\hat{E}_{\|\mathcal{U}_t^+}\{x(t+h) \mid \mathcal{P}_t^-\} = \hat{E}_{\|\mathcal{U}_t^+}\{x(t+h) \mid \mathcal{X}_t^{+/-}\}, \qquad h = 1, 2, \cdots$$

where $\mathcal{X}_t^{+/-}$ is the predictor space defined by (10.23).

Proof. Rewriting the formula (10.25) for $t \to t+h$ yields

$$\begin{bmatrix} \hat{y}(t+h \mid t+h) \\ \hat{y}(t+h+1 \mid t+h) \\ \vdots \\ \hat{y}(t+h+k-1 \mid t+h) \end{bmatrix} = \mathcal{O}_k x(t+h) + \Psi_k u_+(t+h) \tag{10.26}$$

where $\hat{y}(l \mid t+h), l = t+h, \cdots$ denotes the optimal estimate of $y(l)$ based on the observations up to time $t+h-1$ and the inputs after $t+h$. Also, $k \to k+h$ in (10.25),

$$\begin{bmatrix} \hat{y}(t \mid t) \\ \vdots \\ \hat{y}(t+h-1 \mid t) \\ \hat{y}(t+h \mid t) \\ \vdots \\ \hat{y}(t+h+k-1 \mid t) \end{bmatrix} = \mathcal{O}_{h+k} x(t) + \Psi_{h+k} \begin{bmatrix} u(t) \\ \vdots \\ u(t+h-1) \\ u(t+h) \\ \vdots \\ u(t+h+k-1) \end{bmatrix} \tag{10.27}$$

Since $\mathcal{O}_k$ has full column rank, the last k block rows of $\mathcal{O}_{h+k} \in \mathbb{R}^{kp \times n}$ is written as

$$\mathcal{O}_{h|k} = \mathcal{O}_k A_h, \qquad A_h \in \mathbb{R}^{n \times n}$$

Also, the last k block rows of Ψ_{h+k} is expressed as

$$\Psi_{h|k} = \begin{bmatrix} G_h & \cdots & \cdots & G_0 & & & 0 \\ G_{h+1} & G_h & \cdots & \cdots & G_0 & & \\ \vdots & & \ddots & \ddots & & \ddots & \\ G_{h+k-1} & \cdots & \cdots & G_h & \cdots & \cdots & G_0 \end{bmatrix}$$

Hence, we can write the last k block rows of (10.27) as

$$\begin{bmatrix} \hat{y}(t+h \mid t) \\ \hat{y}(t+h+1 \mid t) \\ \vdots \\ \hat{y}(t+h+k-1 \mid t) \end{bmatrix} = \mathcal{O}_k A_h x(t) + \Psi_{h|k} \begin{bmatrix} u(t) \\ u(t+1) \\ \vdots \\ u(t+h+k-1) \end{bmatrix} \tag{10.28}$$

From the definition of $\hat{f}(t \mid t)$ and the property of oblique projection, it can be shown that

$$\hat{E}_{\|\mathcal{U}_t^+}\{\hat{y}(t+h+l \mid t+h) \mid \mathcal{P}_t^-\} = \hat{E}_{\|\mathcal{U}_t^+}\{\hat{y}(t+h+l \mid t) \mid \mathcal{P}_t^-\}$$

holds for $l = 0, 1, \cdots$. Thus, by applying the operator $\hat{E}_{\|\mathcal{U}_t^+}\{\cdot \mid \mathcal{P}_t^-\}$ on both sides of (10.26) and (10.28), we have

$$\mathcal{O}_k \hat{E}_{\|\mathcal{U}_t^+}\{x(t+h) \mid \mathcal{P}_t^-\} = \mathcal{O}_k A_h \hat{E}_{\|\mathcal{U}_t^+}\{x(t) \mid \mathcal{P}_t^-\} = \mathcal{O}_k A_h x(t)$$

Since $\mathcal{O}_k$ has full rank, it follows that

$$\hat{E}_{\|\mathcal{U}_t^+}\{x(t+h) \mid \mathcal{P}_t^-\} = A_h x(t) \in \mathcal{X}_t^{+/-}$$

Also, applying the operator $\hat{E}_{\|\mathcal{U}_t^+}\{\cdot \mid \mathcal{X}_t^{+/-}\}$ to the above equation yields

$$\hat{E}_{\|\mathcal{U}_t^+}\left\{\hat{E}_{\|\mathcal{U}_t^+}\{x(t+h) \mid \mathcal{P}_t^-\} \mid \mathcal{X}_t^{+/-}\right\} = A_h x(t) = \hat{E}_{\|\mathcal{U}_t^+}\{x(t+h) \mid \mathcal{P}_t^-\}$$

where the left-hand side of the above equation reduces to $\hat{E}_{\|\mathcal{U}_t^+}\{x(t+h) \mid \mathcal{X}_t^{+/-}\}$ by using $\mathcal{X}_t^{+/-} \subset \mathcal{P}_t^-$. This completes the proof. □

Remark 10.1. The state vector defined by (10.22) is based on the conditional CCA, so that it is different from the state vector defined for stationary processes in Subsection 8.5.1. According to the discussion therein, it may be more natural to define the state vector as

$$x^c(t) = \Sigma^{1/2}\alpha(t) = \mathcal{C}_k \Sigma_{pp|u}^{-1} \hat{E}(p \mid u_+^{\perp})(t)$$

in terms of the conditional canonical vector $\alpha(t)$ of (10.20), where $\text{cov}\{x^c(t)\} = \Sigma$. It should, however, be noted that we cannot derive a causal state space model by using the above $x^c(t)$. In fact, defining the subspace $\tilde{\mathcal{P}}_t^- := \hat{E}\{\mathcal{P}_t^- \mid (\mathcal{U}_t^+)^\perp\}$, it follows that $\mathcal{P}_t^- \vee \mathcal{U}_t^+ = \tilde{\mathcal{P}}_t^- \oplus \mathcal{U}_t^+$. Thus we obtain the following orthogonal decomposition

$$\begin{aligned}\hat{f}(t \mid t) &= \hat{E}\{f(t) \mid \mathcal{P}_t^- \vee \mathcal{U}_t^+\} = \hat{E}\{f(t) \mid \tilde{\mathcal{P}}_t^- \oplus \mathcal{U}_t^+\} \\ &= \hat{E}\{f(t) \mid \tilde{\mathcal{P}}_t^-\} + \hat{E}\{f(t) \mid \mathcal{U}_t^+\}\end{aligned}$$

From $\tilde{\mathcal{P}}_t^- = \text{span}\{\hat{E}(p \mid u_+^{\perp})(t)\}$, the first term in the right-hand side of the above equation becomes

$$\begin{aligned}\hat{E}\{f(t) \mid \tilde{\mathcal{P}}_t^-\} &= E\{f(t)[\hat{E}(p \mid u_+^{\perp})(t)]^{\mathrm{T}}\} \\ &\qquad \times (\text{cov}\{\hat{E}(p \mid u_+^{\perp})(t)\})^{-1}\hat{E}(p \mid u_+^{\perp})(t) \\ &= \Sigma_{fp|u}(\Sigma_{pp|u})^{-1}\hat{E}(p \mid u_+^{\perp})(t) = \mathcal{O}_k x^c(t)\end{aligned}$$

Thus, though similar to (10.25), we have a different optimal predictor

$$\hat{f}(t \mid t) = \mathcal{O}_k x^c(t) + \tilde{\Psi}_k u_+(t) \tag{10.29}$$

where $\tilde{\Psi}_k$ is a non-causal operator defined by

$$\begin{aligned}\tilde{\Psi}_k u_+(t) :&= \hat{E}\{f(t) \mid \mathcal{U}_t^+\} \\ &= E\{f(t)u_+(t)^{\mathrm{T}}\}(E\{u_+(t)u_+(t)^{\mathrm{T}}\})^{-1}u_+(t)\end{aligned}$$

This implies that being not a causal predictor, $x^c(t)$ of (10.29) cannot be a state vector of a causal model. □

We are now in a position to derive the innovation representation for a stochastic system with exogenous inputs.

10.4 Innovation Representation

In this section, by means of the state vector x of (10.22) and the optimal predictor $\hat{f}(t \mid t)$ of (10.25), we derive a forward innovation model for the output process y.

Let $\mathcal{U}_t = \text{span}\{u(t)\}$ be the subspace spanned by $u(t)$. From Lemma 10.3, showing the causality of the predictor, the first p rows of (10.25) just give the one-step prediction of $y(t)$ based on $\mathcal{P}_t^- \vee \mathcal{U}_t$, so that we have

$$\begin{aligned}\hat{y}(t) &= \hat{E}\{y(t) \mid \mathcal{P}_t^- \vee \mathcal{U}_t^+\} \\ &= \hat{E}\{y(t) \mid \mathcal{P}_t^- \vee \mathcal{U}_t\} = Cx(t) + Du(t) \qquad (10.30)\end{aligned}$$

where $C \in \mathbb{R}^{p\times n}$ and $D \in \mathbb{R}^{p\times m}$ are constant matrices. Since, from the proof of Theorem 10.1, $\mathcal{P}_t^- \cap \mathcal{U}_t = \{0\}$, the right-hand side of (10.30) is a unique direct sum decomposition. Define the prediction error as

$$e(t) := y(t) - \hat{E}\{y(t) \mid \mathcal{P}_t^- \vee \mathcal{U}_t\} \qquad (10.31)$$

Then, the output equation is given by

$$y(t) = Cx(t) + Du(t) + e(t) \qquad (10.32)$$

Since the projection $\hat{E}\{y(t) \mid \mathcal{P}_t^- \vee \mathcal{U}_t\}$ is based on the infinite past and $(u,\, y)$ are jointly stationary, the prediction error e is also stationary. Moreover, from (10.31), the prediction error $e(t)$ is uncorrelated with the past output $\{y(t-1),\, y(t-2),\, \cdots\}$ and the present and past inputs $\{u(t),\, u(t-1),\, \cdots\}$, so that $e(t) \perp (\mathcal{Y}_t^- \vee \mathcal{U}_{t+1}^-)$. Since $e(t+1) \perp (\mathcal{Y}_{t+1}^- \vee \mathcal{U}_{t+2}^-)$, it follows from (10.31) that

$$e(t) \in \mathcal{P}_{t+1}^- = (\mathcal{Y}_{t+1}^- \vee \mathcal{U}_{t+1}^-) \subset (\mathcal{Y}_{t+1}^- \vee \mathcal{U}_{t+2}^-)$$

This implies that $e(t+1) \perp e(t)$, and hence e is a white noise.

Now we compute the dynamics satisfied by $x(t)$. To this end, we define

$$w(t) := x(t+1) - \hat{E}\{x(t+1) \mid \mathcal{X}_t^{+/-} \vee \mathcal{U}_t\} \qquad (10.33)$$

where $\mathcal{X}_t^{+/-} = \text{span}\{x(t)\}$. Since $\mathcal{X}_t^{+/-} \subset \mathcal{P}_t^-$, and hence $\mathcal{X}_t^{+/-} \cap \mathcal{U}_t = \{0\}$, the second term of the right-hand side of the above equation can be expressed as a direct sum of two oblique projections. Thus there are constant matrices $A \in \mathbb{R}^{n\times n}$ and $B \in \mathbb{R}^{n\times m}$ satisfying

$$\hat{E}\{x(t+1) \mid \mathcal{X}_t^{+/-} \vee \mathcal{U}_t\} = Ax(t) + Bu(t) \qquad (10.34)$$

Thus the state equation is given by

$$x(t+1) = Ax(t) + Bu(t) + w(t)$$

Finally, we show that w in the right-hand side of the above equation is expressed in terms of the innovation process e.

Lemma 10.5. *The prediction error $w(t)$ is a function of $e(t)$. In fact, there exists a matrix $K \in \mathbb{R}^{n\times p}$ such that*

$$w(t) = Ke(t) \qquad (10.35)$$

where $K = E\{w(t)e^{\mathrm{T}}(t)\}(\text{cov}\{e(t)\})^{-1}$.

Proof. Since $x(t+1)$ is a function of $\{y(t), u(t), y(t-1), u(t-1), \cdots\}$, it follows from (10.34) that

$$\begin{aligned} Ax(t) + Bu(t) &= \hat{E}\{x(t+1) \mid \mathcal{X}_t^{+/-} \vee \mathcal{U}_t\} \\ &= \hat{E}_{\|\mathcal{U}_t}\{x(t+1) \mid \mathcal{X}_t^{+/-}\} + \hat{E}_{\|\mathcal{X}_t^{+/-}}\{x(t+1) \mid \mathcal{U}_t\} \end{aligned}$$

From Lemma 10.4 (with $h = 1$), the first term in the right-hand side of the above equation is given by

$$\hat{E}_{\|\mathcal{U}_t}\{x(t+1) \mid \mathcal{X}_t^{+/-}\} = \hat{E}_{\|\mathcal{U}_t}\{x(t+1) \mid \mathcal{P}_t^-\} = Ax(t) \tag{10.36}$$

Define $\hat{E}_{\|\mathcal{P}_t^-}\{x(t+1) \mid \mathcal{U}_t\} := B_1 u(t)$. Then, it follows that

$$\hat{E}\{x(t+1) \mid \mathcal{P}_t^- \vee \mathcal{U}_t\} = Ax(t) + B_1 u(t)$$

Since $(\mathcal{X}_t^{+/-} \vee \mathcal{U}_t) \subset (\mathcal{P}_t^- \vee \mathcal{U}_t)$, we obtain

$$\hat{E}\{x(t+1) \mid \mathcal{X}_t^{+/-} \vee \mathcal{U}_t\} = \hat{E}\Big\{\hat{E}\{x(t+1) \mid \mathcal{P}_t^- \vee \mathcal{U}_t\} \Big| \mathcal{X}_t^{+/-} \vee \mathcal{U}_t\Big\}$$

The left-hand side is $Ax(t) + Bu(t)$, while the right-hand side is

$$\hat{E}\{Ax(t) + B_1 u(t) \mid \mathcal{X}_t^{+/-} \vee \mathcal{U}_t\} = Ax(t) + B_1 u(t)$$

so that $B_1 u(t) = Bu(t)$ holds for any $u(t)$, implying that $B_1 = B$. Thus we have

$$\hat{E}_{\|\mathcal{X}_t^{+/-}}\{x(t+1) \mid \mathcal{U}_t\} = Bu(t) = \hat{E}_{\|\mathcal{P}_t^-}\{x(t+1) \mid \mathcal{U}_t\} \tag{10.37}$$

Combining (10.36) and (10.37) yields

$$\begin{aligned} \hat{E}\{x(t+1) \mid \mathcal{X}_t^{+/-} \vee \mathcal{U}_t\} &= \hat{E}_{\|\mathcal{U}_t}\{x(t+1) \mid \mathcal{P}_t^-\} + \hat{E}_{\|\mathcal{P}_t^-}\{x(t+1) \mid \mathcal{U}_t\} \\ &= \hat{E}\{x(t+1) \mid \mathcal{P}_t^- \vee \mathcal{U}_t\} \\ &= \hat{E}\{x(t+1) \mid \mathcal{Y}_t^- \vee \mathcal{U}_{t+1}^-\} \end{aligned}$$

Thus from (10.33), $w(t)$ is also expressed as

$$w(t) = x(t+1) - \hat{E}\{x(t+1) \mid \mathcal{Y}_t^- \vee \mathcal{U}_{t+1}^-\}$$

so that $w(t) \in \mathcal{P}_{t+1}^-$ is orthogonal to $\mathcal{Y}_t^- \vee \mathcal{U}_{t+1}^-$.

On the other hand, the subspace $\mathcal{P}_{t+1}^-$ can be expressed as

$$\begin{aligned} \mathcal{P}_{t+1}^- &:= \overline{\text{span}}\{y(t),\, y(t-1),\, \cdots;\, u(t),\, u(t-1),\, \cdots\} \\ &= \overline{\text{span}}\{e(t),\, y(t-1),\, \cdots;\, u(t),\, u(t-1),\, \cdots\} \\ &= \text{span}\{e(t)\} \oplus \left(\mathcal{Y}_t^- \vee \mathcal{U}_{t+1}^-\right) \end{aligned}$$

It therefore follows that $w(t)$ is expressed as a function of $e(t)$. The matrix K is obtained by $w(t) = \hat{E}\{w(t) \mid e(t)\} = Ke(t)$. □

Summarizing the above results, we have the main theorem in this chapter.

Theorem 10.2. *Suppose that Assumptions 10.1 and 10.2 are satisfied. If the rank condition* $\text{rank}(\Sigma_{fp|u}) = n$ *holds, then the output* y *is described by the following state space model*

$$x(t+1) = Ax(t) + Bu(t) + Ke(t) \tag{10.38a}$$

$$y(t) = Cx(t) + Du(t) + e(t) \tag{10.38b}$$

This is a forward innovation model with the exogenous input u*, and the state vector* x *is an* n*-dimensional basis vector of the predictor space* $\mathcal{X}_t^{+/-}$ *given by (10.23).* □

Thus, it follows from (10.38) that the input and output relation is expressed as in Figure 10.3, where

$$P(z) = D + C(zI - A)^{-1}B$$

$$H(z) = I_p + C(zI - A)^{-1}K$$

Since the plant $P(z)$ and the noise model $H(z)$ have the same poles, we cannot parametrize these models independently. This result is due to the present approach itself based on the conditional CCA with exogenous inputs.

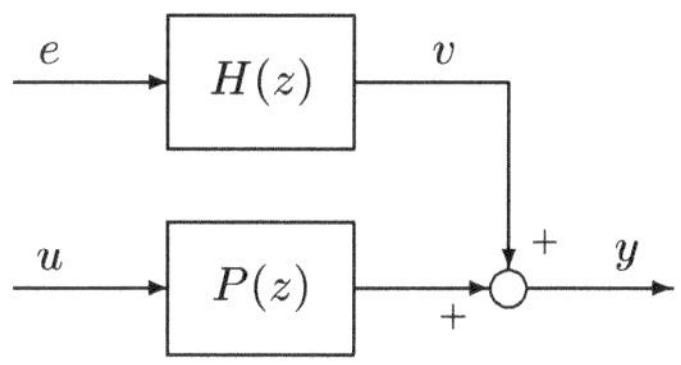

Figure 10.3. Transfer matrix model

From the Kalman filtering theory, we see that all other minimal representations for the output y are given by

$$z(t+1) = Az(t) + Bu(t) + Fv(t) \tag{10.39a}$$

$$y(t) = Cz(t) + Du(t) + Jv(t) \tag{10.39b}$$

where v is a zero mean white noise with covariance matrix I_q $(q \geq p)$. The matrices $F \in \mathbb{R}^{n \times q}$ and $J \in \mathbb{R}^{p \times q}$ are constant, and A, B, C, D are the same as those given in (10.38)[1]. Also, the state vector x of the innovation model of (10.38) is the minimum variance estimate for the state vector $z(t)$ of (10.39), *i.e.*,

$$x(t) = \hat{E}\{z(t) \mid \mathcal{P}_t^- \vee \mathcal{U}_t^+\} = \hat{E}\{z(t) \mid \mathcal{P}_t^-\}$$

The relation between F and J in (10.39) and K and $R := E\{e(t)e^{\mathrm{T}}(t)\}$ is already explained in Section 5.4.

[1] See Subsection 7.3.1, where the size of spectral factors associated with Markov models are discussed.

10.5 Stochastic Realization Based on Finite Data

In practice, the computations must be performed based on finite input-output data, and the construction of the innovation model should be based on the predictor of the future outputs obtained by the available finite data.

Suppose that $\tau \le t \le T$ and let t be the present time. Let $w(\tau), \cdots, w(t-1)$ be the truncated past vectors, and define the stacked vector

$$p_\tau(t) := \begin{bmatrix} w(t-1) \\ w(t-2) \\ \vdots \\ w(\tau) \end{bmatrix} \in \mathbb{R}^{(t-\tau)d\times 1}$$

and let $\mathcal{P}_{[\tau,t)}$ denote the past data space spanned by the above vector $p_\tau(t)$. The symbol $\mathcal{U}_{[t,T]}$ denotes the (finite) future input history after time t.

From these data we can form the finite-memory predictor at time t as

$$\begin{aligned} \hat{f}_\tau(t \mid t) &= \hat{E}\{f(t) \mid \mathcal{P}_{[\tau,t)} \vee \mathcal{U}_{[t,T]}\} \\ &= \hat{E}\{\hat{f}(t \mid t) \mid \mathcal{P}_{[\tau,t)} \vee \mathcal{U}_{[t,T]}\} \\ &= \hat{E}_{\|\mathcal{U}_{[t,T]}}\{\hat{f}(t \mid t) \mid \mathcal{P}_{[\tau,t)}\} + \hat{E}_{\|\mathcal{P}_{[\tau,t)}}\{\hat{f}(t \mid t) \mid \mathcal{U}_{[t,T]}\} \end{aligned}$$

where $\hat{f}(t \mid t)$ is defined by (10.9). The following result, which we shall state without proof, explains the role of the transient Kalman filter in finite data modeling; see also Theorem 6 in [107] and Theorem 3 in [165].

Theorem 10.3. *Suppose that Assumptions 10.1 and 10.2 are satisfied. If $\Sigma_{fp|u}$ has rank n, the process y admits a finite interval realization of the form*

$$\hat{x}_\tau(t+1) = A\hat{x}_\tau(t) + Bu(t) + K(t)\hat{e}_\tau(t) \tag{10.40a}$$

$$y(t) = C\hat{x}_\tau(t) + Du(t) + \hat{e}_\tau(t) \tag{10.40b}$$

where the state vector $\hat{x}_\tau(t)$ is a basis in the finite memory predictor space

$$\hat{\mathcal{X}}_t := \hat{E}\{\mathcal{X}_t^{+/-} \mid \mathcal{P}_{[\tau,t)} \vee \mathcal{U}_{[t,T]}\}$$

and the process $\{\hat{e}_\tau(t), \tau \le t \le T\}$ is the transient innovation of the output process $\{y(t), \tau \le t \le T\}$ with respect to $\mathcal{P}_{[\tau,t)} \vee \mathcal{U}_{[t,T]}$.

Proof. The result is proved by applying the Kalman filter algorithm to (10.39). See the finite interval realization of [106, 107]. □

We briefly make a comment on the non-stationary realization stated in Theorem 10.3. We see that any basis $\hat{x}_\tau(t) \in \hat{\mathcal{X}}_t$ has a representation

$$\hat{x}_\tau(t) = \hat{E}\{x(t) \mid \mathcal{P}_{[\tau,t)} \vee \mathcal{U}_{[t,T]}\}, \qquad t \ge \tau \tag{10.41}$$

where $x(t)$ is a basis in the stationary predictor space $\mathcal{X}_t^{+/-}$, and hence $\hat{x}_\tau(t)$ is also the transient Kalman filter estimate of $z(t)$, the state vector of (10.39), given the data $\mathcal{P}_{[\tau,t)} \vee \mathcal{U}_{[t,T]}$. The initial state for (10.40a) is

$$\hat{x}_\tau(\tau) = \hat{E}\{x(\tau) \mid \mathcal{U}_{[\tau,T]}\}$$

and the matrices $A,\ B,\ C,\ D$ are the same as those in (10.38).

We define the error covariance matrix of the state vector $z(t)$ of the system (10.39) as

$$P(t) = E\{[z(t) - \hat{x}_\tau(t)]\,[z(t) - \hat{x}_\tau(t)]^{\mathrm{T}}\}$$

It thus follows from (5.66) that $P(t)$ satisfies the Riccati equation

$$\begin{aligned} P(t+1) = AP(t)A^{\mathrm{T}} - (AP(t)C^{\mathrm{T}} + FJ^{\mathrm{T}})(CP(t)C^{\mathrm{T}} + JJ^{\mathrm{T}})^{-1} \\ \times (CP(t)A^{\mathrm{T}} + JF^{\mathrm{T}}) + FF^{\mathrm{T}} \end{aligned}$$

where $P(\tau) = \mathrm{cov}\{z(\tau) - \hat{x}_\tau(\tau)\}$, and the transient Kalman gain is given by

$$K(t) = (AP(t)C^{\mathrm{T}} + FJ^{\mathrm{T}})(CP(t)C^{\mathrm{T}} + JJ^{\mathrm{T}})^{-1}$$

Also, if $\tau \to -\infty$, the state vector $\hat{x}_\tau(t)$ of the transient innovation model of (10.40) converges to $x(t)$. Moreover, $P(t)$ converges to a unique stabilizing solution of the ARE

$$\begin{aligned} P = APA^{\mathrm{T}} - (APC^{\mathrm{T}} + FJ^{\mathrm{T}})(CPC^{\mathrm{T}} + JJ^{\mathrm{T}})^{-1} \\ \times (CPA^{\mathrm{T}} + JF^{\mathrm{T}}) + FF^{\mathrm{T}} \end{aligned} \qquad (10.42)$$

and hence $K(t)$ converges to $K = (APC^{\mathrm{T}} + FJ^{\mathrm{T}})(CPC^{\mathrm{T}} + JJ^{\mathrm{T}})^{-1}$.

Remark 10.2. The conditional CCA procedure of Section 10.4 applied to finite past data provides an approximate state vector $\hat{x}_\tau(t)$ differing from $x(t)$ by an additive initial condition term which tends to zero as $\tau \to -\infty$. In fact, from (10.41),

$$\hat{x}_\tau(t) = \hat{E}_{\| \mathcal{U}_{[t,T]}}\{x(t) \mid \mathcal{P}_{[\tau,t)}\} + \hat{E}_{\| \mathcal{P}_{[\tau,t)}}\{x(t) \mid \mathcal{U}_{[t,T]}\} \qquad (10.43)$$

Recall from (10.24) that $x(t) = \mathcal{O}^{\dagger}\hat{E}_{\| \mathcal{U}_{[t,T]}}\{f(t) \mid \mathcal{P}_t^-\}$ holds. Then the first term in the right-hand side of the above equation is an oblique projection which can be obtained by the conditional CCA of the finite future and past data.

Since $x(t) \in \mathcal{P}_t^-$ holds, the second term in the right-hand side of (10.43) tends to zero for $\tau \to -\infty$ by the absence of feedback, and hence the oblique projection of $x(t)$ onto the future $\mathcal{U}_{[t,T]}$ along the past $\mathcal{P}_{[\tau,t)}$ tends to the oblique projection along $\mathcal{P}_t^-$, which is clearly zero. □

10.6 CCA Method

In this section, a procedure of computing matrices Π_k and Ψ_k based on finite input-output data is developed. A basic procedure is to compute approximate solutions of the discrete Wiener-Hopf equations of (10.10), from which we have

$$\Pi_k = \Sigma_{fp|u}(\Sigma_{pp|u})^{-1} \in \mathbb{R}^{kp \times kd} \tag{10.44}$$

$$\Psi_k = \Sigma_{fu|p}(\Sigma_{uu|p})^{-1} \in \mathbb{R}^{kp \times km} \tag{10.45}$$

Once we obtain Π_k and Ψ_k, subspace identification methods of computing the system parameters A, B, C, D, K are easily derived.

Suppose that finite input-output data $u(t)$, $y(t)$ for $t = 0, 1, \cdots, N + 2k - 2$ are given with N sufficiently large and k positive. We assume that the time series $\{u(t),\ y(t)\}$ are sample values of the jointly stationary processes $(u,\ y)$ satisfying the assumptions of the previous sections, in particular the finite dimensionality and the feedback-free conditions. In addition, we assume throughout this section that the sample averages converge to the "true" expected values as $N \to \infty$.

Recall that $d = p + m$, the dimension of the joint process (u, y). Define the $kd \times N$ block Toeplitz matrix with N columns

$$\check{W}_{0|k-1} := \begin{bmatrix} w(k-1) & w(k) & \cdots & w(k+N-2) \\ w(k-2) & w(k-1) & \cdots & w(k+N-3) \\ \vdots & \vdots & & \vdots \\ w(0) & w(1) & \cdots & w(N-1) \end{bmatrix} \in \mathbb{R}^{kd \times N}$$

where $\check{W}_{0|k-1}$ denotes the past input-output data. Also, define block Hankel matrices

$$U_{k|2k-1} := \begin{bmatrix} u(k) & u(k+1) & \cdots & u(k+N-1) \\ u(k+1) & u(k+2) & \cdots & u(k+N) \\ \vdots & \vdots & & \vdots \\ u(2k-1) & u(2k) & \cdots & u(N+2k-2) \end{bmatrix} \in \mathbb{R}^{km \times N}$$

and

$$Y_{k|2k-1} := \begin{bmatrix} y(k) & y(k+1) & \cdots & y(k+N-1) \\ y(k+1) & y(k+2) & \cdots & y(k+N) \\ \vdots & \vdots & & \vdots \\ y(2k-1) & y(2k) & \cdots & y(N+2k-2) \end{bmatrix} \in \mathbb{R}^{kp \times N}$$

where $U_{k|2k-1}$ and $Y_{k|2k-1}$ denote the future input and the future output data, respectively.

In the following, we assume that the integer k is chosen so that $k > n$, where n is the dimension of the underlying stochastic system generating the data. Also, we assume that the input is PE with order $2k$, so that $U_{0|2k-1}$ has full row rank. Consider the following LQ decomposition:

$$\frac{1}{\sqrt{N}}\begin{bmatrix} U_{k|2k-1} \\ \check{W}_{0|k-1} \\ Y_{k|2k-1} \end{bmatrix} = \begin{bmatrix} R_{11} & 0 & 0 \\ R_{21} & R_{22} & 0 \\ R_{31} & R_{32} & R_{33} \end{bmatrix} \begin{bmatrix} Q_1^{\mathrm{T}} \\ Q_2^{\mathrm{T}} \\ Q_3^{\mathrm{T}} \end{bmatrix} =: \bar{R}Q^{\mathrm{T}} \tag{10.46}$$

where $R_{11} \in \mathbb{R}^{km\times km}$, $R_{22} \in \mathbb{R}^{kd\times kd}$, $R_{33} \in \mathbb{R}^{kp\times kp}$ are block lower triangular, and matrices Q_i are orthogonal with $Q_i^{\mathrm{T}} Q_j = I\delta_{ij}$.

By using (10.46), we have

$$\begin{bmatrix} \Sigma_{uu} & \Sigma_{up} & \Sigma_{uf} \\ \Sigma_{pu} & \Sigma_{pp} & \Sigma_{pf} \\ \Sigma_{fu} & \Sigma_{fp} & \Sigma_{ff} \end{bmatrix} := \frac{1}{N}\begin{bmatrix} U_{k|2k-1} \\ \check{W}_{0|k-1} \\ Y_{k|2k-1} \end{bmatrix}\begin{bmatrix} U_{k|2k-1} \\ \check{W}_{0|k-1} \\ Y_{k|2k-1} \end{bmatrix}^{\mathrm{T}} = \bar{R}\bar{R}^{\mathrm{T}}$$

It therefore follows that $\Sigma_{uu} = R_{11}R_{11}^{\mathrm{T}}$, $\Sigma_{pu} = R_{21}R_{11}^{\mathrm{T}}$, $\Sigma_{fu} = R_{31}R_{11}^{\mathrm{T}}$, and

$$\Sigma_{pp} = R_{21}R_{21}^{\mathrm{T}} + R_{22}R_{22}^{\mathrm{T}}, \qquad \Sigma_{fp} = R_{31}R_{21}^{\mathrm{T}} + R_{32}R_{22}^{\mathrm{T}}$$
$$\Sigma_{ff} = R_{31}R_{31}^{\mathrm{T}} + R_{32}R_{32}^{\mathrm{T}} + R_{33}R_{33}^{\mathrm{T}}$$

From the definition of conditional expectation of Lemma 10.1, we get

$$\Sigma_{ff|u} = \Sigma_{ff} - \Sigma_{fu}\Sigma_{uu}^{-1}\Sigma_{uf} = R_{32}R_{32}^{\mathrm{T}} + R_{33}R_{33}^{\mathrm{T}} \tag{10.47a}$$

$$\Sigma_{pp|u} = \Sigma_{pp} - \Sigma_{pu}\Sigma_{uu}^{-1}\Sigma_{up} = R_{22}R_{22}^{\mathrm{T}} \tag{10.47b}$$

$$\Sigma_{fp|u} = \Sigma_{fp} - \Sigma_{fu}\Sigma_{uu}^{-1}\Sigma_{up} = R_{32}R_{22}^{\mathrm{T}} \tag{10.47c}$$

$$\begin{aligned}\Sigma_{fu|p} &= \Sigma_{fu} - \Sigma_{fp}\Sigma_{pp}^{-1}\Sigma_{pu} \\ &= R_{31}R_{11}^{\mathrm{T}} - (R_{31}R_{21}^{\mathrm{T}} + R_{32}R_{22}^{\mathrm{T}})\Sigma_{pp}^{-1}R_{21}R_{11}^{\mathrm{T}}\end{aligned} \tag{10.47d}$$

$$\begin{aligned}\Sigma_{uu|p} &= \Sigma_{uu} - \Sigma_{up}\Sigma_{pp}^{-1}\Sigma_{pu} \\ &= R_{11}R_{11}^{\mathrm{T}} - R_{11}R_{21}^{\mathrm{T}}\Sigma_{pp}^{-1}R_{21}R_{11}^{\mathrm{T}}\end{aligned} \tag{10.47e}$$

It should be noted here that Σ_{uu} is positive definite by the PE condition. Also, we assume that Σ_{pp} and $\Sigma_{pp|u}$ are positive definite.

Lemma 10.6. *In terms of R_{ij} of (10.46), Π_k and Ψ_k are respectively expressed as*

$$\Pi_k = R_{32}R_{22}^{-1} \tag{10.48}$$

$$\Psi_k = (R_{31} - R_{32}R_{22}^{-1}R_{21})R_{11}^{-1} \tag{10.49}$$

Proof. Since $R_{22}^{\mathrm{T}}(R_{22}R_{22}^{\mathrm{T}})^{-1} = R_{22}^{-1}$, (10.48) is obvious from (10.44). We show (10.49). It follows from (10.47d) and (10.47e) that

$$\begin{aligned}\Sigma_{fu|p} &= R_{31}R_{11}^{\mathrm{T}} - (R_{31}R_{21}^{\mathrm{T}} + R_{32}R_{22}^{\mathrm{T}})\Sigma_{pp}^{-1}R_{21}R_{11}^{\mathrm{T}} \\ &= R_{31}(I_{km} - R_{21}^{\mathrm{T}}\Sigma_{pp}^{-1}R_{21})R_{11}^{\mathrm{T}} - R_{32}R_{22}^{\mathrm{T}}\Sigma_{pp}^{-1}R_{21}R_{11}^{\mathrm{T}}\end{aligned}$$

and $\Sigma_{uu|p} = R_{11}(I_{km} - R_{21}^{\mathrm{T}}\Sigma_{pp}^{-1}R_{21})R_{11}^{\mathrm{T}}$, respectively. From (10.45),

$$\begin{aligned}
\Psi_k &= \Sigma_{fu|p}(\Sigma_{uu|p})^{-1} \\
&= R_{31}R_{11}^{-1} - R_{32}R_{22}^{\mathrm{T}}\Sigma_{pp}^{-1}R_{21}(I_{km} - R_{21}^{\mathrm{T}}\Sigma_{pp}^{-1}R_{21})^{-1}R_{11}^{-1} \\
&= R_{31}R_{11}^{-1} - R_{32}R_{22}^{\mathrm{T}}(I_{km} - \Sigma_{pp}^{-1}R_{21}R_{21}^{\mathrm{T}})^{-1}\Sigma_{pp}^{-1}R_{21}R_{11}^{-1} \\
&= R_{31}R_{11}^{-1} - R_{32}R_{22}^{\mathrm{T}}(\Sigma_{pp} - R_{21}R_{21}^{\mathrm{T}})^{-1}R_{21}R_{11}^{-1} \\
&= (R_{31} - R_{32}R_{22}^{\mathrm{T}}(R_{22}R_{22}^{\mathrm{T}})^{-1}R_{21})R_{11}^{-1}
\end{aligned}$$

The right-hand side is equal to that of (10.49). □

Comparing the LQ decompositions of (10.46) and (6.59), we see that $\Pi_k \check{W}_{0|k-1}$ and Ψ_k obtained in Lemma 10.6 are the same as ξ of (6.65) and Ψ_k of (6.66), respectively. Thus, the present method based on the CCA technique is closely related to the N4SID method. The following numerical procedure is, however, different from that of the N4SID in the way of using the SVD to get the extended observability matrix.

In the following algorithm, it is assumed that the conditional covariance matrices $\Sigma_{ff|u}$, $\Sigma_{pp|u}$, $\Sigma_{fp|u}$ of (10.47) have already been obtained.

Subspace Identification of Stochastic System – CCA Method

Step 1: Compute the square roots of conditional covariance matrices[2]

$$\Sigma_{ff|u} = LL^{\mathrm{T}}, \qquad \Sigma_{pp|u} = MM^{\mathrm{T}}$$

Step 2: Compute the normalized SVD [see (10.19)]

$$L^{-1}\Sigma_{fp|u}M^{-\mathrm{T}} = USV^{\mathrm{T}} \simeq \hat{U}\hat{S}\hat{V}^{\mathrm{T}}$$

and then we get

$$\Sigma_{fp|u} \simeq L\hat{U}\hat{S}\hat{V}^{\mathrm{T}}M^{\mathrm{T}}$$

where $\hat{S}$ is obtained by neglecting smaller singular values, so that the dimension of the state vector equals $\dim(\hat{S})$.

Step 3: Define the extended observability and reachability matrices by [see (10.21)]

$$\mathcal{O}_k = L\hat{U}\hat{S}^{1/2}, \qquad \mathcal{C}_k = \hat{S}^{1/2}\hat{V}^{\mathrm{T}}M^{\mathrm{T}}$$

Algorithm A: Realization-based Approach

Step A4: Compute A and C by

$$\begin{aligned}
A &= \mathcal{O}_k(p+1:kp, 1:n)^{\dagger}\mathcal{O}_k(1:(k-1)p, 1:n) \\
C &= \mathcal{O}_k(1:p, 1:n)
\end{aligned}$$

---

[2]In general, there is a possibility that $\Sigma_{ff|u}$ and/or $\Sigma_{pp|u}$ are nearly rank deficient. Thus we use **svd** to compute L and M rather than **chol**, and the inverses are replaced by the pseudo-inverses.

Step A5: Given A, C and Ψ_k of (10.49), compute B and D by the least-squares method

$$\begin{bmatrix} I_p & 0_{p\times n} \\ 0 & \mathcal{O}_{k-1} \\ \hline I_p & 0_{p\times n} \\ 0 & \mathcal{O}_{k-2} \\ \hline \vdots & \vdots \\ \hline I_p & 0_{p\times n} \end{bmatrix} \begin{bmatrix} D \\ B \end{bmatrix} = \begin{bmatrix} \Psi_k(1:kp, 1:m) \\ \hline \Psi_k(p+1:kp, m+1:2m) \\ \hline \vdots \\ \hline \Psi_k((k-1)p+1:kp, (k-1)m+1:km) \end{bmatrix}$$

where $\mathcal{O}_j = \mathcal{O}_k(1:jp, 1:n)$, $j \leq k$.

Remark 10.3. Comparing the LQ decompositions of (10.46) and (9.48), we observe that since the data are the same except for arrangement, R_{32} of (10.46) corresponds to $[L_{42} \;\; L_{43}]$ of (9.48). Thus the construction of the extended observability matrix is quite similar to that of PO-MOESP [171]; see also Remark 9.3. It may be noted that a difference in two methods lies in the use of a normalized SVD of the conditional covariance matrix. □

We next present a subspace identification algorithm based on the use of the state estimates. The algorithm until *Step* 3 is the same as that of Algorithm A.

Algorithm B: Regression Approach Using State Vector

Step B4: The estimate of the state vector is given by [see (10.22)]

$$\bar{X}_k = \mathcal{C}_k \Sigma_{pp|u}^{-1} \check{W}_{0|k-1} = \hat{S}^{1/2} \hat{V}^{\mathrm{T}} M^{-1} \check{W}_{0|k-1} \in \mathbb{R}^{n\times N}$$

and compute matrices with $N-1$ columns

$$\hat{X}_{k+1} = \bar{X}_k(:, 2:N) \qquad \hat{X}_k = \bar{X}_k(:, 1:N-1)$$

$$\hat{Y}_{k|k} = Y_{k|k}(:, 1:N-1) \qquad \hat{U}_{k|k} = U(:, 1:N-1)$$

where $\hat{X}_{k+1}$, the state vector at time $k+1$, is obtained by shifting $\hat{X}_k$ under the assumption that k is sufficiently large.

Step B5: Compute the estimate of the system matrices $(A,\ B,\ C,\ D)$ by applying the least-squares method to the following overdetermined equations

$$\begin{bmatrix} \hat{X}_{k+1} \\ \hat{Y}_{k|k} \end{bmatrix} = \begin{bmatrix} A & B \\ C & D \end{bmatrix} \begin{bmatrix} \hat{X}_k \\ \hat{U}_{k|k} \end{bmatrix} + \begin{bmatrix} \rho_w \\ \rho_e \end{bmatrix}$$

where $\rho_w \in \mathbb{R}^{n\times(N-1)}$ and $\rho_e \in \mathbb{R}^{p\times(N-1)}$ are residuals.

Step B6: Compute the error covariance matrices

$$\begin{bmatrix} \Sigma_{ww} & \Sigma_{we} \\ \Sigma_{ew} & \Sigma_{ee} \end{bmatrix} = \frac{1}{N-1} \begin{bmatrix} \rho_w \rho_w^{\mathrm{T}} & \rho_w \rho_e^{\mathrm{T}} \\ \rho_e \rho_w^{\mathrm{T}} & \rho_e \rho_e^{\mathrm{T}} \end{bmatrix}$$

and solve the Kalman filter ARE [see (8.83)]

$$P = APA^{\mathrm{T}} - (APC^{\mathrm{T}} + \Sigma_{we})(CPC^{\mathrm{T}} + \Sigma_{ee})^{-1}(APC^{\mathrm{T}} + \Sigma_{we})^{\mathrm{T}} + \Sigma_{ww}$$

Then, by using the stabilizing solution, the Kalman gain is given by

$$K = (APC^{\mathrm{T}} + \Sigma_{we})(CPC^{\mathrm{T}} + \Sigma_{ee})^{-1}$$

where the matrix $A - KC$ is stable.

A program of Algorithm B is listed in Table D.6. □

The above procedure is correct for infinitely long past data and $N \to \infty$. For, it follows from Lemma 10.5 that the exact relations $\Sigma_{ee} = \Lambda = E\{e(t)e^{\mathrm{T}}(t)\}$, $\Sigma_{ww} = K\Sigma_{ee}K^{\mathrm{T}}$, $\Sigma_{we} = K\Sigma_{ee}$ should hold, so that the unique stabilizing solution of the Kalman filter ARE above exists and is actually $P = 0$.

For the finite data case, these exact relations do not hold and the sample covariance matrices computed in *Step B5* vary with k and N. However, under the assumption that the data are generated by a true system of order n, if N and k are chosen large enough with $N \gg k$, the procedure provides consistent estimates. It should be noted that the Kalman filter ARE has a unique stabilizing solution $P \geq 0$ from which we can estimate K. This is so, because by construction of the extended observability matrix $\mathcal{O}_k$, the pair (C, A) is observable and the covariance matrix of residuals is generically nonnegative definite.

10.7 Numerical Examples

We show some numerical results obtained by the CCA method, together with results by the ORT and PO-MOESP methods. We employ Algorithm B, which is based on the use of the estimate of the state vector.

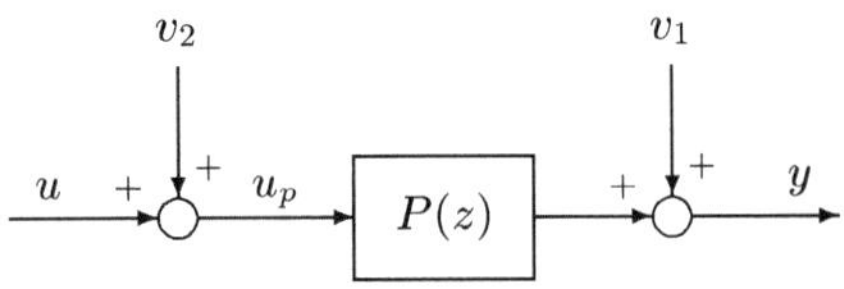

Figure 10.4. Simulation model

We consider the simulation model shown in Figure 10.4, from which the input-output relation of the system is expressed as

$$y(t) = P(z)u_p(t) + v_1(t), \qquad u_p(t) = u(t) + v_2(t) \tag{10.50}$$

where v_1 and v_2 are zero mean noises additively acting on the input and output signals, respectively. The plant $P(z)$ is the same as the one used in Section 9.8, and is given by

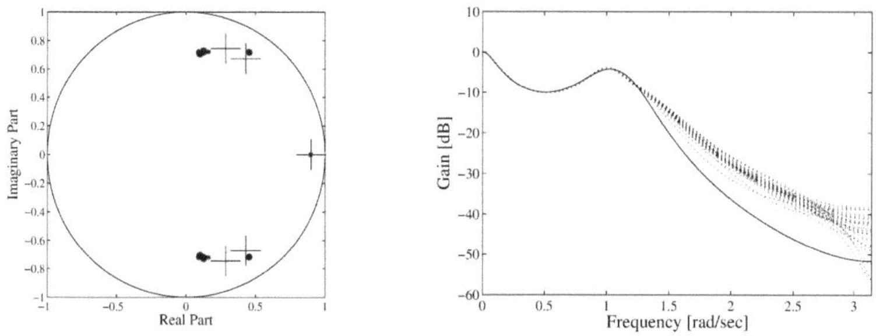

(a) (v_1 = colored, $v_2 = 0$) (b) (v_1 = colored, $v_2 = 0$)

Figure 10.5. Identification results by CCA

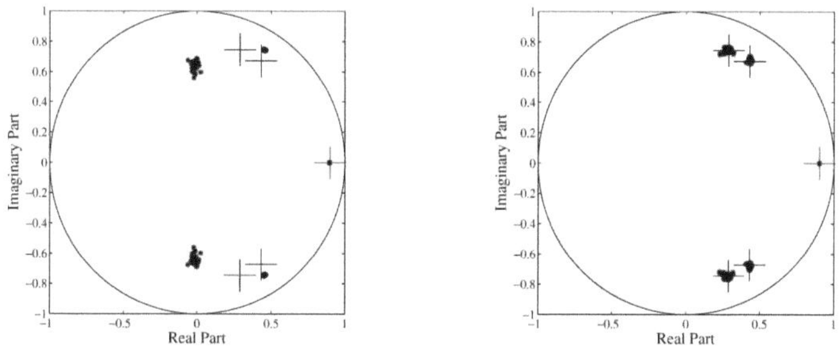

(a) (v_1 = colored, $v_2 = 0$) (b) (v_1 = colored, $v_2 = 0$)

Figure 10.6. Identification results: (a) MOESP and (b) ORT

$$P(z) = \frac{0.0275z^{-4} + 0.0551z^{-5}}{1 - 2.3443z^{-1} + 3.081z^{-2} - 2.5274z^{-3} + 1.2415z^{-4} - 0.3686z^{-5}}$$

Case 1: We consider the case where a colored noise is acting on the output, *i.e.*, v_1 is colored noise and $v_2 = 0$. The plant input u is a white noise with mean zero and variance $\sigma_u^2 = 1$, where the colored noise v_1 is an ARMA process generated by $v_1 = H(z)e$, where the noise model is given by

$$H(z) = \frac{1 - 0.2z^{-1} - 0.48z^{-2}}{1 + 0.4z^{-1} + 0.8z^{-2}}$$

and e is a zero mean white noise with variance σ_e^2, whose value is adjusted so that the variance of the colored noise becomes nearly $\sigma_1^2 = 0.01$.

The Bode gain plots of transfer functions $P(z)$ and $H(z)$ are displayed in Figure 9.8 in Section 9.8[3]. In the CCA and PO-MOESP methods, which are based on the innovation models, it is implicitly assumed that the plant and noise models have the same poles. However, note that the plant and noise models have different poles in

[3]The simulation conditions of the present example are the same as those of the example in Section 9.8; see Figures 9.9(b) and 9.10(b).

the simulation model of Figure 10.4; this is not consistent with the premise of the CCA and PO-MOESP methods. In fact, as shown below, results by the CCA and PO-MOESP methods have biases in the identification results.

As mentioned in Chapter 9, however, the ORT method conforms with this simulation model, since the ORT is based on the state space model with independent parametrizations for the plant and noise models. Hence, we expect that the ORT will provide better results than the CCA and PO-MOESP methods.

Taking the number of data $N = 1000$ and the number of block rows $k = 15$, we performed 30 simulation runs. Figure 10.5(a) displays the poles of the identified plant by the CCA method, where $+$ denotes the true poles of the plant, and $*$ denote the poles identified by 30 simulation runs. Figure 10.5(b) displays the Bode plots of the identified plant transfer functions, where the true gain is shown by the solid curve.

For comparison, Figure 10.6(a) and 10.6(b) display the poles of the identified plants by the PO-MOESP and ORT methods, respectively. In Figures 10.5(a) and 10.6(a), we observe rather large biases in the estimates of the poles, but, as shown in Figure 10.6(b), we do not observe biases in the results by the ORT method. Moreover, we see that the results by the CCA method are somewhat better than those by the PO-MOESP method.

Case 2: We consider the case where both v_1 and v_2 are mutually uncorrelated white Gaussian noises in Figure 10.4, so that $H(z) = 1$.

First we show that in this case the model of Figure 10.4 is reduced to an innovation model with the same form as the one derived in Theorem 10.2. It is clear that the effect of the noise v_2 on the output is given by $P(z)v_2$, so that the input-output relation of (10.50) is described by

$$y(t) = P(z)u(t) + [1 \quad P(z)] \begin{bmatrix} v_1(t) \\ v_2(t) \end{bmatrix} \tag{10.51}$$

where the noise model is a 1×2 transfer matrix $L(z) = [1 \quad P(z)]$; thus the poles of the plant and the noise model are the same. Hence, the transfer matrix model of (10.51) is a special case of the innovation model (see Figure 10.3)

$$y(t) = P(z)u(t) + H(z)e(t)$$

where $H(z)$ is a minimum phase transfer matrix satisfying

$$\sigma_e^2 |H(e^{j\omega})|^2 = L(e^{j\omega}) \begin{bmatrix} \sigma_1^2 & 0 \\ 0 & \sigma_2^2 \end{bmatrix} L^{\mathrm{T}}(e^{j\omega}) = \sigma_1^2 + \sigma_2^2 |P(e^{j\omega})|^2$$

with $H(\infty) = 1$.

The transfer matrix $H(z)$ can be obtained by a technique of spectral factorization. In fact, deriving a state space model for the noise model $L(z)v$, and solving the ARE associate with it, we obtain an innovation model, from which we have the desired transfer function $H(z)$. Thus, in this case, the model of Figure 10.4 is compatible with the CCA method.

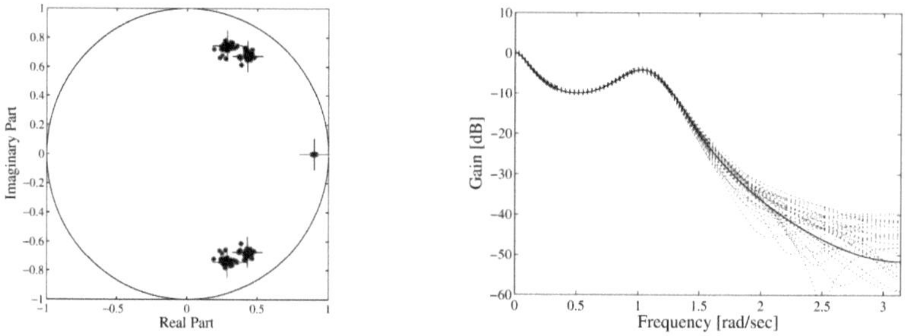

(a) CCA (v_1, v_2 = white) (b) CCA (v_1, v_2 = white)

Figure 10.7. Identification results by CCA

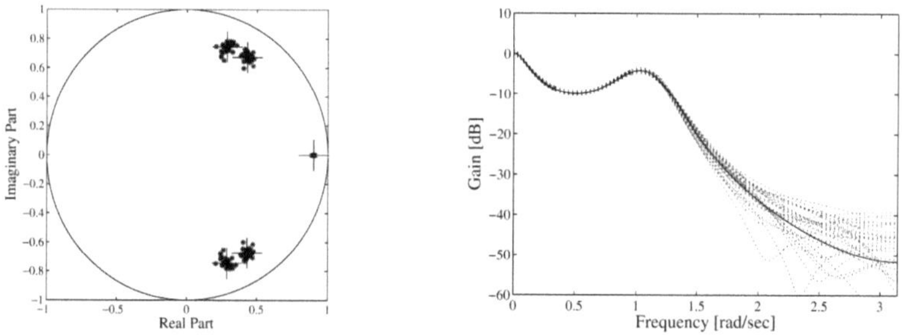

(a) ORT (v_1, v_2 = white) (b) ORT (v_1, v_2 = white)

Figure 10.8. Identification results by ORT

Numerical results by the CCA and ORT methods are displayed in Figures 10.7 and 10.8, respectively, where it is assumed that $\sigma_1^2 = 0.01$, $\sigma_2^2 = 0.09$, $\sigma_u^2 = 1$, and the number of data $N = 1000$, the number of block rows $k = 15$. Since both v_1 and v_2 are white noises, we do not see appreciable biases in the poles of the identified plants, though there are some variations in the estimates. This is due to the fact that the present simulation model is fitted in with the CCA method as well as with the ORT method.

In the simulations above, we have fixed the number of data N and the number of block rows k. In the next case, we present some simulation results by the CCA and ORT methods by changing the number of block rows k.

Case 3: We present some simulation results by the CCA and ORT methods by changing the number of block rows as $k = 8, 10, 15, 20$, while the number of columns of data matrices is fixed as $N = 4000$. The simulation model is the same as in Case 2, where there exist both input and output white noises, and the noise variances are fixed as $\sigma_1^2 = 1$ and $\sigma_2^2 = 0.09$.

We see from Figure 10.9 that the performance of identification by the CCA method is rather independent of the values of k. Though it is generally said that a sufficiently large $k\,(> n)$ is recommended, the present results show that the recom-

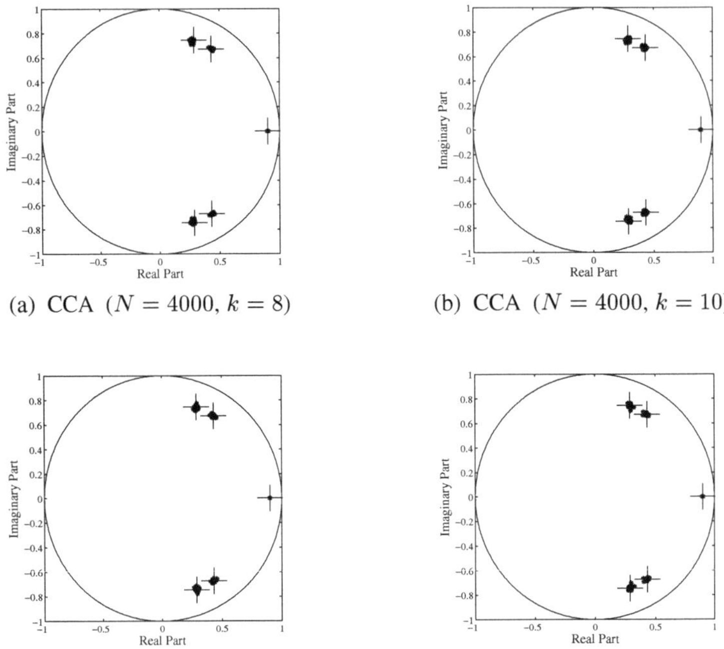

(a) CCA ($N = 4000$, $k = 8$)
(b) CCA ($N = 4000$, $k = 10$)
(c) CCA ($N = 4000$, $k = 15$)
(d) CCA ($N = 4000$, $k = 20$)

Figure 10.9. Identification results by CCA

mendation is not always true. We can thus safely say that the number of block rows k should be chosen in relation to the number of columns N.

Figure 10.10 displays the results of identification by the ORT method. In contrast to the results by CCA method, the performance is improved by taking larger k, while for small k we see some variations in the poles of identified plant. These results may be due to the fact that computation of the deterministic components by the LQ decomposition is not very accurate if k gets smaller.

In this section, we have compared simulation results by using the CCA and ORT methods. We observe that the performance of CCA method is slightly better than that of PO-MOESP, where both methods are based on the innovation models. We also conclude that the performance of the ORT method is better than that of the CCA method, especially if we use a general noise model.

10.8 Notes and References

- In this chapter, we have described some stochastic realization results in the presence of exogenous inputs based on Katayama and Picci [90]. As in Chapter 9, we have assumed that there is no feedback from the output to the input, and the input has a PE condition of sufficiently high order.

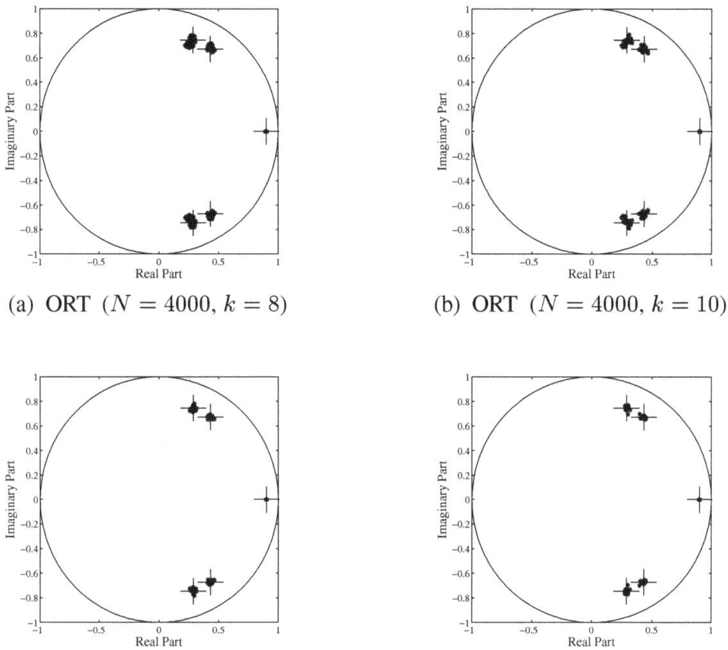

(a) ORT $(N = 4000, k = 8)$

(b) ORT $(N = 4000, k = 10)$

(c) ORT $(N = 4000, k = 15)$

(d) ORT $(N = 4000, k = 20)$

Figure 10.10. Identification results by ORT

- In Section 10.1, after stating the stochastic realization problem in the presence of an exogenous input, we considered a multi-stage Wiener prediction problem of estimating the future outputs in terms of the past input-output and the future inputs. This problem is solved by using the oblique projection, and the optimal predictor for the future outputs is derived in Section 10.2.
- In Section 10.3, by defining the conditional CCA, we have obtained a state vector for a stochastic system that includes the information contained in the past data needed to predict the future. In Section 10.4, the state vector so defined is employed to derive a forward innovation model for the system with an exogenous input.
- In Section 10.5, we have provided a theoretical foundation to adapt the stochastic realization theory to finite input-output data, and derived a non-stationary innovation model based on the transient Kalman filter. In Section 10.6, by means of the LQ decomposition and SVD, we have derived two subspace identification methods. A relation of the CCA method to the N4SID method is also clarified. In Section 10.7, some simulation results are included.

In the following, some comments are provided for the CCA method developed in this chapter and the ORT method in Chapter 9, together with some other methods.

- The earlier papers dealt with subspace identification methods based on the CCA are [100, 101, 128]. The method of Larimore, called the CVA method, is based on the solution of k-step prediction problem, and has been applied to identification of many industrial plants; see [102] and references therein.
- In the ORT method developed in Chapter 9, as shown in Figure 10.11, we start with the decomposition of the output y into a deterministic component $y_d \in \mathcal{U}$ and a stochastic component $y_s \in \mathcal{U}^\perp$, thereby dividing the problem into two identification problems for deterministic and stochastic subsystems. Hence, from the point of view of identifying the plant, the ORT method is similar to deterministic subspace identification methods, and the identification of the noise model is a version of the standard stochastic subspace identification method for stationary processes.

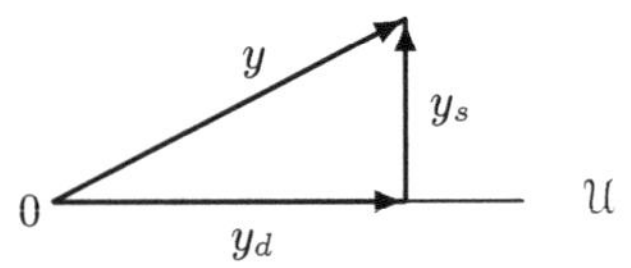

Figure 10.11. Orthogonal decomposition

- On the other hand, the CCA method is based on the conditional canonical correlations between the future and the past after deleting the effects of the future inputs, so that this method is regarded as an extension of the CCA method due to Akaike [2, 3], Desai *et al.* [42, 43], and Larimore [100, 101].

11

Identification of Closed-loop System

This chapter discusses the identification of closed-loop systems based on the subspace identification methods developed in the previous chapters. First we explain three main approaches to the closed-loop identification. Then, in the framework of the joint input-output approach, we consider the stochastic realization problem of the closed-loop system by using the CCA method, and derive a subspace method of identifying the plant and controller. Also, we consider the same problem based on the ORT method, deriving a subspace method of identifying the plant and controller by using the deterministic component of the joint input-output process. Further, a model reduction method is introduced to get lower order models. Some simulation results are included. In the appendix, under the assumption that the system is open-loop stable, we present simple methods of identifying the plant, controller and the noise model from the deterministic and stochastic components of the joint input-output process, respectively.

11.1 Overview of Closed-loop Identification

The identification problem for linear systems operating in closed-loop has received much attention in the literature, since closed-loop experiments are necessary if the open-loop plant is unstable, or the feedback is an inherent mechanism of the system [48, 145, 158]. Also, safety and maintaining high-quality production may prohibit experiments in open-loop setting.

The identification of multivariable systems operating in closed-loop by subspace methods has been the topic of active research in the past decade. For example, in [161], the joint input-output approach is used for deriving the state space models of subsystems, followed by a balanced model reduction. Also, based on a subspace method, a technique of identifying the state space model of a plant operating in closed-loop has been studied by reformulating it as an equivalent open-loop identification problem [170]. In addition, modifying the N4SID method [165], a closed-loop subspace identification method has been derived under the assumption that a finite

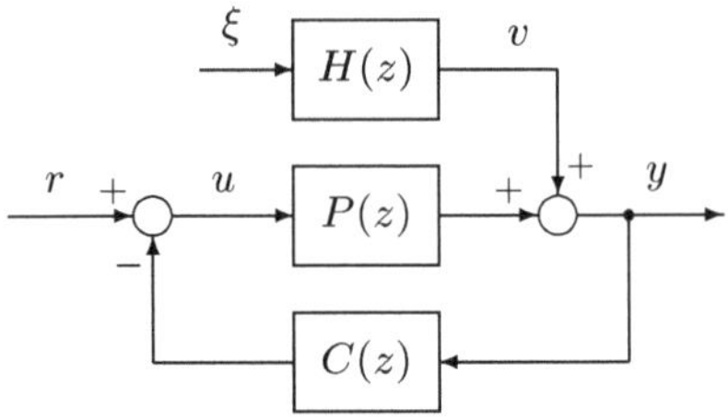

Figure 11.1. A feedback system [48]

number of Markov parameters of the controller are known [167]. And, a subspace-based closed-loop identification of linear state space model has been treated by using the CCA technique [34].

Figure 11.1 shows a typical feedback control system, where $P(z)$, $C(z)$, and $H(z)$ denote respectively the plant, the controller and the noise model, and where r is the exogenous input, u the control input, and y the plant output, v the unmeasurable disturbance. A standard closed-loop identification problem is to identify the plant based on the measurable exogenous input r and the plant input u and output y.

A fundamental difficulty with closed-loop identification is due to the existence of correlations between the external unmeasurable noise v and the control input u. In fact, if there is a correlation between u and v, it is well known that the least-squares method provides a biased estimate of the plant[1]. This is also true for subspace identification methods. Recall that we have assumed in Chapters 9 and 10 that there is no feedback from the output y to the input u, which is a basic condition for the open-loop system identification.

We review three approaches to closed-loop identification [48, 109]. The area of closed-loop identification methods can be classified into three groups.

1. **Direct Approach** Ignoring the existence of the feedback loop, we directly apply open-loop identification methods to the measurable input-output data (u, y) for identifying the plant $P(z)$.
2. **Indirect Approach** Suppose that the exogenous input r is available for identification, and that the controller transfer function $C(z)$ is known. We first identify the transfer function $T_{yr}(z)$ from r to the output y, and then compute the plant transfer function by using the formula

$$P(z) = \frac{T_{yr}(z)}{1 - C(z)T_{yr}(z)} \tag{11.1}$$

3. **Joint Input-Output Approach** Suppose that there exists an input r that can be utilized for system identification. We first identify the transfer functions $T_{ur}(z)$ and $T_{yr}(z)$ from the exogenous input r to the joint input-output (u, y), and then compute the plant transfer function using the algebraic relation

[1]This situation corresponds to the case where Assumption A1) in Section A.1 is violated.

$$P(z) = \frac{T_{yr}(z)}{T_{ur}(z)} \tag{11.2}$$

We shall now provide some comments on the basic approaches to closed-loop identification stated above.

- It is clear that the direct approach provides biased estimates. However, since the procedure is very simple, this approach is practical if the bias is not significant. In order to overcome the difficulty associated with the biases, modified methods called two stage least-square methods and the projection method are developed in [49, 160]. The basic idea is to identify the sensitivity function of the closed-loop system by using ARMA or finite impulse response (FIR) models, by which the estimate $\hat{u}$ of the input u is generated removing the noise effects. Then, the estimated input $\hat{u}$ and the output y are employed to identify the plant transfer function using a standard open-loop identification technique.
- For the indirect approach, the knowledge of the controller transfer function is needed. However due to possible deterioration of the controller characteristics and/or inclusion of some nonlinearities like limiter and dead zone, the quality of the estimates will be degraded. Moreover, the estimate of $P(z)$ obtained by (11.1) is of higher order, which is typically the sum of orders of $T_{yr}(z)$ and $C(z)$, so that we need some model reduction procedures. There are also related methods of using the dual Youla parametrization, which parametrizes all the plants stabilized by a given controller. By using the dual Youla parametrization, the closed-loop identification problem is converted into an open-loop identification problem; see [70, 141, 159] for details.
- The advantage of the joint input-output approach is that the knowledge of the controller is not needed. However, the joint input-output approach has the same disadvantage as the indirect approach that the estimated plant transfer functions are of higher order. It should also be noted that in this approach we should deal with vector processes even if we consider the identification of scalar systems. In this sense, the joint input-output approach should be best studied in the framework of subspace methods.

11.2 Problem Formulation

11.2.1 Feedback System

We consider the problem of identifying a closed-loop system based on input-output measurements. The configuration of the system is shown in Figure 11.2, where $y \in \mathbb{R}^p$ is the output vector of the plant, and $u \in \mathbb{R}^m$ the input vector. The noise models $H(z)$ and $F(z)$ are minimum phase square rational transfer matrices with $H(\infty) = I_p$ and $F(\infty) = I_m$, where the inputs to the noise models are respectively white noises $\nu \in \mathbb{R}^p$ and $\eta \in \mathbb{R}^m$ with means zero and positive definite covariance matrices. The inputs $r_1 \in \mathbb{R}^p$ and $r_2 \in \mathbb{R}^m$ may be interpreted as the exogenous reference signal and a probing input (dither) or a measurable disturbance.

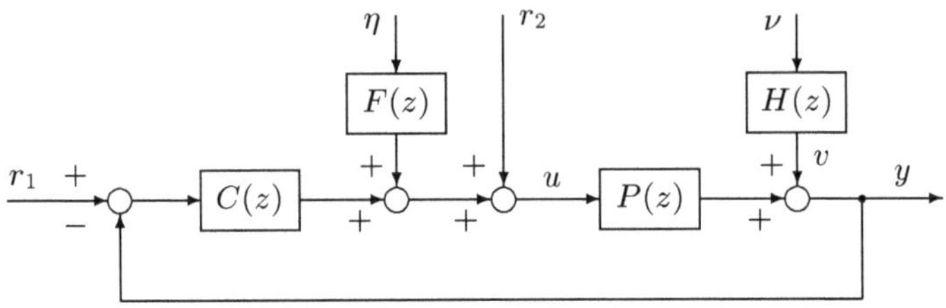

Figure 11.2. Closed-loop system

Let the plant be a finite dimensional LTI system described by

$$y(t) = P(z)u(t) + H(z)\nu(t) \tag{11.3}$$

where $P(z)$ is the $(p \times m)$-dimensional transfer matrix of the plant. Also, the control input is generated by

$$u(t) = r_2(t) + C(z)[r_1(t) - y(t)] + F(z)\eta(t) \tag{11.4}$$

where $C(z)$ denotes the $(m \times p)$-dimensional transfer matrix of the LTI controller.

We introduce the following assumptions on the closed-loop system, exogenous inputs, and noises.

Assumption 11.1. *A1) The closed-loop system is well-posed in the sense that* $(u,\ y)$ *are uniquely determined by the states of the plant and controller and by the exogenous inputs and noises. This generic condition is satisfied if* $I_p + P(\infty)C(\infty)$ *and* $I_m + C(\infty)P(\infty)$ *are nonsingular. For the sake of simplicity, it is assumed that the plant is strictly proper, i.e.,* $P(\infty) = 0$.

A2) The controller internally stabilizes the closed-loop system.

A3) The exogenous input $r := \begin{bmatrix} r_1 \\ r_2 \end{bmatrix} \in \mathbb{R}^d$ $(d = p + m)$ *satisfies PE condition, and is uncorrelated with the noise* $\chi := \begin{bmatrix} \nu \\ \eta \end{bmatrix} \in \mathbb{R}^d$*; thus* $r_1(t),\ r_2(s),\ \nu(\tau),\ \eta(\sigma)$ *are uncorrelated for all* $t,\ s,\ \tau,\ \sigma \in \mathbb{Z}$.

In the following, we consider the problem of identifying the deterministic part of the closed-loop system, or the plant $P(z)$ and controller $C(z)$, using the measurable finite data $\{r_1(t),\ r_2(t),\ u(t),\ y(t),\ t = 0, 1, \cdots, N-1\}$.

Remark 11.1. The identification of controller $C(z)$ may not be needed in applications. However, if the identified controller agrees well with the known controller transfer function, this will be an evidence that the identification results are plausible. Also, there are many chemical plants which contain recycle paths of energy and materials, so that the identification of closed-loop systems is very important from both theoretical and practical points of view. □

The objective of this chapter is to obtain state space models of the plant $P(z)$ and the controller $C(z)$ based on finite measurement data $\{r_1(t),\ r_2(t),\ u(t),\ y(t)\}$ by using subspace identification methods. In the following, we present two closed-loop identification algorithms based on the CCA and ORT methods. The first one, based on the CCA method, is rather close to that of Verhaegen [170]. The second one, based on the ORT method, is quite different from existing closed-loop idnetification algorithms, including that of [170].

11.2.2 Identification by Joint Input-Output Approach

In order to obtain state space models of the plant and controller in closed-loop, we use the joint input-output approach.

Define the joint input-output process

$$w := \begin{bmatrix} y \\ u \end{bmatrix} \in \mathbb{R}^d \tag{11.5}$$

It then follows from Figure 11.2 that these signals are related by

$$w(t) = T_{wr}(z)r(t) + T_{w\chi}(z)\chi(t) \tag{11.6}$$

where $T_{wr}(z)$ and $T_{w\chi}(z)$ are the closed-loop system transfer matrices defined by

$$T_{wr}(z) = \begin{bmatrix} T_{yr_1}(z) & T_{yr_2}(z) \\ T_{ur_1}(z) & T_{ur_2}(z) \end{bmatrix} = \begin{bmatrix} P(z)S_i(z)C(z) & P(z)S_i(z) \\ S_i(z)C(z) & S_i(z) \end{bmatrix} \tag{11.7}$$

and

$$T_{w\chi}(z) = \begin{bmatrix} T_{y\nu}(z) & T_{y\eta}(z) \\ T_{u\nu}(z) & T_{u\eta}(z) \end{bmatrix} = \begin{bmatrix} S_o(z)H(z) & P(z)S_i(z)F(z) \\ -C(z)S_o(z)H(z) & S_i(z)F(z) \end{bmatrix}$$

and where

$$S_i(z) = (I_m + C(z)P(z))^{-1}, \qquad S_o(z) = (I_p + P(z)C(z))^{-1}$$

are the input and output sensitivity matrices, respectively.

Recall that the feedback system is internally stable if and only if the four transfer matrices in (11.7) are stable. Since r and χ are uncorrelated in (11.6), there is no feedback from w to r; hence we can employ open-loop identification techniques to estimate the transfer matrix $T_{wr}(z) = [T_{wr_1}(z)\ \ T_{wr_2}(z)]$, using measurements of the input r and the output w.

In order to deal with a well-posed estimation problem, these transfer matrices should be uniquely obtainable from the overall transfer matrix $T_{wr}(z)$. It follows from (11.7) that $P(z)$ and $C(z)$ are identifiable from

$$P(z) = T_{yr_2}(z)T_{ur_2}^{-1}(z), \qquad C(z) = T_{ur_2}^{-1}(z)T_{ur_1}(z) \tag{11.8}$$

where the inverse exists because $S_i(z)$ is invertible. Hence, contrary to the indirect approach [167], we do not need the knowledge of the controller, nor the auxiliary input needed in the method based on the dual-Youla parametrization approach. It should, however, be noted that in order that both $P(z)$ and $C(z)$ be uniquely identifiable from the data, in general we need to have both signals r_1 and r_2 acting on the system[2].

In addition to Assumption 11.1 A1) $\sim$ A3), we need the following.

Assumption 11.2. *There is no feedback from the joint input-output process w to the exogenous input r.* □

11.3 CCA Method

In this section, we apply the CCA method developed in Chapter 10 to the closed-loop identification problem of identifying the plant and controller based on the joint input-output approach.

11.3.1 Realization of Joint Input-Output Process

It follows from Theorem 10.2 that the innovation model for the joint input-output process w with the input r has the following form

$$x(t+1) = Ax(t) + [B_1 \;\; B_2] \begin{bmatrix} r_1(t) \\ r_2(t) \end{bmatrix} + [K_1 \;\; K_2] \begin{bmatrix} e_1(t) \\ e_2(t) \end{bmatrix} \tag{11.9a}$$

$$\begin{bmatrix} y(t) \\ u(t) \end{bmatrix} = \begin{bmatrix} C_1 \\ C_2 \end{bmatrix} x(t) + \begin{bmatrix} 0 & 0 \\ D_{21} & D_{22} \end{bmatrix} \begin{bmatrix} r_1(t) \\ r_2(t) \end{bmatrix} + \begin{bmatrix} e_1(t) \\ e_2(t) \end{bmatrix} \tag{11.9b}$$

where the dimension of the state vector is generically the sum of the orders of the plant and controller ($n = n_p + n_c$), and where $D_{11} = 0$, $D_{12} = 0$ from the condition $P(\infty) = 0$.

We see from (11.9) that the transfer matrices from r_1 to u and from r_2 to y, u are given by

$$T_{ur_1} = \left[\begin{array}{c|c} A & B_1 \\ \hline C_2 & D_{21} \end{array}\right], \quad T_{yr_2} = \left[\begin{array}{c|c} A & B_2 \\ \hline C_1 & 0 \end{array}\right], \quad T_{ur_2} = \left[\begin{array}{c|c} A & B_2 \\ \hline C_2 & D_{22} \end{array}\right] \tag{11.10}$$

Thus we have the following result.

Lemma 11.1. *Suppose that a realization of the joint input-output process w is given by (11.9). Then, realizations of the plant and controller are respectively computed by*

$$P(z) = \left[\begin{array}{c|c} A - B_2 D_{22}^{-1} C_2 & B_2 D_{22}^{-1} \\ \hline C_1 & 0 \end{array}\right] \tag{11.11}$$

and

[2] The case where one of the two signals is absent is discussed in [89].

$$C(z) = \left[\begin{array}{c|c} A - B_2 D_{22}^{-1} C_2 & B_1 - B_2 D_{22}^{-1} D_{21} \\ \hline D_{22}^{-1} C_2 & D_{22}^{-1} D_{21} \end{array}\right] \tag{11.12}$$

Proof. Since $P = T_{yr_2} T_{ur_2}^{-1}$, it follows from (11.10) that

$$\begin{aligned} P(z) &= \left[\begin{array}{c|c} A & B_2 \\ \hline C_1 & 0 \end{array}\right] \left[\begin{array}{c|c} A & B_2 \\ \hline C_2 & D_{22} \end{array}\right]^{-1} \\ &= \left[\begin{array}{c|c} A & B_2 \\ \hline C_1 & 0 \end{array}\right] \left[\begin{array}{c|c} A - B_2 D_{22}^{-1} C_2 & B_2 D_{22}^{-1} \\ \hline -D_{22}^{-1} C_2 & D_{22}^{-1} \end{array}\right] \\ &= \left[\begin{array}{cc|c} A & -B_2 D_{22}^{-1} C_2 & B_2 D_{22}^{-1} \\ 0 & A - B_2 D_{22}^{-1} C_2 & B_2 D_{22}^{-1} \\ \hline C_1 & 0 & 0 \end{array}\right] \end{aligned}$$

By the coordinate transform $T = \begin{bmatrix} I & I \\ 0 & I \end{bmatrix}$, we obtain (11.11). Also, from the relation $C = T_{ur_2}^{-1} T_{ur_1}$, we can prove (11.12). □

It may be noted that the matrix D_{22} should be nonsingular to compute the inverse matrix above. This implies that the exogenous input r_2 must satisfy the PE condition.

Let the state space models of the plant and controller be given by

$$x_p(t+1) = A_p x_p(t) + B_p u(t) \tag{11.13a}$$

$$y(t) = C_p x_p(t) \tag{11.13b}$$

and

$$x_c(t+1) = A_c x_c(t) + B_c[r_1(t) - y(t)] \tag{11.14a}$$

$$u(t) = r_2(t) + C_c x_c(t) + D_c[r_1(t) - y(t)] \tag{11.14b}$$

where $x_p \in \mathbb{R}^{n_p}$ and $x_c \in \mathbb{R}^{n_c}$ are the state vectors of the plant and controller, respectively. We show that the models of (11.11) and (11.12) are not minimal.

Lemma 11.2. *Suppose that realizations of the plant and controller are respectively given by (11.13) and (11.14). Then, the following realizations*

$$P(z) = \left[\begin{array}{cc|c} A_p & 0 & B_p \\ -B_c C_p & A_c & 0 \\ \hline C_p & 0 & 0 \end{array}\right] \tag{11.15}$$

and

$$C(z) = \left[\begin{array}{cc|c} A_p & 0 & 0 \\ -B_p C_p & A_c & B_c \\ \hline -D_c C_p & C_c & D_c \end{array}\right] \tag{11.16}$$

are input-output equivalent to realizations of (11.11) and (11.12), respectively. Hence, the reachable and observable part of non-minimal realizations are the state space realizations of the plant and controller, respectively.

Proof. Combining (11.13) and (11.14) yields

$$T_{wr}(z) = \left[\begin{array}{cc|cc} A_p - B_pD_cC_p & B_pC_c & B_pD_c & B_p \\ -B_cC_p & A_c & B_c & 0 \\ \hline C_p & 0 & 0 & 0 \\ -D_cC_p & C_c & D_c & I_m \end{array}\right] \tag{11.17}$$

For simplicity, we define

$$\bar{A} = \begin{bmatrix} A_p & 0 \\ -B_cC_p & A_c \end{bmatrix}, \quad [\bar{B}_1 \ \bar{B}_2] = \begin{bmatrix} B_pD_c & B_p \\ B_c & 0 \end{bmatrix}, \quad \begin{bmatrix} \bar{C}_1 \\ \bar{C}_2 \end{bmatrix} = \begin{bmatrix} C_p & 0 \\ -D_cC_p & C_c \end{bmatrix}$$

Then, we have

$$T_{yr_2}(z) = \left[\begin{array}{c|c} \bar{A} + \bar{B}_2\bar{C}_2 & \bar{B}_2 \\ \hline \bar{C}_1 & 0 \end{array}\right], \qquad T_{ur_2}(z) = \left[\begin{array}{c|c} \bar{A} + \bar{B}_2\bar{C}_2 & \bar{B}_2 \\ \hline \bar{C}_2 & I_m \end{array}\right]$$

so that $T_{ur_2}^{-1}(z) = \left[\begin{array}{c|c} \bar{A} & -\bar{B}_2 \\ \hline \bar{C}_2 & I_m \end{array}\right]$. Thus, it follows that

$$P(z) = T_{yr_2}(z)T_{ur_2}^{-1}(z) = \left[\begin{array}{cc|c} \bar{A} & 0 & -\bar{B}_2 \\ \bar{B}_2\bar{C}_2 & \bar{A} + \bar{B}_2\bar{C}_2 & \bar{B}_2 \\ \hline 0 & \bar{C}_1 & 0 \end{array}\right] =: \left[\begin{array}{c|c} A_s & B_s \\ \hline C_s & 0 \end{array}\right]$$

Let $S = \begin{bmatrix} I & 0 \\ -I & I \end{bmatrix}$, and $S^{-1} = \begin{bmatrix} I & 0 \\ I & I \end{bmatrix}$. Then, we obtain

$$S^{-1}A_sS = \begin{bmatrix} \bar{A} & 0 \\ 0 & \bar{A} + \bar{B}_2\bar{C}_2 \end{bmatrix}, \quad S^{-1}B_s = \begin{bmatrix} -\bar{B}_2 \\ 0 \end{bmatrix}, \quad C_sS = [-\bar{C}_1 \ \ \bar{C}_1]$$

It therefore follows that

$$P(z) = \left[\begin{array}{cc|c} \bar{A} & 0 & -\bar{B}_2 \\ 0 & \bar{A} + \bar{B}_2\bar{C}_2 & 0 \\ \hline -\bar{C}_1 & \bar{C}_1 & 0 \end{array}\right] = \left[\begin{array}{c|c} \bar{A} & \bar{B}_2 \\ \hline \bar{C}_1 & 0 \end{array}\right]$$

The right-hand side is equal to (11.15). Similarly, for a proof of (11.16), we can use $T_{ur_1}(z) = \left[\begin{array}{c|c} \bar{A} + \bar{B}_2\bar{C}_2 & \bar{B}_1 \\ \hline \bar{C}_2 & D_c \end{array}\right]$ obtained from (11.17) and $C(z) = T_{yr_2}(z)T_{ur_1}^{-1}(z)$.

For the realization (11.15), it follows from Theorems 3.4 (ii) and 3.7 (ii) that the rank conditions

$$\operatorname{rank} \begin{bmatrix} zI - A_p & 0 & B_p \\ B_cC_p & zI - A_c & 0 \end{bmatrix} < n_p + n_c, \qquad z \in \lambda(A_c)$$

and

$$\operatorname{rank} \begin{bmatrix} zI - A_p & 0 \\ B_c C_p & zI - A_c \\ C_p & 0 \end{bmatrix} < n_p + n_c, \qquad z \in \lambda(A_c)$$

hold, implying that there are n_c pole-zero cancellations in the realization of $P(z)$. Thus, this realization is unreachable and unobservable. Hence the reachable and observable part of the realization (11.15) will be a relevant state space realization of the plant. For the realization (11.16) of the controller, it can be shown that there exist n_p pole-zero cancellations. □

Since a strict pole-zero cancellation does not exist in the realizations of (11.11) and (11.12), which are identified by using finite data, we see that the dimension of the state space realizations are of higher dimension with $n := n_p + n_c$. It is therefore necessary to obtain lower order models from higher order models by using a model reduction procedure. This problem is treated in Section 11.5.

11.3.2 Subspace Identification Method

We describe a subspace identification method based on the results of Section 10.6. Let $r_1(t),\ r_2(t),\ u(t),\ y(t),\ \ t = 0, 1, \cdots, N + 2k - 2$ be a set of given finite data, where N is sufficiently large and $k > n$. Recall that the exogenous inputs and the joint input-output w are defined as $r(t) = \begin{bmatrix} r_1(t) \\ r_2(t) \end{bmatrix} \in \mathbb{R}^d$ and $w(t) = \begin{bmatrix} y(t) \\ u(t) \end{bmatrix} \in \mathbb{R}^d$, where $d = p + m$.

Let k be the present time. Define the block Toeplitz matrix formed by the past data as

$$\check{P}_{0|k-1} := \begin{bmatrix} w(k-1) & w(k) & \cdots & w(k+N-2) \\ r(k-1) & r(k) & \cdots & r(k+N-2) \\ \vdots & \vdots & \ddots & \vdots \\ w(0) & w(1) & \cdots & w(N-1) \\ r(0) & r(1) & \cdots & r(N-1) \end{bmatrix} \in \mathbb{R}^{2kd \times N}$$

Similarly, the block Hankel matrices formed by the future of r and w are respectively defined as

$$R_{k|2k-1} := \begin{bmatrix} r(k) & r(k+1) & \cdots & r(k+N-1) \\ r(k+1) & r(k+2) & \cdots & r(k+N) \\ \vdots & \vdots & & \vdots \\ r(2k-1) & r(2k) & \cdots & r(N+2k-2) \end{bmatrix} \in \mathbb{R}^{kd \times N}$$

and

$$W_{k|2k-1} := \begin{bmatrix} w(k) & w(k+1) & \cdots & w(k+N-1) \\ w(k+1) & w(k+2) & \cdots & w(k+N) \\ \vdots & \vdots & & \vdots \\ w(2k-1) & w(2k) & \cdots & w(N+2k-2) \end{bmatrix} \in \mathbb{R}^{kd \times N}$$

We consider the LQ decomposition

$$\frac{1}{\sqrt{N}}\begin{bmatrix} R_{k|2k-1} \\ \check{P}_{0|k-1} \\ W_{k|2k-1} \end{bmatrix} = \begin{bmatrix} R_{11} & 0 & 0 \\ R_{21} & R_{22} & 0 \\ R_{31} & R_{32} & R_{33} \end{bmatrix}\begin{bmatrix} Q_1^{\mathrm{T}} \\ Q_2^{\mathrm{T}} \\ Q_3^{\mathrm{T}} \end{bmatrix} =: \bar{R}Q^{\mathrm{T}} \tag{11.18}$$

where $R_{11} \in \mathbb{R}^{kd\times kd}$, $R_{22} \in \mathbb{R}^{2kd\times 2kd}$, $R_{33} \in \mathbb{R}^{kd\times kd}$ are lower block triangular and Q_i are orthogonal. Then, the conditional covariance matrices are given by

$$\Sigma_{ww|r} = R_{32}R_{32}^{\mathrm{T}} + R_{33}R_{33}^{\mathrm{T}}, \qquad \Sigma_{pp|r} = R_{22}R_{22}^{\mathrm{T}}, \qquad \Sigma_{wp|r} = R_{32}R_{22}^{\mathrm{T}}$$

The following closed-loop subspace identification algorithm is derived by using Algorithm B of Section 10.6.

Closed-loop Identification – CCA Method

Step 1: Compute the square root matrices such that

$$\Sigma_{ww|r} = LL^{\mathrm{T}}, \qquad \Sigma_{pp|r} = MM^{\mathrm{T}}$$

Step 2: Compute the SVD of a normalized covariance matrix by

$$L^{-1}\Sigma_{wp|r}M^{-\mathrm{T}} = USV^{\mathrm{T}} \simeq \hat{U}\hat{S}\hat{V}^{\mathrm{T}}$$

where $\hat{S}$ is obtained by deleting smaller singular values of S.

Step 3: Define the extended observability and reachability matrices as

$$\mathcal{O}_k = L\hat{U}\hat{S}^{1/2}, \qquad \mathcal{C}_k = \hat{S}^{1/2}\hat{V}^{\mathrm{T}}M^{\mathrm{T}}$$

Step 4: Compute the estimate of state vector by

$$\bar{X}_k = \mathcal{C}_k\Sigma_{pp|r}^{-1}\check{P}_{0|k-1} = \hat{S}^{1/2}\hat{V}^{\mathrm{T}}M^{-1}\check{P}_{0|k-1}$$

and form the following matrices with $N-1$ columns

$$\hat{X}_{k+1} = \bar{X}_k(:, 2:N), \qquad \hat{X}_k = \bar{X}_k(:, 1:N-1)$$
$$\hat{W}_{k|k} = W_{k|k}(:, 1:N-1)$$

Step 5: Compute the estimates of the matrices $(A,\ B,\ C,\ D)$ by applying the least-squares method to the regression model

$$\begin{bmatrix} \hat{X}_{k+1} \\ \hat{W}_{k|k} \end{bmatrix} = \begin{bmatrix} A & B \\ C & D \end{bmatrix}\begin{bmatrix} \hat{X}_k \\ R_{k|k} \end{bmatrix} + \begin{bmatrix} \rho_w \\ \rho_e \end{bmatrix}$$

Step 6: Partition the matrices $B,\ C,\ D$ as

$$B = [B_1 \quad B_2], \qquad C = \begin{bmatrix} C_1 \\ C_2 \end{bmatrix}, \qquad D = \begin{bmatrix} 0 & 0 \\ D_{21} & D_{22} \end{bmatrix}$$

and compute the higher order models $P(z)$ and $C(z)$ of the plant and controller by the formulas (11.11) and (11.12), respectively.

Step 7: Compute lower dimensional models by using a model reduction algorithm (see Section 11.5).

11.4 ORT Method

In this section, we develop a closed-loop subspace identification method based on the ORT method derived in Chapter 9.

11.4.1 Orthogonal Decomposition of Joint Input-Output Process

As usual, we introduce Hilbert spaces generated by the exogenous inputs and by the joint input-output process, which are respectively denoted by

$$\mathcal{R} = \overline{\text{span}}\{r(\tau) \mid \tau = 0, \pm 1, \cdots\}, \qquad \mathcal{W} = \overline{\text{span}}\{w(\tau) \mid \tau = 0, \pm 1, \cdots\}$$

We also define Hilbert subspaces spanned by the infinite past and infinite future of the various processes at the present time t as

$$\mathcal{R}_t^- := \overline{\text{span}}\{r(\tau) \mid \tau < t\}, \qquad \mathcal{W}_t^- := \overline{\text{span}}\{w(\tau) \mid \tau < t\}$$

and

$$\mathcal{R}_t^+ := \overline{\text{span}}\{r(\tau) \mid \tau \geq t\}, \qquad \mathcal{W}_t^+ := \overline{\text{span}}\{w(\tau) \mid \tau \geq t\}$$

These are all subspaces of the ambient Hilbert space $\mathcal{H} := \mathcal{R} \vee \mathcal{W}$ spanned by the observable input and output processes (r, w).

Since there is no feedback from w to r, the future of r is conditionally uncorrelated with the past of w given the past of r. From Theorem 9.1 (ii), this feedback-free condition is written as

$$\hat{E}\{w(t) \mid \mathcal{R}\} = \hat{E}\{w(t) \mid \mathcal{R}_{t+1}^-\}, \qquad t = 0, \pm 1, \cdots \tag{11.19}$$

implying that the smoothed estimate of w based on r is causal.

It follows from (11.19) that

$$\begin{aligned} w_s(t) &= w(t) - \hat{E}\{w(t) \mid \mathcal{R}_{t+1}^-\} \\ &= w(t) - \hat{E}\{w(t) \mid \mathcal{R}\} = \hat{E}\{w(t) \mid \mathcal{R}^\perp\} \end{aligned}$$

where $\mathcal{R}^\perp$ is the orthogonal complement of $\mathcal{R}$ in $\mathcal{H}$, and w_s is called the stochastic component of w. Similarly,

$$w_d(t) = \hat{E}\{w(t) \mid \mathcal{R}\}$$

is called the deterministic component of w. The deterministic component w_d is the part of w that can be linearly expressed in terms of the exogenous input r.

As in Section 9.4, we obtain the orthogonal decomposition of the joint input-output process $w = w_d + w_s$, i.e.,

$$\begin{bmatrix} y(t) \\ u(t) \end{bmatrix} = \begin{bmatrix} y_d(t) \\ u_d(t) \end{bmatrix} + \begin{bmatrix} y_s(t) \\ u_s(t) \end{bmatrix} \tag{11.20}$$

where the deterministic and stochastic components are mutually uncorrelated, so that we see from Lemma 9.3 that

$$E\{w_s(t)w_d^{\mathrm{T}}(\tau)\} = 0, \qquad \forall\, t,\, \tau = 0,\, \pm 1,\, \cdots$$

Applying this orthogonal decomposition to the feedback system shown in Figure 11.2, we have equations satisfied by the deterministic and stochastic components.

Lemma 11.3. *The deterministic and stochastic components respectively satisfy the decoupled equations*

$$y_d(t) = P(z)u_d(t) \tag{11.21a}$$

$$u_d(t) = r_2(t) + C(z)[r_1(t) - y_d(t)] \tag{11.21b}$$

and

$$y_s(t) = P(z)u_s(t) + H(z)\nu(t) \tag{11.22a}$$

$$u_s(t) = -C(z)y_s(t) + F(z)\eta(t) \tag{11.22b}$$

Proof. From (11.3), (11.4) and (11.20), we have

$$y_d(t) + y_s(t) = P(z)[u_d(t) + u_s(t)] + H(z)\nu(t)$$

$$u_d(t) + u_s(t) = r_2(t) + C(z)[r_1(t) - y_d(t) - y_s(t)] + F(z)\eta(t)$$

Since ν, η, y_s, u_s are orthogonal to $\mathcal{R}$, the orthogonal projection of the above equations onto $\mathcal{R}$ and $\mathcal{R}^{\perp}$ yields (11.21) and (11.22), respectively. □

We can easily see from (11.21) that

$$\begin{bmatrix} y_d(t) \\ u_d(t) \end{bmatrix} = \begin{bmatrix} P(z)S_i(z)C(z) & P(z)S_i(z) \\ S_i(z)C(z) & S_i(z) \end{bmatrix} \begin{bmatrix} r_1(t) \\ r_2(t) \end{bmatrix} \tag{11.23}$$

Since the transfer matrices in the right-hand side of (11.23) are the same as those of (11.7), the transfer matrices of the plant and the controller can be obtained from a state space realization of the deterministic component w_d.

We can draw some interesting observations from Lemma 11.3 for the decoupled deterministic and stochastic components.

1. We see that the realizations of deterministic and stochastic components can be decoupled, since the two components are mutually uncorrelated. It should be, however, noted that though true for infinite data case, the above observation is not true practically. This is because, in case of finite input-output data, the estimate of the stochastic component w_s is influenced by the unknown initial condition associated with the estimate of the deterministic component w_d as discussed in Section 9.6. However, the effect due to unknown initial conditions surely decreases for a sufficiently long data.

2. Suppose that $P(z)$ and $C(z)$ are stable. Then, we can apply the ORT method to the deterministic part (11.21) to obtain state space realizations of $P(z)$ and $C(z)$; see Appendix of Section 11.8. In this case, we also show that the noise models $H(z)$ and $F(z)$ can be obtained from the stochastic part (11.22).
3. If $P(z)$ and/or $C(z)$ is open-loop unstable or marginally stable, we cannot follow the above procedure, since the deterministic (or stochastic) subspace method applied to (11.21) yields erroneous results. For, it is impossible to connect the second-order stationary processes u_d and y_d (or u_s and y_s) by an unstable transfer matrix $P(z)$. Moreover, controllers in practical control systems are often marginally stable due to the existence of integrators implemented. In this case, we need the joint input-output approach as show below.

11.4.2 Realization of Closed-loop System

Suppose that for each t the input space $\mathcal{R}$ admits the direct sum decomposition

$$\mathcal{R} = \mathcal{R}_t^+ + \mathcal{R}_t^-, \qquad \mathcal{R}_t^+ \cap \mathcal{R}_t^- = 0$$

An analogous condition is that the spectral density matrix of r is strictly positive definite on the unit circle, *i.e.*, $\Phi_{rr}(\omega) > cI_d$, $\exists c > 0$ or all canonical angles between the past and future subspaces of r are strictly positive. As already mentioned, in practice, it suffices to assume that r satisfies a sufficiently high order PE condition, and that the "true" system is finite dimensional.

Let $\hat{\mathcal{W}}$ be spanned by deterministic component w_d. Let $\hat{\mathcal{W}}_t^+$ denote the subspace generated by the future $w_d(\tau)$, $\tau = t, t+1, \cdots$. According to Subsection 9.5.2, we define the oblique predictor subspace as

$$\mathcal{X}_t^{+/-} := \hat{E}_{\|\mathcal{R}_t^+}\{\hat{\mathcal{W}}_t^+ \mid \mathcal{R}_t^-\} \tag{11.24}$$

This is the oblique projection of $\hat{\mathcal{W}}_t^+$ onto the past $\mathcal{R}_t^-$ along the future $\mathcal{R}_t^+$, so that $\mathcal{X}_t^{+/-}$ is the state space for the deterministic component. Clearly, if r is a white noise process, (11.24) reduces to the orthogonal projection onto $\mathcal{R}_t^-$.

Let the dimension of the state space $\mathcal{X}_t^{+/-}$ be n, which in general equals the sum of the orders of the plant and the controller. From Theorem 9.3, any basis vector $x_d(t) \in \mathcal{X}_t^{+/-}$ yields a state space representation of $w_d(t)$, *i.e.*,

$$x_d(t+1) = Ax_d(t) + [B_1 \quad B_2]\, r(t) \tag{11.25a}$$

$$\begin{bmatrix} y_d(t) \\ u_d(t) \end{bmatrix} = \begin{bmatrix} C_1 \\ C_2 \end{bmatrix} x_d(t) + \begin{bmatrix} 0 & 0 \\ D_{21} & D_{22} \end{bmatrix} r(t) \tag{11.25b}$$

where $A \in \mathbb{R}^{n \times n}$. Since $P(z)$ is assumed to be strictly proper, we have $D_{11} = 0$, $D_{12} = 0$. Also, from the configuration of Figure 11.2, $D_{22} = I_m$ and $D_{21} = D_c$ hold. It therefore follows from (11.25) that the transfer matrices of the closed-loop system are given by

$$\begin{bmatrix} T_{yr_1}(z) & T_{yr_2}(z) \\ T_{ur_1}(z) & T_{ur_2}(z) \end{bmatrix} = \left[\begin{array}{c|cc} A & B_1 & B_2 \\ \hline C_1 & 0 & 0 \\ C_2 & D_{21} & D_{22} \end{array}\right]$$

Hence, from (11.7), we have $T_{yr_2}(z) = P(z)T_{ur_2}(z)$ and $T_{ur_1}(z) = T_{ur_2}(z)C(z)$, so that the plant and the controller are computed by $P(z) = T_{yr_2}(z)T_{ur_2}^{-1}(z)$ and $C(z) = T_{ur_2}^{-1}(z)T_{ur_1}(z)$, respectively.

Lemma 11.4. *Let* $\det D_{22} \neq 0$. *Then, the (non-minimal) realizations of the plant and controller are respectively given by*

$$P(z) = \left[\begin{array}{c|c} A - B_2 D_{22}^{-1} C_2 & B_2 D_{22}^{-1} \\ \hline C_1 & 0 \end{array}\right] \tag{11.26}$$

and

$$C(z) = \left[\begin{array}{c|c} A - B_2 D_{22}^{-1} C_2 & B_1 - B_2 D_{22}^{-1} D_{21} \\ \hline D_{22}^{-1} C_2 & D_{22}^{-1} D_{21} \end{array}\right] \tag{11.27}$$

Proof. A proof is similar to that of Lemma 11.1. □

Remark 11.2. Lemma 11.4 is seemingly the same as Lemma 11.1. However, the subspace identification algorithm derived from Lemma 11.4 is different from the one derived from Lemma 11.1. For it is clear that the way of computing state space realizations is quite different in two methods. □

Since the transfer matrices $P(z)$ and $C(z)$ obtained from the realization of the deterministic component of (11.25) are of higher order, we apply a model reduction technique to get lower order models. This will be discussed in detail in Section 11.5.

11.4.3 Subspace Identification Method

In this section, we present a subspace identification method based on finite data. The notation used here is the same as that of Subsection 11.3.2. Suppose that finite input-output data $r_1(t)$, $r_2(t)$, $u(t)$, $y(t)$ for $t = 0, 1, \cdots, N + 2k - 2$ are given with N sufficiently large and $k > n$. We assume that they are samples from jointly stationary processes with means zero and finite covariance matrices.

Let $R_{0|k-1}$, $R_{k|2k-1} \in \mathbb{R}^{kd \times N}$ be the block Hankel matrices generated by the past and the future exogenous inputs, and similarly for $W_{0|k-1}$, $W_{k|2k-1} \in \mathbb{R}^{kd \times N}$. Moreover, we define the block Hankel matrices

$$R_{0|2k-1} := \begin{bmatrix} R_{0|k-1} \\ R_{k|2k-1} \end{bmatrix}, \qquad W_{0|2k-1} := \begin{bmatrix} W_{0|k-1} \\ W_{k|2k-1} \end{bmatrix}$$

and then the subspaces $\mathcal{R}_{0|2k-1}$ and $\mathcal{W}_{0|2k-1}$ generated by $R_{0|2k-1}$ and $W_{0|2k-1}$, respectively.

The first step of subspace identification is to obtain the deterministic component w_d by means of the orthogonal projection

$$\hat{W}^d_{0|2k-1} = \hat{E}\left\{W_{0|2k-1} \mid \mathcal{R}_{0|2k-1}\right\} \tag{11.28}$$

The following development is based on the argument of Section 9.7.

To derive the matrix input-output equation satisfied by $W^d_{k|2k-1}$ from (11.25), we define $B := [B_1 \;\; B_2] \in \mathbb{R}^{n \times d}$, $C := \begin{bmatrix} C_1 \\ C_2 \end{bmatrix} \in \mathbb{R}^{d \times n}$, $D := \begin{bmatrix} 0 & 0 \\ D_{21} & D_{22} \end{bmatrix} \in \mathbb{R}^{d \times d}$ and the extended observability matrix

$$\mathcal{O}_k = \begin{bmatrix} C \\ CA \\ \vdots \\ CA^{k-1} \end{bmatrix} \in \mathbb{R}^{kd \times n}, \qquad k > n$$

and the lower triangular block Toeplitz matrix

$$\Psi_k = \begin{bmatrix} D & & & 0 \\ CB & D & & \\ CAB & CB & D & \\ \vdots & \vdots & \ddots & \ddots \\ CA^{k-2}B & CA^{k-3}B & \cdots & CB \;\; D \end{bmatrix} \in \mathbb{R}^{kd \times kd} \tag{11.29}$$

Then, it follows from (11.25) that

$$W^d_{k|2k-1} = \mathcal{O}_k X^d_k + \Psi_k R_{k|2k-1} \tag{11.30}$$

where

$$X^d_k = [x_d(k) \;\; x_d(k+1) \;\; \cdots \;\; x_d(k+N-1)] \in \mathbb{R}^{n \times N}$$

By using Lemma 9.8, the matrix $\hat{W}^d_{k|2k-1}$, a part of $\hat{W}^d_{0|2k-1}$ defined by (11.28), satisfies the same equation as (11.30), *i.e.*,

$$\hat{W}^d_{k|2k-1} = \mathcal{O}_k \hat{X}^d_k + \Psi_k R_{k|2k-1} \tag{11.31}$$

[see (9.44)], where the state vector is given by $\hat{X}^d_k := \hat{E}\{X^d_k \mid \mathcal{R}_{0|2k-1}\}$.

Motivated by the above discussion, we consider the following LQ decomposition

$$\begin{bmatrix} R_{k|2k-1} \\ R_{0|k-1} \\ W_{0|k-1} \\ W_{k|2k-1} \end{bmatrix} = \begin{bmatrix} L_{11} & 0 & 0 & 0 \\ L_{21} & L_{22} & 0 & 0 \\ L_{31} & L_{32} & L_{33} & 0 \\ L_{41} & L_{42} & L_{43} & L_{44} \end{bmatrix} \begin{bmatrix} Q_1^{\mathrm{T}} \\ Q_2^{\mathrm{T}} \\ Q_3^{\mathrm{T}} \\ Q_4^{\mathrm{T}} \end{bmatrix} \tag{11.32}$$

where $L_{11}, L_{22}, L_{33}, L_{44} \in \mathbb{R}^{kd \times kd}$ are block lower triangular, and Q_i are orthogonal. Then, from (11.28), the deterministic component can be given by

$$\hat{W}^d_{0|2k-1} = \hat{E}\{W_{0|2k-1} \mid \mathcal{R}_{0|2k-1}\} = \begin{bmatrix} L_{31} & L_{32} \\ L_{41} & L_{42} \end{bmatrix} \begin{bmatrix} Q_1^{\mathrm{T}} \\ Q_2^{\mathrm{T}} \end{bmatrix}$$

Thus from the above equation and (11.31),

$$\hat{W}^d_{k|2k-1} = L_{41}Q_1^{\mathrm{T}} + L_{42}Q_2^{\mathrm{T}} = \Psi_k L_{11} Q_1^{\mathrm{T}} + \mathcal{O}_k \hat{X}^d_k \tag{11.33}$$

By using the orthogonality $Q_1^{\mathrm{T}}Q_2 = 0$, we see from (11.33) that

$$L_{42} = \mathcal{O}_k \hat{X}^d_k Q_2$$

In the following algorithm, it is assumed that the LQ decomposition of (11.32) is already given.

Closed-loop Identification – ORT Method

Step 1: Compute the SVD of L_{42}, *i.e.*,

$$L_{42} = [\hat{U} \quad \bar{U}] \begin{bmatrix} \hat{S} & 0 \\ 0 & \bar{S} \end{bmatrix} \begin{bmatrix} \hat{V}^{\mathrm{T}} \\ \bar{V}^{\mathrm{T}} \end{bmatrix} \simeq \hat{U}\hat{S}\hat{V}^{\mathrm{T}} \tag{11.34}$$

where $\hat{S}$ is obtained by neglecting sufficiently small singular values. Thus the dimension of the state vector is the same as the dimension of $\hat{S}$, so that we have

$$\mathcal{O}_k \hat{X}^d_k Q_2 = L_{42} \simeq \hat{U}\hat{S}\hat{V}^{\mathrm{T}} = \left(\hat{U}\hat{S}^{1/2}\right)\left(\hat{S}^{1/2}\hat{V}^{\mathrm{T}}\right)$$

Under the assumption that $\hat{X}^d_k Q_2$ has full rank, the extended observability matrix is given by

$$\mathcal{O}_k = \hat{U}\hat{S}^{1/2}$$

Step 2: Compute the matrices A and C by

$$A = \mathcal{O}^{\dagger}_{k-1}\overline{\mathcal{O}}_k, \qquad C = \mathcal{O}_k(1:d,:)$$

where $\overline{\mathcal{O}}_k$ is obtained by deleting the first d rows from $\mathcal{O}_k$.

Step 3: Given the estimates of A and C, compute the least-squares estimates of B and D from

$$\bar{U}^{\mathrm{T}}\Psi_k(B, D) = \bar{U}^{\mathrm{T}}L_{41}L_{11}^{-1}$$

where L_{11} and L_{41} are obtained by (11.32), and $\bar{U}$ of (11.34) satisfies $\bar{U}^{\mathrm{T}}\mathcal{O}_k = 0$, and with $D_{11} = 0$, $D_{12} = 0$.

Step 4: Partition the obtained matrices B, C, D as

$$B = [B_1 \;\; B_2], \qquad C = \begin{bmatrix} C_1 \\ C_2 \end{bmatrix}, \qquad D = \begin{bmatrix} 0 & 0 \\ D_{21} & D_{22} \end{bmatrix}$$

and compute the state space realizations of $P(z)$ and $C(z)$ from (11.26) and (11.27), respectively.

Step 5: Compute lower order models of $P(z)$ and $C(z)$ by using a model reduction method. This will be explained in the next section.

11.5 Model Reduction

As mentioned already, all the identified transfer matrices have higher orders than the true one. To recover reduced order models from Lemma 11.1 (or Lemma 11.4), it is therefore necessary to delete nearly unreachable and/or unobservable modes. Since the open-loop plant is possibly unstable, we need the model reduction technique that can be applied to both stable and unstable transfer matrices [168, 186].

In this section, we employ a direct model reduction method introduced in Lemma 3.7. The technique starts with a given balanced realization, but higher order models in question are not necessarily balanced nor minimal. Hence a desired model reduction procedure should have the following property.

(a) Applicable to non-minimal and non-balanced realizations.

(b) Numerically reliable.

Let $G(z) := (A, B, C, D)$ be a realization to be reduced, where we assume that $A \in \mathbb{R}^{n\times n}$ is stable. Let P and Q be reachability and observability Gramians, respectively, satisfying

$$P = APA^{\mathrm{T}} + BB^{\mathrm{T}}, \qquad Q = A^{\mathrm{T}}QA + C^{\mathrm{T}}C \tag{11.35}$$

For the computation of Gramians for unstable A, see Lemma 3.9.

A similarity transform of the state vector by a matrix Z yields

$$\left[\begin{array}{c|c} Z^{-1}AZ & Z^{-1}B \\ \hline CZ & D \end{array}\right] = \left[\begin{array}{cc|c} A_{11} & A_{12} & B_1 \\ A_{21} & A_{22} & B_2 \\ \hline C_1 & C_2 & D \end{array}\right] \tag{11.36}$$

Define $Z = [T \;\; \bar{T}]$ and $Z^{-1} = \begin{bmatrix} L \\ \bar{L} \end{bmatrix}$. Then, we have

$$\begin{bmatrix} LAT & LA\bar{T} \\ \bar{L}AT & \bar{L}A\bar{T} \end{bmatrix} = \begin{bmatrix} A_{11} & A_{12} \\ A_{21} & A_{22} \end{bmatrix}, \quad \begin{bmatrix} LB \\ \bar{L}B \end{bmatrix} = \begin{bmatrix} B_1 \\ B_2 \end{bmatrix}, \quad [CT \;\; C\bar{T}] = [C_1 \;\; C_2]$$

so that we get $A_{11} = LAT$, $B_1 = LB$ and $C_1 = CT$.

The requirement (a) mentioned above is fulfilled by computing the matrices T and L without actually forming the matrices Z and Z^{-1}. Also, the requirement (b) is attained by using the SVD-based computation. The following algorithm satisfies these requirements.

SR Algorithm

Step 1: Obtain the Gramians P and Q by solving (11.35).

Step 2: Compute the factorizations

$$P = S^{\mathrm{T}}S, \qquad Q = R^{\mathrm{T}}R$$

Note that **chol** in MATLAB® does not work unless P and Q are positive definite.

Step 3: Compute the SVD of $SR^{\mathrm{T}} \in \mathbb{R}^{n \times n}$ as

$$SR^{\mathrm{T}} = U\Sigma V^{\mathrm{T}} = [U_1 \quad U_2] \begin{bmatrix} \Sigma_1 & 0 \\ 0 & \Sigma_2 \end{bmatrix} \begin{bmatrix} V_1^{\mathrm{T}} \\ V_2^{\mathrm{T}} \end{bmatrix} \tag{11.37}$$

where $\Sigma = \mathrm{diag}(\sigma_1 \geq \cdots \geq \sigma_r > \sigma_{r+1} \geq \cdots \geq \sigma_n \geq 0)$, which are the Hankel singular values of the system.

Step 4: Partition $\Sigma = \mathrm{diag}\{\Sigma_1,\ \Sigma_2\}$, where

$$\Sigma_1 = \mathrm{diag}(\sigma_1, \cdots, \sigma_r), \qquad \Sigma_2 = \mathrm{diag}(\sigma_{r+1}, \cdots, \sigma_n)$$

and define the matrices T and L as

$$T = S^{\mathrm{T}} U_1 \Sigma_1^{-1/2}, \qquad L = \Sigma_1^{-1/2} V_1^{\mathrm{T}} R \tag{11.38}$$

Then, a reduced order model is obtained by

$$G_r(z) = (LAT,\ LB,\ CT,\ D) \tag{11.39}$$

Using Lemma 3.7, we can prove that $G_r(z)$ is a reduced order model.

Lemma 11.5. *A reduced order model is given by $G_r(z)$ of (11.39). In general, $G_r(z)$ is not balanced, but if we take the parameter r so that $\Sigma_1 > 0,\ \Sigma_2 = 0$, then $G_r(z)$ is balanced and minimal.*

Proof. By the definition of Hankel singular values,

$$\begin{aligned}\sqrt{\lambda_i(PQ)} &= \sqrt{\lambda_i(S^{\mathrm{T}}SR^{\mathrm{T}}R)} = \sqrt{\lambda_i(RS^{\mathrm{T}}SR^{\mathrm{T}})} \\ &= \sqrt{\sigma_i(SR^{\mathrm{T}})^2} = \sigma_i(SR^{\mathrm{T}})\end{aligned}$$

This shows that the diagonal elements of Σ obtained in *Step 3* of SR algorithm are the Hankel singular values. Pre-multiplying the first equation of (11.35) by $\Sigma^{-1/2}V^{\mathrm{T}}R$ and post-multiplying by $R^{\mathrm{T}}V\Sigma^{-1/2}$ yield

$$\begin{aligned}\Sigma^{-1/2}V^{\mathrm{T}}RPR^{\mathrm{T}}V\Sigma^{-1/2} &= \Sigma^{-1/2}V^{\mathrm{T}}R(APA^{\mathrm{T}})R^{\mathrm{T}}V\Sigma^{-1/2} \\ &\quad + \Sigma^{-1/2}V^{\mathrm{T}}R(BB^{\mathrm{T}})R^{\mathrm{T}}V\Sigma^{-1/2} \\ &=: I_1 + I_2\end{aligned} \tag{11.40}$$

From $P = S^{\mathrm{T}}S$ and (11.37), the left-hand side of (11.40) becomes

$$\Sigma^{-1/2}V^{\mathrm{T}}RS^{\mathrm{T}}SR^{\mathrm{T}}V\Sigma^{-1/2} = \Sigma^{-1/2}V^{\mathrm{T}}V\Sigma U^{\mathrm{T}}U\Sigma V^{\mathrm{T}}V\Sigma^{-1/2} = \Sigma \tag{11.41}$$

To compute the right-hand side of (11.40), we note that

$$\Sigma^{-1/2}V^{\mathrm{T}}R = \begin{bmatrix} \Sigma_1^{-1/2} & 0 \\ 0 & \Sigma_2^{-1/2} \end{bmatrix} \begin{bmatrix} V_1^{\mathrm{T}} \\ V_2^{\mathrm{T}} \end{bmatrix} R =: \begin{bmatrix} L \\ \bar{L} \end{bmatrix} \tag{11.42}$$

By using the fact that $UU^{\mathrm{T}} = I_n$, we get

$$\begin{aligned} P &= S^{\mathrm{T}} U \Sigma^{-1/2} \Sigma \Sigma^{-1/2} U^{\mathrm{T}} S \\ &= S^{\mathrm{T}} [U_1 \ \ U_2] \begin{bmatrix} \Sigma_1^{-1/2} & 0 \\ 0 & \Sigma_2^{-1/2} \end{bmatrix} \begin{bmatrix} \Sigma_1 & 0 \\ 0 & \Sigma_2 \end{bmatrix} \\ &\quad \times \begin{bmatrix} \Sigma_1^{-1/2} & 0 \\ 0 & \Sigma_2^{-1/2} \end{bmatrix} \begin{bmatrix} U_1^{\mathrm{T}} \\ U_2^{\mathrm{T}} \end{bmatrix} S \\ &= [T \ \ \bar{T}] \begin{bmatrix} \Sigma_1 & 0 \\ 0 & \Sigma_2 \end{bmatrix} \begin{bmatrix} T^{\mathrm{T}} \\ \bar{T}^{\mathrm{T}} \end{bmatrix} = T \Sigma_1 T^{\mathrm{T}} + \bar{T} \Sigma_2 \bar{T}^{\mathrm{T}} \end{aligned} \tag{11.43}$$

where T is defined by (11.38) and $\bar{T} := S^{\mathrm{T}} U_2 \Sigma_2^{-1/2}$. Hence, from (11.40), (11.42) and (11.43),

$$I_1 = \begin{bmatrix} L \\ \bar{L} \end{bmatrix} A (T \Sigma_1 T^{\mathrm{T}} + \bar{T} \Sigma_2 \bar{T}^{\mathrm{T}}) A^{\mathrm{T}} [L^{\mathrm{T}} \ \ \bar{L}^{\mathrm{T}}]$$

$$I_2 = \begin{bmatrix} L \\ \bar{L} \end{bmatrix} B B^{\mathrm{T}} [L^{\mathrm{T}} \ \ \bar{L}^{\mathrm{T}}]$$

Thus from (11.41), the $(1,1)$-block of (11.40) is given by

$$\begin{aligned} \Sigma_1 &= (LAT) \Sigma_1 (LAT)^{\mathrm{T}} + (LB)(LB)^{\mathrm{T}} + (LA\bar{T}) \Sigma_2 (LA\bar{T})^{\mathrm{T}} \\ &= A_{11} \Sigma_1 A_{11}^{\mathrm{T}} + B_1 B_1^{\mathrm{T}} + A_{12} \Sigma_2 A_{12}^{\mathrm{T}} \end{aligned} \tag{11.44}$$

Similarly, from the second equation of (11.35), we have

$$\Sigma_1 = A_{11}^{\mathrm{T}} \Sigma_1 A_{11} + C_1^{\mathrm{T}} C_1 + A_{21}^{\mathrm{T}} \Sigma_2 A_{21} \tag{11.45}$$

Equations (11.44) and (11.45) derived above are the same as (3.41a) and (3.43), respectively. Thus $G_r(z)$ is a reduced order model of $G(z)$, but is not balanced.

Putting $\Sigma_2 = 0$ in (11.44) and (11.45) gives

$$\Sigma_1 = A_{11} \Sigma_1 A_{11}^{\mathrm{T}} + B_1 B_1^{\mathrm{T}}, \qquad \Sigma_1 = A_{11}^{\mathrm{T}} \Sigma_1 A_{11} + C_1^{\mathrm{T}} C_1$$

implying that $G_r(z)$ is balanced.

The minimality of $G_r(z)$ is proved similarly to Lemma 3.7. □

In the SR algorithm derived above, it is assumed that A is stable. For the case where A is unstable, defining the Gramians P and Q as in Definition 3.10, it is possible to compute them by the algorithm of Lemma 3.9. Hence, there needs to be no change in the SR algorithm except for *Step 1*.

11.6 Numerical Results

Some numerical results for closed-loop system identification are presented. The first model is a closed-loop system with a 2nd-order plant and a 1st-order controller, for

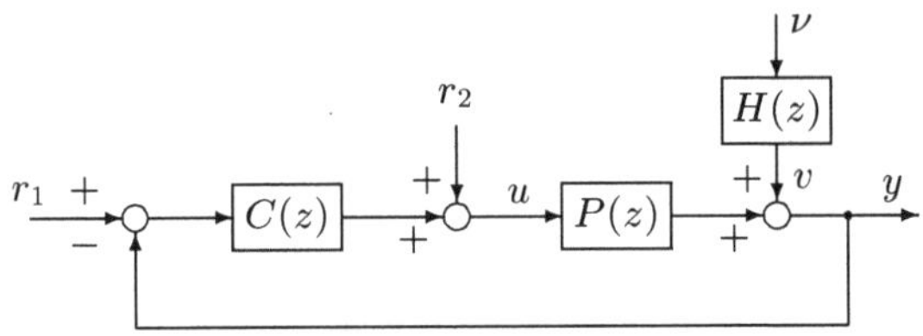

Figure 11.3. A feedback control system

which results obtained by the ORT and CCA methods are compared. In the second example, we present identification results for a 5th-order plant with a 4th-order controller by means of the ORT method. A feedback control system used in the present simulation is displayed in Figure 11.3.

11.6.1 Example 1

Suppose that the transfer functions of plant and controller are given by

$$P(z) = \frac{z^{-1}}{1 - 1.6z^{-1} + 0.89z^{-2}}, \qquad C(z) = \frac{z - 0.8}{z}$$

where the closed-loop poles are located at $z = 0,\ 0.3,\ 0.3$. We assume that the noise v is an ARMA process generated by

$$H(z) = \frac{1 - 1.56z^{-1} + 1.045z^{-2} - 0.3338z^{-3}}{1 - 2.35z^{-1} + 2.09z^{-2} - 0.6675z^{-3}}$$

This model is a slightly modified version of the one used in [160], in which only the probing input r_2 is used to identify the plant, but here we include r_1 and r_2 as reference inputs in order to identify both the plant and controller. The reference input r_1 is a composite sinusoid of the form

$$r_1(t) = \rho \sum_{j=1}^{30} A_j \sin(\omega_j t + \phi_j), \qquad t = 0,\ 1, \cdots, N + 2k - 2$$

where a magnitude ρ is adjusted so that $\sigma_1^2 = 1$, and A_j is a white noise with $\mathcal{N}(0,\ 1)$. The parameters ω_j and ϕ_j are uniformly distributed over $(0,\ \pi)$, so that r_1 has PE condition of order 60. The r_2 and ν are Gaussian white noises with variances $\sigma_2^2 = (0.2)^2$ and $\sigma_\nu^2 = 1/9$, respectively.

For the ORT method, since the sum of the orders of plant and controller is three, 3rd-order state-space models are fitted to the input-output data (r, w). Then, the 3rd-order plant and controller models so identified are reduced to the second- and the first-order models, respectively.

On the other hand, for the CCA method, 6th-order models are fitted to $(r,\ w)$, because the sum of orders of the plant, controller and noise model is six, and because the state space model cannot be divided into separate deterministic and stochastic

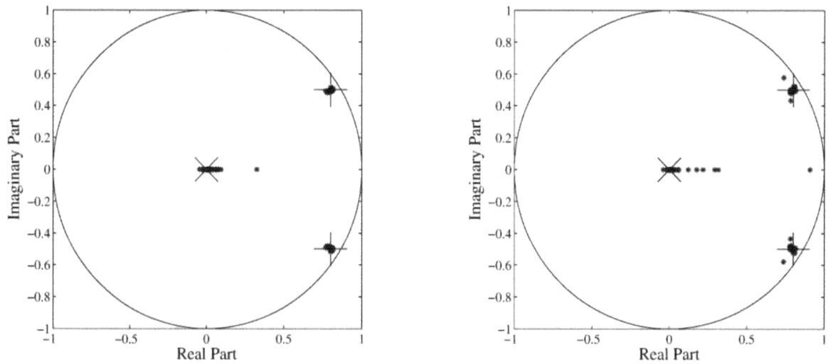

(a) ORT method (b) CCA method

Figure 11.4. Estimates of poles, $(+)$: plant, $(\times)$: controller

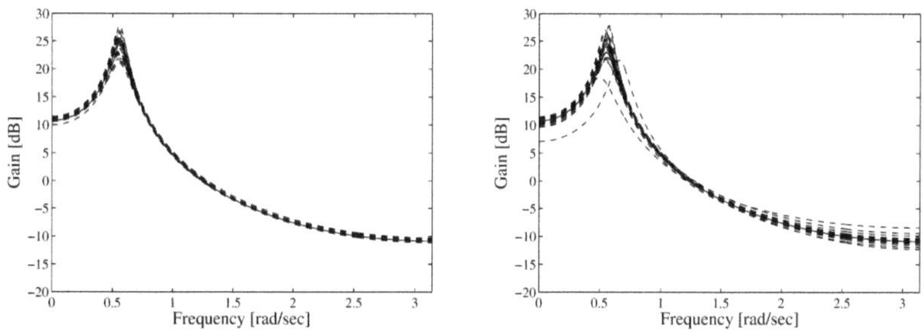

(a) ORT method (b) CCA method

Figure 11.5. Bode plots of $P(z)$

components. Thus, in this case, the identified 6th-order models of plant and controller are reduced to the second- and the first-order models, respectively.

Case 1: We take the number of data points $N = 2000$ and the number of block rows $k = 15$, and generated 30 data sets, each with different samples for r_1, r_2 and ν. Figures 11.4(a) and 11.4(b) respectively display the poles of the plant and controller identified by the ORT and CCA methods, where $+$ and $\times$ denote the true poles of plant and controller, respectively. Figures 11.5(a) and 11.5(b) respectively display the Bode plots of the estimated plant, and Figures 11.6(a) and 11.6(b) the Bode plots of the estimated controller. We see from these figures that the identification results by the ORT method are quite good, but the results by the CCA method are somewhat degraded compared with the results by the ORT method.

The Bode plots of the plant identified by the ORT and CCA methods based on the direct approach are shown in Figures 11.7(a) and 11.7(b), respectively. We clearly see biases in the estimates of Bode magnitude, where the ORT provides somewhat larger biases than the CCA method.

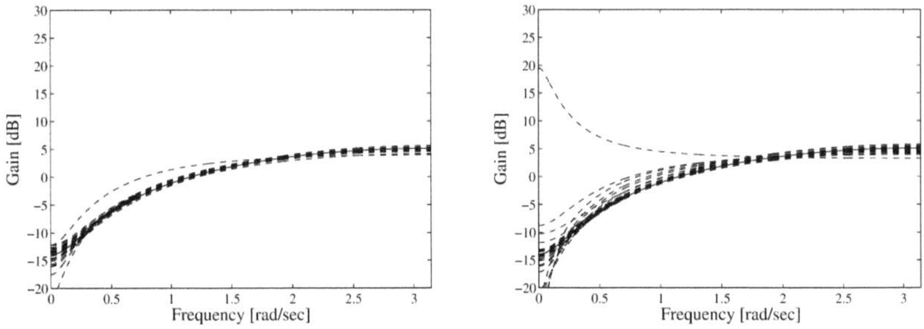

(a) ORT method (b) CCA method

Figure 11.6. Bode plots of $C(z)$

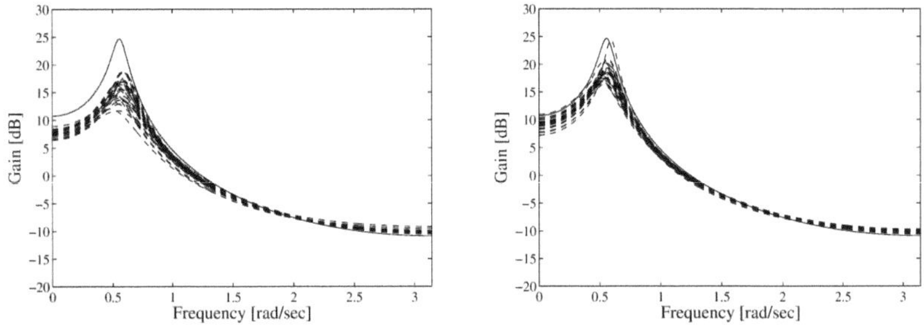

(a) ORT method (b) CCA method

Figure 11.7. Identification results by direct method

Case 2: As the second experiment, we consider the effect of the number of data on the performance of identification. For the plant parameter vector $\theta := (-1.6 \;\; 0.89 \;\; 1 \;\; 0) \in \mathbb{R}^4$, the performance is measured by the norm of the estimation error

$$I_N = \frac{1}{M} \sum_{i=1}^{M} ||\theta - \hat{\theta}(i, N)||^2$$

where $\hat{\theta}(i, N) \in \mathbb{R}^4$ denotes the estimate of θ at ith run, and where the number of data is $N = 200, 500, 1000, 2000, 5000$, and the number of runs is $M = 30$ in each case. Figure 11.8 compares the performance of the identification of plant transfer function by the ORT and CCA methods. This figure clearly shows the advantage of ORT-based algorithm over the CCA-based algorithm.

As mentioned before, if the exogenous inputs r_1, r_2 are white noise, then the two algorithm present quite similar identification results. However, if at least one of the exogenous inputs is colored, then we can safely say that the ORT method outperforms the CCA method in the performance.

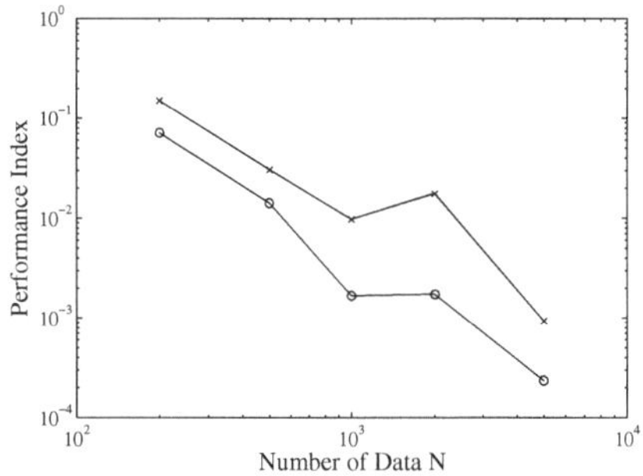

Figure 11.8. Comparison of identification results: the ORT (○ - - - ○) and CCA (× - - - ×) methods

11.6.2 Example 2

We assume that the plant is a discrete-time model of laboratory plant setup of two circular plates rotated by an electrical servo-motor with flexible shafts [169], where the transfer function of the plant $P(z)$ is given by

$$P(z) = \frac{10^{-3}\left(0.98z^4 + 12.990z^3 + 18.589z^2 + 3.2987z - 0.02\right)}{z^5 - 4.3986z^4 + 8.0852z^3 - 7.8233z^2 + 3.9954z - 0.8588}$$

and where a stabilizing controller is chosen as

$$C(z) = \frac{0.6300z^4 - 2.0830z^3 + 2.8222z^2 - 1.8650z + 0.4978}{z^4 - 2.6500z^3 + 3.1100z^2 - 1.7500z + 0.3900}$$

The configuration of the plant and controller is the same as the one depicted in Figure 11.3, where the output noise process $v = \nu$ is a Gaussian white noise sequence with $E\left\{\nu^2(t)\right\} = 1/9$. Both the reference signals r_1 and r_2 are Gaussian white noises with variances $\sigma_1^2 = 1$ and $\sigma_2^2 = 0.5$, respectively. Note that $P(z)$ has poles at $z = 1$, $0.9674 \pm 0.1493j$, $0.7319 \pm 0.6005j$; thus the plant has one integrator and therefore is marginally stable. Also, the controller $C(z)$ has the poles at $z = 0.7169 \pm 0.6678j$, $0.6081 \pm 0.1910j$; thus the controller is stable. We take the number of data points $N = 4000$ and the number of block rows $k = 15$. We generated 30 data sets, each with the same reference inputs r_1 and r_2, but with a different noise sequence ν.

In this experiment, we have employed the ORT method. Figure 11.9 shows the estimated eigenvalues of the matrix $A_2 - B_{22}D_{12}^{-1}C_{12}$ [see Lemma 11.4], where $+$ denotes the true poles of the plant, $\times$ those of the controller and $*$ the estimated eigenvalues. From Figure 11.9 we can see that the nine poles of plant and controller are identified very well.

The estimated transfer function $P(z)$ of 5th-order is displayed in Figure 11.10. Figure 11.10(a) shows the estimated poles of the plant, where $+$ denotes the true poles and $*$ denotes the estimated poles over 30 experiments. The Bode plot of the estimated transfer function of the plant is depicted in Figure 11.10(b), where the

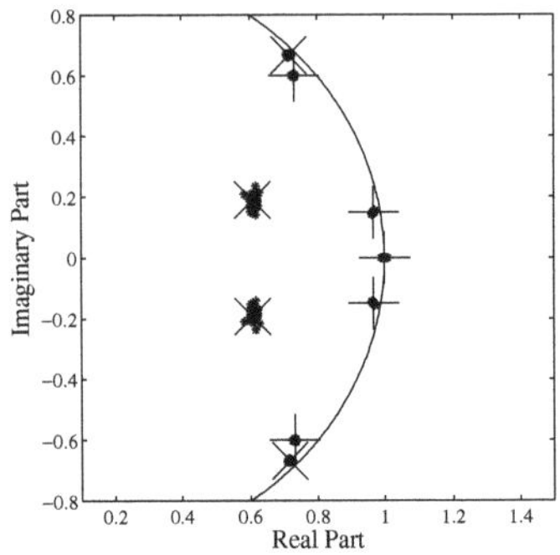

Figure 11.9. ORT method : poles of $A - B_2 D_{22}^{-1} C_2$ [(11.26)]; +: plant, ×: controller

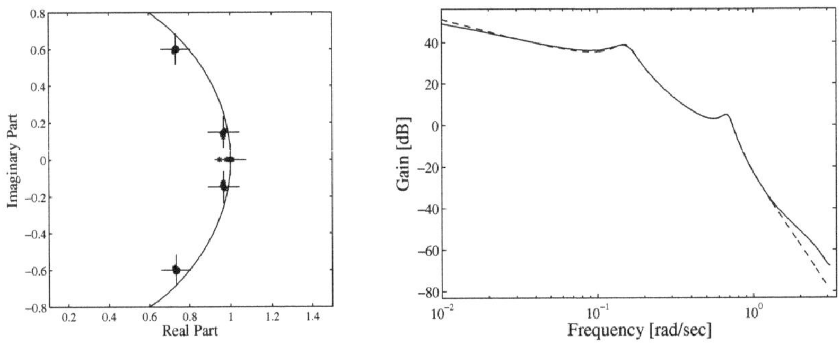

(a) Estimates of poles of $P(z)$ (b) Bode plots of $P(z)$

Figure 11.10. Identification by ORT method

dashed line depicts the true transfer function of the plant and the solid line depicts the average over 30 experiments. From these figures, we can see that the ORT method-based algorithm performs very well in the identification of the plant. Since $C(z)$ is stable, there are no unstable pole-zero cancellations in the reduction of the estimated plant; thus it seems that the model reduction is performed nicely.

Furthermore, the estimation results of the controller are depicted in Figures 11.11 and 11.12. As in the case of the plant estimation, the estimation of the controller needs the model reduction by approximate pole-zero cancellations. It should be noted that in order to estimate the controller having the same order as the true one, we need to perform an unstable pole-zero cancellation at $z = 1$. Figure 11.11 depicts the estimated controller as a 4th-order model, which is the same order as the true one, where Figure 11.11(a) shows the pole estimation, where $\times$ denotes the true poles and $*$ denotes the estimated ones, and Figure 11.11(b) shows the Bode plots of the true transfer function (dashed line) and the average transfer function over 30 experiments (solid line). We can see from these figures that there are many incorrect poles around

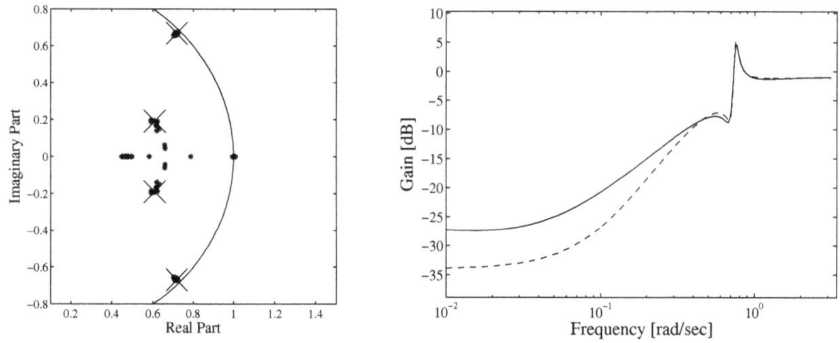

(a) Pole plots of $C(z)$ (b) Bode plots of $C(z)$

Figure 11.11. Identification of controller (4th order model)

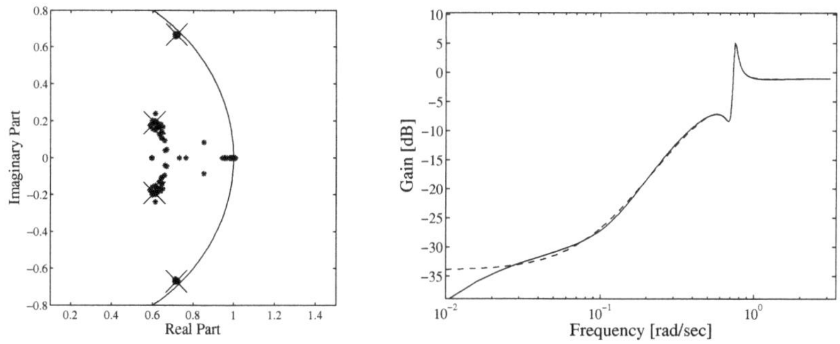

(a) Pole plots of $C(z)$ (b) Bode plots of $C(z)$

Figure 11.12. Identification of controller (5th-order model)

positive real axis including $z = 1$, and the Bode plot is biased in low frequency range.

On the other hand, Figure 11.12 displays the estimated controller as a 5th-order model, *i.e.*, the estimated 9th-order models are reduced to 5th order. In this case, though Figure 11.12(a) shows that there are many incorrect poles around real axis, we can see from Figure 11.12(b) that the Bode gain of controller is estimated very well by using a 5th-order transfer function.

11.7 Notes and References

- In this chapter, based on Katayama *et al.* [87,88], we have developed two closed-loop subspace identification methods based on the ORT and CCA methods derived in Chapters 9 and 10, in the framework of the joint input-output approach. See also Katayama *et al.* [89], in which the role of input signal in closed-loop identification is discussed in detail.

- The importance and the basic approaches of closed-loop identification are reviewed in Section 11.1 [48, 109, 145]. In Section 11.2, the problem is formulated and the fundamental idea of the joint input-output approach is explained. The present problem is virtually the same as the one treated in [170].
- Section 11.3 is devoted to the realization of feedback system and the derivation of subspace identification method based on the CCA method. Section 11.4 derives a subspace identification method based on the ORT method, and shows that the plant and controller can be identified by a realization of deterministic component of the joint input-output process.
- Since the transfer matrices derived by the joint input-output approach are necessarily of higher order than the true one, we have presented a model reduction technique called the SR algorithm in Section 11.5.
- Section 11.6 shows the procedure of closed-loop identification methods through some numerical results. The performance of closed-loop identification depends on the basic open-loop identification techniques; numerical results show that performance of the ORT based method is somewhat superior to that of CCA based method. Some related numerical results are also found in [89].
- Under the assumption that the plant is stable, a simple closed-loop identification method based on the orthogonal decomposition of the joint input-output process is described in Appendix below.

11.8 Appendix: Identification of Stable Transfer Matrices

In this section, as Appendix to this chapter, we present a simple closed-loop identification procedure by using the result of Lemma 11.3 under the assumption that all the open-loop transfer matrices in Figure 11.2 are stable. In the following, Assumptions 1 and 2 stated in Section 11.2 are satisfied.

11.8.1 Identification of Deterministic Parts

From (11.21), we have two deterministic equations

$$y_d(t) = P(z)u_d(t) \tag{11.46}$$

and

$$\tilde{u}_d(t) = -C(z)\tilde{y}_d(t) \tag{11.47}$$

where

$$\tilde{y}_d(t) = y_d(t) - r_1(t), \qquad \tilde{u}_d(t) = u_d(t) - r_2(t)$$

It should be noted that above relations are satisfied by deterministic components (u_d, y_d) and $(\tilde{u}_d, \tilde{y}_d)$, since the noise components are removed in these relations. Thus Figure 11.13 displays two independent open-loop systems for the plant and controller, so that we can use (11.46) and (11.47) to identify the open-loop plant

Figure 11.13. Plant and controller in terms of deterministic components

$P(z)$ and the controller $C(z)$ independently. The present idea is somewhat related to the two-stage method [160] and the projection method [49].

Identification Algorithm of Plant and Controller

Step 1: By using LQ decomposition, we compute the deterministic components of the joint input-output process $(y_d,\ u_d)$ and then compute $(\tilde{y}_d,\ \tilde{u}_d)$.

Step 2: We apply the ORT (or CCA) method to the input-output data $(u_d,\ y_d)$ to obtain

$$x_p(t+1) = A_p x_p(t) + B_p u_d(t) \tag{11.48a}$$

$$y_d(t) = C_p x_p(t) \tag{11.48b}$$

Then the plant transfer matrix is given by $P(z) = (A_p,\ B_p,\ C_p)$.

Step 3: We apply the ORT (or CCA) method to the input-output data $(\tilde{u}_d,\ \tilde{y}_d)$ to obtain

$$x_c(t+1) = A_c x_c(t) + B_c \tilde{y}_d(t) \tag{11.49a}$$

$$-\tilde{u}_d(t) = C_c x_d(t) + D_c \tilde{y}_d(t) \tag{11.49b}$$

Then the controller transfer matrix is given by $C(z) = (A_c,\ B_c,\ C_c,\ D_c)$.

For numerical results based on the above technique, see Katayama *et al.* [92].

11.8.2 Identification of Noise Models

We have not discussed the identification of noise models in this chapter. But, they can easily be identified, if both the plant and controller are open-loop stable. It should be noted that the noise models are located outside the closed-loop, so that the identification of noise models is actually an open-loop identification problem.

Under the assumption that $P(z)$ and $C(z)$ are stable, we compute

$$\tilde{y}_s(t) := y_s(t) - P(z)u_s(t) = H(z)\nu(t) \tag{11.50}$$

and

$$\tilde{u}_s(t) := u_s(t) + C(z)y_s(t) = F(z)\eta(t) \tag{11.51}$$

Figure 11.14 shows the block diagrams for noise models.

Since $(\tilde{u}_s,\ \tilde{y}_s)$ are second-order jointly stationary processes, we can identify noise models $H(z)$ and $F(z)$ by applying the CCA method (or a stochastic subspace identification technique).

Figure 11.14. Noise models in terms of stochastic components

In the following, we assume that the plant and controller in Figure 11.13 are already identified by the procedure stated above.

Identification Algorithm of Noise Models

Step 1: By using (11.50) and (11.51), we compute $\tilde{y}_s$ and $\tilde{u}_s$.

Step 2: Applying the CCA method developed in Chapter 8 to the data $\tilde{y}_s$, we identify

$$x_h(t+1) = A_h x_h(t) + K_h e_h(t) \tag{11.52a}$$

$$\tilde{y}_s(t) = C_h x_h(t) + e_h(t) \tag{11.52b}$$

Then, the plant noise model is given by $H(z) = (A_h, C_h, K_h, I_p)$.

Step 3: Applying the CCA method developed in Chapter 8 to the data $\tilde{u}_s$, we identify

$$x_f(t+1) = A_f x_f(t) + K_f e_f(t) \tag{11.53a}$$

$$\tilde{u}_s(t) = C_f x_f(t) + e_f(t) \tag{11.53b}$$

Then, the controller noise model is given by $F(z) = (A_f, C_f, K_f, I_m)$.

Appendix

A

Least-Squares Method

We briefly review the least-squares method for a linear regression model, together with its relation to the LQ decomposition.

A.1 Linear Regressions

Suppose that there exists a linear relation between the output variable $y(t)$ and the d-dimensional regression vector $\varphi(t) = [\varphi_1(t)\ \varphi_2(t)\ \cdots\ \varphi_d(t)]^{\mathrm{T}}$. We assume that N observations $\{y(t),\ \varphi(t),\ t = 0,\ 1,\ \cdots,\ N-1\}$ are given. Then, it follows that

$$y(t) = \varphi_1(t)\theta_1 + \cdots + \varphi_d(t)\theta_d + e(t), \qquad t = 0,\ 1,\ \cdots,\ N-1 \tag{A.1}$$

where $e(t)$ denotes the measurement noise, or the variation in $y(t)$ that cannot be explained by means of $\varphi_1(t),\ \cdots,\ \varphi_d(t)$. We also assume that $\varphi_1(t),\ \cdots,\ \varphi_d(t)$ have no uncertainties[1].

For simplicity, we define the stacked vectors

$$\theta = \begin{bmatrix} \theta_1 \\ \theta_2 \\ \vdots \\ \theta_d \end{bmatrix}, \qquad y = \begin{bmatrix} y(0) \\ y(1) \\ \vdots \\ y(N-1) \end{bmatrix}, \qquad e = \begin{bmatrix} e(0) \\ e(1) \\ \vdots \\ e(N-1) \end{bmatrix}$$

and the matrix

$$\Phi = \begin{bmatrix} \varphi_1(0) & \varphi_2(0) & \cdots & \varphi_d(0) \\ \varphi_1(1) & \varphi_2(1) & \cdots & \varphi_d(1) \\ \vdots & \vdots & & \vdots \\ \varphi_1(N-1) & \varphi_2(N-1) & \cdots & \varphi_d(N-1) \end{bmatrix} = \begin{bmatrix} \varphi^{\mathrm{T}}(0) \\ \varphi^{\mathrm{T}}(1) \\ \vdots \\ \varphi^{\mathrm{T}}(N-1) \end{bmatrix}$$

where $\Phi \in \mathbb{R}^{N \times d}$. Then (A.1) can be written as

[1]If φ are also subject to noises, (A.1) is called an errors-in-variables model.

$$y = \Phi\theta + e$$

This is referred to as a linear regression model. The regression analysis involves the estimation of unknown parameters and the analysis of residuals.

The basic assumptions needed for the least-squares method are listed below.

A1) The error vector e is uncorrelated with Φ and θ.

A2) The error vector is a random vector with mean zero.

A3) The covariance matrix of the error vector is $\sigma^2 I_N$ with $\sigma^2 > 0$.

A4) The column vectors of Φ are linearly independent, *i.e.*, $\text{rank}(\Phi) = d$.

Under the above assumptions, we consider the least-squares problem minimizing the quadratic performance index

$$J(\theta) := \sum_{t=0}^{N-1} [y(t) - \varphi^{\mathrm{T}}(t)\theta]^2 = \|y - \Phi\theta\|^2$$

Setting the gradient of $J(\theta)$ with respect to θ to zero yields

$$\left(\sum_{t=0}^{N-1} \varphi(t)\varphi^{\mathrm{T}}(t)\right)\theta = \sum_{t=0}^{N-1} \varphi(t)y(t) \quad \Rightarrow \quad (\Phi^{\mathrm{T}}\Phi)\theta = \Phi^{\mathrm{T}}y \tag{A.2}$$

This is a well-known normal equation.

From Assumption A4), we see that $\Phi^{\mathrm{T}}\Phi \in \mathbb{R}^{d\times d}$ is nonsingular. Thus, solving (A.2), the least-squares estimate is given by

$$\hat{\theta}_{\mathrm{LS}} := \left(\sum_{t=0}^{N-1} \varphi(t)\varphi^{\mathrm{T}}(t)\right)^{-1} \sum_{t=0}^{N-1} \varphi(t)y(t) = (\Phi^{\mathrm{T}}\Phi)^{-1}\Phi^{\mathrm{T}}y \tag{A.3}$$

Also, from Assumptions A1) and A2),

$$\begin{aligned} E\{\hat{\theta}_{\mathrm{LS}}\} &= E\{(\Phi^{\mathrm{T}}\Phi)^{-1}\Phi^{\mathrm{T}}(\Phi\theta + e)\} \\ &= \theta + (\Phi^{\mathrm{T}}\Phi)^{-1}\Phi^{\mathrm{T}}E\{e\} = \theta \end{aligned} \tag{A.4}$$

so that the least-squares estimate $\hat{\theta}_{\mathrm{LS}}$ is unbiased. It follows from Assumption A3) that the error covariance matrix of the estimate $\hat{\theta}_{\mathrm{LS}}$ is

$$\mathrm{cov}\{\hat{\theta}_{\mathrm{LS}}\} = E\{[\theta - \hat{\theta}_{\mathrm{LS}}][\theta - \hat{\theta}_{\mathrm{LS}}]^{\mathrm{T}}\} = \sigma^2(\Phi^{\mathrm{T}}\Phi)^{-1}$$

Moreover, define the residual vector as $\varepsilon := y - \Phi\hat{\theta}_{\mathrm{LS}}$. Then, it follows that

$$\varepsilon = [I_N - \Phi(\Phi^{\mathrm{T}}\Phi)^{-1}\Phi^{\mathrm{T}}]y = [I_N - \Phi(\Phi^{\mathrm{T}}\Phi)^{-1}\Phi^{\mathrm{T}}]e \tag{A.5}$$

It should be noted that $\Pi := \Phi(\Phi^{\mathrm{T}}\Phi)^{-1}\Phi^{\mathrm{T}}$ satisfies $\Pi^2 = \Pi$ and $\Pi = \Pi^{\mathrm{T}}$, so that Π is an orthogonal projection onto $\mathrm{Im}(\Phi)$. Also, $Q := I_N - \Pi$ is an orthogonal projection onto the orthogonal complement $(\operatorname{Im} \Phi)^{\perp} = \operatorname{Ker}(\Phi^{\mathrm{T}})$. Then, $||\varepsilon||^2 = y^{\mathrm{T}}(I_N - \Pi)y$ denotes the square of the minimum distance between the point y and the space $\operatorname{Im}(\Phi^{\mathrm{T}})$.

We compute the variance of the residual. From (A.5),

$$\begin{aligned} E\{||\varepsilon||^2\} &= \operatorname{trace} E\{\varepsilon\varepsilon^{\mathrm{T}}\} \\ &= \operatorname{trace}\left([I_N - \Phi(\Phi^{\mathrm{T}}\Phi)^{-1}\Phi^{\mathrm{T}}]E\{ee^{\mathrm{T}}\}[I_N - \Phi(\Phi^{\mathrm{T}}\Phi)^{-1}\Phi^{\mathrm{T}}]\right) \\ &= \sigma^2 \operatorname{trace}[I_N - \Phi(\Phi^{\mathrm{T}}\Phi)^{-1}\Phi^{\mathrm{T}}] \\ &= \sigma^2[\operatorname{trace}(I_N) - \operatorname{trace}(\Phi(\Phi^{\mathrm{T}}\Phi)^{-1}\Phi^{\mathrm{T}})] = \sigma^2(N-d) \end{aligned}$$

Hence, the unbiased estimate of the variance σ^2 is given by

$$s^2 := \frac{1}{N-d}\sum_{t=0}^{N-1}\varepsilon^2(t) = \frac{1}{N-d}||\varepsilon||^2$$

In practice, the above assumptions A1) $\sim$ A4) are not completely satisfied. If either A1) or A2) is not satisfied, then a bias arises in the least-squares estimate. In fact, in the computation of (A.4), we have

$$E\{\hat{\theta}_{\mathrm{LS}}\} = \theta + E\{(\Phi^{\mathrm{T}}\Phi)^{-1}\Phi^{\mathrm{T}}e\} \neq \theta$$

Suppose that $E\{ee^{\mathrm{T}}\} = R > 0$, so that A3) does not hold. In this case, we consider a weighted least-squares problem of minimizing

$$J(\theta) = ||y - \Phi\theta||^2_{R^{-1}} = (y - \Phi\theta)^{\mathrm{T}}R^{-1}(y - \Phi\theta)$$

By using the same technique of deriving the least-squares estimate $\hat{\theta}_{\mathrm{LS}}$, we can show that the optimal estimate is given by

$$\hat{\theta}_{\mathrm{GLS}} := (\Phi^{\mathrm{T}}R^{-1}\Phi)^{-1}\Phi^{\mathrm{T}}R^{-1}y$$

where $\hat{\theta}_{\mathrm{GLS}}$ is called the generalized least-squares estimate. The corresponding error covariance matrix becomes

$$\operatorname{cov}\{\hat{\theta}_{\mathrm{GLS}}\} = (\Phi^{\mathrm{T}}R^{-1}\Phi)^{-1}$$

We now turn to Assumption A4). In real problems, we often encounter the case where there exist some "approximate" linear relations among regression vectors (column vectors of Φ); this is called a multicolinearity problem in econometrics. In this case, one or more eigenvalues of $\Phi^{\mathrm{T}}\Phi$ get closer to zero, so that the condition number $\kappa(\Phi)$ becomes very large, leading to unreliable least-squares estimates. An SVD-based method of solving a least-squares problem under ill-conditioning is introduced in Section 2.7. There are also other methods to solve ill-conditioned least-squares problems, including regularization methods, the ridge regression, *etc.*

Example A.1. Consider the normal equation of (A.2):

$$(\Phi^{\mathrm{T}}\Phi)\theta = \Phi^{\mathrm{T}}y, \qquad \Phi \in \mathbb{R}^{N\times d}, \qquad y \in \mathbb{R}^{N} \tag{A.6}$$

We show that (A.6) has always a solution for any $y \in \mathbb{R}^N$. It is well known that (A.6) is solvable if and only if the vector $\Phi^{\mathrm{T}}y$ belongs to $\mathrm{Im}(\Phi^{\mathrm{T}}\Phi)$. However, this is easily verified by noting that $\Phi^{\mathrm{T}}y \in \mathrm{Im}(\Phi^{\mathrm{T}}) = \mathrm{Im}(\Phi^{\mathrm{T}}\Phi)$.

By direct manipulation, we can show that $\theta = \Phi^{\dagger}y$ is a solution of the normal equation, where $\Phi^{\dagger}$ is the pseudo-inverse defined in Lemma 2.10. Indeed, we have

$$(\Phi^{\mathrm{T}}\Phi)\Phi^{\dagger}y = \Phi^{\mathrm{T}}\Phi\Phi^{\dagger}y = \Phi^{\mathrm{T}}(\Phi^{\dagger})^{\mathrm{T}}\Phi^{\mathrm{T}}y = \Phi^{\mathrm{T}}(\Phi^{\mathrm{T}})^{\dagger}\Phi^{\mathrm{T}}y = \Phi^{\mathrm{T}}y$$

where the Moore-Penrose condition (iii) is used (see Problem 2.9). Also, the general solution

$$\theta = \Phi^{\dagger}y + (I_N - \Phi^{\dagger}\Phi)z, \qquad \forall\, z \in \mathbb{R}^d$$

satisfies the normal equation. □

Let $\boldsymbol{\Theta}$ be the set of minimizers

$$\boldsymbol{\Theta} := \{\,\theta \mid \|y - \Phi\theta\| = \min\}$$

Then, we can show that

1. If θ is a minimizer, *i.e.* $\theta \in \boldsymbol{\Theta}$, then $\Phi^{\mathrm{T}}(y - \Phi\theta) = 0$, and *vice versa.*
2. If $\mathrm{rank}(\Phi) = d$, then $\boldsymbol{\Theta} = \{\hat{\theta}_{\mathrm{LS}}\}$, a singleton.
3. The set $\boldsymbol{\Theta}$ is convex.
4. The set $\boldsymbol{\Theta}$ has a unique minimum norm solution $\theta = \Phi^{\dagger}y$.

We apply the regression analysis technique to an ARX model[2], leading to a least-squares identification method, which is one of the simplest methods for a realistic identification problem.

Example A.2. Consider an ARX model

$$A(z)y(t) = B(z)u(t) + e(t) \tag{A.7}$$

where the unknown parameters are $\theta := (a_1 \quad \cdots \quad a_n \quad b_1 \quad \cdots \quad b_m)^{\mathrm{T}}$ and the noise variance σ^2. This is also called an equation error model, which is most easily identified by using the least-squares method. It should be noted that the ARX model of (A.7) is derived from (1.1) by setting $H(z) = 1/A(z)$.

From (A.7), the prediction error is given by

$$\varepsilon(t,\theta) := A(z,\theta)y(t) - B(z,\theta)u(t) = y(t) - \varphi^{\mathrm{T}}(t)\theta \tag{A.8}$$

where $\varphi(t)$ is the regression vector defined by

[2] ARX = AutoRegressive with eXogenous input.

$$\varphi(t) := [-y(t-1) \ \cdots \ -y(t-n) \ u(t-1) \ \cdots \ u(t-m)]^{\mathrm{T}} \in \mathbb{R}^{d \times 1}$$

Also, the unknown parameter vector is given by

$$\theta := (a_1 \cdots a_n \ b_1 \cdots b_m)^{\mathrm{T}}$$

Thus, it follows from (1.3) and (A.8) that

$$V_N(\theta) = \frac{1}{N} \sum_{t=0}^{N-1} [y(t) - \varphi^{\mathrm{T}}(t)\theta]^2$$

This implies that the PEM as applied to ARX models reduces to the least-squares method, so that the optimal estimate is given by

$$\hat{\theta}_{\mathrm{LS}}(N) = \left(\frac{1}{N} \sum_{t=0}^{N-1} \varphi(t)\varphi^{\mathrm{T}}(t) \right)^{-1} \frac{1}{N} \sum_{t=0}^{N-1} \varphi(t)y(t) \tag{A.9}$$

Suppose that the actual observations are expressed as

$$y(t) = \varphi^{\mathrm{T}}(t)\theta_0 + v_0(t) \tag{A.10}$$

where v_0 is a noise, and θ_0 is the "true" parameter. Substituting the above equation into (A.9) yields

$$\hat{\theta}_{\mathrm{LS}}(N) = \theta_0 + \left(\frac{1}{N} \sum_{t=0}^{N-1} \varphi(t)\varphi^{\mathrm{T}}(t) \right)^{-1} \frac{1}{N} \sum_{t=0}^{N-1} \varphi(t)v_0(t)$$

Suppose that

LS1) $$\lim_{N\to\infty} \frac{1}{N} \sum_{t=0}^{N-1} \varphi(t)\varphi^{\mathrm{T}}(t) = E\{\varphi(t)\varphi^{\mathrm{T}}(t)\} = \text{nonsingular}$$

LS2) $$\lim_{N\to\infty} \frac{1}{N} \sum_{t=0}^{N-1} \varphi(t)v_0(t) = E\{\varphi(t)v_0(t)\} = 0$$

hold[3]. Then we can show that

$$\lim_{N\to\infty} \hat{\theta}_{\mathrm{LS}}(N) = \theta_0$$

Thus the least-squares estimate is consistent. □

For convergence results based on laws of large numbers, see [109, 145]. If the above condition LS2) is not satisfied, then the least-squares estimate becomes biased. In order to obtain an unbiased estimate, we can employ a vector sequence correlated with the regressor vector $\varphi(t)$ but uncorrelated with the external noise $v_0(t)$.

[3]The second condition is surely satisfied if v_0 is a filtered white noise and $\varphi(t)$ is a bounded sequence [109].

Example A.3. (IV estimate) Let $\zeta(t) \in \mathbb{R}^d$ be a vector sequence. Pre-multiplying (A.10) by $\zeta(t)$ and summing over $[0, N-1]$ yield

$$\frac{1}{N}\sum_{t=0}^{N-1}\zeta(t)y(t) = \left(\frac{1}{N}\sum_{t=0}^{N-1}\zeta(t)\varphi(t)\right)\theta_0 + \frac{1}{N}\sum_{t=0}^{N-1}\zeta(t)v_0(t)$$

Suppose that a vector $\zeta(t)$ satisfies two conditions

$$\text{IV1)}\quad \lim_{N\to\infty}\frac{1}{N}\sum_{t=0}^{N-1}\zeta(t)\varphi^{\mathrm{T}}(t) = E\{\zeta(t)\varphi^{\mathrm{T}}(t)\} = \text{nonsingular}$$

$$\text{IV2)}\quad \lim_{N\to\infty}\frac{1}{N}\sum_{t=0}^{N-1}\zeta(t)v_0(t) = E\{\zeta(t)v_0(t)\} = 0$$

Then, we obtain a consistent estimate

$$\hat{\theta}_{\mathrm{IV}}(N) = \left(\frac{1}{N}\sum_{t=0}^{N-1}\zeta(t)\varphi^{\mathrm{T}}(t)\right)^{-1}\frac{1}{N}\sum_{t=0}^{N-1}\zeta(t)y(t) \tag{A.11}$$

This estimate is usually called an instrumental variable (IV) estimate, and the vectors $\zeta(t)$ satisfying the conditions IV1) and IV2) are called IV vectors. □

Detailed discussions on the IV estimate, including the best choice of the IV vector and convergence results, are found in [109, 145].

A.2 LQ Decomposition

We consider the relation between the least-squares method and LQ decomposition, which is a key technique in subspace identification methods.

Consider an FIR (finite impulse response) model

$$y(t) = \sum_{i=0}^{k-1} g_i u(t-i) + e(t) \tag{A.12}$$

where e is a white noise with mean zero and variance σ^2. The problem is to identify the impulse responses $\theta := (g_{k-1} \ \cdots \ g_1 \ g_0)^{\mathrm{T}}$ based on the input-output data $\{u(t),\, y(t),\, t = 0,\, 1,\, \cdots,\, N+k-2\}$. We define a data matrix

$$\left[\begin{array}{c} U_{0|k-1} \\ \hline Y_{k-1|k-1} \end{array}\right] := \left[\begin{array}{cccc} u(0) & u(1) & \cdots & u(N-1) \\ u(1) & u(2) & \cdots & u(N) \\ \vdots & \vdots & & \vdots \\ u(k-1) & u(k) & \cdots & u(N+k-2) \\ \hline y(k-1) & y(k) & \cdots & y(N+k-2) \end{array}\right] \in \mathbb{R}^{(k+1)\times N}$$

where we assume that $U_{0|k-1}$ has full row rank, so that $\text{rank}(U_{0|k-1}) = k$.

We temporarily assume that $\sigma^2 = 0$. Then from (A.12), we get

$$[g_{k-1} \;\; g_{k-2} \;\; \cdots \;\; g_0 \;\; -1] \begin{bmatrix} u(0) & u(1) \cdots & u(N-1) \\ u(1) & u(2) \cdots & u(N) \\ \vdots & \vdots & \vdots \\ u(k-1) & u(k) \cdots & u(N+k-2) \\ \hline y(k-1) & y(k) \cdots & y(N+k-2) \end{bmatrix} = 0$$

or this can be simply written as

$$[\theta^{\mathrm{T}} \;\; -1] \begin{bmatrix} U_{0|k-1} \\ Y_{k-1|k-1} \end{bmatrix} = 0 \tag{A.13}$$

As shown in Example 6.2, this problem can be solved by using the SVD of the data matrix. In fact, let $\begin{bmatrix} U_{0|k-1} \\ Y_{k-1|k-1} \end{bmatrix} = USV^{\mathrm{T}}$. Since the last singular value is zero due to (A.13), *i.e.* $\sigma_{k+1} = 0$, the $(k+1)$th left singular vector u_{k+1} satisfies

$$u_{k+1}^{\mathrm{T}} \begin{bmatrix} U_{0|k-1} \\ Y_{k-1|k-1} \end{bmatrix} = 0$$

Thus, normalizing the vector u_{k+1} so that the last element becomes -1, we obtain an estimate of the vector θ.

Now we assume that $\sigma^2 > 0$, where no θ exists satisfying (A.13), so that we must take a different route to estimate the vector θ. The LQ decomposition of the data matrix yields

$$\begin{bmatrix} U_{0|k-1} \\ Y_{k-1|k-1} \end{bmatrix} = \begin{bmatrix} L_{11} & 0 \\ L_{21} & L_{22} \end{bmatrix} \begin{bmatrix} Q_1^{\mathrm{T}} \\ Q_2^{\mathrm{T}} \end{bmatrix} \tag{A.14}$$

where $L_{11} \in \mathbb{R}^{k \times k}$, $L_{22} \in \mathbb{R}^{1 \times 1}$, $L_{21} \in \mathbb{R}^{1 \times k}$, and matrices $Q_1 \in \mathbb{R}^{N \times k}$, $Q_2 \in \mathbb{R}^{N \times 1}$ are orthogonal. By the rank condition for $U_{0|k-1}$, we see that $\det(L_{11}) \neq 0$, so that

$$Y_{k-1|k-1} = L_{21}Q_1^{\mathrm{T}} + L_{22}Q_2^{\mathrm{T}} = L_{21}L_{11}^{-1}U_{0|k} + L_{22}Q_2^{\mathrm{T}}$$

Since $Q_1^{\mathrm{T}} Q_2 = 0$, two terms in the right-hand side of the above equation are uncorrelated. Define

$$(g_{k-1} \;\; \cdots \;\; g_{k-1} \;\; g_0) := L_{21}L_{11}^{-1}$$

and

$$[e(k-1) \;\; e(k) \;\; \cdots \;\; e(N+k-2)] := L_{22}Q_2^{\mathrm{T}}$$

Then, for $t = k-1, k, \cdots, N+k-2$, we have

$$y(t) = g_0 u(t) + g_1 u(t-1) + \cdots + g_{k-1} u(t-k+1) + e(t)$$

This is the same FIR model as (A.12), implying that $\theta^{\mathrm{T}} = L_{21}L_{11}^{-1} \in \mathbb{R}^{1 \times k}$ is the least-squares estimates of impulse response parameters.

We show that the above result is also derived by solving the normal equation. The identification problem for the FIR model (A.12) can be cast into a least-squares problem

$$\min J(\theta) = ||Y_{k-1|k-1}^{\mathrm{T}} - U_{0|k-1}^{\mathrm{T}}\theta||^2$$

Thus, from (A.3) and (A.14), the least-squares estimate is given by

$$\begin{aligned}\hat{\theta} &= (U_{0|k-1}U_{0|k-1}^{\mathrm{T}})^{-1}U_{0|k-1}Y_{k-1|k-1}^{\mathrm{T}} \\ &= (L_{11}L_{11}^{\mathrm{T}})^{-1}[L_{11}Q_1^{\mathrm{T}}(L_{21}Q_1^{\mathrm{T}} + L_{22}Q_2^{\mathrm{T}})^{\mathrm{T}}] = (L_{21}L_{11}^{-1})^{\mathrm{T}}\end{aligned}$$

This is exactly the same as the least-squares estimate of θ obtained above by using the LQ decomposition. Thus we conclude that the least-squares problem can be solved by using the LQ decomposition.

B

Input Signals for System Identification

The selection of input signals has crucial effects on identification results. In this section, several input signals used for system identification are described, including step signals, sinusoids as well as random signals. One of the most important concepts related to input signals is the persistently exciting (PE) condition.

Let $u(t)$, $t = 0,\ 1,\ \cdots$ be a deterministic function. Then we define the mean and auto-covariance function as

$$\mu_u = \lim_{N\to\infty} \frac{1}{N} \sum_{t=0}^{N-1} u(t) \tag{B.1}$$

and for $l = 0,\ \pm 1,\ \cdots,$

$$\Lambda_{uu}(l) = \lim_{N\to\infty} \frac{1}{N} \sum_{t=0}^{N-1} [u(t+l) - \mu_u][u(t) - \mu_u] \tag{B.2}$$

Example B.1. (a) A step function is defined by

$$u(t) = \begin{cases} u_0, & t = 0,\ 1, \cdots \\ 0, & t = -1,\ -2,\ \cdots \end{cases}$$

In this case, we have $\Lambda_{uu}(l) = 0$ for $l = 0, \pm 1, \cdots$.

(b) Consider a sinusoid defined by

$$u(t) = a \sin(\omega t + \phi), \qquad t = 0,\ 1, \cdots \tag{B.3}$$

where $\omega > 0$ denotes the angular frequency, and $a > 0$ and $0 < \phi < \pi$ are the amplitude and phase, respectively. Let

$$S_N = \frac{1}{N} \sum_{t=0}^{N-1} \sin(\omega t + \phi), \qquad C_N = \frac{1}{N} \sum_{t=0}^{N-1} \cos(\omega t + \phi)$$

Since $\lim_{N\to\infty} S_N = 0$ and $\lim_{N\to\infty} C_N = 0$ hold, we have

$$\begin{aligned}\Lambda_{uu}(l) &= \lim_{N\to\infty} \frac{1}{N} \sum_{t=0}^{N-1} a^2 \sin(\omega(t+l)+\phi)\sin(\omega t+\phi) \\ &= \frac{a^2}{2}\cos(\omega l), \qquad l = 0, \pm 1, \cdots\end{aligned}$$

where the formula: $\sin\alpha \sin\beta = [\cos(\alpha-\beta) - \cos(\alpha+\beta)]/2$ is used.

Also, consider a composite sinusoid

$$u(t) = \sum_{j=1}^{p} a_j \sin(\omega_j t + \phi_j), \qquad t = 0,\ 1, \cdots \tag{B.4}$$

where $0 < \omega_1 < \cdots < \omega_p$ denote the angular frequencies, and $\{a_j\}$ and $\{\phi_j\}$ denote the amplitudes and phases, respectively. Then, it can be shown that

$$\Lambda_{uu}(l) = \sum_{j=1}^{p} \frac{a_j^2}{2}\cos(\omega_j l), \qquad l = 0, \pm 1, \cdots$$

(c) In system identification, a pseudo-random binary signal (PRBS) shown in Figure B.1 is often employed as test inputs. The PRBS is a periodic sequence with the maximum period $N = 2^p - 1$ where p is an integer greater than three, and is easily generated by p-stage shift registers. It is shown [145] that the mean and auto-covariance of a PRBS taking values on $\pm b$ are given by

$$\mu_u = \frac{1}{N}\sum_{t=1}^{N} u(t) = \frac{b}{N} \tag{B.5}$$

$$\Lambda_{uu}(l) = \begin{cases} b^2\left(1 - \dfrac{1}{N^2}\right), & l = 0 \pmod N \\ -\dfrac{b^2}{N}\left(1 + \dfrac{1}{N}\right), & l \neq 0 \pmod N \end{cases} \tag{B.6}$$

The auto-covariance function are shown in Figure B.2. □

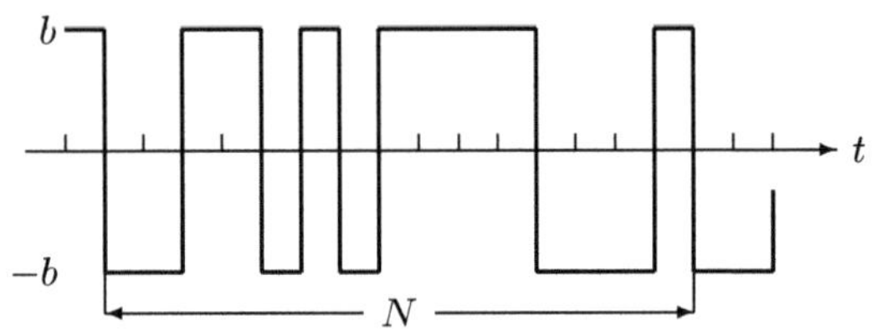

Figure B.1. A PRBS with the maximum period $N = 15$

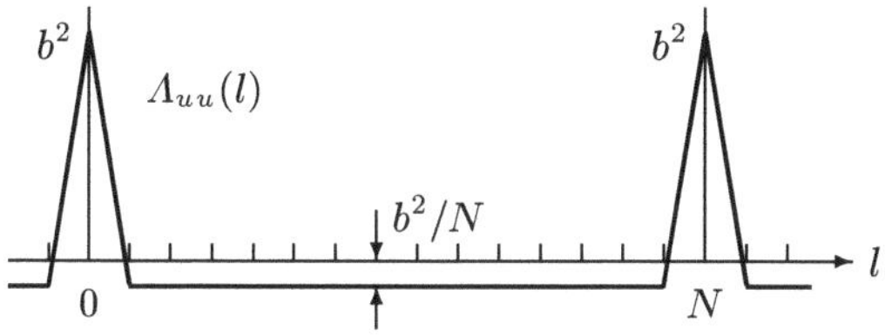

Figure B.2. The auto-covariance sequence of PRBS

In order to explain the PE condition of input signals, we consider the same FIR model as (A.12):

$$y(t) = \sum_{i=0}^{k-1} g_i u(t-i) + e(t) \tag{B.7}$$

where e is a zero mean white noise with variance σ^2. We deal with the identification of the impulse responses $\theta = (g_{k-1} \ \cdots \ g_1 \ g_0)^{\mathrm{T}}$ of the FIR model based on input-output data $\{u(t),\ y(t),\ t = 0,\ 1,\ \cdots,\ N-1\}$. For notational simplicity, we define the stacked vectors

$$y_{N-1} = \begin{bmatrix} y(k-1) \\ y(k) \\ \vdots \\ y(N-1) \end{bmatrix}, \qquad e_{N-1} = \begin{bmatrix} e(k-1) \\ e(k) \\ \vdots \\ e(N-1) \end{bmatrix} \in \mathbb{R}^{(N-k+1)\times 1}$$

and the matrix

$$U_{N-1} = \begin{bmatrix} u(0) & u(1) & \cdots & u(k-1) \\ u(1) & u(2) & \cdots & u(k) \\ \vdots & \vdots & \ddots & \vdots \\ u(N-k) & u(N-k+1) & \cdots & u(N-1) \end{bmatrix} \in \mathbb{R}^{(N-k+1)\times k}$$

Then, from (B.7), we have a linear regression model of the form

$$y_{N-1} = U_{N-1}\theta + e_{N-1} \tag{B.8}$$

The least-squares estimate of θ for (B.8) is obtained by solving

$$\min_{\theta} \|y_{N-1} - U_{N-1}\theta\|$$

Recall that Conditions A1) $\sim$ A4) in Section A.1 are required for solving the least-squares estimation problems. In particular, to get a unique solution, it is necessary to assume that $\text{rank}(U_{N-1}) = k$. This condition is equivalent to the fact that

$$\text{rank} \begin{bmatrix} u(0) & u(1) & \cdots & u(N-k) \\ u(1) & u(2) & \cdots & u(N-k+1) \\ \vdots & \vdots & & \vdots \\ u(k-1) & u(k) & \cdots & u(N-1) \end{bmatrix} = k \tag{B.9}$$

It may be noted that the data length is finite for this case.

Definition B.1. *[165] A deterministic sequence u with length N is PE of order k if (B.9) holds. If the input is a vector process $u \in \mathbb{R}^m$, then the rank condition (B.9) is replaced by* $\operatorname{rank}(U_{N-1}) = km$. □

For a zero mean stationary process $u \in \mathbb{R}^m$, we define the covariance matrix by

$$\bar{\Lambda}_{uu}(k) = \lim_{N\to\infty} \frac{1}{N} U_N^{\mathrm{T}} U_N$$
$$= \begin{bmatrix} \Lambda_{uu}(0) & \Lambda_{uu}^{\mathrm{T}}(1) & \cdots & \Lambda_{uu}^{\mathrm{T}}(k-1) \\ \Lambda_{uu}(1) & \Lambda_{uu}(0) & \cdots & \Lambda_{uu}^{\mathrm{T}}(k-2) \\ \vdots & \vdots & \ddots & \vdots \\ \Lambda_{uu}(k-1) & \Lambda_{uu}(k-2) & \cdots & \Lambda_{uu}(0) \end{bmatrix} \tag{B.10}$$

Then the PE condition for a stationary stochastic process is defined as follows.

Definition B.2. *[109, 145] If $\bar{\Lambda}_{uu}(k)$ of (B.10) is positive definite, then we say that u has the PE condition of order k.* □

The following example shows that when we deal with finite data, there always exist some ambiguities regarding how we treat boundary data.

Example B.2. Consider the step function treated in Example B.1. It can be shown that the step function is not PE since we have $\operatorname{rank}(U_{N-1}) = 1$.

However, in practice, step signals are often used for system identification. To consider this problem, we express (B.7) as

$$\begin{aligned} y(0) &= g_0 u(0) + g_1 u(-1) + \cdots + g_{k-1} u(-k+1) + e(0) \\ y(1) &= g_0 u(1) + g_1 u(0) + \cdots + g_{k-1} u(-k+2) + e(1) \\ &\vdots \\ y(k-1) &= g_0 u(k-1) + g_1 u(k-2) + \cdots + g_{k-1} u(0) + e(k-1) \\ y(k) &= g_0 u(k) + g_1 u(k-1) + \cdots + g_{k-1} u(1) + e(k) \\ &\vdots \end{aligned}$$

Suppose that the system is at rest for $t < 0$. Then we have $u(t) = 0$, $t = -1, -2, \cdots, -k+1$. Rearranging the above equations and assuming that $e(t) = 0$ for $t = 0, 1, \cdots$, we get

$$\begin{aligned} &[y(0)\ y(1)\ \cdots\ y(N-1)] \\ &= [g_{k-1}\ \cdots\ g_0] \begin{bmatrix} 0 & \cdots & 0 & u(0) & \cdots & u(N-k) \\ \vdots & \iddots & \iddots & u(1) & \cdots & u(N-k+1) \\ 0 & \iddots & \iddots & \vdots & & \vdots \\ u(0) & u(1) & \cdots & u(k-1) & \cdots & u(N-1) \end{bmatrix} \end{aligned}$$

Thus if $u(t) = 1,\ t = 0,\ 1, \cdots$, the wide rectangular matrix in the right-hand side of the above equation has rank k. Hence, by using the least-squares method, we can identify the impulse responses $g_0,\ g_1,\ \cdots, g_{k-1}$. However, in this case, it should be understood that the estimate is obtained by using the additional information that $u(t) = 0,\ t = -1,\ -2,\ \cdots,\ -k+1$. □

Example B.3. We consider the order of PE condition for simple signals based on Definition B.2.

(a) Let $u(t)$ be a zero mean white noise with variance σ^2. Then, for all $k > 0$, we see that $\bar{\Lambda}_{uu}(k) = \sigma^2 I_k$ is positive definite. Thus the white noise satisfies the PE condition of order infinity.

(b) Consider a sinusoid $u(t) = A\sin(\lambda_0 t),\ 0 < \lambda_0 < \pi$. Then, the auto-covariance function is given by $\Lambda_{uu}(k) = (A^2/2)\cos(\lambda_0 k)$, so that

$$\bar{\Lambda}_{uu}(2) = \frac{A^2}{2}\begin{bmatrix} 1 & \cos\lambda_0 \\ \cos\lambda_0 & 1 \end{bmatrix}$$

$$\bar{\Lambda}_{uu}(3) = \frac{A^2}{2}\begin{bmatrix} 1 & \cos\lambda_0 & \cos 2\lambda_0 \\ \cos\lambda_0 & 1 & \cos\lambda_0 \\ \cos 2\lambda_0 & \cos\lambda_0 & 1 \end{bmatrix}$$

We see that $\mathrm{rank}[\bar{\Lambda}_{uu}(2)] = 2$, and $\mathrm{rank}[\bar{\Lambda}_{uu}(k)] = 2$ for $k = 3,\ 4,\ \cdots$. Hence the sinusoid has PE condition of order two. This is obvious because a sinusoid has two independent parameters, a magnitude and a phase shift. □

Lemma B.1. *The PE conditions for some familiar stochastic processes are provided.*

(i) ARMA processes have the PE condition of order infinity.

(ii) The composite sinusoid of (B.4) satisfies the PE condition of order 2p.

Proof. [145] (i) Let u be a zero mean ARMA process with the spectral density function $\Phi_{uu}(\omega)$. Define $h = (h(0),\ h(1),\ \cdots,\ h(l-1))^{\mathrm{T}}$, and

$$H(z) = \sum_{i=0}^{l-1} h(i) z^{-i}$$

Consider a process defined by $y = H(z)u$. Then we easily see that y is a zero mean second-order stationary process, so that the variance of y is given by

$$\sigma_y^2 = E\left\{\left|\sum_{i=0}^{l-1} h(i)u(t-i)\right|^2\right\} = \sum_{i,j=0}^{l-1} \Lambda_{uu}(i-j)h(i)h(j) = h^{\mathrm{T}}\bar{\Lambda}_{uu}(l)h$$

It follows from Lemma 4.4 that

$$h^{\mathrm{T}}\bar{\Lambda}_{uu}(l)h = \frac{1}{2\pi}\int_{-\pi}^{\pi} |H(e^{j\omega})|^2 \Phi_{uu}(\omega)d\omega \tag{B.11}$$

Suppose that u does not satisfy the PE condition of order l. Then there exists a nonzero vector $h \in \mathbb{R}^l$ such that $h^{\mathrm{T}} \bar{\Lambda}_{uu}(l) h = 0$. Since the integrand of (B.11) is nonnegative, we have $|H(e^{j\omega})|^2 \Phi_{uu}(\omega) = 0$ for all ω[1]. However, from (4.35), the spectral density function of the ARMA process is positive except for at most finite points. It therefore follows that $H(e^{j\omega}) = 0$ (a.e.), and hence $h = 0$. This is a contradiction, implying that the ARMA process satisfies the PE condition of order l. Since l is arbitrary, the ARMA process satisfies the PE condition of infinite order.

(ii) Since, as shown in Example B.3, a sinusoid has the PE condition of order two, the composite sinusoid of (B.4) has the PE condition of order $2p$. □

From Lemma B.1 (i), we can say that for a stationary process u, if

$$\Phi_{uu}(\omega) > 0, \qquad -\pi < \omega < \pi$$

is satisfied, then u is PE of order infinity. This condition has already been mentioned in Chapters 9 and 10.

[1]The equality holds for $\omega \in (-\pi, \pi)$ almost everywhere (a.e.).

C

Overlapping Parametrization

In this section, we derive an overlapping parametrization for a stationary process; see also Example 1.2. From Theorems 4.3 and 4.4 (see Section 4.5), a zero mean regular full rank process $y \in \mathbb{R}^p$ can uniquely be expressed as

$$y(t) = \sum_{i=0}^{\infty} H_i \, e(t-i) = \sum_{i=-\infty}^{t} H_{t-i} \, e(i) \tag{C.1}$$

where e is the innovation process with mean 0 and covariance matrix $R > 0$, and where H_i, $i = 0, 1, \cdots$ are impulse response matrices satisfying

$$\sum_{i=0}^{\infty} \|H_i\|^2 < \infty; \qquad H_0 = I_p$$

Define the transfer matrix by

$$H(z) = \sum_{i=0}^{\infty} H_i z^{-i}$$

Moreover, define

$$\mathcal{Y}_t^- = \overline{\text{span}}\{y(t-1), \, y(t-2), \, \cdots\}$$
$$\mathcal{E}_t^- = \overline{\text{span}}\{e(t-1), \, e(t-2), \, \cdots\}$$

Then, it follows that $\mathcal{Y}_t^- = \mathcal{E}_t^-$, $t = 0, \pm 1, \cdots$. In the following, we assume that both $H(z)$ and $H^{-1}(z)$ are stable.

Let t be the present time. Then, from (C.1),

$$y(t+k) = \sum_{i=t}^{t+k} H_{t+k-i} \, e(i) + \sum_{i=-\infty}^{t-1} H_{t+k-i} \, e(i), \quad k = 0, 1, \cdots \tag{C.2}$$

Thus we see that the first term in the right-hand side of the above equation is a linear combination of the future innovations $e(t), \cdots, e(t+k)$ and that the second term is a linear combination of the past innovations $e(t-1),\ e(t-2),\ \cdots$. Since $\mathcal{Y}_t^- = \mathcal{E}_t^-$, the second term is also expressed as a linear combination of the past outputs $y(t-1),\ y(t-2),\ \cdots$, and hence it belongs to $\mathcal{Y}_t^-$. Thus it follows that the optimal predictor for $y(t+k)$ based on the past $\mathcal{Y}_t^-$ is given by (see Example 4.10)

$$\hat{y}(t+k \mid t-1) = \sum_{i=-\infty}^{t-1} H_{t+k-i}\, e(i), \qquad k = 0,\ 1,\ \cdots \tag{C.3}$$

Repeated use of this relation yields

$$\begin{bmatrix} \hat{y}(t \mid t-1) \\ \hat{y}(t+1 \mid t-1) \\ \hat{y}(t+2 \mid t-1) \\ \vdots \end{bmatrix} = \begin{bmatrix} H_1 & H_2 & H_3 & \cdots \\ H_2 & H_3 & H_4 & \cdots \\ H_3 & H_4 & H_5 & \cdots \\ \vdots & \vdots & \vdots & \ddots \end{bmatrix} \begin{bmatrix} e(t-1) \\ e(t-2) \\ e(t-3) \\ \vdots \end{bmatrix} \tag{C.4}$$

It should be noted that this is a free response of the system with the initial state resulting from the past inputs e up to time $t-1$ (see also Section 6.2).

Let the block Hankel operator be

$$H = \begin{bmatrix} H_1 & H_2 & H_3 & \cdots \\ H_2 & H_3 & H_4 & \cdots \\ H_3 & H_4 & H_5 & \cdots \\ \vdots & \vdots & \vdots & \ddots \end{bmatrix}$$

where it is assumed that $\operatorname{rank}(H) = n < \infty$. As shown in Section 8.3, the predictor space is defined by

$$\mathcal{X}_t^{+/-} := \hat{E}\{\mathcal{Y}_t^+ \mid \mathcal{Y}_t^-\} = \overline{\operatorname{span}}\{\hat{y}(t+k \mid t-1) \mid k = 0,\ 1,\ \cdots\}$$

Thus, we can find n independent vectors from the infinite components

$$\{\hat{y}_i(t+k \mid t-1), \quad i = 1,\ \cdots,\ p, \quad k = 0,\ 1, \cdots\} \tag{C.5}$$

where the n independent vectors form a basis of the predictor space $\mathcal{X}_t^{+/-}$.

Suppose that $\hat{y}(t \mid t-1)$ has full rank, *i.e.* $\operatorname{cov}\{\hat{y}(t \mid t-1)\} > 0$. Then,

$$\hat{y}_1(t \mid t-1),\ \hat{y}_2(t \mid t-1),\ \cdots,\ \hat{y}_p(t \mid t-1) \tag{C.6}$$

are linearly independent, and hence we see that the first p rows of H are linearly independent.

Let $\overline{n} = (n_1,\ \cdots,\ n_p)$ be a set of p positive integers such that $n_1 + \cdots + n_p = n$. We pick n elements including the p components of (C.6) from the infinite components defined by (C.5). Let such vectors be given by

$$\begin{array}{ccc}
\hat{y}_1(t \mid t-1), & \hat{y}_1(t+1 \mid t-1), \cdots, & \hat{y}_1(t+n_1-1 \mid t-1) \\
\hat{y}_2(t \mid t-1), & \hat{y}_2(t+1 \mid t-1), \cdots, & \hat{y}_2(t+n_2-1 \mid t-1) \\
 & \vdots \qquad\qquad & \vdots \\
\hat{y}_p(t \mid t-1), & \hat{y}_p(t+1 \mid t-1), \cdots, & \hat{y}_p(t+n_p-1 \mid t-1)
\end{array}$$

Note that, for example, if $n_1 = 1$, then only $\hat{y}_1(t \mid t-1)$ is selected from the first row. If these n vectors are linearly independent, we call $\overline{n} = (n_1, \cdots, n_p)$ a multi-index; see [54, 68, 109] for more details.

By using the above linearly independent components, we define a state vector of the system by

$$x(t) := \begin{bmatrix} \hat{y}_1(t \mid t-1) \\ \vdots \\ \hat{y}_1(t+n_1-1 \mid t-1) \\ \vdots \\ \hat{y}_p(t \mid t-1) \\ \vdots \\ \hat{y}_p(t+n_p-1 \mid t-1) \end{bmatrix} \in \mathbb{R}^n \tag{C.7}$$

From (C.3), we get

$$\begin{aligned}
\hat{y}(t+k \mid t) &= \sum_{i=-\infty}^{t} H_{t+k-i}\, e(i) = \sum_{i=-\infty}^{t-1} H_{t+k-i}\, e(i) + H_k\, e(t) \\
&= \hat{y}(t+k \mid t-1) + H_k\, e(t), \qquad k = 0, 1, \cdots
\end{aligned}$$

In terms of the components, this can be written as

$$\hat{y}_i(t+k \mid t) = \hat{y}_i(t+k \mid t-1) + h_{ik} e(t), \qquad k = 0, 1, \cdots \tag{C.8}$$

where $i = 1, \cdots, p$, and $h_{ik} = [h_{ik}(1) \ \cdots \ h_{ik}(p)] \in \mathbb{R}^{1\times p}$ is the ith row of H_k. Also, from (C.7), the state vector at $t+1$ is expressed as

$$x(t+1) = \begin{bmatrix} \hat{y}_1(t+1 \mid t) \\ \vdots \\ \hat{y}_1(t+n_1 \mid t) \\ \vdots \\ \hat{y}_p(t+1 \mid t) \\ \vdots \\ \hat{y}_p(t+n_p \mid t) \end{bmatrix} \in \mathbb{R}^n$$

Thus from (C.8), we can verify that

$$x(t+1) = \begin{bmatrix} \hat{y}_1(t+1 \mid t-1) \\ \vdots \\ \hat{y}_1(t+n_1 \mid t-1) \\ \vdots \\ \hat{y}_p(t+1 \mid t-1) \\ \vdots \\ \hat{y}_p(t+n_p \mid t-1) \end{bmatrix} + \begin{bmatrix} h_{11}(1) & \cdots & h_{11}(p) \\ \vdots & & \vdots \\ h_{1n_1}(1) & \cdots & h_{1n_1}(p) \\ \vdots & & \vdots \\ h_{p1}(1) & \cdots & h_{p1}(p) \\ \vdots & & \vdots \\ h_{pn_p}(1) & \cdots & h_{pn_p}(p) \end{bmatrix} e(t) \tag{C.9}$$

Note that the first term in the right-hand side of (C.9) belongs to the space $\mathcal{X}_t^{+/-}$. In particular, we see that $\hat{y}_i(t+n_i \mid t-1)$, $i = 1, \cdots, p$ are expressed in terms of a linear combination of the components of the basis vector $x(t)$. Thus, we have

$$\hat{y}_i(t+n_i \mid t-1) = \sum_{j=1}^{p} \sum_{k=1}^{n_j} \alpha_{ij}^k \hat{y}_j(t+k-1 \mid t-1), \quad i = 1, \cdots, p \tag{C.10}$$

Other components $\hat{y}_i(t+l \mid t-1)$, $l = 1, \cdots, n_i - 1$ are already contained in the vector $x(t)$ as its elements, so that they are expressed in terms of shift operations. Moreover, putting $k = 0$ in (C.8) and noting that $\hat{y}(t \mid t) = y(t)$ and $H_0 = I_p$ yield

$$y_i(t) = \hat{y}_i(t \mid t-1) + e_i(t), \qquad i = 1, \cdots, p \tag{C.11}$$

where $\hat{y}_i(t \mid t-1)$ belongs to $\mathcal{X}_t^{+/-}$.

For simplicity, we consider a 3-dimensional process y with 9-dimensional state vector, and assume that $n_1 = 3, n_2 = 4, n_3 = 2$ with $n_1 + n_2 + n_3 = 9$. Then, by using (C.8) and (C.11), we have the following A- and C-matrix:

$$A = \left[\begin{array}{ccc|cccc|cc} 0 & 1 & 0 & 0 & 0 & 0 & 0 & 0 & 0 \\ 0 & 0 & 1 & 0 & 0 & 0 & 0 & 0 & 0 \\ \alpha_{11}^1 & \alpha_{11}^2 & \alpha_{11}^3 & \alpha_{12}^1 & \alpha_{12}^2 & \alpha_{12}^3 & \alpha_{12}^4 & \alpha_{13}^1 & \alpha_{13}^2 \\ \hline 0 & 0 & 0 & 0 & 1 & 0 & 0 & 0 & 0 \\ 0 & 0 & 0 & 0 & 0 & 1 & 0 & 0 & 0 \\ 0 & 0 & 0 & 0 & 0 & 0 & 1 & 0 & 0 \\ \alpha_{21}^1 & \alpha_{21}^2 & \alpha_{21}^3 & \alpha_{22}^1 & \alpha_{22}^2 & \alpha_{22}^3 & \alpha_{22}^4 & \alpha_{23}^1 & \alpha_{23}^2 \\ \hline 0 & 0 & 0 & 0 & 0 & 0 & 0 & 0 & 1 \\ \alpha_{31}^1 & \alpha_{31}^2 & \alpha_{31}^3 & \alpha_{32}^1 & \alpha_{32}^2 & \alpha_{32}^3 & \alpha_{32}^4 & \alpha_{33}^1 & \alpha_{33}^2 \end{array}\right]$$

$$(p = 3, n = 9, n_1 = 3, n_2 = 4, n_3 = 2)$$

and

$$C = \left[\begin{array}{ccc|cccc|cc} 1 & 0 & 0 & 0 & 0 & 0 & 0 & 0 & 0 \\ \hline 0 & 0 & 0 & 1 & 0 & 0 & 0 & 0 & 0 \\ \hline 0 & 0 & 0 & 0 & 0 & 0 & 0 & 1 & 0 \end{array}\right]$$

We can easily infer the forms of A- and C-matrix for general cases. Thus, we have the following state space equation

$$x(t+1) = Ax(t) + Ke(t) \tag{C.12a}$$

$$y(t) = Cx(t) + e(t) \tag{C.12b}$$

where $K \in \mathbb{R}^{n \times p}$ is the coefficient matrix for $e(t)$ of (C.9). We see that the number of unknown parameters in this Markov model is $2np$, since K has no particular structure.

From the property of block Hankel matrix, we have the following lemma.

Lemma C.1. *[54, 109] Any n-dimensional stochastic LTI state space system can be expressed by means of a state space model (C.12) with a particular multi-index $\overline{n}$. In other words, the state space model (C.12) with a particular multi-index $\overline{n}$ can describe almost all n-dimensional stochastic LTI systems.* □

More precisely, let $M_{\overline{n}}(p)$ be the model structure of (C.12) with a multi-index $\overline{n}$. Also, let the sum of $M_{\overline{n}}(p)$ over possible multi-indices be

$$\overline{\mathcal{M}}(p) = \bigcup_{\overline{n}} \text{Im } M_{\overline{n}}(p)$$

Then, the set $\overline{\mathcal{M}}(p)$ denotes the set of all n-dimensional linear stochastic system with p outputs. Of course, $M_{\overline{n}}(p)$ may overlap, but $\overline{\mathcal{M}}(p)$ contains all the n-dimensional linear systems $M_{\overline{n}}(p)$.

The state space model of (C.12) is called an overlapping parametrization with $2np$ independent parameters. Thus, we can employ the PEM to identify the $2np$ unknown parameters, but we need some complicated algorithms for switching from a particular $\bar{n}^1$ to another $\bar{n}^2$ during the parameter identification, since we do not know the multi-index $\bar{n}$ prior to identification.

In general, a p-dimensional process y with state dimension n is called generic if the state vector x is formed as in (C.7) using some multi-index $\bar{n} = (n_1, \cdots, n_p)$. The next example shows that there exist non-generic processes.

Example C.1. [54] Let $p = 2$ and $n = 3$, and consider the following matrices:

$$C = \begin{bmatrix} 1 & 0 & 0 \\ 0 & 1 & 0 \end{bmatrix}, \qquad A = \begin{bmatrix} \alpha & \beta & 0 \\ 0 & 0 & 1 \\ 0 & 1 & 0 \end{bmatrix}, \qquad K = \begin{bmatrix} 0 & 0 \\ 1 & 0 \\ 0 & 1 \end{bmatrix}$$

where $\alpha\beta \neq 0$. Then, since $H_j = CA^{j-1}K$, $j = 1, \cdots$, we get

$$H_1 = \begin{bmatrix} 0 & 0 \\ 1 & 0 \end{bmatrix}, \quad H_2 = \begin{bmatrix} \beta & 0 \\ 0 & 1 \end{bmatrix}, \quad H_3 = \begin{bmatrix} \alpha\beta & \beta \\ 1 & 0 \end{bmatrix}, \quad H_4 = \begin{bmatrix} \alpha^2\beta + \beta & \alpha\beta \\ 0 & 1 \end{bmatrix}, \cdots$$

Thus the first 3×2 block submatrix of H is given by

$$H_{3,2} := \begin{bmatrix} H_1 & H_2 \\ H_2 & H_3 \\ H_3 & H_4 \end{bmatrix} = \begin{bmatrix} 0 & 0 & \beta & 0 \\ 1 & 0 & 0 & 1 \\ \beta & 0 & \alpha\beta & \beta \\ 0 & 1 & 1 & 0 \\ \alpha\beta & \beta & \alpha^2\beta + \beta & \alpha\beta \\ 1 & 0 & 0 & 1 \end{bmatrix}$$

It is easy to see that the first two rows of the block Hankel matrix are linearly independent, but the 3rd row is linearly dependent on the first two rows. Thus we observe that the selection $\bar{n} = (2, 1)$ $(n_1 = 2, n_2 = 1)$ does not yield a basis. Actually, in this case, we should pick the first two rows and the fourth row to form a basis. □

D

List of Programs

In Appendix D, some of MATLAB® programs used in this book are included.

D.1 Deterministic Realization Algorithm

Table D.1 displays a program for the Ho-Kalman's algorithm of Lemma 6.1, where it is assumed that $k,\ l > n := \text{rank}(H)$.

Table D.1. Ho-Kalman's algorithm

```
% Function  zeiger.m
% Lemma 6.1
function[A,B,C] = zeiger(H,p,m,n)
% p = dim(y); m = dim(u); n = dim(x)
% (p, m) are known
% kp × lm Hankel matrix
% k, l > n; H must be finite rank
kp = size(H,1); lm = size(H',1);
[U,S,V] = svd(H);                                  % Eq. (6.14)
n=rank(S); % if n is known, this is redundant.
S1 = sqrtm(S(1:n,1:n));
% T = identity matrix                              % Eq. (6.15)
Ok = U(:,1:n)*S1;
Cl = S1*V(:,1:n)';
A = Ok(1:kp-p,:)\Ok(p+1:kp,:);                     % Eq. (6.16)
B = Cl(:,1:m);
C = Ok(1:p,:);
```

Table D.2. MOESP method

```
% Function moeps.m
% Lemma 6.6
% m = dim(u), p = dim(y), n = dim(x); k = number of block rows
% U = km x N input data matrix
% Y = kp x N output data matrix
function [A,B,C,D] = moesp(U,Y,m,p,n,k)
km = size(U,1); kp = size(Y,1);
L = triu(qr([U;Y]'))'; % LQ decomposition
L11 = L(1:km,1:km);
L21 = L(km+1:km+kp,1:km);
L22 = L(km+1:km+kp,km+1:km+kp);
[UU,SS,VV] = svd(L22);                                  % Eq. (6.39)
U1 = UU(:,1:n); % n is known
Ok = U1*sqrtm(SS(1:n,1:n));
% Matrices A and C
C = Ok(1:p,1:n);                                        % Eq. (6.41)
A = pinv(Ok(1:p*(k-1),1:n))*Ok(p+1:p*k,1:n);            % Eq. (6.42)
% Matrices B and D
U2 = UU(:,n+1:size(UU',1));
Z = U2'*L21/L11;
XX = []; RR = [];
for j = 1:k
XX = [XX; Z(:,m*(j-1)+1:m*j)];
  Okj = Ok(1:p*(k-j),:);
  Rj = [zeros(p*(j-1),p) zeros(p*(j-1),n);
     eye(p) zeros(p,n); zeros(p*(k-j),p) Okj];
RR = [RR; U2'*Rj];
end
DB = pinv(RR)*XX;                                       % Eq. (6.44)
D = DB(1:p,:);
B = DB(p+1:size(DB,1),:);
```

D.2 MOESP Algorithm

Table D.2 displays a program for the basic MOESP method developed in [172, 173]. A formation of data matrices is omitted in this program, but Table D.3 contains a related method of constructing data matrices.

It should be noted that way of computing matrices A and C is different in each method, but the computing method of B and D in the MOESP method in Table D.2 is commonly used in many other subspace identification methods (not always). Thus we can say that differences in algorithms of subspace system identification methods are attributed to the way of computing A and C, or the image of extended observability matrix $\mathrm{Im}(\mathcal{O}_k)$.

D.3 Stochastic Realization Algorithms

We show two algorithms of stochastic realization based on Lemma 7.9 in Section 7.7 and Algorithm A in Section 8.7. It will be instructive to understand the difference between the two stochastic realization algorithms.

Table D.3. Stochastic realization algorithm

```
% Function  stochastic.m
% Lemma 7.9
% function [A,C,Cb,K,R] = stochastic(y,n,k)
% y = [y(1),y(2),...,y(Ndat)];  p × Ndat  matrix
% n = dim(x);  k = number of block rows
function [A,C,Cb,K,R] = stochastic(y,n,k)
[p,Ndat] = size(y);  N = Ndat-2*k;
ii = 0;
for i = 1:p:2*k*p-p+1
   ii = ii+1;
   Y(i:i+p-1,:) = y(:,ii:ii+N-1);
end;
% Data matrix
Ypp = Y(1:k*p,:);
for i = 1:k
   j = (k-i)*p+1;
   Yp(j:j+p-1,:) = Ypp((i-1)*p+1:i*p,:); % Yp := Y check
end
Yf = Y(k*p+1:2*k*p,:);
Rfp = (Yf*Yp')/N;  % Covariance matrix
[U,S,V] = svd(Rfp);                                   % Eq. (7.81)
S2 = sqrtm(S(1:n,1:n));
Ok = U(:,1:n)*S2;                                     % Eq. (7.82)
Ck = S2*V(:,1:n)';
A = Ok(1:k*p-p,:)\ Ok(p+1:k*p,:);                     % Eq. (7.83)
C = Ok(1:p,:);
Cb = Ck(1:n,1:p)';
RR = (Yf*Yf')/N;
R0 = RR(1:p,1:p);  % Variance of output
[P,L,G,Rept] = dare(A',C',zeros(n,n),-R0,-Cb');       % ARE (7.84)
K = G'
R = R-C*P*C';
```

Table D.3 displays the stochastic realization algorithm of Lemma 7.9, in which ARE is solved by using the function **dare**. This function **dare** can solve the ARE appearing in stochastic realization as well as the one appearing in Kalman filtering. For details, see the manual of the function **dare**.

Table D.4. Balanced stochastic realization – Algorithm A

```
% Function stocha_bal.m
% Algorithm A in Section 8.7
% y = [y(1),y(2),...,y(Ndat)]; p × Ndat matrix
% n = dim(x); k = number of block rows
function [A,C,Cb,K,R] = stocha_bal (y,n,k)
[p,Ndat] = size(y); N = Ndat-2*k;
ii = 0;
for i = 1:p:2*k*p-p+1
   ii = ii+1;    Y(i:i+p-1,:) = y(:,ii:ii+N-1);
end
Yp = Y(1:k*p,:); Yf = Y(k*p+1:2*k*p,:);
% LQ decomposition
H = [Yp; Yf]; [Q,L] = qr(H',0); L = L'/sqrt(N);          % Eq. (8.76)
L11 = L(1:k*p,1:k*p); L21 = L(k*p+1:2*k*p,1:k*p);
L22 = L(k*p+1:2*k*p,k*p+1:2*k*p);
% Covariance matrices
Rff = (L21*L21'+L22*L22');
Rfp = L21*L11'; Rpp = L11*L11';
% Square roots & inverses
[Uf,Sf,Vf] = svd(Rff); [Up,Sp,Vp] = svd(Rpp);
Sf = sqrtm(Sf); Sp = sqrtm(Sp);
L = Uf*Sf*Vf'; M = Up*Sp*Vp';                              % Eq. (8.77)
Sfi = inv(Sf); Spi = inv(Sp);
Linv = Vf*Sfi*Uf'; Minv = Vp*Spi*Up';
OC = Linv*Rfp*Minv';
[UU,SS,VV] = svd(OC);                                      % Eq. (8.78)
Lambda = Rpp(1:p,1:p); % Covariance matrix of output
S = SS(1:n,1:n);
Ok = L*UU(:,1:n)*sqrtm(S);                                 % Eq. (8.79)
Ck = sqrtm(S)*VV(:,1:n)'*M';
A = Ok(1:k*p-p,:)\Ok(p+1:k*p,:);                           % Eq. (8.80)
C = Ok(1:p,:);   Cb = Ck(:,(k-1)*p+1:k*p)';
R = Lambda-C*S*C'; K = (Cb'-A*S*C')/R;                     % Eq. (8.81)
```

Table D.4 shows a program for Algorithm A of Section 8.7. The form of data matrix Y_p in Table D.4 is slightly different from Y_p in Table D.3, since in Table D.3, after generating Y_p, we formed $\check{Y}_p$ by re-ordering the elements. Thus a way of computing $\bar{C}^{\mathrm{T}}$ in Table D.4 is different from that in Table D.3. There is no theoretical difference, but numerical results may be slightly different.

The program of Table D.4 is very simple since the solution of ARE is not employed, but there are possibilities that $A - BK$ is unstable. Also, it should be noted that we compute L^{-1} and M^{-1} by using pseudo-inverses. For, if the function **chol** is used for computing the matrix square roots, the program stops unless Σ_{ff} and Σ_{pp} are positive definite, but these matrices may be rank deficient.

D.4 Subspace Identification Algorithms

The programs for the ORT and CCA methods derived in Sections 9.7 and 10.6 are displayed in Tables D.5 and D.6, respectively. Also, a program of the PO-MOESP is included in Table D.7. Comparing the programs in Tables D.5 and D.7, we can easily understand the difference in algorithms of the ORT and PO-MOESP; both use the same LQ decomposition, but the way of utilizing L factors is different. For identifying B and D, the ORT uses the same method as the PO-MOESP.

Table D.5. Subspace identification of deterministic subsystem – ORT

```
% Function ort_pk.m
% Subsection 9.7.1
function [A,B,C,D] = ort_pk(U,Y,m,p,n,k);
% ORT method by Picci and Katayama
km = size(U,1)/2; kp = size(Y,1)/2;
% LQ decomposition                                  % Eq. (9.48)
L = triu(qr([U;Y]'))';
L11 = L(1:km,1:km);
L41 = L(2*km+kp+1:2*km+2*kp,1:km);
L42 = L(2*km+kp+1:2*km+2*kp,km+1:2*km);
% SVD                                               % Eq. (9.52)
[UU,SS,VV] = svd(L42);
U1 = UU(:,1:n);
Ok = U1*sqrtm(SS(1:n,1:n));
C = Ok(1:p,1:n);
A = pinv(Ok(1:p*(k-1),1:n))*Ok(p+1:k*p,1:n);        % Eq. (9.53)
% Matrices B and D
U2 = UU(:,n+1:size(UU',1));
Z = U2'*L41/L11;                                    % Eq. (9.54)
% The program for computing B and D is the same
% as that of MOESP of Table D.2.
XX = [];
RR = [];
for j = 1:k
XX = [XX; Z(:,m*(j-1)+1:m*j)];
Okj = Ok(1:p*(k-j),:);
Rj = [zeros(p*(j-1),p),zeros(p*(j-1),n);
eye(p), zeros(p,n);
zeros(p*(k-j),p),Okj];
RR = [RR;U2'*Rj];
end
DB = pinv(RR)*XX;
D = DB(1:p,:);
B = DB(p+1:size(DB,1),:);
```

Table D.6. Stochastic subspace identification – CCA

```
% Function cca.m
% Section 10.6 CCA Algorithm B
% y = [y(1),y(2),...,y(Ndat)]; p×Ndat matrix
% u = [u(1),u(2),...,u(Ndat)]; m×Ndat matrix
% n = dim(x); k = number of block rows
% Written by H. Kawauchi; modified by T. Katayama
function [A,B,C,D,K] = cca(y,u,n,k)
[p,Ndat] = size(y); [m,Ndat] = size(u); N = Ndat-2*k;
ii = 0;
for i = 1:m:2*k*m-m+1
  ii = ii+1;   U(i:i+m-1,:) = u(:,ii:ii+N-1); % Data matrix
end
ii = 0;
for i = 1:p:2*k*p-p+1
  ii = ii+1;
  Y(i:i+p-1,:) = y(:,ii:ii+N-1); % Data matrix
end
Uf = U(k*m+1:2*k*m,:); Yf = Y(k*p+1:2*k*p,:);
Up = U(1:k*m,:); Yp = Y(1:k*p,:); Wp = [Up; Yp];
H = [Uf; Up; Yp; Yf];
[Q,L] = qr(H',0); L = L'; % LQ decomposition
L22 = L(k*m+1:k*(2*m+p),k*m+1:k*(2*m+p));
L32 = L(k*(2*m+p)+1:2*k*(m+p),k*m+1:k*(2*m+p));
L33 = L(k*(2*m+p)+1:2*k*(m+p),k*(2*m+p)+1:2*k*(m+p));
Rff = L32*L32'+L33*L33'; Rpp = L22*L22'; Rfp = L32*L22';
[Uf,Sf,Vf] = svd(Rff); [Up,Sp,Vp] = svd(Rpp);
Sf = sqrtm(Sf); Sfi = inv(Sf); Sp = sqrtm(Sp); Spi = inv(Sp);
Lfi = Vf*Sfi*Uf'; Lpi = Vp*Spi*Up'; % Lf = Uf*Sf*Vf'; Lp = Up*Sp*Vp'
OC = Lfi*Rfp*Lpi';
[UU,SS,VV] = svd(OC); % Normalized SVD
S1 = SS(1:n,1:n); U1 = UU(:,1:n); V1 = VV(:,1:n);
X = sqrtm(S1)*V1'*Lpi*Wp; XX = X(:,2:N); X = X(:,1:N-1);
U = Uf(1:m,1:N-1); Y = Yf(1:p,1:N-1);
ABCD = [XX;Y]/[X;U]; % System matrices
A = ABCD(1:n,1:n); B = ABCD(1:n,n+1:n+m);
C = ABCD(n+1:n+p,1:n); D = ABCD(n+1:n+p,n+1:n+m);
W = XX-A*X-B*U;  E = Y-C*X-D*U;
SigWE = [W;E]*[W;E]'/(N-1);
QQ = SigWE(1:n,1:n); RR = SigWE(n+1:n+p,n+1:n+p);
SS = SigWE(1:n,n+1:n+p);
[P,L,G,Rept] = dare(A',C',QQ,RR,SS); % Kalman filter ARE
K = G'; % Kalman gain
```

The CCA method – Algorithm B – in Table D.6 is based on the use of estimates of state vectors. It may be noted that the LQ decomposition in the above table is different from the one defined by (10.46); in fact, in the above program, the past input-output data $\begin{bmatrix} U_p \\ Y_p \end{bmatrix}$ is employed for $\check{W}_{0|k-1}$, since the row spaces of both data matrices are the same.

The following table shows a program of the PO-MOESP algorithm [171].

Table D.7. PO-MOESP algorithm

```
% Function po_moesp.m
function [A,B,C,D] = po_moesp(U,Y,m,p,n,k);
% cf. Remark 9.3
% m=dim(u), p=dim(y), n=dim(x)
% k=number of block rows; U=2km x N matrix; Y=2kp x N matrix
km=k*m;
kp=k*p;
% LQ decomposition
L = triu(qr([U;Y]'))';
L11 = L(1:km,1:km);
L21 = L(km+1:2*km,1:km);
L22 = L(km+1:2*km,km+1:2*km);
L31 = L(2*km+1:2*km+kp,1:km);
L32 = L(2*km+1:2*km+kp,km+1:2*km);
L41 = L(2*km+kp+1:2*km+2*kp,1:km);
L42 = L(2*km+kp+1:2*km+2*kp,km+1:2*km);
L43 = L(2*km+kp+1:2*km+2*kp,2*km+1:2*km+kp);
[UU,SS,VV]=svd([L42 L43]);
U1 = UU(:,1:n);
Ok = U1*sqrtm(SS(1:n,1:n));
C = Ok(1:p,1:n);
A = pinv(Ok(1:p*(k-1),1:n))*Ok(p+1:k*p,1:n);
% Matrices B and D
U2 = UU(:,n+1:size(UU',1));
Z = U2'*[L31 L32 L41]/[L21 L22 L11];
% The rest is the same as that of MOESP of Table D.2.
% The subsequent part is omitted.
```

E

Solutions to Problems

Chapter 2

2.1 (a) Suppose that $\mathrm{rank}(A) = r$. Let $A = U\Sigma V^{\mathrm{T}}$, where $\Sigma = \mathrm{diag}(\Sigma_r, 0)$, and $\Sigma_r \in \mathbb{R}^{r\times r} > 0$. Also, partition $U = [U_r \ \ \tilde{U}_r]$ and $V = [V_r \ \ \tilde{V}_r]$. From Lemma 2.9 (i), we see that $\mathrm{Im}(A) = \mathrm{Im}(U_r)$, $\mathrm{Ker}(A^{\mathrm{T}}) = \mathrm{Im}(\tilde{U}_r)$, and $\mathrm{Im}(A^{\mathrm{T}}) = \mathrm{Im}(V_r)$, $\mathrm{Ker}(A) = \mathrm{Im}(\tilde{V}_r)$. Item (a) is proved by using

$$\mathrm{Im}(U_r) \oplus \mathrm{Im}(\tilde{U}_r) = \mathbb{R}^m, \qquad \mathrm{Im}(U_r) \perp \mathrm{Im}(\tilde{U}_r)$$

$$\mathrm{Im}(V_r) \oplus \mathrm{Im}(\tilde{V}_r) = \mathbb{R}^n, \qquad \mathrm{Im}(V_r) \perp \mathrm{Im}(\tilde{V}_r)$$

(b) These are the restatement of the relations in (a).
(c) We can prove the first relation of (c) as

$$\mathrm{Im}(AA^{\mathrm{T}}) = \mathrm{Im}(U_r\Sigma_r^2 U_r^{\mathrm{T}}) = \mathrm{Im}(U_r) = \mathrm{Im}(A)$$

Also, the second relation is proved as follows:

$$A\mathrm{Im}(B) = \{Ax \mid x = B\eta,\ \eta \in \mathbb{R}^p\} = \{AB\eta \mid \eta \in \mathbb{R}^p\} = \mathrm{Im}(AB)$$

2.2 Compute the product of three matrices in the right-hand side.

2.3 (a) It suffices to compute the determinant of both sides of the relations in Problem 2.2. (b) This is obvious from (a). (c) Pre-multiplying the right-hand side of the formula by $\begin{bmatrix} A & B \\ C & D \end{bmatrix}$ yields the identity. (d) Comparing the $(1,1)$-blocks of the formula in (c) gives

$$[A - BD^{-1}C]^{-1} = A^{-1} + A^{-1}B[D - CA^{-1}B]^{-1}CA^{-1}$$

By changing the sign of D, we get the desired result.

2.4 Let $Px = \lambda x$, $x \neq 0$. Then, $P(Px) = P(\lambda x) = \lambda^2 x$ holds. Hence, from $P^2 = P$, we have $Px = \lambda^2 x$. It thus follows that $\lambda x = \lambda^2 x$ for $x \neq 0$, implying that $\lambda = 0$ or $\lambda = 1$. Suppose that $\lambda_1 = \cdots = \lambda_r = 1$ and $\lambda_{r+1} = \cdots = \lambda_n = 0$.

We see that trace$(P) = \sum_{i=1}^n \lambda_i = r$. Let the SVD of P be given by $P = U\Sigma V^{\mathrm{T}}$, where $\Sigma = \mathrm{diag}(\sigma_1, \cdots, \sigma_n)$. Since, in this case, $\sigma_i = \lambda_i$, we see that rank$(P) =$ rank$(\Sigma) = r$.

2.5 Suppose that $P^2 = P$ holds. Then, from Lemma 2.4 and Corollary 2.1, we have (2.17) and (2.18). Thus (a) implies (b).

We show (b) $\to$ (c). As in the proof of Lemma 2.5, we define $\mathcal{V} = \mathrm{Im}(P)$ and $\mathcal{W} = \mathrm{Im}(I_n - P)$. Note that for the dimensions of subspaces, we have

$$\dim(\mathcal{V} \vee \mathcal{W}) = \dim(\mathcal{V}) + \dim(\mathcal{W}) - \dim(\mathcal{V} \cap \mathcal{W})$$

Since $x = Px + (I_n - P)x$, it follows that $\mathbb{R}^n = \mathcal{V} \vee \mathcal{W}$ and $n = \dim(\mathcal{V} \vee \mathcal{W})$. Also, from (b), we get $\dim(\mathcal{V}) + \dim(\mathcal{W}) = n$, and hence $\dim(\mathcal{V} \cap \mathcal{W}) = 0$. This implies that $\mathcal{V} \cap \mathcal{W} = \{0\}$, so that (c) holds.

Finally, we show (c) $\to$ (a). Post-multiplying $I_n = P + (I_n - P)$ by P yields $P = P^2 + (I_n - P)P$, so that we have $P(I_n - P) = (I_n - P)P$. Thus

$$\mathrm{Im}\, P(I_n - P) \subset \mathrm{Im}(P), \qquad \mathrm{Im}\,(I_n - P)P \subset \mathrm{Im}(I_n - P)$$

hold. If (c) holds, we get $\mathrm{Im}(P) \cap \mathrm{Im}(I_n - P) = \{0\}$, implying that $\mathrm{Im}[P(I_n - P)] = \{0\}$ follows. Hence, we have $P^2 = P$. This completes the proof.

2.6 Since $LT = I_r$, we get $P^2 = TLTL = TL = P$. Also, T and L are of full rank, so that $\mathrm{Im}(P) = \mathrm{Im}(TL) = \mathrm{Im}(T)$ and $\mathrm{Ker}(P) = \mathrm{Ker}(TL) = \mathrm{Ker}(L)$. This implies that P is the oblique projection on $\mathrm{Im}(T)$ along $\mathrm{Ker}(L)$. Similarly, we can prove that Q is a projection.

2.7 Define $L = [L_1 \;\; L_2]$ and $V = [V_1 \;\; V_2]$. Since $\begin{bmatrix} L \\ V \end{bmatrix} [T \;\; U] = \begin{bmatrix} I_r & 0 \\ 0 & I_{n-r} \end{bmatrix}$, we have

$$\begin{bmatrix} L_1 & L_2 \\ V_1 & V_2 \end{bmatrix} \begin{bmatrix} I_r & -X \\ 0 & I_{n-r} \end{bmatrix} = \begin{bmatrix} I_r & 0 \\ 0 & I_{n-r} \end{bmatrix}$$

This implies that $L_1 = I_r$, $L_2 = X$, $V_1 = 0$, $V_2 = I_{n-r}$, and hence

$$P = \begin{bmatrix} I_r & X \\ 0 & 0 \end{bmatrix}$$

2.8 (a) Let $P = V_r V_r^{\mathrm{T}}$. Then, $P^2 = P$ and $P^{\mathrm{T}} = P$ hold, so that P is an orthogonal projection. Also, from Lemma 2.9, we have $\mathrm{Im}(A^{\mathrm{T}}) = \mathrm{Im}(V_r) = \mathrm{Im}(V_r V_r^{\mathrm{T}})$. Similarly, we can prove (b), (c), (d).

2.9 Let $A = U\Sigma V^{\mathrm{T}}$, where $U \in \mathbb{R}^{m\times m}$, $V \in \mathbb{R}^{n\times n}$ are orthogonal and $\Sigma = \begin{bmatrix} \Sigma_s & 0 \\ 0 & 0 \end{bmatrix} \in \mathbb{R}^{m\times n}$ with $\Sigma_s \in \mathbb{R}^{r\times r}$ diagonal, where $r := \mathrm{rank}(A)$. Then, we get

$$(AA^{\mathrm{T}})^\dagger = (U\Sigma\Sigma^{\mathrm{T}}U^{\mathrm{T}})^\dagger = U(\Sigma\Sigma^{\mathrm{T}})^\dagger U^{\mathrm{T}}$$

where

$$(\Sigma\Sigma^{\mathrm{T}})^\dagger = \begin{bmatrix} \Sigma_s^{-2} & 0 \\ 0 & 0 \end{bmatrix} \in \mathbb{R}^{m\times m}$$

Thus

$$A^{\mathrm{T}}(AA^{\mathrm{T}})^{\dagger} = V \begin{bmatrix} \Sigma_s & 0 \\ 0 & 0 \end{bmatrix}^{\mathrm{T}} \begin{bmatrix} \Sigma_s^{-2} & 0 \\ 0 & 0 \end{bmatrix} U^{\mathrm{T}} = V \begin{bmatrix} \Sigma_s^{-1} & 0 \\ 0 & 0 \end{bmatrix} U^{\mathrm{T}} = A^{\dagger}$$

That $(A^{\mathrm{T}}A)^{\dagger}A^{\mathrm{T}} = A^{\dagger}$ is proved similarly.

2.10 Let $A = U\Sigma V^{\mathrm{T}}$, where $\Sigma = \mathrm{diag}(\sigma_1, \cdots, \sigma_n)$, and $U \in \mathbb{R}^{m\times n}$, $V \in \mathbb{R}^{n\times n}$. Then, we have $Q = UV^{\mathrm{T}}$ and $\Pi = V\Sigma V^{\mathrm{T}}$. Note that $VV^{\mathrm{T}} = V^{\mathrm{T}}V = I_n$.

Chapter 3

3.1 Since $|g_k| = 1/k$, $k = 1, 2, \cdots$, we have

$$\sum_{k=1}^{\infty} |g_k| = \sum_{k=1}^{\infty} \frac{1}{k} = \infty$$

This implies that the system is not stable.

3.2 To apply the Routh-Hurwitz test for a continuous-time LTI system to a discrete-time LTI system, let $z = (s+1)/(s-1)$. Then, we see that $|z| < 1 \Leftrightarrow \mathfrak{Re}[s] < 0$. From $f((s+1)/(s-1)) = 0$, we get

$$(1 + a_1 + a_2)s^2 + 2(1 - a_2)s + (1 - a_1 + a_2) = 0$$

Thus the stability condition for $z^2 + a_1 z + z_2$ is given by

$$1 + a_1 + a_2 > 0, \qquad 1 - a_1 + a_2 > 0, \qquad 1 - a_2 > 0 \tag{E.1}$$

3.3 From a diagonal system of Figure 3.3, A and B are given by

$$A = \begin{bmatrix} \lambda_1 & 0 & 0 \\ 0 & \lambda_2 & 0 \\ 0 & 0 & \lambda_3 \end{bmatrix}, \qquad B = \begin{bmatrix} b_1 \\ b_2 \\ b_3 \end{bmatrix}$$

Thus, from Theorem 3.4 (ii), it suffices to find the condition such that

$$\mathrm{rank} \begin{bmatrix} \lambda_1 - z & 0 & 0 & b_1 \\ 0 & \lambda_2 - z & 0 & b_2 \\ 0 & 0 & \lambda_3 - z & b_3 \end{bmatrix} = 3, \qquad z = \lambda_1, \lambda_2, \lambda_3$$

holds. Hence, the reachability condition becomes

$$b_1 b_2 b_3 \neq 0, \qquad (\lambda_1 - \lambda_2)(\lambda_2 - \lambda_3)(\lambda_3 - \lambda_1) \neq 0$$

3.4 Note that $\mathcal{C} = [b \;\; Ab \;\; \cdots \;\; A^{n-1}b]$. From (2.3), we have

$$A^n = -(\alpha_1 A^{n-1} + \cdots + \alpha_{n-1}A + \alpha_n I)$$

Hence,

$$\begin{aligned}A\mathcal{C} &= [Ab\ \ A^2b\ \ \cdots\ \ A^nb] \\ &= [Ab\ \ A^2b\ \ \cdots\ \ -(\alpha_1 A^{n-1} + \cdots + \alpha_{n-1}A + \alpha_n I)b] \\ &= [b\ \ Ab\ \ \cdots\ \ A^{n-1}b]\begin{bmatrix} 0 & & & -\alpha_n \\ 1 & & & -\alpha_{n-1} \\ & \ddots & & \vdots \\ & & 1 & -\alpha_1 \end{bmatrix} = \mathcal{C}\bar{A}\end{aligned}$$

and

$$b = [b\ \ Ab\ \ \cdots\ \ A^{n-1}b]\begin{bmatrix}1 \\ 0 \\ \vdots \\ 0\end{bmatrix} = \mathcal{C}\bar{b}$$

3.5 We can show that

$$\text{rank}[A + BK - \lambda I \quad B] = n \quad \Leftrightarrow \quad \text{rank}[A - \lambda I \quad B] = n$$

and

$$\text{rank}\begin{bmatrix} A + LC - \lambda I \\ C \end{bmatrix} = n \quad \Leftrightarrow \quad \text{rank}\begin{bmatrix} A - \lambda I \\ C \end{bmatrix} = n$$

The results follow from Theorems 3.4 $\sim$ 3.9.

3.6 Define $\bar{A} := A/(\rho(A) + \varepsilon)$. Then, the spectral radius of $\bar{A}$ is strictly less than 1, and hence $\bar{A}^k \to 0$ as $k \to \infty$. Thus, in particular, the elements of the sequence $\{\bar{A}^k,\ k = 1, 2, \cdots\}$ are bounded, so that we have $|(\bar{A}^k)_{ij}| \leq C$, $C > 0$ for $k = 1, 2, \cdots$ and $i, j = 1, \cdots, n$. Since $(\bar{A}^k)_{ij} = (A^k)_{ij}/(\rho(A) + \varepsilon)^k$, we get the desired result.

3.7 Before proving this assertion, it will be helpful to look at the proof of a basic convergence result for the Césaro sum in Problem 4.3 (a).

The solution $x(t)$ is given by

$$x(t) = A^t x(0) + \sum_{k=0}^{t-1} A^{t-k-1} f(k)$$

By assumption, $\rho(A) < 1$. Thus we can take $\varepsilon > 0$ such that $\rho(A) + \varepsilon =: a < 1$. From Problem 3.6,

$$|(A^k)_{ij}| \leq C(\rho(A) + \varepsilon)^k \quad \Rightarrow \quad \|A^k\|_\alpha \leq C_1 a^k, \quad k = 1, 2, \cdots$$

where $C_1 > 0$, and $\|\cdot\|_\alpha$ is a matrix norm (see Section 2.3). By using the above estimate,

$$\|x(t)\| \leq C_1 a^t \|x(0)\| + C_1 \sum_{k=0}^{t-1} a^{t-k-1} \|f(k)\|$$

Since the first term tends to zero as $t \to \infty$, it suffices to show that the second term tends to zero as $t \to \infty$. Let the second term be $g(t)$. Then, we get

$$g(t) = C_1 a^{t-1} \sum_{k=0}^{t-1} a^{-k} \beta(k), \qquad \beta(k) = \|f(k)\|$$

By hypothesis, $\lim_{k\to\infty} \beta(k) = 0$, so that for any $\varepsilon_1 > 0$, there exists $N_0 > 0$ such that $\beta(k) < \varepsilon_1(1-a)/(aC_1)$ for all $k > N_0$. Thus, for a sufficiently large t,

$$\begin{aligned} g(t) &= C_1 a^{t-1} \left[\sum_{k=0}^{N_0} a^{-k}\beta(k) + \sum_{k=N_0+1}^{t-1} a^{-k}\beta(k) \right] \\ &\le C_1 a^{t-1} \left[\sum_{k=0}^{N_0} a^{-k}\beta(k) + \frac{\varepsilon(1-a)}{aC_1} \sum_{k=N_0+1}^{t-1} a^{-k} \right] \\ &= C_1 a^{t-1} \left[\sum_{k=0}^{N_0} a^{-k}\beta(k) + \frac{\varepsilon_1(1-a)}{aC_1} \left(\frac{a^{-N_0-1} - a^{-t+1}}{1-a^{-1}} \right) \right] \\ &\le C_1 a^{t-1} \left[\sum_{k=0}^{N_0} a^{-k}\beta(k) \right] + \varepsilon_1 \left[1 - a^{t-N_0-2} \right] \end{aligned}$$

The first term in the right-hand side of the above inequality tends to zero as $t \to \infty$, while the second term is smaller than ε_1. This completes the proof.

3.8 It can be shown that

$$\begin{bmatrix} zI - A & B \\ -C & D \end{bmatrix} = \begin{bmatrix} I_n & 0 \\ -C(zI-A)^{-1} & I_p \end{bmatrix} \begin{bmatrix} zI - A & B \\ 0 & G(z) \end{bmatrix}$$

where $G(z) = C(zI-A)^{-1}B$. Since rank $\begin{bmatrix} I_n & 0 \\ -C(zI-A)^{-1} & I_p \end{bmatrix} = n + p$, we get

$$\mathrm{rank}_z S(z) = \mathrm{rank}_z (zI - A) + \mathrm{rank}_z G(z) = n + \mathrm{rank}_z G(z)$$

3.9 ([51], vol. 2, pp. 206–207) Suppose that $R(z) = b(z)/a(z)$ is rational, and that the series expansion

$$\frac{b(z)}{a(z)} = \frac{h_1}{z} + \frac{h_2}{z^2} + \cdots \tag{E.2}$$

converges for $|z| > \rho$ for some $\rho > 0$. Suppose that polynomials $a(z)$ and $b(z)$ are given by

$$a(z) = z^m + a_1 z^{m-1} + \cdots + a_m, \quad b(z) = b_1 z^{m-1} + b_2 z^{m-2} + \cdots + b_m$$

Multiplying (E.2) by $a(z)$ yields

$$b_1 z^{m-1} + b_2 z^{m-2} + \cdots + b_m = (z^m + a_1 z^{m-1} + \cdots + a_m) \left(\frac{h_1}{z} + \frac{h_2}{z^2} + \cdots \right)$$

Equating the coefficients of equal powers of z on both sides, we obtain

$$\begin{aligned} b_1 &= h_1 \\ b_2 &= h_2 + a_1 h_1 \\ &\vdots \\ b_m &= h_m + a_1 h_{m-1} + \cdots + a_{m-1} h_1 \end{aligned}$$

and for $j = m+1, \cdots,$

$$0 = h_j + a_1 h_{j-1} + \cdots + a_m h_{j-m}$$

This implies that (2.40) holds with $r = m$, so that the Hankel matrix (2.35) has finite rank.

Conversely, if H has finite rank, then (2.40) holds from Lemma 2.14. Hence, by using $a_1, \cdots, a_r$ of (2.40) and the above relations, we can define $b_1, \cdots, b_r$. Thus we see that $b(z)/a(z)$ is a desired rational function, which equals $R(z) = h_1/z + h_2/z^2 + \cdots$.

3.10 Note that the following power series expansion:

$$\log(1 + z^{-1}) = z^{-1} - \frac{1}{2} z^{-2} + \frac{1}{3} z^{-3} - \cdots, \qquad |z| > 1$$

Thus the right-hand side converges to a non-rational transfer function, implying that the impulse response cannot be realized by a state space model.

Chapter 4

4.1 Putting $i - j = k$, we change variables from (i, j) to (j, k). Then, k is bounded by $-N+1 \leq k \leq N-1$, and j is bounded by $1 \leq j \leq N-k$ if $k \geq 0$ and by $-k+1 \leq j \leq N$ if $k < 0$. Thus we get

$$\begin{aligned} \sum_{i=1}^{N} \sum_{j=1}^{N} \phi(i-j) &= \sum_{k=0}^{N-1} \sum_{j=1}^{N-k} \phi(k) + \sum_{k=-N+1}^{-1} \sum_{j=-k+1}^{N} \phi(k) \\ &= \sum_{k=0}^{N-1} (N-k)\phi(k) + \sum_{k=-N+1}^{-1} (N+k)\phi(k) \\ &= \sum_{k=-N+1}^{N-1} (N - |k|)\phi(k) \end{aligned}$$

4.2 Define $k = t - s$. Then, applying the formula in Problem 4.1, we have

$$\sum_{i=-N}^{N} \sum_{j=-N}^{N} \Lambda(t-s) = \sum_{k=-2N}^{2N} (2N + 1 - |k|)\Lambda(k)$$

Thus dividing the above equation by $2N+1$ gives (4.13).

4.3 (a) Let $\varepsilon > 0$ be a small number. From the assumption, there exists an integer $p > 0$ such that $|a_k| < \varepsilon$ for $k > p$. Let $M = \max\{|a_1|, \cdots, |a_p|\}$. Then,

$$\left|\frac{a_1 + \cdots + a_n}{n}\right| < \frac{pM + \varepsilon(n-p)}{n} < \frac{pM}{n} + \varepsilon$$

Taking the limit $n \to \infty$, we have $pM/n \to 0$. Since $\varepsilon > 0$ is arbitrary, the assertion is proved.

(b) Define $B_n = (1/n)\sum_{k=1}^{n} a_k$ with $B_0 = 0$. Then, $\lim_{n\to\infty} |B_n| = 0$ by hypothesis. Noting that

$$ka_k = k^2 B_k - (k-1)^2 B_{k-1} - (k-1)B_{k-1}$$

we have

$$I_n := \frac{1}{n}\sum_{k=1}^{n}\left(1 - \frac{k}{n}\right) a_k = B_n - \frac{1}{n^2}\sum_{k=1}^{n} ka_k = \frac{1}{n^2}\sum_{k=1}^{n}(k-1)B_{k-1}$$

Thus

$$|I_n| \le \frac{1}{n}\sum_{k=1}^{n}\left|\frac{k-1}{n}\right| |B_{k-1}| \le \frac{1}{n}\sum_{k=1}^{n}|B_{k-1}| \to 0$$

since $\lim_{k\to\infty} |B_k| = 0$.

(c) Define $C_n = \sum_{k=n}^{\infty} a_k$. By assumption, $\lim_{n\to\infty} C_n = 0$. It can be shown that

$$\left|\sum_{k=1}^{\infty} a_k - \sum_{k=1}^{n}\left(1 - \frac{k}{n}\right) a_k\right| \le \left|\sum_{k=n+1}^{\infty} a_k\right| + \frac{1}{n}\left|\sum_{k=1}^{n} ka_k\right|$$

$$= |C_{n+1}| + \frac{1}{n}\left|\sum_{k=1}^{n} ka_k\right|$$

Since the first term in the right-hand side of the above equation converges to zero, it remains to prove the convergence of the second term. By the definition of C_n,

$$\frac{1}{n}\sum_{k=1}^{n} ka_k = \frac{1}{n}\sum_{k=1}^{n} k(C_k - C_{k+1}) = \frac{1}{n}\sum_{k=1}^{n}\big([kC_k - (k+1)C_{k+1}] + C_{k+1}\big)$$

$$= \frac{C_1}{n} - \frac{n+1}{n}C_{n+1} + \frac{1}{n}\sum_{k=1}^{n} C_{k+1}$$

We see that the first and second terms of the right-hand side of the above equation converge to zero, and the third term also converges to zero by (a).

4.4 For zero mean Gaussian random variables $a,\ b,\ c,\ d$, we have (see *e.g.* [145])

$$E\{abcd\} = E\{ab\}E\{cd\} + E\{ac\}E\{bd\} + E\{ad\}E\{bc\} \tag{E.3}$$

By using (E.3), it follows from (4.17) that

$$\begin{aligned}
\Lambda_{\xi\xi}(k) &= E\{x(t+l+k)x(t+k)x(t+l)x(t)\} - \mu_\xi^2 \\
&= \Lambda_{xx}(l)\Lambda_{xx}(l) + \Lambda_{xx}(k)\Lambda_{xx}(k) + \Lambda_{xx}(l+k)\Lambda_{xx}(l-k) - \mu_\xi^2 \\
&= \Lambda_{xx}^2(k) + \Lambda_{xx}(l+k)\Lambda_{xx}(l-k)
\end{aligned}$$

By the Schwartz inequality,

$$\begin{aligned}
\left|\sum_{k=0}^{N} \Lambda_{\xi\xi}(k)\right| &\le \sum_{k=0}^{N} \Lambda_{xx}^2(k) + \left|\sum_{k=0}^{N} \Lambda_{xx}(k+l)\Lambda_{xx}(k-l)\right| \\
&\le \sum_{k=0}^{N} \Lambda_{xx}^2(k) + \left(\sum_{k=0}^{N} \Lambda_{xx}^2(k+l) \sum_{k=0}^{N} \Lambda_{xx}^2(k-l)\right)^{1/2}
\end{aligned} \tag{E.4}$$

Since (4.20) holds, it follows that

$$\lim_{N\to\infty} \frac{1}{N+1} \sum_{k=0}^{N} \Lambda_{xx}^2(k \pm l) = 0, \qquad l = 0,\ 1,\ \cdots$$

and hence from (E.4)

$$\lim_{N\to\infty} \frac{1}{N+1} \sum_{k=0}^{N} \Lambda_{\xi\xi}(k) = 0$$

This implies that (4.19) holds from Problem 4.3 (c).

4.5 Similarly to the calculation in Problem 4.2, we have

$$\begin{aligned}
I_N(\omega) &= \frac{1}{2N+1} \sum_{l=-N}^{N} \sum_{k=-N}^{N} E\{x(l)x(k)\}e^{-j\omega(l-k)} \\
&= \sum_{\tau=-2N}^{2N} \left(1 - \frac{|\tau|}{2N+1}\right) \Lambda(\tau)e^{-j\omega\tau}
\end{aligned} \tag{E.5}$$

Note that

$$\lim_{N\to\infty} \sum_{\tau=-2N}^{2N} \Lambda(\tau)e^{-j\omega\tau} = \Phi(\omega)$$

exists. It therefore from Problem 4.3 (c) that the limit of the right-hand side of (E.5) converges to $\Phi(\omega)$.

4.6 A proof is similar to that of Lemma 4.4. Post-multiplying

$$y(t) = \sum_{k=0}^{\infty} g_k u(t-k)$$

by $u(t-l)$ and taking the expectation yield

$$\Lambda_{yu}(l) = \sum_{k=0}^{\infty} g_k E\{u(t-k)u(t-l)\} = \sum_{k=0}^{\infty} g_k \Lambda_{uu}(l-k)$$

Post-multiplying the above equation by $e^{-j\omega l}$ and summing up with respect to l yield

$$\begin{aligned}\Phi_{yu}(\omega) &= \sum_{l=-\infty}^{\infty} \sum_{k=0}^{\infty} g_k e^{-j\omega k} \Lambda_{uu}(l-k) e^{-j\omega(l-k)} \\ &= \sum_{k=0}^{\infty} g_k e^{-j\omega k} \Phi_{uu}(\omega) = G(e^{j\omega})\Phi_{uu}(\omega)\end{aligned}$$

4.7 Since $\Phi_{yy}(\omega) = 2 - 2\cos\omega = 4\sin^2(\omega/2)$,

$$\begin{aligned}\int_{-\pi}^{\pi} \log \Phi_{yy}(\omega) d\omega &= 2\int_0^{\pi} \log\left[4\sin^2(\omega/2)\right] d\omega \\ &= 4\pi \log 2 + 4\int_0^{\pi} \log \sin(\omega/2) d\omega \\ &= 4\pi \log 2 + 8\int_0^{\pi/2} \log \sin\theta \, d\theta = 0 > -\infty\end{aligned}$$

where $\int_0^{\pi/2} \log\sin\theta d\theta = -(\pi/2)\log 2$ (Euler) is used.

4.8 The form of $\Phi(\omega)$ implies that y is a one-dimensional ARMA process, so that

$$y(t) + ay(t-1) = e(t) + ce(t-1)$$

Thus from (4.35), the spectral density function of y becomes

$$\Phi(\omega) = \sigma^2 \left| \frac{1 + ce^{j\omega}}{1 + ae^{j\omega}} \right|^2 = \sigma^2 \frac{1 + c^2 + 2c\cos\omega}{1 + a^2 + 2a\cos\omega}$$

Comparing the coefficients, we have $a = -0.9$, $c = 0.5$.

4.9 Since $H(z) = (z+c)/(z+a)$, we have

$$z^m H(z) = \frac{z^m(z+c)}{z+a} = (z^m + cz^{m-1})(1 + (-a)z^{-1} + (-a)^2 z^{-2} + \cdots)$$

Computing $[z^m H(z)]_+$ yields

$$\begin{aligned}[z^m H(z)]_+ &= (-a)^m + (-a)^{m+1} z^{-1} + \cdots + c[(-a)^{m-1} + (-a)^m z^{-1} + \cdots] \\ &= (-a)^{m-1}(c-a)(1 + (-a)z^{-1} + (-a)^2 z^{-1} + \cdots) \\ &= \frac{(-a)^{m-1}(c-a)z}{z+a}\end{aligned}$$

Thus from (4.57), the optimal predictor is given by

$$G(z) = \frac{(-a)^{m-1}(c-a)z}{z+a}\,\frac{z+a}{z+c} = \frac{(-a)^{m-1}(c-a)z}{z+c}$$

4.10 From (4.58),

$$\begin{bmatrix} x(t+1) \\ y(t+1) \end{bmatrix} = \begin{bmatrix} A(t) & 0 \\ C(t)A(t) & 0 \end{bmatrix} \begin{bmatrix} x(t) \\ y(t) \end{bmatrix} + \begin{bmatrix} B(t) & 0 \\ C(t)B(t) & I \end{bmatrix} \begin{bmatrix} w(t) \\ v(t+1) \end{bmatrix}$$

This is a state space equation, implying that the joint process (x, y) is Markov.

4.11 A proof is by direct substitution.

4.12 By definition,

$$\Phi_{yy}(z) = \sum_{l=-\infty}^{-1} \bar{C}(A^{\mathrm{T}})^{-l-1}C^{\mathrm{T}}z^{-l} + \Lambda_{yy}(0) + \sum_{l=1}^{\infty} CA^{l-1}\bar{C}^{\mathrm{T}}z^{-l}$$

Since $\bar{C}^{\mathrm{T}} = A\Pi C^{\mathrm{T}} + S$, we compute the terms that include S. Thus,

$$\begin{aligned} I_S := & \sum_{l=-\infty}^{-1} S^{\mathrm{T}}(A^{\mathrm{T}})^{-l-1}C^{\mathrm{T}}z^{-l} + \sum_{l=1}^{\infty} CA^{l-1}Sz^{-l} \\ &= S^{\mathrm{T}}\left(\sum_{l=1}^{\infty}(A^{\mathrm{T}})^{l-1}z^{l}\right)C^{\mathrm{T}} + C\left(\sum_{l=1}^{\infty}A^{l-1}z^{-l}\right)S \\ &= S^{\mathrm{T}}(z^{-1}I - A^{\mathrm{T}})^{-1}C^{\mathrm{T}} + C(zI - A)^{-1}S \\ &= S^{\mathrm{T}}W^{\mathrm{T}}(z^{-1}) + W(z)S \end{aligned}$$

Adding I_S to the right-hand side of (4.80) yields (4.81).

Chapter 5

5.1 This is a special case of Lemma 5.1.

5.2 Let $K_\alpha(t)$ and $P_\alpha(t)$ respectively be the Kalman gain and the error covariance matrices corresponding to $\alpha Q(t)$, $\alpha S(t)$, $\alpha R(t)$, $\alpha P(0)$. We use the algorithm of Theorem 5.1. For $t = 0$, it follows from (5.41a) that

$$\begin{aligned} K_\alpha(0) &= [A(0)\alpha P(0)C^{\mathrm{T}}(0) + B(0)\alpha S(0)][C(0)\alpha P(0)C^{\mathrm{T}}(0) + \alpha R(0)]^{-1} \\ &= K(0) \end{aligned}$$

Also, from (5.42a),

$$\begin{aligned} P_\alpha(1) = A(0)\alpha P(0)A^{\mathrm{T}}(0) - K_\alpha(0)[C(0)\alpha P(0)C^{\mathrm{T}}(0) + \alpha R(0)]K_\alpha^{\mathrm{T}}(0) \\ + B(0)\alpha Q(0)B^{\mathrm{T}}(0) = \alpha P(1) \end{aligned}$$

Similarly, for $t = 1$, we have $K_\alpha(1) = K(1)$, $P_\alpha(2) = \alpha P(2)$, and hence inductively $K_\alpha(t) = K(t)$, $t = 2, 3, \cdots$.

5.3 It follows that

$$x(t) - \mu_x(t) = \tilde{x}(t \mid t-1) + (\hat{x}(t \mid t-1) - \mu_x(t))$$

where $\tilde{x}(t \mid t-1) \perp \hat{x}(t \mid t-1) - \mu_x(t)$. Thus we have

$$\varPi(t) = P(t \mid t-1) + \varSigma(t)$$

Since $\varSigma(t) \geq 0$ and $P(t \mid t-1) \geq P(t \mid t) \geq 0$, we get

$$\varPi(t) \geq P(t \mid t-1) \geq P(t \mid t) \geq 0, \qquad \varPi(t) \geq \varSigma(t)$$

5.4 Follow the hint.

5.5 The derivation is straightforward.

5.6 Substituting $A = \varPhi + SR^{-1}C$ into (5.68), we have

$$\begin{aligned} K &= [(\varPhi + SR^{-1}C)PC^{\mathrm{T}} + S](CPC^{\mathrm{T}} + R)^{-1} \\ &= \varPhi PC^{\mathrm{T}}(CPC^{\mathrm{T}} + R)^{-1} + SR^{-1} = \varGamma + SR^{-1} \end{aligned}$$

Thus we get $A - KC = \varPhi - \varGamma C$.

It follows from (5.67) that

$$\begin{aligned} P &= APA^{\mathrm{T}} - K(CPC^{\mathrm{T}} + R)K^{\mathrm{T}} + Q \\ &= APA^{\mathrm{T}} - (\varGamma + SR^{-1})(CPC^{\mathrm{T}} + R)(\varGamma + SR^{-1})^{\mathrm{T}} + Q \\ &= (\varPhi + SR^{-1}C)P(\varPhi + SR^{-1}C)^{\mathrm{T}} \\ &\quad - (\varGamma + SR^{-1})(CPC^{\mathrm{T}} + R)(\varGamma + SR^{-1})^{\mathrm{T}} + Q \end{aligned}$$

From the definition of $\varGamma$,

$$\begin{aligned} P &= \varPhi P \varPhi^{\mathrm{T}} - \varGamma(CPC^{\mathrm{T}} + R)\varGamma^{\mathrm{T}} + Q + SR^{-1}CPF^{\mathrm{T}} \\ &\quad + \varPhi PC^{\mathrm{T}}R^{-1}S^{\mathrm{T}} + SR^{-1}CPC^{\mathrm{T}}R^{-1}S^{\mathrm{T}} \\ &\quad - \varGamma(CPC^{\mathrm{T}} + R)R^{-1}S^{\mathrm{T}} - SR^{-1}(CPC^{\mathrm{T}} + R)\varGamma^{\mathrm{T}} \\ &\quad - SR^{-1}(CPC^{\mathrm{T}} + R)R^{-1}S^{\mathrm{T}} \\ &= \varPhi P \varPhi^{\mathrm{T}} - \varGamma(CPC^{\mathrm{T}} + R)\varGamma^{\mathrm{T}} + (Q - SR^{-1}S^{\mathrm{T}}) \end{aligned}$$

This proves (5.70) since $M = Q - SR^{-1}S^{\mathrm{T}}$.

5.7 Equation (5.90) is given by

$$\varSigma = A\varSigma A^{\mathrm{T}} + (\bar{C}^{\mathrm{T}} - A\varSigma C^{\mathrm{T}})(\varLambda(0) - C\varSigma C^{\mathrm{T}})^{-1}(\bar{C} - C\varSigma A^{\mathrm{T}}) \tag{E.6}$$

Using $A = F + \bar{C}^{\mathrm{T}}\varLambda^{-1}(0)C$, the first term in the right-hand side of (E.6) is

$$\begin{aligned} I_1 :&= (F + \bar{C}^{\mathrm{T}}\varLambda^{-1}(0)C)\varSigma(F + \bar{C}^{\mathrm{T}}\varLambda^{-1}(0)C)^{\mathrm{T}} \\ &= F\varSigma F^{\mathrm{T}} + \bar{C}^{\mathrm{T}}\varLambda^{-1}(0)C\varSigma F^{\mathrm{T}} + F\varSigma C^{\mathrm{T}}\varLambda^{-1}(0)\bar{C} \\ &\quad + \bar{C}^{\mathrm{T}}\varLambda^{-1}(0)C\varSigma C^{\mathrm{T}}\varLambda^{-1}(0)\bar{C} \end{aligned}$$

Also, we have

$$\bar{C}^{\mathrm{T}} - A\Sigma C^{\mathrm{T}} = -F\Sigma C^{\mathrm{T}} + \bar{C}^{\mathrm{T}}\Lambda^{-1}(0)(\Lambda(0) - C\Sigma C^{\mathrm{T}})$$

so that the second term in the right-hand side of (E.6) becomes

$$\begin{aligned} I_2 := &(-F\Sigma C^{\mathrm{T}} + \bar{C}^{\mathrm{T}}\Lambda^{-1}(0)[\Lambda(0) - C\Sigma C^{\mathrm{T}}])(\Lambda(0) - C\Sigma C^{\mathrm{T}})^{-1} \\ &\times (-F\Sigma C^{\mathrm{T}} + \bar{C}^{\mathrm{T}}\Lambda^{-1}(0)[\Lambda(0) - C\Sigma C^{\mathrm{T}}])^{\mathrm{T}} \\ = &F\Sigma C^{\mathrm{T}}(\Lambda(0) - C\Sigma C^{\mathrm{T}})^{-1}C\Sigma F^{\mathrm{T}} - F\Sigma C^{\mathrm{T}}\Lambda^{-1}(0)\bar{C} \\ &- \bar{C}^{\mathrm{T}}\Lambda^{-1}(0)C\Sigma F^{\mathrm{T}} + \bar{C}^{\mathrm{T}}\Lambda^{-1}(0)(\Lambda(0) - C\Sigma C^{\mathrm{T}})\Lambda^{-1}(0)\bar{C} \end{aligned}$$

Computing $I_1 + I_2$, we get (5.91).

5.8 Since

$$N = \begin{bmatrix} F^{\mathrm{T}} & 0 \\ -\bar{C}^{\mathrm{T}}\Lambda^{-1}(0)\bar{C} & I_n \end{bmatrix}, \qquad L = \begin{bmatrix} I_n & -C^{\mathrm{T}}\Lambda^{-1}(0)C \\ 0 & F \end{bmatrix}$$

we have

$$\begin{aligned} L\hat{J}L^{\mathrm{T}} &= \begin{bmatrix} I_n & -C^{\mathrm{T}}\Lambda^{-1}(0)C \\ 0 & F \end{bmatrix} \begin{bmatrix} 0 & I_n \\ -I_n & 0 \end{bmatrix} \begin{bmatrix} I_n & -C^{\mathrm{T}}\Lambda^{-1}(0)C \\ 0 & F \end{bmatrix}^{\mathrm{T}} \\ &= \begin{bmatrix} 0 & F^{\mathrm{T}} \\ -F & 0 \end{bmatrix} = N\hat{J}N^{\mathrm{T}} \end{aligned}$$

Consider the following two eigenvalue problems:

$$\text{(A)} \qquad Nx = \lambda Lx \qquad\qquad \text{(B)} \qquad L^{\mathrm{T}}x = \mu N^{\mathrm{T}}x$$

Let $\lambda \neq 0$ be an eigenvalue of Problem (A). Since

$$\det(L^{\mathrm{T}} - \mu N^{\mathrm{T}}) = \det(L - \mu N) = 0$$

we see that $\mu = 1/\lambda$ is an eigenvalue of Problem (B). Also, pre-multiplying $L^{\mathrm{T}}x = \mu N^{\mathrm{T}}x$ by $N\hat{J}$ yields

$$N\hat{J}L^{\mathrm{T}}x = \mu N\hat{J}N^{\mathrm{T}}x = \mu L\hat{J}L^{\mathrm{T}}x \quad \Rightarrow \quad Nz = \mu Lz, \quad z = \hat{J}L^{\mathrm{T}}x$$

Thus $\mu = 1/\lambda$ is also an eigenvalue of Problem (A).

Chapter 6

6.1 (a) Since $g_k = k,\ k = 1, \cdots, g_0 = 0$, we have

$$H_{44} = \begin{bmatrix} 1 & 2 & 3 & 4 \\ 2 & 3 & 4 & 5 \\ 3 & 4 & 5 & 6 \\ 4 & 5 & 6 & 7 \end{bmatrix}, \qquad \text{rank } H_{44} = 2$$

By using the MATLAB® program in Table D.1, we get

$$A = \begin{bmatrix} 1.2182 & -0.2182 \\ 0.2182 & 0.7818 \end{bmatrix}, \quad B = \begin{bmatrix} -1.3039 \\ 0.8368 \end{bmatrix}, \quad C = [-1.3039 \ \ -0.8368]$$

Thus the transfer function is given by $G(z) = z/(z-1)^2$.

(b) In this case, the Hankel matrix becomes

$$H_{66} = \begin{bmatrix} 1 & 0 & -1 & 0 & 0 & 1 \\ 0 & -1 & 0 & 0 & 1 & 0 \\ -1 & 0 & 0 & 1 & 0 & -1 \\ 0 & 0 & 1 & 0 & -1 & 0 \\ 0 & 1 & 0 & -1 & 0 & 1 \\ 1 & 0 & -1 & 0 & 1 & 0 \end{bmatrix}, \qquad \text{rank}\, H_{66} = 4$$

so that we have

$$A = \begin{bmatrix} 0.1450 & 0.8808 & -0.3327 & -0.3239 \\ -0.8808 & 0.3551 & 0.3533 & -0.0115 \\ 0.3327 & -0.3533 & -0.6187 & -0.5087 \\ 0.3239 & -0.0115 & 0.5087 & -0.8814 \end{bmatrix}, \qquad B = \begin{bmatrix} -1.0016 \\ 0.1151 \\ -0.2418 \\ 0.2200 \end{bmatrix}$$

$$C = [-1.0016 \ \ -0.1151 \ \ -0.2418 \ \ -0.2200]$$

Thus the transfer function is given by $G(z) = (z^3 + z^2)/(z^4 + z^3 + z^2 + z + 1)$.

6.2 Let P be the reachability Gramian. Substituting $A = SA^{\mathrm{T}}S$, $B = SC^{\mathrm{T}}$ into (3.34) yields

$$P = APA^{\mathrm{T}} + BB^{\mathrm{T}} = SA^{\mathrm{T}}SPSAS + SC^{\mathrm{T}}CS$$

Since $SS = I$, we get $SPS = A^{\mathrm{T}}(SPS)A + C^{\mathrm{T}}C$. Thus the observability Gramian is expressed as $Q = SPS$. Though (A, B, C) are not balanced, both Gramians have the same eigenvalues. Note that Σ_s (with $T = I$) is diagonal, *i.e.*,

$$\Sigma_s = \mathcal{C}_k \mathcal{C}_k^{\mathrm{T}} = \sum_{i=0}^{k-1} A^i BB^{\mathrm{T}} (A^{\mathrm{T}})^i \ \left(\neq \sum_{i=0}^{\infty} A^i BB^{\mathrm{T}} (A^{\mathrm{T}})^i = P \right)$$

6.3 Since the orthogonal projection is expressed as $\hat{E}\{A \mid B\} = KB$, $K \in \mathbb{R}^{p \times q}$, the optimality condition is reduced to $A - KB \perp B$. Hence we have

$$(A - KB)B^{\mathrm{T}} = 0 \quad \Rightarrow \quad K = (AB^{\mathrm{T}})(BB^{\mathrm{T}})^{\dagger}$$

showing that $\hat{E}\{A \mid B\} = (AB^{\mathrm{T}})(BB^{\mathrm{T}})^{\dagger} B$.

6.4 Since $Q_1^{\mathrm{T}} Q_2 = 0$, two terms in the right-hand side of $A = L_{21} Q_1^{\mathrm{T}} + L_{22} Q_2^{\mathrm{T}}$ are orthogonal. From $B = L_{11} Q_1^{\mathrm{T}}$ with B full row rank, we see that L_{11} is nonsingular and Q_1^{T} forms a basis of the space spanned by the row vectors of B. It therefore follows that $\hat{E}\{A \mid B\} = L_{21} Q_1^{\mathrm{T}} = L_{21} L_{11}^{-1} B$. Also, from $AQ_1 = L_{21}$, we get $\hat{E}\{A \mid B\} = A(Q_1 Q_1^{\mathrm{T}})$. Since $L_{22}Q_2^{\mathrm{T}}$ is orthogonal to the row space of B, it follows that $\hat{E}\{A \mid B^{\perp}\} = L_{22}Q_2^{\mathrm{T}}$.

6.5 Let $D = \begin{bmatrix} B \\ C \end{bmatrix}$. Then, D has full row rank. Thus from Problem 6.3,

$$\hat{E}\{A \mid D\} = A[B^{\mathrm{T}}\ \ C^{\mathrm{T}}] \begin{bmatrix} BB^{\mathrm{T}} & BC^{\mathrm{T}} \\ CB^{\mathrm{T}} & CC^{\mathrm{T}} \end{bmatrix}^{-1} \begin{bmatrix} B \\ C \end{bmatrix}$$

We see that the above equation is expressed as

$$\begin{aligned} \hat{E}\{A \mid D\} &= A[B^{\mathrm{T}}\ \ C^{\mathrm{T}}] \begin{bmatrix} BB^{\mathrm{T}} & BC^{\mathrm{T}} \\ CB^{\mathrm{T}} & CC^{\mathrm{T}} \end{bmatrix}^{-1} \begin{bmatrix} B \\ 0 \end{bmatrix} \\ &\quad + A[B^{\mathrm{T}}\ \ C^{\mathrm{T}}] \begin{bmatrix} BB^{\mathrm{T}} & BC^{\mathrm{T}} \\ CB^{\mathrm{T}} & CC^{\mathrm{T}} \end{bmatrix}^{-1} \begin{bmatrix} 0 \\ C \end{bmatrix} \end{aligned}$$

Since $\mathrm{span}\{B\} \cap \mathrm{span}\{C\} = \{0\}$, the first term of the right-hand side of the above equation is the oblique projection of the row vectors of A onto the space spanned by the row vectors of B along the row vectors of C. Thus we have

$$\hat{E}_{\|C}\{A \mid B\} = A[B^{\mathrm{T}}\ \ C^{\mathrm{T}}] \begin{bmatrix} BB^{\mathrm{T}} & BC^{\mathrm{T}} \\ CB^{\mathrm{T}} & CC^{\mathrm{T}} \end{bmatrix}^{-1} \begin{bmatrix} B \\ 0 \end{bmatrix}$$

6.6 Note that $R_{22} = \begin{bmatrix} L_{22} & 0 \\ L_{32} & L_{33} \end{bmatrix}$ and $R_{32} = [L_{42}\ \ L_{43}]$. Let $\begin{bmatrix} \eta \\ \xi \end{bmatrix} \in \mathrm{Ker}(R_{22})$. Then, $L_{22}\eta = 0$ and $L_{32}\eta + L_{33}\xi = 0$ hold. However, since L_{22} is nonsingular, we have $\eta = 0$, so that $L_{33}\xi = 0$. Thus it suffices to show that $L_{33}\xi = 0$ implies $L_{43}\xi = 0$. Consider the following vectors

$$\begin{bmatrix} L_{13} \\ L_{23} \\ L_{33} \\ L_{43} \end{bmatrix} \xi = \begin{bmatrix} 0 \\ 0 \\ L_{33}\xi \\ L_{43}\xi \end{bmatrix} = \begin{bmatrix} 0 \\ 0 \\ 0 \\ L_{43}\xi \end{bmatrix}$$

Lemmas 6.4 and 6.5 show that the above vector is also an input-output pair. However, since the past input-output and future inputs are zero, the future outputs must be zero, implying that $L_{43}\xi = 0$. This completes the proof.

Chapter 7

7.1 Let $Z(z) = B(z)/A(z)$. Let $z = e^{j\omega}$. Then, we have

$$Z(e^{j\omega}) = \frac{B(e^{j\omega})}{A(e^{j\omega})} = \frac{c(\omega) + jd(\omega)}{a(\omega) + jb(\omega)}$$

It thus follows that

$$\mathfrak{Re}\, Z(e^{j\omega}) = \frac{a(\omega)c(\omega) + b(\omega)d(\omega)}{a(\omega)^2 + b(\omega)^2} \geq 0, \qquad -\pi < \omega \leq \pi$$

Hence $Z(z)$ is positive real, if $A(z)$ is stable and

$$a(\omega)c(\omega) + b(\omega)d(\omega) \geq 0, \qquad -\pi < \omega \leq \pi \tag{E.7}$$

From the given first-order transfer function, we have

$$Z(e^{j\omega}) = \frac{c + b\cos\omega + jb\sin\omega}{a + \cos\omega + j\sin\omega}$$

Thus from (E.7), the positivity is satisfied if $z + a$ is stable and if

$$ac + b + (ab + c)\cos\omega \geq 0, \qquad -\pi < \omega \leq \pi$$

It therefore follows that $|ab + c| \leq ac + b$ and $ac + b > 0$. Hence, we have

$$|a| < 1, \qquad |c| \leq b, \qquad b \geq 0$$

7.2 It can be shown that

$$\begin{aligned} \mathfrak{Re}[A(e^{j\omega})] &= 1 + a_1\cos\omega + a_2\cos 2\omega \\ &= 2a_2\cos^2\omega + a_1\cos\omega - a_2 + 1 \end{aligned} \tag{E.8}$$

For $a_2 = 0$, we see that the positive real condition is reduced to

$$a_1\cos\omega + 1 \geq 0, \qquad -\pi < \omega \leq \pi$$

This is satisfied if and only if $-1 \leq a_1 \leq 1$.

In the following, we assume that $a_2 \neq 0$, and define

$$f(x) := 2x^2 + (a_1/a_2)x + 1/a_2 - 1, \quad -1 \leq x \leq 1$$

1. Suppose that $a_2 < 0$. Then, from (E.8), the positive real condition becomes

$$f(x) \leq 0, \qquad -1 \leq x \leq 1 \tag{E.9}$$

Since $f(0) = 1/a_2 - 1 < 0$, (E.9) is satisfied if and only if $f(-1) \leq 0$ and $f(1) \leq 0$. Thus we have

$$a_2 + a_1 + 1 \geq 0, \qquad a_2 - a_1 + 1 \geq 0, \qquad a_2 \leq 0$$

2. Suppose that $a_2 > 0$. In this case, the positive real condition becomes

$$f(x) \geq 0, \qquad -1 \leq x \leq 1$$

Let $x_1 = -a_1/4a_2$. According to the location of x_1, we have three cases:

a) If $x_1 \leq -1$, then $f(-1) \geq 0$. This implies that

$$0 < a_2 \leq a_1/4, \qquad a_2 \geq a_1 - 1$$

b) If $-1 \leq x_1 \leq 1$, then $f(x_1) \geq 0$, so that

$$a_2 \geq \frac{a_1}{4}, \qquad a_2 \geq -\frac{a_1}{4}, \qquad \frac{a_1^2}{2} + 4\left(a_2 - \frac{1}{2}\right)^2 \leq 1 \tag{E.10}$$

c) If $x_1 \geq 1$, then $f(1) \geq 0$. Hence, we have

$$0 < a_2 \leq -a_1/4, \qquad a_2 \geq -a_1 - 1,$$

Thus, the region $D = \{(a_1, a_2) \mid \mathfrak{Re}[A(e^{j\omega})] \geq 0\}$ is a convex set enclosed by the two lines $a_2 = \pm a_1 - 1$ and a portion of the ellipsoid in (E.10) [see Figure E.1].

$$a_2 \geq a_1 - 1, \qquad a_2 \geq -a_1 - 1, \qquad \frac{a_1^2}{2} + 4\left(a_2 - \frac{1}{2}\right)^2 \leq 1 \tag{E.11}$$

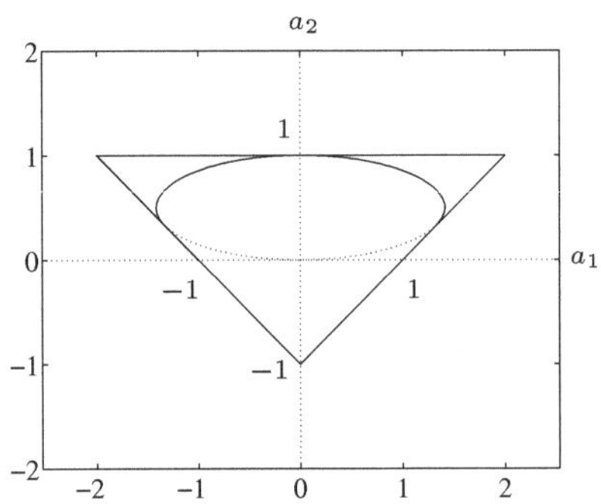

Figure E.1. Region of positive realness in (a_1, a_2)-plane

7.3 It is easy to see that $Z(z)$ is positive real if and only if

$$f(x) := 1 - a_1^2 - a_2^2 + 2a_1a_2x \geq 0, \qquad -1 \leq x \leq 1$$

Thus the condition is given by

$$|a_1 - a_2| \leq 1, \qquad |a_1 + a_2| \leq 1 \tag{E.12}$$

Remark E.1. It will be instructive to compare the positive real conditions (E.11) and (E.12) above and the stability condition (E.1). □

7.4 Using the Frobenius norm, we have

$$\begin{aligned}
\left\| \begin{bmatrix} A & B \\ C & D \end{bmatrix} \right\|_F^2 &= \text{trace}\left(\begin{bmatrix} A & B \\ C & D \end{bmatrix} \begin{bmatrix} A & B \\ C & D \end{bmatrix}^{\mathrm{T}} \right) \\
&= \text{trace}(AA^{\mathrm{T}} + BB^{\mathrm{T}} + CC^{\mathrm{T}} + DD^{\mathrm{T}}) \\
&= \|A\|_F^2 + \|B\|_F^2 + \|C\|_F^2 + \|D\|_F^2 \\
&\leq (\|A\|_F + \|B\|_F + \|C\|_F + \|D\|_F)^2
\end{aligned}$$

Taking the square root of the above relation, we get the desired result.

For the 2-norm, we define $X = \begin{bmatrix} A \\ C \end{bmatrix}$, $Y = \begin{bmatrix} B \\ D \end{bmatrix}$. By the definition of 2-norm,

$$\begin{aligned} \|X\|_2^2 &= \overline{\sigma}(X)^2 = \max \lambda \left(A^{\mathrm{T}} A + C^{\mathrm{T}} C\right) \\ &\le \max \lambda(A^{\mathrm{T}} A) + \max \lambda(C^{\mathrm{T}} C) = \|A\|_2^2 + \|C\|_2^2 \le (\|A\|_2 + \|C\|_2)^2 \end{aligned}$$

Thus we get $\|X\|_2 \le \|A\|_2 + \|C\|_2$. Similarly, we get $\|Y\|_2 \le \|B\|_2 + \|D\|_2$. Thus combining these results,

$$\begin{aligned} \|[X \;\; Y]\|_2^2 &= \max \lambda(XX^{\mathrm{T}} + YY^{\mathrm{T}}) \\ &\le \max \lambda(XX^{\mathrm{T}}) + \max \lambda(YY^{\mathrm{T}}) \\ &= \|X\|_2^2 + \|Y\|_2^2 \le (\|X\|_2 + \|Y\|_2)^2 \end{aligned}$$

Hence we have

$$\|[X \;\; Y]\|_2 \le \|X\|_2 + \|Y\|_2 \le \|A\|_2 + \|B\|_2 + \|C\|_2 + \|D\|_2$$

7.5 $M(\Pi)$ is easily derived. Let $\Pi = 3$. Since $M(\Pi) = \begin{bmatrix} 8/3 & 0 \\ 0 & 0 \end{bmatrix} \ge 0$, we see that $\Pi = 3$ satisfies the LMI. Now suppose that $\Pi < 3$. It then follows from (7.35) that

$$\frac{8}{9}\Pi - \frac{1}{3}\left(1 - \frac{1}{3}\Pi\right) \ge 0 \quad \Rightarrow \quad \Pi \ge 1/3$$

Hence we have $\Pi_* = 1/3$ and $\Pi^* = 3$, implying that the solutions of LMI satisfy $1/3 \le \Pi \le 3$. Note that in this case $F := A - \bar{C}^{\mathrm{T}} \Lambda^{-1}(0) C = 0$; see (5.91).

7.6 By the definition of $\mathcal{C}_k$ and $T_-(k)$,

$$\begin{aligned} \Omega_{k+1} &= \mathcal{C}_{k+1} T_-^{-1}(k+1) \mathcal{C}_{k+1}^{\mathrm{T}} \\ &= [\bar{C}^{\mathrm{T}} \;\; A\mathcal{C}_k] \begin{bmatrix} \Lambda(0) & C\mathcal{C}_k \\ \mathcal{C}_k^{\mathrm{T}} C^{\mathrm{T}} & T_-(k) \end{bmatrix}^{-1} \begin{bmatrix} \bar{C} \\ \mathcal{C}_k^{\mathrm{T}} A^{\mathrm{T}} \end{bmatrix} \end{aligned}$$

Note that this equation has the same form as (7.59). It is easy to see that Ω_k satisfies (7.62) by the following correspondence in (7.60).

$$\bar{\Omega}_k \leftrightarrow \Omega_k, \qquad A \leftrightarrow A^{\mathrm{T}}, \qquad \bar{C} \leftrightarrow C$$

7.7 First from (7.64), we note that

$$K(\Lambda(0) - C\Pi C^{\mathrm{T}}) + A\Pi C^{\mathrm{T}} = \bar{C}^{\mathrm{T}} \tag{E.13}$$

Substituting $A = A_K + KC$ into (7.63) yields

$$\begin{aligned} \Pi &= (A_K + KC)\Pi(A_K + KC)^{\mathrm{T}} + K(\Lambda(0) - C\Pi C^{\mathrm{T}})K^{\mathrm{T}} \\ &= A_K \Pi A_K^{\mathrm{T}} + KC\Pi C^{\mathrm{T}} K^{\mathrm{T}} + KC\Pi A_K^{\mathrm{T}} + A_K \Pi C^{\mathrm{T}} K^{\mathrm{T}} \\ &\quad + K(\Lambda(0) - C\Pi C^{\mathrm{T}})K^{\mathrm{T}} \end{aligned}$$

Again by using $A_K = A - KC$, it follows that

$$\begin{aligned}
\Pi &= A_K \Pi A_K^{\mathrm{T}} + KC\Pi C^{\mathrm{T}} K^{\mathrm{T}} + KC\Pi (A - KC)^{\mathrm{T}} + (A - KC)\Pi C^{\mathrm{T}} K^{\mathrm{T}} \\
&\quad + K(\Lambda(0) - C\Pi C^{\mathrm{T}})K^{\mathrm{T}} \\
&= A_K \Pi A_K^{\mathrm{T}} + KC\Pi A^{\mathrm{T}} + A\Pi C^{\mathrm{T}} K^{\mathrm{T}} - KC\Pi C^{\mathrm{T}} K^{\mathrm{T}} \\
&\quad + K(\Lambda(0) - C\Pi C^{\mathrm{T}})K^{\mathrm{T}} + (K\Lambda(0)K^{\mathrm{T}} - K\Lambda(0)K^{\mathrm{T}})
\end{aligned}$$

Using (E.13) in the right-hand side of the above equation, we readily obtain the desired result, *i.e.*, (7.65).

7.8 In view of Subsection 7.4.2, the constraint is given by $\xi = \mathcal{O}^{\mathrm{T}} \boldsymbol{u}$, so that the Lagrangian becomes

$$\mathcal{L} = \boldsymbol{u}^{\mathrm{T}} \boldsymbol{u} + \lambda^{\mathrm{T}} (\xi - \mathcal{O}^{\mathrm{T}} \boldsymbol{u})$$

Differentiating $\mathcal{L}$ with respect to $\boldsymbol{u}$ yields $2\boldsymbol{u} - \mathcal{O}\lambda = 0$. Thus, from the constraint, we have

$$\xi - \mathcal{O}^{\mathrm{T}} \mathcal{O} \lambda / 2 = 0 \quad \Rightarrow \quad \lambda = 2(\mathcal{O}^{\mathrm{T}} \mathcal{O})^{-1} \xi$$

so that $\boldsymbol{u} = \mathcal{O}(\mathcal{O}^{\mathrm{T}} \mathcal{O})^{-1} \xi$ holds. Hence we have $\min_{\boldsymbol{u}} (\boldsymbol{u}^{\mathrm{T}} \boldsymbol{u}) = \xi^{\mathrm{T}} (\mathcal{O}^{\mathrm{T}} \mathcal{O})^{-1} \xi$.

Chapter 8

8.1 It is easy to show that

$$\begin{bmatrix} L^{\mathrm{T}} & 0 \\ 0 & M^{\mathrm{T}} \end{bmatrix} \begin{bmatrix} \Sigma_{xx} & \Sigma_{xy} \\ \Sigma_{yx} & \Sigma_{yy} \end{bmatrix} \begin{bmatrix} L & 0 \\ 0 & M \end{bmatrix} = \begin{bmatrix} L^{\mathrm{T}} \Sigma_{xx} L & L^{\mathrm{T}} \Sigma_{xy} M \\ M^{\mathrm{T}} \Sigma_{yx} L & M^{\mathrm{T}} \Sigma_{yy} M \end{bmatrix}$$

Thus from (8.9) and (8.10), the result follows. Also, the computation of the determinant is immediate.

8.2 Though this can be proved by using the orthogonality condition $a - Kb \perp b$, we give a different proof. See also Problem 6.3.

Since $I := \|a - Kb\|_{\mathcal{H}}^2 = \mathrm{trace} E\{(a - Kb)(a - Kb)^{\mathrm{T}}\}$,

$$I = \mathrm{trace}\Big(E\{aa^{\mathrm{T}}\} - E\{ab^{\mathrm{T}}\}K^{\mathrm{T}} - KE\{ba^{\mathrm{T}}\} + KE\{bb^{\mathrm{T}}\}K^{\mathrm{T}} \Big)$$

We see that the right-hand side is a quadratic form in $K = (k_{ij})$.

Recall the formulas for the differentiation of trace (*e.g.* see [185]):

$$\frac{\partial}{\partial X} \mathrm{trace}(AX) = A^{\mathrm{T}}, \qquad \frac{\partial}{\partial X} \mathrm{trace}(AX^{\mathrm{T}}) = A$$

$$\frac{\partial}{\partial X} \mathrm{trace}(AXBX^{\mathrm{T}}) = A^{\mathrm{T}} X B^{\mathrm{T}} + AXB$$

Thus it follows that

$$\frac{\partial I}{\partial K} = -2E\{ab^{\mathrm{T}}\} + 2KE\{bb^{\mathrm{T}}\} = 0 \quad \Rightarrow \quad K = E\{ab^{\mathrm{T}}\}(E\{bb^{\mathrm{T}}\})^{-1}$$

8.3 Applying Lemma 4.11 to (8.48), we get

$$\begin{aligned}\Lambda_{yy}(l) &= CA^l \Sigma C^{\mathrm{T}} + CA^{l-1}KR \\ &= CA^{l-1}(A\Sigma C^{\mathrm{T}} + KR) = CA^{l-1}\bar{C}^{\mathrm{T}}\end{aligned}$$

for $l = 1, 2, \cdots$. For $l = 0$, we have $\Lambda_{yy}(0) = C\Sigma C^{\mathrm{T}} + R$, so that we have the desired result. We can show that Theorem 8.5 and Lemma 8.5 give the same result.

8.4 Since y is scalar, we note that $T_+ = T_-$, $L = M$, $H^T = H$ hold. Thus $\bar{H} := L^{-1}HL^{-\mathrm{T}} = U\Sigma V^{\mathrm{T}}$ is symmetric, so that $\bar{H} = U\Sigma V^{\mathrm{T}} = V\Sigma U^{\mathrm{T}}$. Since $\mathrm{Im}(\bar{H}) = \mathrm{Im}(U) = \mathrm{Im}(V)$ holds, there exists a nonsingular matrix $S \in \mathbb{R}^{n\times n}$ such that $U = VS$. Since $I_n = U^{\mathrm{T}}U = S^{\mathrm{T}}V^{\mathrm{T}}VS = S^{\mathrm{T}}S$, we see that $S = V^{\mathrm{T}}U$ is orthogonal. From $U\Sigma V^{\mathrm{T}} = V\Sigma U^{\mathrm{T}}$, we have $\Sigma S^{\mathrm{T}} = S\Sigma$, so that similarly to the proof of Lemma 5.2, we can show that $S = \mathrm{diag}(\pm 1, \cdots, \pm)$ holds. By using $\mathcal{O} = LU\Sigma^{1/2}$, $U = VS$, $\Sigma = S\Sigma S$, we see that

$$\mathcal{C} = \Sigma^{1/2}V^{\mathrm{T}}M^{\mathrm{T}} = \Sigma^{1/2}SU^{\mathrm{T}}L^{\mathrm{T}} = S\Sigma^{1/2}U^{\mathrm{T}}L^{\mathrm{T}} = S\mathcal{O}^{\mathrm{T}}$$

holds, where we used the fact that S and $\Sigma^{1/2}$ are diagonal. Thus,

$$\mathcal{C}^{\leftarrow} = S(\mathcal{O}^{\uparrow})^{\mathrm{T}}, \quad \mathcal{C}^{\dagger} = (\mathcal{O}^{\dagger})^{\mathrm{T}}S$$

Hence, from (8.50), we get

$$A = \mathcal{C}^{\leftarrow}\mathcal{C}^{\dagger} = S(\mathcal{O}^{\uparrow})^{\mathrm{T}}(\mathcal{O}^{\dagger})^{\mathrm{T}}S = S(\mathcal{O}^{\dagger}\mathcal{O}^{\uparrow})^{\mathrm{T}}S = SA^{\mathrm{T}}S$$

Also, from (8.52),

$$\bar{C}^{\mathrm{T}} = \mathcal{C}(1:n, 1:p) = SC^{\mathrm{T}}$$

implying that $\bar{C} = CS$ holds.

8.5 Since $H = L_1U_1\Sigma V_1^{\mathrm{T}}M_1^{\mathrm{T}} = L_2U_2\Sigma V_2^{\mathrm{T}}M_2^{\mathrm{T}}$, and since $\mathrm{Im}(L_1U_1) = \mathrm{Im}(L_2U_2)$, there exists a nonsingular $S \in \mathbb{R}^{n\times n}$ such that $L_2U_2 = L_1U_1S$. Note that $L_1L_1^{\mathrm{T}} = L_2L_2^{\mathrm{T}}$ holds. Thus $Z = L_1^{-1}L_2$ becomes an orthogonal matrix with $ZZ^{\mathrm{T}} = Z^{\mathrm{T}}Z = I$. This implies that ZU_2 becomes an orthogonal matrix, and hence $S = U_1^{\mathrm{T}}(ZU_2)$ becomes an orthogonal matrix. Again, using $L_1U_1\Sigma V_1^{\mathrm{T}}M_1^{\mathrm{T}} = L_2U_2\Sigma V_2^{\mathrm{T}}M_2^{\mathrm{T}}$, and noting that $M_1M_1^{\mathrm{T}} = M_2M_2^{\mathrm{T}}$, it follows that

$$U_1\Sigma V_1^{\mathrm{T}} = L_1^{-1}L_2U_2\Sigma V_2^{\mathrm{T}}M_2^{\mathrm{T}}M_1^{-\mathrm{T}}$$

so that

$$U_1\Sigma^2U_1^{\mathrm{T}} = L_1^{-1}L_2U_2\Sigma^2U_2^{\mathrm{T}}L_2^{\mathrm{T}}L_1^{-\mathrm{T}} = U_1S\Sigma^2S^{\mathrm{T}}U_1^{\mathrm{T}}$$

and hence

$$\Sigma^2 = S\Sigma^2S^{\mathrm{T}} \tag{E.14}$$

It should be noted that Σ^2 is a diagonal matrix with different elements and that S is orthogonal. Thus similarly to Lemma 5.2, we have $S = (\pm 1, \cdots, \pm 1)$. In fact, suppose that

$$S = \begin{bmatrix} S_{n-1} & a \\ b^{\mathrm{T}} & c \end{bmatrix}, \quad S_{n-1} \in \mathbb{R}^{(n-1)\times(n-1)}, \quad a,\, b \in \mathbb{R}^{n-1}, \quad c \in \mathbb{R}$$

Then, from $S^{\mathrm{T}} S = I$, we have $\|a\|^2 + c^2 = 1$ and from (E.14)

$$\Sigma_{n-1}^2 S_{n-1}^{\mathrm{T}} = S_{n-1} \Sigma_{n-1}^2, \qquad \Sigma_{n-1}^2 a = \sigma_n^2 a, \qquad b^{\mathrm{T}} \Sigma_{n-1} = \sigma_n^2 b^{\mathrm{T}}$$

Since σ_n^2 is not an eigenvalue of Σ_{n-1}^2, we see that $a = 0$, $b = 0$, so that $c^2 = 1$.

By using $L_2 U_2 = L_1 U_1 S$ and $U_1 \Sigma V_1^{\mathrm{T}} = L_1^{-1} L_2 U_2 \Sigma V_2^{\mathrm{T}} M_2^{\mathrm{T}} M_1^{-\mathrm{T}}$,

$$\Sigma = S \Sigma V_2^{\mathrm{T}} M_2^{\mathrm{T}} M_1^{-\mathrm{T}} V_1 = \Sigma S V_2^{\mathrm{T}} M_2^{\mathrm{T}} M_1^{-\mathrm{T}} V_1 \quad \Rightarrow \quad V_1^{\mathrm{T}} M_1^{-1} M_2 V_2 S = I_n$$

where we used the fact that $\det \Sigma \neq 0$. Since the right-inverse of $V_1^{\mathrm{T}} M_1^{-1}$ is $M_1 V_1$, we get $M_2 V_2 S = M_1 V_1$, so that $M_2 V_2 = M_1 V_1 S$. Also, from (8.41),

$$\begin{aligned} \mathcal{O}_2 &= L_2 U_2 \Sigma^{1/2} = L_1 U_1 S \Sigma^{1/2} = \mathcal{O}_1 S \\ \mathcal{C}_2 &= \Sigma^{1/2} V_2^{\mathrm{T}} M_2^{\mathrm{T}} = \Sigma^{1/2} S V_1^{\mathrm{T}} M_1^{\mathrm{T}} = S \mathcal{C}_1 \end{aligned}$$

It thus follows from (8.50) that

$$A_2 = \mathcal{C}_2^{\leftarrow} \mathcal{C}_2^{\dagger} = S \mathcal{C}_1^{\leftarrow} \mathcal{C}_1^{\dagger} S = S A_1 S$$

Moreover, from (8.51) and (8.52),

$$\begin{aligned} C_2 &= \mathcal{O}_2(1:p, 1:n) = \mathcal{O}_1(1:p, 1:n) S = C_1 S \\ \bar{C}_2 &= \mathcal{C}_2(1:p, 1:n)^{\mathrm{T}} = \mathcal{C}_1(1:p, 1:n)^{\mathrm{T}} S = \bar{C}_1 S \end{aligned}$$

From (8.53) and (8.54), we have

$$\begin{aligned} R_2 &= \Lambda(0) - C_2 \Sigma C_2^{\mathrm{T}} = \Lambda(0) - C_1 S \Sigma S C_1^{\mathrm{T}} = \Lambda(0) - C_1 \Sigma C_1^{\mathrm{T}} = R_1 \\ K_2 &= (\bar{C}_2^{\mathrm{T}} - A_2 \Sigma C_2^{\mathrm{T}}) R_2^{-1} = (S \bar{C}_1^{\mathrm{T}} - S A_1 S \Sigma S C_1^{\mathrm{T}}) R_1^{-1} \\ &= S(\bar{C}_1^{\mathrm{T}} - A_1 \Sigma C_1^{\mathrm{T}}) R_1^{-1} = S K_1 \end{aligned}$$

This completes the proof.

Glossary

Notation

$\mathbb{R}$, $\mathbb{C}$, $\mathbb{Z}$	real numbers, complex numbers, integers
$\mathbb{R}^n$, $\mathbb{C}^n$	n-dimensional real vectors, complex vectors
$\mathbb{R}^{m\times n}$	$(m \times n)$-dimensional real matrices
$\mathbb{C}^{m\times n}$	$(m \times n)$-dimensional complex matrices
$\dim(x)$	dimension of vector x
$\dim(\mathcal{V})$	dimension of subspace $\mathcal{V}$
$\mathcal{V} \vee \mathcal{W}$	vector sum of subspaces $\mathcal{V}$ and $\mathcal{W}$
$\mathcal{V} + \mathcal{W}$	direct sum of subspaces $\mathcal{V}$ and $\mathcal{W}$
$\text{span}\{v, w, x\}$	subspace generated by vectors v, w, x
A^{T}, A^{H}	transpose of $A \in \mathbb{R}^{m\times n}$, conjugate transpose of $A \in \mathbb{C}^{m\times n}$
A^{-1}, $A^{-\mathrm{T}}$	inverse and transpose of the inverse of A
$A^{\dagger}$	pseudo-inverse of A
$A \geq 0$	symmetric, nonnegative definite
$A > 0$	symmetric, positive definite
$A^{1/2}$	square root of A
$\det(A)$	determinant of A
$\text{trace}(A)$	trace of A
$\text{rank}(A)$	rank of A
$\lambda(A)$, $\lambda_i(A)$	eigenvalue, ith eigenvalue of A
$\rho(A)$	spectral radius, *i.e.*, $\max_i \lvert\lambda_i(A)\rvert$
$\sigma(A)$, $\sigma_i(A)$	singular value, ith singular value of A
$\underline{\sigma}(A)$, $\overline{\sigma}(A)$	minimum singular value, maximum singular value of A
$\text{Im}(A)$	image (or range) of A
$\text{Ker}(A)$	kernel (or null space) of A
$\lVert x\rVert_2$, $\lVert x\rVert_\infty$	2-norm, ∞-norm of x
$\lVert A\rVert_2$, $\lVert A\rVert_F$	2-norm, Frobenius norm of A
$\left[\begin{array}{c\|c} A & B \\ \hline C & D \end{array}\right]$	transfer matrix $G(z) = D + C(zI - A)^{-1}B$

$E\{x\}$	mathematical expectation of random vector x
$\mathrm{cov}\{x, y\}$	(cross-) covariance matrix of random vectors x and y
$\mathcal{N}(\mu, \Sigma)$	Gaussian (normal) distribution with mean μ and covariance matrix Σ
$E\{x \mid y\}$	conditional expectation of x given y
$(x, y)_{\mathcal{H}}$	inner product of x and y in Hilbert space $\mathcal{H}$
$\|x\|_{\mathcal{H}}$	norm of x in Hilbert space $\mathcal{H}$
$\overline{\mathrm{span}}\{\cdots\}$	closed Hilbert subspace generated by infinite elements $\{\cdots\}$
$\hat{E}\{x \mid \mathcal{Y}\}$	orthogonal projection of x onto subspace $\mathcal{Y}$
$\hat{E}_{\parallel \mathcal{Z}}\{x \mid \mathcal{Y}\}$	oblique projection of x onto $\mathcal{Y}$ along $\mathcal{Z}$
$a := b$	a is defined by b
$a =: b$	b is defined by a
$\mathfrak{Z}$	z-transform operator
z	complex variable, shift operator $zf(t) := f(t+1)$
$\mathfrak{Re}$	real part
$\mathrm{Ric}(\cdot)$	Riccati operator; (7.34)

Abbreviations

AIC	Akaike Information Criterion; see Section 1.1
AR	AutoRegressive; (4.33)
ARMA	AutoRegressive Moving Average; (4.34)
ARMAX	AutoRegressive Moving Average with eXogenous input; (1.4)
ARX	AutoRegressive with eXogenous input; (A.7)
ARE	Algebraic Riccati Equation; (5.67)
ARI	Algebraic Riccati Inequality; (7.35)
BIBO	Bounded-Input, Bounded-Output; see Section 3.2
CCA	Canonical Correlation Analysis; see Section 8.1
CVA	Canonical Variate Analysis; see Section 10.8
FIR	Finite Impulse Response; (A.12)
IV	Instrumental Variable; see Section A.1
LMI	Linear Matrix Inequality; see (7.26)
LTI	Linear Time-Invariant; see Section 3.2
MA	Moving Average; (4.44)
MIMO	Multi-Input, Multi-Output; see Section 1.3
ML	Maximum Likelihood; see Section 1.1
MOESP	Multivariable Output Error State sPace; see Section 6.5
N4SID	Numerical algorithms for Subspace State Space System IDentification; see Section 6.6
ORT	ORThogonal decomposition based; see Section 9.7
PE	Persistently Exciting; see Sections 6.3 and Appendix B
PEM	Prediction Error Method; see Sections 1.2 and 1.3
PO-MOESP	Past Output MOESP; see Section 6.6
SISO	Single-Input, Single-Output; see Section 3.2
SVD	Singular Value Decomposition; see (2.26)

References

1. H. Akaike, "Stochastic theory of minimal realization," *IEEE Trans. Automatic Control*, vol. AC-19, no. 6, pp. 667–674, 1974.
2. H. Akaike, "Markovian representation of stochastic processes by canonical variables," *SIAM J. Control*, vol. 13, no. 1, pp. 162–173, 1975.
3. H. Akaike, "Canonical correlation analysis of time series and the use of an information criterion," In *System Identification: Advances and Case Studies* (R. Mehra and D. Lainiotis, eds.), Academic, pp. 27–96, 1976.
4. H. Akaike, "Comments on 'On model structure testing in system identification'," *Int. J. Control*, vol. 27, no. 2, pp. 323–324, 1978.
5. U. M. Al-Saggaf and G. F. Franklin, "An error bound for a discrete reduced order model of a linear multivariable system," *IEEE Trans. Automatic Control*, vol. 32, no. 9, pp. 815–819, 1987.
6. U. M. Al-Saggaf and G. F. Franklin, "Model reduction via balanced realizations: An extension and frequency weighting techniques," *IEEE Trans. Automatic Control*, vol. 33, no. 7, pp. 687–692, 1988.
7. B. D. O. Anderson, "A system theory criterion for positive real matrices," *SIAM J. Control*, vol. 5, no. 2, pp. 171–182, 1967.
8. B. D. O. Anderson, "An algebraic solution to the spectral factorization problem," *IEEE Trans. Automatic Control*, vol. AC-12, no. 4, pp. 410–414, 1967.
9. B. D. O. Anderson, "The inverse problem of stationary covariance generation," *J. Statistical Physics*, vol. 1, no. 1, pp. 133–147, 1969.
10. B. D. O. Anderson, K. L. Hitz and N. D. Diem, "Recursive algorithm for spectral factorization," *IEEE Trans. Circuits Systems*, vol. CAS-21, no. 6, pp. 742–750, 1974.
11. B. D. O. Anderson and J. B. Moore, *Optimal Filtering*, Prentice-Hall, 1979.
12. B. D. O. Anderson and M. R. Gevers, "Identifiability of linear stochastic systems operating under linear feedback," *Automatica*, vol. 18, no. 2, pp. 195–213, 1982.
13. T. W. Anderson, *The Statistical Analysis of Time Series*, Wiley, 1971.
14. T. W. Anderson, *An Introduction to Multivariable Statistical Analysis* (2nd ed.), Wiley, 1984.
15. M. Aoki, *State Space Modeling of Time Series* (2nd ed.), Springer, 1990.
16. M. Aoki and A. M. Havenner (eds.), *Applications of Computer Aided Time Series Modeling* (Lecture Notes in Statistics 119), Springer, 1997.
17. K. J. Åström and T. Bohlin, "Numerical identification of linear dynamic systems for normal operating records," *Proc. 2nd IFAC Symp. Theory of Self-Adaptive Systems*, Teddington, pp. 96–111, 1965.

18. K. S. Arun and S. Y. Kung, "Balanced approximation of stochastic systems," *SIAM J. Matrix Analysis and Applications*, vol. 11, no. 1, pp. 42–68, 1990.
19. D. Bauer and M. Jansson, "Analysis of the asymptotic properties of the MOESP type of subspace algorithms," *Automatica*, vol. 36, no. 4, pp. 497–509, 2000.
20. S. Bittanti, A. J. Laub and J. C. Willems (eds.), *The Riccati Equation*, Springer, 1991.
21. A. Björck and G. H. Golub, "Numerical methods for computing angles between linear subspaces," *Mathematics of Computation*, vol. 27, pp. 579–594, July 1973.
22. G. E. P. Box and G. M. Jenkins, *Time Series Analysis - Forecasting and Control*, Holden-Day, 1970.
23. R. S. Bucy, *Lectures on Discrete Time Filtering*, Springer, 1994.
24. P. E. Caines and C. W. Chan, "Feedback between stationary stochastic processes," *IEEE Trans. Automatic Control*, vol. AC-20, no. 4, pp. 498–508, 1975.
25. P. E. Caines and C. W. Chan, "Estimation, identification and feedback," In *System Identification: Advances and Case Studies* (R. Mehra and D. Lainiotis, eds.), Academic, pp. 349–405, 1976.
26. P. E. Caines, "Weak and strong feedback free processes," *IEEE Trans. Automatic Control*, vol. AC-21, no. 5, pp. 737–739, 1976.
27. C. T. Chen, *Linear System Theory and Design*, Holt-Saunders, 1984.
28. H. -F. Chen, P. K. Kumar and J. H. van Schuppen, "On Kalman filtering for conditionally Gaussian systems with random matrices," *Systems and Control Letters*, vol. 13, pp. 397–404, 1989.
29. A. Chiuso and G. Picci, "Subspace identification by orthogonal decomposition," *Proc. 14th IFAC World Congress*, Beijing, vol. H, pp. 241–246, 1999.
30. A. Chiuso and G. Picci, "Some algorithmic aspects of subspace identification with inputs," *Int. J. Applied Math. and Computer Science*, vol. 11, no. 1, pp. 55–75, 2001.
31. A. Chiuso and G. Picci, "On the ill-conditioning of subspace identification with inputs," *Automatica*, vol. 40, no. 4, pp. 575–589, 2004.
32. A. Chiuso and G. Picci, "Numerical conditioning and asymptotic variance of subspace estimates," *Automatica*, vol. 40, no. 4, pp. 677–683, 2004.
33. T. Chonavel, *Statistical Signal Processing*, Springer, 2002.
34. C. T. Chou and M. Verhaegen, "Closed-loop identification using canonical correlation analysis," *Proc. 5th European Control Conference*, Karlsruhe, F-162, 1999.
35. N. L. C. Chui and J. M. Maciejowski, "Realization of stable models with subspace methods," *Automatica*, vol. 32, no. 11, pp. 1587–1595, 1996.
36. R. V. Churchill and J. W. Brown, *Complex Variables and Applications* (4th ed.), McGraw-Hill, 1984.
37. B. Codrons, B. D. O. Anderson and M. Gevers, "Closed-loop identification with an unstable or nonminimum phase controller," *Automatica*, vol. 38, no. 12, pp. 2127–2137, 2002.
38. A. Dahlén, A. Lindquist and J. Mari, "Experimental evidence showing that stochastic subspace identification methods may fail," *Systems and Control Letters*, vol. 34, no. 2, pp. 302–312, 1998.
39. K. De Cock, *Principal Angles in System Theory, Information Theory and Signal Processing*, Ph.D. Thesis, Katholieke Universiteit Lueven, Belgium, 2002.
40. B. De Moor, "The singular value decomposition and long and short spaces of noisy matrices," *IEEE Trans. Signal Processing*, vol. 41, no. 9, pp. 2826–2838, 1993.
41. B. De Moor, M. Moonen, L. Vandenberghe and J. Vandewalle, "Identification of linear state space models with SVD using canonical correlation analysis," In *Singular Value Decomposition and Signal Processing* (E. Deprettere, ed.), North-Holland, pp. 161–169, 1988.

42. U. B. Desai and D. Pal, "A transformation approach to stochastic model reduction," *IEEE Trans. Automatic Control*, vol. AC-29, no. 12, pp. 1097–1100, 1984.
43. U. B. Desai, D. Pal and R. D. Kirkpatrick, "A realization approach to stochastic model reduction," *Int. J. Control*, vol. 42, no. 4, pp. 821–838, 1985.
44. J. L. Doob, *Stochastic Processes*, Wiley, 1953.
45. P. Faurre, "Réalisations markoviennes de processus stationnaires," Technical Report No. 13, INRIA, March 1973.
46. P. L. Faurre, "Stochastic realization algorithms," In *System Identification: Advances and Case Studies* (R. Mehra and D. Lainiotis, eds.), Academic, pp. 1–25, 1976.
47. P. Faurre, M. Clerget and F. Germain, *Opérateurs Rationnels Positifs*, Dunod, 1979.
48. U. Forssell and L. Ljung, "Closed-loop identification revisited," *Automatica*, vol. 35, no. 7, pp. 1215–1241, 1999.
49. U. Forssell and L. Ljung, "A projection method for closed-loop identification," *IEEE Trans. Automatic Control*, vol. 45, no. 11, pp. 2101–2106, 2000.
50. S. Fujishige, H. Nagai and Y. Sawaragi, "System-theoretic approach to model reduction and system-order determination," *Int. J. Control*, vol. 22, no. 6, pp. 807–819, 1975.
51. F. R. Gantmacher, *The Theory of Matrices* (2 vols.), Chelsea, 1959.
52. I. M. Gel'fand and A. M. Yaglom, "Calculation of the amount of information about a random function contained in another such function," *American Math. Soc. Transl.*, vol. 12, pp. 199–246, 1959.
53. M. Gevers and B. D. O. Anderson, "On joint stationary feedback-free stochastic processes," *IEEE Trans. Automatic Control*, vol. AC-27, no. 2, pp. 431–436, 1982.
54. M. Gevers and V. Wertz, "Uniquely identifiable state-space and ARMA parametrizations for multivariable linear systems," *Automatica*, vol. 20, no. 3, pp. 333–347, 1984.
55. M. Gevers, "A personal view on the development of system identification," *Proc. 13th IFAC Symp. System Identification*, Rotterdam, pp. 773–784, 2003.
56. E. G. Gilbert, "Controllability and observability in multivariable control systems," *SIAM J. Control*, vol. 1, no. 2, pp. 128–151, 1963.
57. M. Glover and J. C. Willems, "Parametrizations of linear dynamical systems: Canonical forms and identifiability," *IEEE Trans. Automatic Control*, vol. AC-19, no. 6, pp. 640–646, 1974.
58. I. Goethals, T. Van Gestel, J. Suykens, P. Van Dooren and B. De Moor, "Identification of positive real models in subspace identification by using regularization," *IEEE Trans. Automatic Control*, vol. 48, no. 10, pp. 1843–1847, 2003.
59. G. H. Golub and C. F. van Loan, *Matrix Computations* (3rd ed.), The Johns Hopkins University Press, 1996.
60. G. H. Golub and H. A. van der Vorst, "Eigenvalue computation in the 20th century," *J. Computational and Applied Mathematics*, vol.123, pp. 35–65, 2000.
61. G. C. Goodwin and R. L. Payne, *Dynamic System Identification: Experiment Design and Data Analysis*, Academic, 1977.
62. B. Gopinath, "On the identification of linear time-invariant systems from input-output data," *Bell Systems Technical Journal*, vol. 48, no. 5, pp. 1101–1113, 1969.
63. C. W. J. Granger, "Economic processes involving feedback," *Information and Control*, vol. 6, pp. 28–48, 1963.
64. M. Green, "All-pass matrices, the positive real lemma, and unit canonical correlations between future and past," *J. Multivariate Analysis*, vol. 24, pp. 143–154, 1988.
65. U. Grenander and G. Szegö, *Toeplitz Forms and Their Applications*, Chelsea, 1958.
66. M. S. Grewal and A. P. Andrews, *Kalman Filtering – Theory and Practice*, Prentice-Hall, 1993.

67. R. Guidorzi, "Canonical structures in the identification of multivariable systems," *Automatica*, vol. 11, no. 4, pp. 361–374, 1975.
68. E. J. Hannan and M. Deistler, *The Statistical Theory of Linear Systems*, Wiley, 1988.
69. E. J. Hannan and D. S. Poskit, "Unit canonical correlations between future and past," *Annals of Statistics*, vol. 16, pp. 784–790, 1988.
70. F. R. Hansen, G. F. Franklin and R. Kosut, "Closed-loop identification via the fractional representation: Experiment design," *Proc. 1989 American Control Conference*, Pittsburgh, pp. 1422–1427, 1989.
71. D. Hinrichsen and A. J. Prichard, "An improved error estimate for reduced-order models of discrete-time systems," *IEEE Trans. Automatic Control*, vol. AC-35, no. 3, pp. 317–320, 1990.
72. B. L. Ho and R. E. Kalman, "Effective construction of linear state-variable models from input/output functions," *Regelungstechnik*, vol. 14, no. 12, pp. 545–548, 1966.
73. R. A. Horn and C. A. Johnson, *Matrix Analysis*, Cambridge University Press, 1985.
74. H. Hotelling, "Analysis of a complex of statistical variables into principal components," *J. Educational Psychology*, vol. 24, pp. 417–441 and pp. 498–520, 1933.
75. H. Hotelling, "Relations between two sets of variates," *Biometrica*, vol. 28, pp. 321–377, 1936.
76. M. Jansson and B. Wahlberg, "On consistency of subspace methods for system identification," *Automatica*, vol. 34, no. 12, pp. 1507–1519, 1998.
77. E. A. Jonckheere and J. W. Helton, "Power spectrum reduction by optimal Hankel norm approximation of the phase of the outer spectral factor," *IEEE Trans. Automatic Control*, vol. AC-30, no. 12, pp. 1192–1201, 1985.
78. T. Kailath, "A view of three decades of linear filtering theory," *IEEE Trans. Information Theory*, vol. IT-20, no. 2, pp. 146–181, 1974.
79. T. Kailath, *Lectures on Linear Least-Squares Estimation*, Springer, 1976.
80. T. Kailath, *Linear Systems*, Prentice-Hall, 1980
81. R. E. Kalman, "A new approach to linear filtering and prediction problems," *Trans. ASME, J. Basic Engineering*, vol. 82D, no.1, pp. 34–45, 1960.
82. R. E. Kalman, "Mathematical description of linear dynamical systems," *SIAM J. Control*, vol. 1, no. 2, pp. 152–192, 1963.
83. R. E. Kalman, "Algebraic aspects of the generalized inverse of a rectangular matrix," In *Generalized Inverses and Applications* (Z. Nashed, ed.), pp. 111–121, Academic, 1976.
84. R. E. Kalman and R. S. Bucy, "New results in linear filtering and prediction theory," *Trans. ASME, J. Basic Engineering*, vol. 83D, no.1, pp. 95–108, 1961.
85. R. E. Kalman, P. L. Falb and M. A. Arbib, *Topics in Mathematical System Theory*, McGraw-Hill, 1969.
86. T. Katayama, "Subspace-based system identification – A view from realization theory," *J. Systems, Control and Information Engineers*, vol. 41, no. 9, pp. 380–387, 1997 (in Japanese).
87. T. Katayama, H. Kawauchi and G. Picci, "A subspace identification of deterministic part of state space model operating in closed loop," *Proc. 6th European Control Conference*, Porto, pp. 2505–2510, 2001.
88. T. Katayama, H. Kawauchi and G. Picci, "Subspace identification of closed loop systems by stochastic realization," *CD-ROM Preprints 15th IFAC World Congress*, Barcelona, Paper # T-Mo-M02-2, 2002.
89. T. Katayama, H. Kawauchi and G. Picci, "Subspace identification of closed loop systems by the orthogonal decomposition method," *Automatica*, vol. 41, no. 5, pp. 863–872, 2005.

90. T. Katayama and G. Picci, "Realization of stochastic systems with exogenous inputs and subspace identification methods," *Automatica*, vol. 35, no. 10, pp. 1635–1652, 1999.
91. T. Katayama and S. Sugimoto (eds.), *Statistical Methods in Control and Signal Processing*, Marcel Dekker, 1997.
92. T. Katayama, H. Tanaka and T. Enomoto, "A simple subspace identification method of closed-loop systems using orthogonal decomposition," *Preprints 16th IFAC World Congress*, Prague, July 2005.
93. H. Kawauchi, A. Chiuso, T. Katayama and G. Picci, "Comparison of two subspace identification methods for combined deterministic-stochastic systems," *Proc. 31st ISCIE Symp. Stochastic Systems Theory and Applications,* Yokohama, pp. 7–12, 1999.
94. V. Klema and A. J. Laub, "The singular value decomposition: Its computation and some applications," *IEEE Trans. Automatic Control*, vol. AC-25, no. 2, pp. 164–176, 1980.
95. L. H. Koopmans, *The Spectral Analysis of Time Series*, Academic, 1974.
96. T. C. Koopmans (ed.), *Statistical Inference in Dynamic Economic Models*, Wiley, 1950.
97. V. Kucèra, "The discrete Riccati equation of optimal control," *Kybernetika*, vol. 8, no. 5, pp. 430–447, 1972.
98. H. Kwakernaak and R. Sivan, *Modern Signals and Systems*, Prentice-Hall, 1991.
99. P. Lancaster and L. Rodman, *Algebraic Riccati Equations*, Oxford Science Publications, 1995.
100. W. E. Larimore, "System identification, reduced-order filtering and modeling via canonical variate analysis," *Proc. 1983 American Control Conference*, San Francisco, pp. 445–451, 1983.
101. W. E. Larimore, "Canonical variate analysis in identification, filtering, and adaptive control," *Proc. 29th IEEE Conference on Decision and Control*, Honolulu, pp. 596–604, 1990.
102. W. E. Larimore, "The ADAPTx software for automated and real-time multivariable system identification," *Proc. 13th IFAC Symp. System Identification*, Rotterdam, pp. 1496–1501, 2003.
103. A. J. Laub, "A Schur method for solving algebraic Riccati equations," *IEEE Trans. Automatic Control*, vol. AC-24, no. 6, pp. 913–921, 1979.
104. M. J. Levin, "Estimation of a system pulse transfer function in the presence of noise," *IEEE Trans. Automatic Control*, vol. 9, no. 2, pp. 229–235, 1964.
105. A. Lindquist and G. Picci, "A geometric approach to modelling and estimation of linear stochastic systems," *J. Math. Systems, Estimation and Control*, vol. 1, no. 3, pp. 241–333, 1991.
106. A. Lindquist and G. Picci, "Canonical correlation analysis, approximate covariance extension, and identification of stationary time series," *Automatica*, vol. 32, no. 5, pp. 709–733, 1996.
107. A. Lindquist and G. Picci, "Geometric methods for state space identification," In *Identification, Adaptation, Learning: The Science of Learning Models from Data* (S. Bittanti and G. Picci, eds.), Springer, pp. 1–69, 1996.
108. Y. Liu and B. D. O. Anderson, "Singular perturbation approximation of balanced systems," *Int. J. Control*, vol. 50, no. 4, pp. 1379–1405, 1989.
109. L. Ljung, *System Identification - Theory for The User* (2nd ed.), Prentice-Hall, 1999.
110. L. Ljung and T. McKelvey, "Subspace identification from closed loop data," *Signal Processing*, vol. 52, no. 2, pp. 209–216, 1996.
111. D. Luenberger, *Optimization by Vector Space Methods*, Wiley, 1969.
112. J. M. Maciejowski, "Parameter estimation of multivariable systems using balanced realizations," In *Identification, Adaptation, Learning: The Science of Learning Models from Data* (S. Bittanti and G. Picci, eds.), Springer, pp. 70–119, 1996.

113. T. McKelvey, H. Ackay and L. Ljung, "Subspace-based multivariable identification from frequency response data," *IEEE Trans. Automatic Control*, vol. AC-41, no. 7, pp. 960–979, 1996.
114. T. McKelvey, A. Helmersson and T. Ribarits, "Data driven local coordinates for multivariable linear systems and their application to system identification," *Automatica*, vol. 40, no. 9, pp. 1629–1635, 2004.
115. P. Masani, "The prediction theory of multivariable stochastic processes - III," *Acta Mathematica*, vol. 104, pp. 141–162, 1960.
116. R. K. Mehra and C. T. Leondes (eds.), *System Identification - Advances and Case Studies*, Academic, 1976.
117. B. P. Molinari, "The stabilizing solution of the discrete algebraic Riccati equation," *IEEE Trans. Automatic Control*, vol. AC-20, no. 3, pp. 396–399, 1975.
118. M. Moonen, B. De Moor, L. Vandenberghe and J. Vandewalle, "On- and off-line identification of linear state-space models," *Int. J. Control*, vol. 49, no. 1, pp. 219–232, 1989.
119. M. Moonen and J. Vandewalle, "QSVD approach to on- and off-line state-space identification," *Int. J. Control*, vol. 51, no. 5, pp. 1133–1146, 1990.
120. A. Ohsumi, K. Kameyama and K. Yamaguchi, "Subspace identification for continuous-time stochastic systems via distribution-based approach," *Automatica*, vol. 38, no. 1, pp. 63–79, 2002.
121. A. V. Oppenheim and R. W. Schafer, *Digital Signal Processing*, Prentice-Hall, 1975.
122. C. C. Paige, "Properties of numerical algorithms related to computing controllability," *IEEE Trans. Automatic Control*, vol. AC-26, no. 1, pp. 130–138, 1981.
123. A. Papoulis and S. U. Pillai, *Probability, Random Variables and Stochastic Processes* (4th ed.), McGraw-Hill, 2002.
124. T. Pappas, A. J. Laub and N. R. Sandell, Jr., "On the numerical solution of the discrete time algebraic Riccati equation," *IEEE Trans. Automatic Control*, vol. AC-25, no. 4, pp. 631–641, 1980.
125. R. Patel, A. J. Laub and P. M. Van Dooren (eds.), *Numerical Linear Algebra Techniques for Systems and Control*, IEEE Press, 1994.
126. M. Pavon, "Canonical correlations of past inputs and future outputs for linear stochastic systems," *Systems and Control Letters*, vol. 4, pp. 209–215, 1984.
127. L. Pernebo and L. M. Silverman, "Model reduction via balanced state space representations," *IEEE Trans. Automatic Control*, vol. AC-27, no. 2, pp. 382–387, 1982.
128. K. Peternell, W. Scherrer and M. Deistler, "Statistical analysis of novel subspace identification methods," *Signal Processing*, vol. 52, no. 2, pp. 161–177, 1996.
129. G. Picci, "Stochastic realization and system identification," In *Statistical Methods in Control and Signal Processing* (T. Katayama and S. Sugimoto, eds.), Marcel Dekker, pp. 1–36, 1997.
130. G. Picci and T. Katayama, "Stochastic realization with exogenous inputs and 'subspace methods' identification," *Signal Processing*, vol. 52, no. 2, pp. 145–160, 1996.
131. G. Picci and T. Katayama, "A simple 'subspace' identification method with exogenous inputs," *Proc. IFAC 13th World Congress*, San Francisco, Vol. I, pp. 175–180, 1996.
132. R. Pintelon and J. Schoukens, *System Identification – A Frequency Domain Approach*, IEEE Press, 2001.
133. V. F. Pisarenko, "The retrieval of harmonics from a covariance function," *Geophysical J. Royal Astronomical Society*, vol. 33, pp. 347–366, 1973.
134. M. B. Priestley, *Spectral Analysis and Time Series, vol. 1: Univariate Series; vol. 2: Multivariate Series, Prediction and Control*, Academic, 1981.
135. A. Rantzer, "On the Kalman-Yakubovich-Popov lemma," *Systems and Control Letters*, vol. 28, no. 1, pp. 7–10, 1996.

136. C. R. Rao, *Linear Statistical Inference and Its Applications* (2nd ed.), Wiley, 1973.
137. J. Rissanen, "Recursive identification of linear systems," *SIAM J. Control*, vol. 9, no. 3, pp. 420–430, 1971.
138. Yu. A. Rozanov, *Stationary Random Processes*, Holden-Day, 1967.
139. A. H. Sayed, *Fundamentals of Adaptive Filtering*, Wiley, 2003.
140. R. O. Schmidt, "Multiple emitter location and signal parameter estimation," *IEEE Trans. Antennas and Propagation*, vol. 34, no. 3, pp. 276–280, 1986.
141. R. Schrama, "Control oriented approximate closed loop identification via fractional representation," *Proc. 1991 American Control Conference*, Boston, pp. 719–720, 1991.
142. R. H. Shumway, *Applied Statistical Time Series Analysis*, Prentice-Hall, 1988.
143. L. M. Silverman, "Realization of linear dynamical systems," *IEEE Trans. Automatic Control*, vol. AC-16, no. 6, pp. 554–567, 1971.
144. T. Söderström, *Discrete-time Stochastic Systems*, Springer, 2002.
145. T. Söderström and P. Stoica, *System Identification*, Prentice-Hall, 1989.
146. L. H. Son and B. D. O. Anderson, "Design of Kalman filters using signal-model output statistics," *Proc. IEE*, vol. 120, no. 2, pp. 312–318, 1973.
147. E. D. Sontag, *Mathematical Control Theory*, Springer, 1990.
148. G. W. Stewart, "On the early history of the singular value decomposition," *SIAM Review*, vol. 35, no. 4, pp. 551–566, 1993.
149. S. M. Stigler, *The History of Statistics: The Measurement of Uncertainty before 1900*, The Belknap Press of Harvard University Press, 1986.
150. P. Stoica and R. Moses, *Introduction to Spectral Analysis*, Prentice-Hall, 1997.
151. H. Tanaka and T. Katayama, "A stochastic realization in a Hilbert space based on 'LQ decomposition' with application to subspace identification," *Proc. 13th IFAC Symp. System Identification*, Rotterdam, pp. 899–904, 2003.
152. H. Tanaka and T. Katayama, "Stochastic realization on a finite interval via 'LQ decomposition' in Hilbert space," *Proc. 7th European Control Conference*, Cambridge, Paper No. SI-2-2, 2003.
153. H. Tanaka and T. Katayama, "A stochastically balanced realization on a finite interval," *Proc. 16th Int. Symp. on Mathematical Theory of Network and Systems*, THP8 (Identification), Leuven, 2004.
154. H. Tanaka and T. Katayama, "Stochastic subspace identification algorithm guarnteeing stability and minimum phase," *Preprints 16th IFAC World Congress*, Prague, July 2005.
155. A. J. Tether, "Construction of minimal linear state-variable models from finite input-output data," *IEEE Trans. Automatic Control*, vol. 15, no. 4, pp. 427–436, 1970.
156. L. Tong and S. Perreau, "Multichannel blind identification: From subspace to maximum likelihood methods," *Proc. IEEE*, vol. 86, no. 10, pp. 1951–1968, 1998.
157. L. N. Trefethen and D. Bau, III, *Numerical Linear Algebra*, SIAM, 1997.
158. P. M. J. Van den Hof, "Closed-loop issues in system identification," *Proc. 11th IFAC Symp. System Identification*, Kitakyushu, Japan, pp. 1651–1664, 1997.
159. P. M. J. Van den Hof and R. A. de Callafon, "Multivariable closed-loop identification: From indirect identification to dual-Youla parametrization," *Proc. 35th IEEE Conference on Decision and Control*, Kobe, Japan, pp. 1397–1402, 1996.
160. P. M. J. Van den Hof and R. J. P. Schrama, "An indirect method for transfer function estimation from closed loop data," *Automatica*, vol. 29, no. 6, pp. 1523–1527, 1993.
161. A. C. Van der Klauw, M. Verhaegen and P. P. J. Van den Bosch, "State space identification of closed loop systems," *Proc. 30th IEEE Conference on Decision and Control*, Brighton, pp. 1327–1332, 1991.

162. A. J. Van der Veen, E. F. Deprettere and A. Swindlehurst, "Subspace-based signal analysis using singular value decomposition," *Proc. IEEE*, vol. 81, no. 9, pp. 1277–1308, 1993.
163. P. Van Overschee and B. De Moor, "Subspace algorithms for the stochastic identification problem," *Automatica*, vol. 29, no. 3, pp. 649–660, 1993.
164. P. Van Overschee and B. De Moor, "N4SID - Subspace algorithms for the identification of combined deterministic - stochastic systems," *Automatica*, vol. 30, no. 1, pp. 75–93, 1994.
165. P. Van Overschee and B. De Moor, *Subspace Identification for Linear Systems*, Kluwer Academic Pub., 1996.
166. P. Van Overschee, B. De Moor, W. Dehandschutter and J. Swevers, "A subspace algorithm for the identification of discrete-time frequency domain power spectra," *Automatica*, vol. 33, no. 12, pp. 2147–2157, 1997.
167. P. Van Overschee and B. De Moor, "Closed loop subspace systems identification," *Proc. 36th IEEE Conference on Decision and Control*, San Diego, pp. 1848–1853, 1997.
168. A. Varga, "On balancing and order reduction of unstable periodic systems," *Preprints 1st IFAC Workshop on Periodic Control Systems*, Como, pp. 177–182, 2001.
169. M. Verhaegen, "Subspace model identification, Part 3: Analysis of the ordinary output-error state-space model identification algorithm," *Int. J. Control*, vol. 58, no. 3, pp. 555–586, 1993.
170. M. Verhaegen, "Application of a subspace model identification technique to identify LTI systems operating in closed loop," *Automatica*, vol. 29, no. 4, pp. 1027–1040, 1993.
171. M. Verhaegen, "Identification of the deterministic part of MIMO state space models given in innovations form from input-output data," *Automatica*, vol. 30, no. 1, pp. 61–74, 1994.
172. M. Verhaegen and P. Dewilde, "Subspace model identification, Part 1: The output-error state-space model identification class of algorithms," *Int. J. Control*, vol. 56, no. 5, pp. 1187–1210, 1992.
173. M. Verhaegen and P. Dewilde, "Subspace model identification, Part 2: Analysis of the elementary output-error state space model identification algorithm," *Int. J. Control*, vol. 56, no. 5, pp. 1211–1241, 1992.
174. M. Viberg, B. Ottersten, B. Wahlberg and L. Ljung, "A statistical perspective on state-space modeling using subspace methods," *Proc. 30th IEEE Conference on Decision and Control*, Brighton, pp. 1337–1342, 1991.
175. M. Viberg, "Subspace-based methods for identification of linear time-invariant systems," *Automatica*, vol. 31, no. 12, pp. 1835–1851, 1995.
176. M. Viberg and P. Stoica, "Editorial note," *Signal Processing*, Special Issue on Subspace Methods, Part I: Array Processing and Subspace Computations, vol. 50, nos. 1-2, pp. 1–3, 1996; Part II: System Identification, vol. 52, no. 2, pp. 99–101, 1996.
177. K. Wada, "What's the subspace identification methods?" *J. Society Instrument and Control Engineers*, vol. 36, no. 8, pp. 569–574, 1997 (in Japanese).
178. P. Whittle, *Prediction and Regulation by Linear Least-Square Methods* (2nd ed.), Basil Blackwell Publisher, 1984.
179. N. Wiener and P. Masani, "The prediction theory of multivariable stochastic processes - I," *Acta Mathematica*, vol. 98, pp. 111–150, 1957.
180. N. Wiener and P. Masani, "The prediction theory of multivariable stochastic processes - II," *Acta Mathematica*, vol. 99, pp. 93–137, 1958.
181. J. Willems, I. Markovsky, P. Rapisarda and B. De Moor, "A note on persistency of excitation," Technical Report 04-54, Department of Electrical Engineering, Katholieke Universiteit Lueven, 2004.

182. W. M. Wonham, "On a matrix Riccati equation of stochastic control," *SIAM J. Control*, vol. 6, no. 4, pp. 681–698, 1968.
183. N. Young, *An Introduction to Hilbert Space*, Cambridge University Press, 1988.
184. H. P. Zeiger and A. J. McEwen, "Approximate linear realization of given dimension via Ho's algorithm," *IEEE Trans. Automatic Control*, vol. AC-19, no. 2, p. 153, 1974.
185. K. Zhou, J. C. Doyle and K. Glover, *Robust Optimal Control*, Prentice-Hall, 1996.
186. K. Zhou, G. Salomon and E. Wu, "Balanced realization and model reduction for unstable systems," *Int. J. Robust and Nonlinear Control*, vol. 9, no. 2, pp. 183–198, 1999.

Index

σ-algebra, 74, 112
z-transform, 41
  inverse transform, 43
  properties of, 43
2-norm, 22, 32, 47

admissible inputs, 123
Akaike's method, 209
algebraic Riccati equation (ARE), 129
algebraic Riccati inequality (ARI), 179
AR model, 84, 121
ARE, 129, 179, 192
  numerical solution of, 134
  stabilizing solution of, 134, 136, 287
ARI, 179, 180
  degenerate solution, 183
ARMA model, 84, 230
ARMAX model, 5

backward Kalman filter, 132
  stationary, 216
backward Markov model, 102, 131
backward process, 188
balanced realization, 58
  stochastically, 223
basis, 20
  orthonormal, 26
bounded-input and bounded-output (BIBO) stable, 45

canonical angles, 11, 246, 275
canonical correlation analysis (CCA), 11, 203
canonical correlations, 205, 218, 230
  between future and past, 216
  conditional, 279
canonical decomposition theorem, 55
canonical variables, 205
canonical vectors, 207, 218
  conditional, 279
Cayley-Hamilton theorem, 18
CCA, 203, 207
CCA method, 288
Cholesky factorization, 217, 279
closed-loop identification, 299
  CCA method, 308
  direct approach, 300
  indirect approach, 300
  joint input-output approach, 300, 303
  ORT method, 314
coercive, 174, 184, 194
condition number, 36
conditional distribution, 108
conditional mean, 109
conditional orthogonality, 241, 277
conditionally orthogonal, 241
controllable, 52
covariance function, 76
  cross-, 78
covariance matrix, 76, 97, 175
  conditional, 273, 274
  of predicted estimate, 127, 130
CVA method, 298

data matrix, 149, 152
detectability, 53
detectable, 53
deterministic component, 245, 246

realization of, 249
deterministic realization algorithm, 145
deterministic realization problem, 142
direct sum, 21
decomposition, 246, 283

eigenvalues, 18
eigenvectors, 18
ergodic process, 79
ergodic theorem
covariance, 81
mean, 80
error covariance matrix, 115

feedback system, 301
feedback-free condition, 242–244, 309
Fibonacci sequence, 145
filtered estimate, 119
filtration, 74
finite impulse response (FIR) model, 334, 339
Fourier transform, 47
full rank, 171

Gauss-Markov process, 96
Gaussian distribution
2-dimensional, 137
multivariate, 107
Gaussian process, 76
generalized eigenvalue problem (GEP), 134, 204

Hankel matrix, 36, 37
block, 55, 65, 227
properties of, 143
Hankel operator, 36, 344
block, 142
Hankel singular values, 58, 216, 316
Hilbert space, 89
Ho-Kalman's method, 142
Householder transform, 23

identification methods
classical, 4
prediction error method (PEM), 5
image (or range), 20
impulse response, 45
matrix, 142
inner product, 17

innovation model
backward, 133
forward, 130, 285
innovation process, 94, 116
backward, 131
innovation representation, 219
innovations, 112
inverse filter, 86

joint distribution, 74
joint input-output process, 242
joint probability density function, 74

Kalman filter, 120
block diagram, 120
Kalman filter with inputs, 123
block diagram, 126
Kalman filtering problem, 113
kernel (or null space), 20

least-squares estimate, 330
generalized, 331
least-squares method, 33, 329
least-squares problem, 33, 329
basic assumptions for, 330
minimum norm solution of, 35
linear matrix inequality (LMI), 176
linear regression model, 109, 330
linear space, 19
linear time-invariant (LTI) system, 44
LMI, 176, 177
LQ decomposition, 155, 162, 258, 288, 334
MATLAB® program, 155
LTI system
external description of, 50
internal description of, 49
Lyapunov equation, 54, 99, 175
Lyapunov stability, 50

Markov model, 101, 212
backward, 101, 176, 222
forward, 101, 176, 222
reduced order, 226
Markov parameters, 49
Markov process, 75
matrix
block, 39
inverse of, 39
Hankel, 36
Hermitian, 18

idempotent, 28, 40
numerical rank of, 33
observability, 56
orthogonal, 17
perturbed, 33
projection, 28
reachability, 51, 56
square root, 19
Toeplitz, 37
matrix input-output equation, 257
matrix inversion lemma, 39, 110
maximum singular value, 32
mean function, 76
mean vector, 76
minimal phase, 85, 244
minimal realization, 66
minimum singular value, 32
minimum variance estimate, 109, 114
unbiased, 115, 118
model reduction
singular perturbation approximation (SPA) method, 62
SR algorithm, 315
model structure, 2
MOESP method, 157, 169
moment function, 75
moving average (MA) representation, 90
multi-index, 345, 347
MUSIC, 11, 170

N4SID algorithm, 166
N4SID method, 161, 170
direct, 8
realization-based, 10
norm
H_∞-, 47
H_2-, 47
l_2-induced, 48
Euclidean, 22
Frobenius, 22, 32
infinity-, 22
operator, 22

oblique projection, 27, 40, 161, 163, 240, 277
observability, 51
observability Gramian, 58
of unstable system, 63
observability matrix, 53
extended, 66, 144
observable, 51
one-step prediction error, 5
optimal predictor, 280
ORT method, 256
orthogonal, 21
orthogonal complement, 21
orthogonal decomposition, 29, 245
orthogonal projection, 29, 40, 240
orthonormal basis vectors, 211
overlapping parametrization, 7, 343

PE condition, 151, 246, 275, 340
PO-MOESP, 291
PO-MOESP algorithm, 166, 259
positive real, 174, 184, 224
strictly, 174, 224
positive real lemma, 183, 185
predicted estimate, 118
prediction error method (PEM), 5
MIMO model, 8
prediction problem, 93, 242
predictor space, 209, 250
backward, 210, 218
basis vector of, 280
finite-memory, 286
forward, 210, 218
oblique splitting, 249
projection, 27
pseudo-canonical form, 7
pseudo-inverse, 34

QR decomposition, 23, 24, 26, 33
quadratic form, 17

random walk, 74, 77
rank, 21
normal, 72
reachability, 51
reachability Gramian, 58
of unstable system, 63
reachability matrix, 51
extended, 66, 144
reachable, 51
realizable, 67
realization, 56
balanced, 58, 59
finite interval, 286
minimal, 56
realization theory, 65

recursive sequence, 67
reduced order model, 59, 316
regularity condition, 91
Riccati equation, 121, 127, 287

second-order process, 76
shaping filter, 86
shift invariant, 143
singular value decomposition (SVD), 30
singular values, 31, 168
  Hankel, 58, 316
singular vectors, 32
  left, 168
spectral analysis, 81
spectral density function, 81
spectral density matrix, 100, 173
  additive decomposition, 173
spectral factor, 177, 179
spectral radius, 22, 72
splitting subspace, 212
  oblique, 249, 250
SR algorithm, 315
stabilizable, 52
stable, 45
  asymptotically, 50
state estimation problem, 114
state space model
  block structure of, 253
state space system, 48
stationary Kalman filter, 129, 182, 214
stationary process, 77
  second-order, 78
stochastic component, 245, 246
  realization of, 248
stochastic linear system, 95
stochastic LTI system, 98
stochastic process, 73
  full rank, 171, 243, 271
  Hilbert space of, 89
  regular, 90, 243, 271
  singular, 90
stochastic realization, 12, 171
  algorithm, 198, 227, 228
  balanced, 219
  based on finite data, 286
  with exogenous inputs, 242
stochastic realization problem, 174, 207, 242
  solution of, 176
  with exogenous inputs, 272
strictly positive real, 174, 184
  conditions, 194
subspace, 20, 148
  Hilbert, 209, 243
  invariant, 21
  noise, 168
  signal, 168
subspace identification
  CCA method, 290
  ORT method, 258, 260
    deterministic subsystem, 258
    stochastic subsystem, 260
subspace method, 8, 10
SVD, 145, 163, 166
  and additive noise, 166
system identification, 1
  flow chart of, 3
  input signals for, 337

Toeplitz matrix, 37
  block, 184, 227

variance, 76, 83
vector sum, 21

white noise, 74, 95, 114
Wiener-Hopf equation, 247, 274, 276
Wiener-Khinchine formula, 82
Wold decomposition theorem, 90, 243

zero-input response, 49, 142, 143, 156
zero-state response, 49

MIX
Papier aus verantwortungsvollen Quellen
Paper from responsible sources
FSC® C105338

If you have any concerns about our products,
you can contact us on
ProductSafety@springernature.com

In case Publisher is established outside the EU,
the EU authorized representative is:
Springer Nature Customer Service Center GmbH
Europaplatz 3, 69115 Heidelberg, Germany

Printed by Libri Plureos GmbH
in Hamburg, Germany